CELL BIOLOGY AND GENETICS

CECIE STARR / RALPH TAGGART

BIOLOGY
THE UNITY AND DIVERSITY OF LIFE
EIGHTH EDITION

WADSWORTH PUBLISHING COMPANY

I⊕P® AN INTERNATIONAL THOMSON PUBLISHING COMPANY

Belmont, CA • Albany, NY • Bonn • Boston
Cincinnati • Detroit • Johannesburg • London • Madrid
Melbourne • Mexico City • New York
Paris • San Francisco • Singapore
Tokyo • Toronto • Washington

BIOLOGY PUBLISHER: Jack C. Carey

ASSISTANT EDITOR: Kristin Milotich

MEDIA PROJECT MANAGER: Pat Waldo

MARKETING MANAGER: Halee Dinsey

DEVELOPMENTAL EDITOR: Mary Arbogast

PROJECT EDITOR: Sandra Craig

EDITORIAL ASSISTANT: Michael Burgreen

PRINT BUYER: Karen Hunt

PRODUCTION: Mary Douglas, Rogue Valley Publications

TEXT AND COVER DESIGN, ART DIRECTION: Gary Head,
Gary Head Design

ART COORDINATOR: Myrna Engler-Forkner

EDITORIAL PRODUCTION: Mary Roybal, Karen Stough,
Susan Gall

PRIMARY ARTISTS: Raychel Ciemma; Precision Graphics
(Jan Troutt, J.C. Morgan)

ADDITIONAL ARTISTS: Robert Demarest, Darwen Hennings,
Vally Hennings, Betsy Palay, Nadine Sokol, Kevin Somerville,
Lloyd Townsend

PHOTO RESEARCH, PERMISSIONS: Stephen Forsling, Roberta Broyer

COVER PHOTOGRAPH: *Minnehaha Falls,* © Richard Hamilton Smith

COMPOSITION: Precision Graphics (Jim Gallagher, Kirsten
Dennison)

COLOR PROCESSING: H&S Graphics (Tom Anderson,
Nancy Dean, John Deady, Rich Stanislawski)

PRINTING AND BINDING: World Color, Versailles

BOOKS IN THE WADSWORTH BIOLOGY SERIES

Biology: Concepts and Applications, Third, Starr
Biology: The Unity and Diversity of Life, Eighth, Starr/Taggart
Human Biology, Second, Starr/McMillan
Laboratory Manual for Biology, Perry and Morton
General Botany, Rost et al.
Introduction to Biotechnology, Barnum
General Ecology, Krohne
Introduction to Microbiology, Ingraham/Ingraham
Living in the Environment, Tenth, Miller
Environmental Science, Seventh, Miller
Sustaining the Earth, Third, Miller
Environment: Problems and Solutions, Miller
Introduction to Cell and Molecular Biology, Wolfe
Molecular and Cellular Biology, Wolfe
Cell Ultrastructure, Wolfe
Marine Life and the Sea, Milne
Essentials of Oceanography, Garrison
Oceanography: An Invitation to Marine Science, Second, Garrison
Oceanography: An Introduction, Fifth, Ingmanson/Wallace
Plant Physiology, Fourth, Salisbury/Ross
Plant Physiology Laboratory Manual, Ross
Plants: An Evolutionary Survey, Second, Scagel et al.
Psychobiology: The Neuron and Behavior, Hoyenga/Hoyenga
Sex, Evolution, and Behavior, Second, Daly/Wilson
Dimensions of Cancer, Kupchella
Evolution: Process and Product, Third, Dodson/Dodson

For more information, contact Wadsworth Publishing Company,
10 Davis Drive, Belmont, California 94002, or electronically at
http://www.thomson.com/wadsworth.html

International Thomson Publishing Europe
Berkshire House 168-173, High Holborn
London, WC1V7AA, England

Thomas Nelson Australia
102 Dodds Street
South Melbourne 3205, Victoria, Australia

Nelson Canada
1120 Birchmount Road
Scarborough, Ontario, Canada M1K 5G4

International Thomson Editores
Campos Eliseos 385, Piso 7
Col. Polanco, 11560 México D.F. México

International Thomson Publishing GmbH
Königswinterer Strasse 418
53227 Bonn, Germany

International Thomson Publishing Asia
221 Henderson Road, #05-10 Henderson Building
Singapore 0315

International Thomson Publishing Japan
Hirakawacho Kyowa Building, 3F
2-2-1 Hirakawacho, Chiyoda-ku, Tokyo 102, Japan

International Thomson Publishing Southern Africa
Building 18, Constantia Park
240 Old Pretoria Road
Halfway House, 1685 South Africa

CONTENTS IN BRIEF

Highlighted chapters are included in CELL BIOLOGY AND GENETICS.

Highlighted chapters are included in CELL BIOLOGY AND GENETICS.

DETAILED CONTENTS

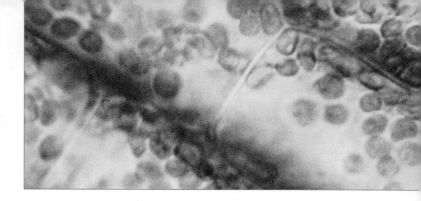

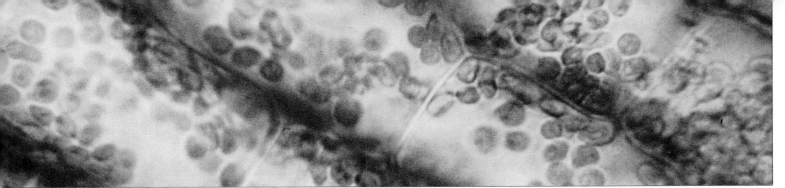

CHLOROPLASTS IN LIVING PLANT CELLS

PREFACE

Not too long from now we will cross the threshold of a new millennium, a rite of passage that invites reflection on where biology has been and where it might be heading. About 500 years ago, during an age of global exploration, naturalists first started to systematically catalog and think about the staggering diversity of organisms all around the world. Less than 150 years ago, just before the start of a civil war that would shred the fabric of a new nation, the naturalist Charles Darwin shredded preconceived notions about life's diversity. It was only about 50 years ago that biologists caught their first glimpse of life's unity at the molecular level. Until that happened a biologist could still hope to be a generalist—someone who viewed life as Darwin did, without detailed knowledge of mechanisms that created it, that perpetuate it, that change it.

No more. Biology grew to encompass hundreds of specialized fields, each focused on one narrow aspect of life and yielding volumes of information about it. Twenty years ago I wondered whether introductory textbooks could possibly keep up with the rapid and divergent splintering of biological inquiry. James Bonner, a teacher and researcher at the California Institute of Technology, turned my thinking around on this. He foresaw that authors and instructors for introductory courses must become the new generalists, the ones who give each generation of students broad perspective on what we know about life and what we have yet to learn.

And we must do this, for the biological perspective remains one of the most powerful of education's gifts. With it, students who travel down specialized roads can sense intuitively that their research and its applications may have repercussions in unexpected places in the world of life. With that perspective, students in general might cut their own intellectual paths through social, medical, and environmental thickets. And they might come to understand the past and to predict possible futures for ourselves and all other organisms.

CONCERNING THE EIGHTH EDITION

Like earlier editions, this book starts with an overview of the basic concepts and scientific methods. Three units on the principles of biochemistry, inheritance, and evolution follow. The principles provide the conceptual background necessary for deeper probes into life's unity and diversity, starting with a richly illustrated evolutionary survey of each kingdom. Units on the comparative anatomy and physiology of plants, then animals, follow. The last unit focuses on the patterns and consequences of organisms interacting with one another and with their environment. Thus the organization parallels the levels of biological organization, from cells through the biosphere. We adhere to this traditional approach for good reason: it works.

As before, we identify and highlight the key concepts, current understandings, and research trends for the major fields of inquiry. Through examples of problem solving and experiments, we give ample evidence of "how we know what we know" and thus demonstrate the power of critical thinking. We explain the structure and functioning of a broad sampling of organisms in enough detail so that students can develop a working vocabulary about life's parts and processes. We also updated the glossary.

CONCEPT SPREADS

In the first chapter, an overview of the levels of biological organization kicks off a story that continues through the rest of the book. Telling such a big, complex story might be daunting unless you remind yourself of the question *"How do you eat an elephant?"* and its answer, *"One bite at a time."* We who have told the story again and again know how the parts fit together, but many students need help to keep the story line in focus within and between chapters. And they need to chew on concepts one at a time.

In every chapter we present each concept on its own table, so to speak. That is, we organize the descriptions, art, and supporting evidence for it on two facing pages, at most. Think of this as a concept spread, as in Figure *A*. Each starts with a numbered tab and ends with boldface statements to summarize the key points. Students can use these cues as reminders to digest one topic before starting on another. Well-crafted transitions between spreads help students focus on where topics fit in the larger story and gently discourage memorization for its own sake. The clear demarcation also gives instructors greater flexibility in assigning or skipping topics within a chapter.

By restricting the space available for each concept, we force ourselves to clear away the clutter of superfluous detail. Within each concept spread, we block out headings and subheadings to rank the importance of its various parts. Any good story has such a hierarchy of information, with background settings, major and minor characters, and high points and an ending where everything comes together. Without a hierarchy, a story has all the excitement, flow, and drama of an encyclopedia. Where details are useful as expansions of concepts, we integrate them into suitable illustrations to keep them from disrupting the text flow.

Not all students are biology majors, and many of them approach biology textbooks with apprehension. If the words don't engage them, they sometimes end up hating the book, and the subject. It comes down to line-by-line judgment calls. During twenty-two years of authorship, we developed a sense of when to leave core material alone and when to loosen it up to give students breathing room. Interrupting, say, an account of mitotic cell division with a distracting anecdote does no good. Plunking a humorous aside into a chapter that ties together the evolution of the Earth and life trivializes a magnificent story. Including an entertaining story is fine, provided that doing so reinforces a key concept. Thus, for example, we include the story of a misguided species introduction that resulted in wild European rabbits running amok through Australia.

BALANCING CONCEPTS WITH APPLICATIONS

Each chapter starts with a lively or sobering application that leads into an adjoining list of key concepts. The list is an advance organizer for the chapter as a whole. At strategic points, examples of applications parallel the core material—not so many as to be distracting, but enough to keep minds perking along with the conceptual development. Many brief applications are integrated in the text. Others are in *Focus* essays, which give more depth on medical, environmental, and social issues but do not interrupt the text flow.

FOUNDATIONS FOR CRITICAL THINKING

To help students develop a capacity for critical thinking, we walk them through experiments that yielded evidence in favor of or against hypotheses being discussed. The main index for the book will give you a sense of the number and types of experiments used (see the entry *Experiments*).

We use certain chapter introductions as well as entire chapters to show students some of the productive results of critical thinking. Among these are the introductions to the chapters on Mendelian genetics (11), DNA structure and function (14), and speciation (19).

Many *Focus on Science* essays provide more detailed, optional examples of how biologists apply critical thinking to problem solving. For example, one of these describes RFLP analysis (Section 16.3) and a few of its more jarring applications. Another essay helps convey to students that biology is not a closed book. Even when new research brings a sweeping story into sharp focus, it also opens up new roads of inquiry.

This edition has *Critical Thinking* questions at the end of chapters. Katherine Denniston of Towson State University developed these thought-provoking questions. Chapters 11 and 12 also include a large selection of *Genetics Problems* that help students grasp the principles of inheritance.

To keep readers focused, we cover each concept on one or two facing pages, starting with a numbered tab . . .

5.1

MEMBRANE STRUCTURE AND FUNCTION

Earlier chapters provided you with a brief look at the structure of cell membranes and the general functions of their component parts. Here, we incorporate some of the background information in a more detailed picture.

The Lipid Bilayer of Cell Membranes

Fluid bathes the two surfaces of a cell membrane and is vital for its functioning. The membrane, too, has a fluid quality; it is not a solid, static wall between cytoplasmic and extracellular fluids. For instance, puncture a cell with a fine needle, and its cytoplasm will not ooze out. The membrane will flow over the puncture site and seal it!

How does a fluid membrane remain distinct from its fluid surroundings? To arrive at the answer, start by reviewing what we have already learned about its most abundant components, the phospholipids. Recall that a **phospholipid** has a phosphate-containing head and two fatty acid tails attached to a glycerol backbone (Figure 5.2a). The head is hydrophilic; it easily dissolves in water. Its tails are hydrophobic; water repels them. Immerse a number of phospholipid molecules in water, and they will interact with water molecules and with one another until they spontaneously cluster in a sheet or film at the water's surface. Their jostlings may even force them to become organized in two layers, with all fatty acid tails sandwiched between all hydrophilic heads. This **lipid bilayer** arrangement, remember, is the structural basis of cell membranes (Section 4.1 and Figure 5.2c).

The organization of each lipid bilayer minimizes the total number of hydrophobic

groups exposed to water, so the fatty acid tails do not have to spend a lot of energy fighting water molecules, so to speak. A "punctured" membrane exhibits sealing behavior precisely because a puncture is energetically unfavorable. It leaves far too many hydrophobic groups exposed to the surrounding fluid.

Ordinarily, few cells get jabbed by fine needles. But the self-sealing behavior of membrane phospholipids is good for more than damage control. Among other things, it functions in vesicle formation. For example, as vesicles bud away from ER or Golgi membranes, phospholipids interact hydrophobically with cytoplasmic water. They get pushed together, and the rupture seals. You will read more about vesicle formation later in the chapter.

Fluid Mosaic Model of Membrane Structure

Figure 5.3 shows a bit of membrane that corresponds to the **fluid mosaic model**. By this model, cell membranes are a mixed composition—a "mosaic"—of phospholipids, glycolipids, sterols, and proteins. The phospholipid heads as well as the length and saturation of the tails are not all the same. (Recall that unsaturated fatty acids have one or more double bonds in their backbone and fully saturated ones have none.) The glycolipids are structurally similar to phospholipids, but their head incorporates one or more sugar monomers. In animal cell membranes, cholesterol is the most abundant sterol (Figure 5.2b). Phytosterols are their equivalent in plant cell membranes.

Also by this model, the membrane is "fluid" owing to the motions and interactions of its component parts.

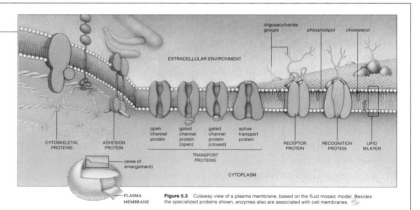

Figure 5.3 Cutaway view of a plasma membrane, based on the fluid mosaic model. Besides the specialized proteins shown, enzymes also are associated with cell membranes.

The hydrophobic interactions that give rise to most of a membrane's structure are weaker than covalent bonds. This means most phospholipids and some proteins are free to drift sideways. Also, the phospholipids can spin about their long axis and flex their tails, which keeps neighboring molecules from packing together in a solid layer. Short or kinked (unsaturated) fatty acid tails also contribute to membrane fluidity.

The fluid mosaic model is a good starting point for exploring cell membranes. But bear in mind, membranes differ in the details of their molecular composition and arrangements, and they are not even the same on both surfaces of their bilayer. For example, oligosaccharides and other carbohydrates are covalently bonded to protein and lipid components of a plasma membrane, but only on its outward-facing surface (Figure 5.3). Moreover, they differ in number and kind from one species to the next, even among the different cells of the same individual.

Overview of Membrane Proteins

The proteins embedded in a lipid bilayer or attached to one of its surfaces carry out most membrane functions. Many are enzyme components of metabolic machinery. Others are **transport proteins** that allow water-soluble substances to move through their interior, which spans the bilayer. They bind molecules or ions on one side of the membrane, then release them on the other side. The **receptor proteins** bind extracellular substances, such as hormones, that trigger changes in cell activities.

For example, certain enzymes that crank up machinery for cell growth and division become switched on when somatotropin, a hormone, binds with receptors for it. Different cells have different combinations of receptors.

Diverse **recognition proteins** at the cell surface are like molecular fingerprints; their oligosaccharide chains identify a cell as being of a specific type. For example, "self" proteins pepper the plasma membrane of your cells. Certain white blood cells chemically recognize the proteins and leave your own cells alone, but they attack invading bacterial cells having "nonself" proteins at their surface. Finally, **adhesion proteins** of multicelled organisms help cells of the same type locate and stick to one another and stay positioned in the proper tissues. They are glycoproteins with oligosaccharides attached. After tissues form, the sites of adhesion may become a type of cell junction, as described earlier in Section 4.10.

A cell membrane has two layers composed mainly of lipids, phospholipids especially. This lipid bilayer is the structural foundation for the membrane and also serves as a barrier to water-soluble substances.

Hydrophilic heads of the phospholipids are dissolved in fluids that bathe the two outer surfaces of the bilayer. Their hydrophobic tails are sandwiched between the heads.

Proteins associated with the bilayer carry out most membrane functions. Many are enzymes, transporters of substances across the bilayer, or receptors for extracellular substances. Other types function in cell-to-cell recognition or adhesion.

Figure 5.2

(a) Structural formula of phosphatidylcholine, a phospholipid that is one of the most common components of the membranes of animal cells. *Orange* indicates its hydrophilic head; *yellow* indicates its hydrophobic tails.

(b) Structural formula of cholesterol, the major sterol in animal tissues.

(c) Diagram showing how lipids that are placed in liquid water may spontaneously organize themselves into a bilayer structure.

FIGURE A   *A concept spread from this edition.*

. . . and ending with one or more summary statements.

VISUAL OVERVIEWS OF MAJOR CONCEPTS

While writing the text, we simultaneously develop the illustrations as inseparable parts of the same story. This integrative approach appeals to students who are visual learners. When they can first work their way through a visual overview of some process, then reading through the corresponding text becomes less intimidating. Over the years, students have repeatedly thanked us for our hundreds of overview illustrations, which contain step-by-step, written descriptions of biological parts and processes. We break down the information into a series of illustrated steps that are more inviting than a complex, "wordless" diagram. Figure *B* is a sample. Notice how simple descriptions, integrated with the art, take students through the stages by which mRNA transcripts become translated into polypeptide chains, one step at a time.

Similarly, we continue to create visual overviews for anatomical drawings. The illustrations integrate structure and function. Students need not jump back and forth from the text, to tables, to illustrations, and back again in order to comprehend how an organ system is put together and what its parts do. Even individual descriptions of parts are hierarchically arranged to reflect the structural and functional organization of that system.

COLOR CODING

In line illustrations, we consistently use the same colors for the same types of molecules and cell structures. Visual consistency makes it easier for students to track complex parts and processes. Figure C is the color coding chart.

ZOOM SEQUENCES

Many illustrations in the book progress from macroscopic to microscopic views of the same subject. Figure 7.2 is an example; this zoom sequence shows where the reactions of photosynthesis proceed, starting with a plant growing by a roadside.

ICONS

Within the text, small diagrams next to an illustration help relate the topic to the big picture. For instance, in Figure *A*, a simple representation of a cell subtly reminds students of the location of the plasma membrane relative to the cytoplasm. Other icons serve as reminders of the location of reactions and processes in cells and how they

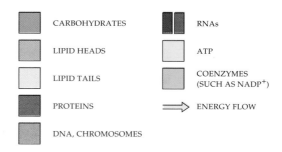

FIGURE C   *Color coding chart for the diagrams of biological molecules and cell structures.*

Step-by-step art with simple descriptions helps students visualize a process before reading text about it.

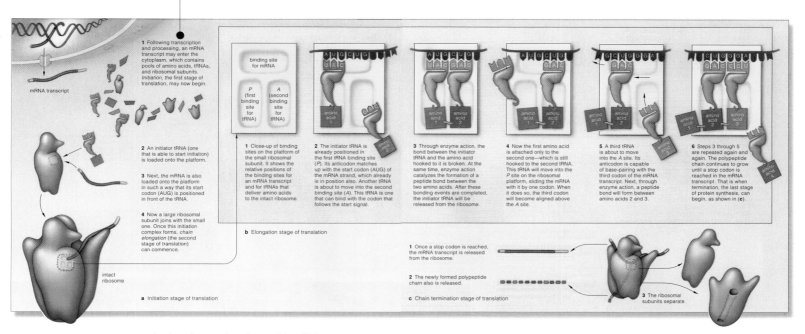

FIGURE B   *A visual overview from this edition.*

interrelate to one another. Still other icons remind students of the evolutionary relationships among groups of organisms.

New to this edition are icons that invite students to use multimedia. One icon directs them to art in the CD-ROM enclosed with each student copy, another to supplemental material on the Web, and a third to InfoTrac:

CD-ROM ICON:

WEB ICON:

INFOTRAC ICON:

END-OF-CHAPTER STUDY AIDS

Figure D shows a sampling of our end-of-chapter study aids, which reinforce the key concepts. Each chapter ends with a summary in list form, review questions, a self-quiz, critical thinking questions, selected key terms, and a list of readings. Italicized page numbers tie the review questions and key terms to relevant text pages.

END-OF-BOOK STUDY AIDS

At the book's end, the detailed classification scheme in Appendix I is helpful for reference purposes. Appendix II includes metric-English conversion charts. Appendix III has detailed answers to the genetics problems in Chapters 11 and 12, and the fourth has answers to the self-quizzes at the end of each chapter. For those interested students and professors who prefer the added detail, Appendix V illustrates the structural formulas for the major metabolic pathways. Appendix VI shows the periodic table of the elements.

A Glossary includes boldfaced terms from the text, with pronunciation guides and word origins to make the formidable words less so. The Appendixes as well as the Glossary are printed on paper that is tinted different colors to preclude frustrating searches for where one ends and the next begins. The Index is detailed enough to help readers find doors to the text more quickly.

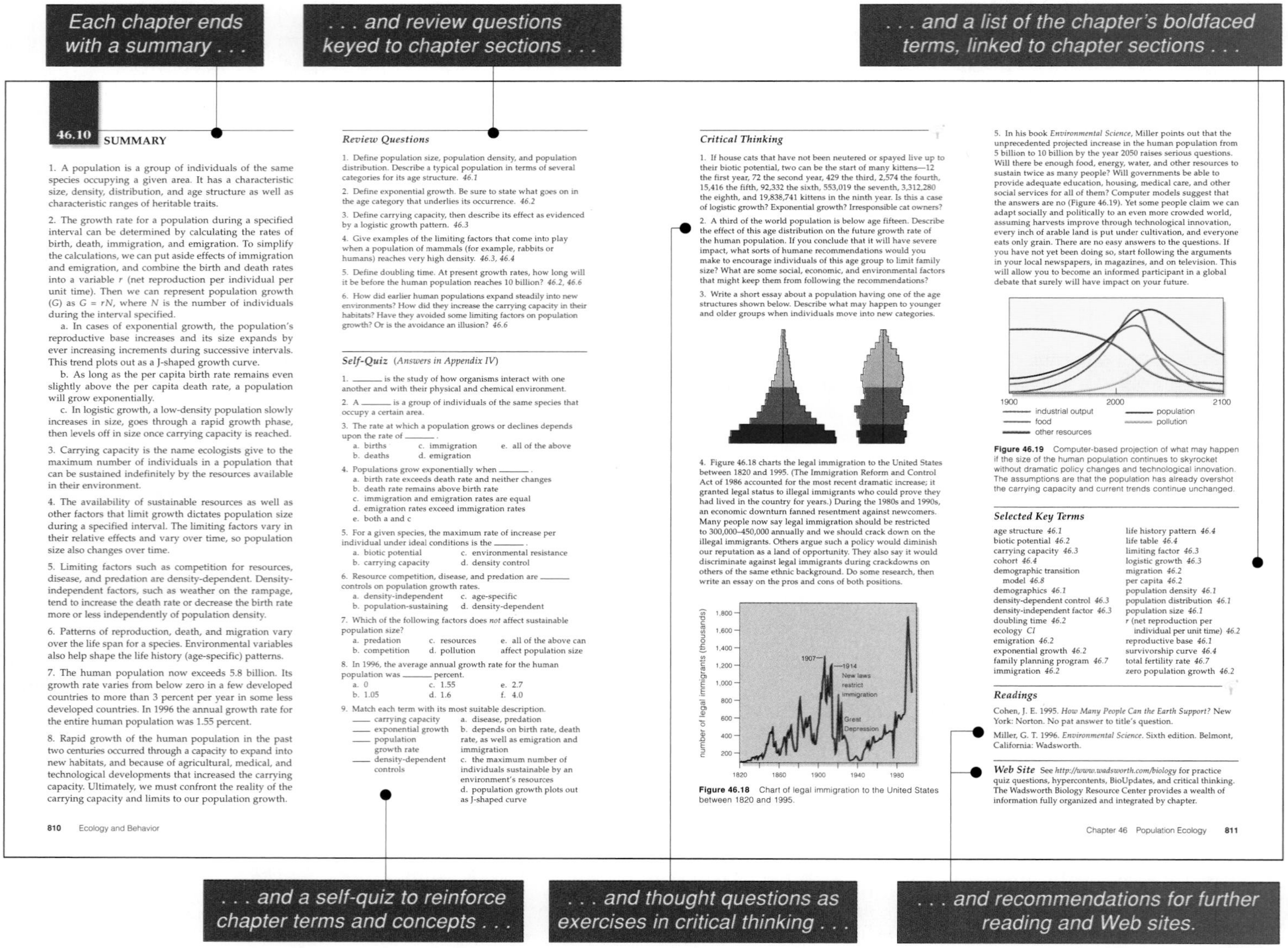

Each chapter ends with a summary . . .

. . . and review questions keyed to chapter sections . . .

. . . and a list of the chapter's boldfaced terms, linked to chapter sections . . .

46.10 SUMMARY

1. A population is a group of individuals of the same species occupying a given area. It has a characteristic size, density, distribution, and age structure as well as characteristic ranges of heritable traits.

2. The growth rate for a population during a specified interval can be determined by calculating the rates of birth, death, immigration, and emigration. To simplify the calculations, we can put aside effects of immigration and emigration, and combine the birth and death rates into a variable r (net reproduction per individual per unit time). Then we can represent population growth (G) as $G = rN$, where N is the number of individuals during the interval specified.

   a. In cases of exponential growth, the population's reproductive base increases and its size expands by ever increasing increments during successive intervals. This trend plots out as a J-shaped growth curve.

   b. As long as the per capita birth rate remains even slightly above the per capita death rate, a population will grow exponentially.

   c. In logistic growth, a low-density population slowly increases in size, goes through a rapid growth phase, then levels off in size once carrying capacity is reached.

3. Carrying capacity is the name ecologists give to the maximum number of individuals in a population that can be sustained indefinitely by the resources available in their environment.

4. The availability of sustainable resources as well as other factors that limit growth dictates population size during a specified interval. The limiting factors vary in their relative effects and vary over time, so population size also changes over time.

5. Limiting factors such as competition for resources, disease, and predation are density-dependent. Density-independent factors, such as weather on the rampage, tend to increase the death rate or decrease the birth rate more or less independently of population density.

6. Patterns of reproduction, death, and migration vary over the life span for a species. Environmental variables also help shape the life history (age-specific) patterns.

7. The human population now exceeds 5.8 billion. Its growth rate varies from below zero in a few developed countries to more than 3 percent per year in some less developed countries. In 1996 the annual growth rate for the entire human population was 1.55 percent.

8. Rapid growth of the human population in the past two centuries occurred through a capacity to expand into new habitats, and because of agricultural, medical, and technological developments that increased the carrying capacity. Ultimately, we must confront the reality of the carrying capacity and limits to our population growth.

Review Questions

1. Define population size, population density, and population distribution. Describe a typical population in terms of several categories for its age structure. 46.1

2. Define exponential growth. Be sure to state what goes on in the age category that underlies its occurrence. 46.2

3. Define carrying capacity, then describe its effect as evidenced by a logistic growth pattern. 46.3

4. Give examples of the limiting factors that come into play when a population of mammals (for example, rabbits or humans) reaches very high density. 46.3, 46.4

5. Define doubling time. At present growth rates, how long will it be before the human population reaches 10 billion? 46.2, 46.6

6. How did earlier human populations expand steadily into new environments? How did they increase the carrying capacity in their habitats? Have they avoided some limiting factors on population growth? Or is the avoidance an illusion? 46.6

Self-Quiz (Answers in Appendix IV)

1. _____ is the study of how organisms interact with one another and with their physical and chemical environment.

2. A _____ is a group of individuals of the same species that occupy a certain area.

3. The rate at which a population grows or declines depends upon the rate of _____.
   a. births   c. immigration   e. all of the above
   b. deaths   d. emigration

4. Populations grow exponentially when _____.
   a. birth rate exceeds death rate and neither changes
   b. death rate remains above birth rate
   c. immigration and emigration rates are equal
   d. emigration rates exceed immigration rates
   e. both a and c

5. For a given species, the maximum rate of increase per individual under ideal conditions is the _____.
   a. biotic potential   c. environmental resistance
   b. carrying capacity   d. density control

6. Resource competition, disease, and predation are _____ controls on population growth rates.
   a. density-independent   c. age-specific
   b. population-sustaining   d. density-dependent

7. Which of the following factors does not affect sustainable population size?
   a. predation   c. resources   e. all of the above can
   b. competition   d. pollution   affect population size

8. In 1996, the average annual growth rate for the human population was _____ percent.
   a. 0   c. 1.55   e. 2.7
   b. 1.05   d. 1.6   f. 4.0

9. Match each term with its most suitable description.
   ____ carrying capacity   a. disease, predation
   ____ exponential growth   b. depends on birth rate, death
   ____ population   rate, as well as emigration and
       growth rate   immigration
   ____ density-dependent   c. the maximum number of
       controls   individuals sustainable by an
                                environment's resources
                              d. population growth plots out
                                as J-shaped curve

810   Ecology and Behavior

Critical Thinking

1. If house cats that have not been neutered or spayed live up to their biotic potential, two can be the start of many kittens—12 the first year, 72 the second year, 429 the third, 2,574 the fourth, 15,416 the fifth, 92,332 the sixth, 553,019 the seventh, 3,312,280 the eighth, and 19,838,741 kittens in the ninth year. Is this a case of logistic growth? Exponential growth? Irresponsible cat owners?

2. A third of the world population is below age fifteen. Describe the effect of this age distribution on the future growth rate of the human population. If you conclude that it will have severe impact, what sorts of humane recommendations would you make to encourage individuals of this age group to limit family size? What are some social, economic, and environmental factors that might keep them from following the recommendations?

3. Write a short essay about a population having one of the age structures shown below. Describe what may happen to younger and older groups when individuals move into new categories.

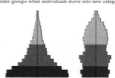

4. Figure 46.18 charts the legal immigration to the United States between 1820 and 1995. (The Immigration Reform and Control Act of 1986 accounted for the most recent dramatic increase; it granted legal status to illegal immigrants who could prove they had lived in the country for years.) During the 1980s and 1990s, an economic downturn fanned resentment against newcomers. Many people see legal immigration should be restricted to 300,000–450,000 annually and we should crack down on the illegal immigrants. Others argue such a policy would diminish our reputation as a land of opportunity. They also say it would discriminate against legal immigrants during crackdowns on others of the same ethnic background. Do some research, then write an essay on the pros and cons of both positions.

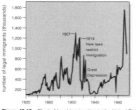

Figure 46.18 Chart of legal immigration to the United States between 1820 and 1995.

5. In his book *Environmental Science*, Miller points out that the unprecedented projected increase in the human population from 5 billion to 10 billion by the year 2050 raises serious questions. Will there be enough food, energy, water, and other resources to sustain twice as many people? Will governments be able to provide adequate education, housing, medical care, and other social services for all of them? Computer models suggest that the answers are no (Figure 46.19). Yet some people claim we can adapt socially and politically to an even more crowded world, assuming harvests improve through technological innovation, every inch of arable land is put under cultivation, and everyone eats only grain. There are no easy answers to the questions. If you have not yet been doing so, start following the arguments in your local newspapers, in magazines, and on television. This will allow you to become an informed participant in a global debate that surely will have impact on your future.

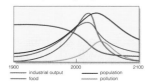

Figure 46.19 Computer-based projection of what may happen if the size of the human population continues to skyrocket without dramatic policy changes and technological innovation. The assumptions are that the population has already overshot the carrying capacity and current trends continue unchanged.

Selected Key Terms

age structure 46.1
biotic potential 46.2
carrying capacity 46.3
cohort 46.4
demographic transition model 46.8
demographics 46.1
density-dependent control 46.3
density-independent factor 46.3
doubling time 46.2
ecology CI
emigration 46.2
exponential growth 46.2
family planning program 46.7
immigration 46.2
life history pattern 46.4
life table 46.4
limiting factor 46.3
logistic growth 46.3
migration 46.2
per capita 46.2
population density 46.1
population distribution 46.1
population size 46.1
r (net reproduction per individual per unit time) 46.2
reproductive base 46.1
survivorship curve 46.4
total fertility rate 46.7
zero population growth 46.2

Readings

Cohen, J. E. 1995. *How Many People Can the Earth Support?* New York: Norton. No pat answer to title's question.

Miller, G. T. 1996. *Environmental Science*. Sixth edition. Belmont, California: Wadsworth.

Web Site See http://www.wadsworth.com/biology for practice quiz questions, hypercontents, BioUpdates, and critical thinking. The Wadsworth Biology Resource Center provides a wealth of information fully organized and integrated by chapter.

Chapter 46   Population Ecology   **811**

. . . and a self-quiz to reinforce chapter terms and concepts . . .

. . . and thought questions as exercises in critical thinking . . .

. . . and recommendations for further reading and Web sites.

FIGURE D   *Study aids at the end of a chapter from this edition.*

CONTENT REVISIONS

Instructors who use *Biology: The Unity and Diversity of Life* may wish to evaluate this overview of key modifications. Overall, conceptual development is more integrated. The writing is still crisp but not too brief, because some topics can confuse students when presented in insufficient detail. New research called for some adjustments in the overall framework. For example, we subscribe to the six-kingdom classification scheme, which now has exceptional support from comparative biochemical studies. Even the end-of-chapter Critical Thinking questions incorporate current material.

INTRODUCTION   This conceptual overview for the book is more focused, starting with introductions to the molecular trinity (DNA to RNA to protein), energy, and the levels of biological organization. An early glimpse of life's diversity reflects the six-kingdom model. We follow it with a simple explanation of evolution by natural selection, a dominant theme throughout the book. We treat scientific methods in more detail. To highlight the power of scientific inquiry, we have a new spread on experimental tests of an alternative to antibiotics. This detailed, timely example builds on an earlier description of selection for antibiotic resistance.

UNIT I. PRINCIPLES OF CELLULAR LIFE   Chapter 1 opens with a new vignette on a chemical element and a current application (phytoremediation, as at Chernobyl). Sections on radioactive decay and on acids, bases, and buffers are rewritten. Chapter 3 opens with a "carbon story" about swings in atmospheric CO_2. It has an essay on pesticides. Chapter 4 has better treatment of cytoskeletal elements and accessory proteins, including myosin. Chapter 5 has improved explanations of membrane proteins, diffusion, osmosis, and membrane transport mechanisms. A rewrite of Chapter 6 makes the ground rules of metabolism more intelligible. The photosynthesis chapter has new icons, new concept spreads on properties of light and photosynthetic pigments, and a visual comparison of carbon-fixing routes. Chapter 8 includes a new illustration of the energy harvest from aerobic respiration, and a more lively, updated treatment of alternative energy sources in humans and other mammals.

UNIT II. PRINCIPLES OF INHERITANCE   Chapter 9 includes a new spread on the cell cycle and chromosome structure. Chapter 11 includes rewrites on pleiotropy, phenotypic variation, and some new, illustrated genetics problems. We strengthened Chapter 12 with reorganized text and new art on linkage, recombination patterns, linkage mapping, and changes in chromosome structure. It has more graphic examples of inheritance patterns. Chapter 13 has a visual overview as well as a detailed look at DNA replication and an essay on DNA mutations and cancer. We improved the Chapter 14 treatment of mutation. Chapter 15 opens with a new vignette on apoptosis, a prelude to an updated section on cancer at the chapter's end. Definitions are clarified. Chapter 16 has a new vignette and a new essay on RFLPs.

MULTIMEDIA SUPPLEMENTS

1. Interactive Concepts in Biology. Packaged free with all student copies, this is the first CD-ROM to address the full sweep of biology. The cross-platform CD-ROM covers all concept spreads in the book. Because students can learn by doing, it encourages them to manipulate the book's art. A combination of text, graphics, photographs, animations, video, and audio enhances each book chapter. The revised CD-ROM has three times as many animations, many more interactive quizzes, and many more interactive exercises.

2. InfoTrac College Edition. This on-line library is available FREE with each copy of *Biology: The Unity and Diversity of Life*. It gives students access to full articles—not abstracts—from more than 600 scholarly and popular periodicals dating back as far as four years. The articles are available through InfoTrac's impressive database that has such periodicals as *Discover*, *Audubon*, and *Health*.

3. Student Guide to InfoTrac College Edition. This guide is on the Wadsworth Biology Resource Center site on the World Wide Web. It has an introduction to InfoTrac and a set of electronic readings for each chapter, updated frequently. It links some of each chapter's critical thinking questions to InfoTrac articles to invite deeper examination of issues.

4. Biology Resource Center. All information is arranged by the eighth edition's chapters. Every month it has new BioUpdates on relevant applications, and hyperlinks and an average of 40 practice quiz questions per chapter. It includes descriptions of degrees and careers in biology, a student feedback site, cool clip art, ideas for teaching on the Web, and a forum where instructors can share ideas on teaching courses. It also includes flashcards for all glossary terms, critical thinking exercises, news groups, surfing lessons including biology surfing, and a variety of search engines, biological games, a BioTutor, and a final Blitz set of practice questions. Perhaps most importantly, it also has Internet exercises for each chapter to guide students to doing more than randomly browse sites. A cool event of the quarter will have an ongoing experiment in which students and instructors can participate. The address for the Wadsworth Biology Resource Center is:

http://www.wadsworth.com/biology

5. Internet Activities for General Biology. This book guides students to more productive activities than browsing the Web. Students learn by interactive dissections, surveys, genetic crosses, lab experiments, notice postings, and other diverse activities. It has tear-out worksheets that may be handed in for evaluation.

6. An Introduction to the Internet. This 100-page booklet helps students learn how to get around the Internet when using a browser such as Netscape, search engines, e-mail, setting up home pages, and related topics. It lists useful biology sites on the Net that correspond to book chapters.

7. The Biology Place (www.biology.com). Wadsworth is an official distributor of this site, created by Peregrine Publishers, Inc., for instructors and students on the World Wide Web. A community of educators who developed and maintain it offer learning activities categorized by topic and type, including interactive study guides, lab investigations, study projects, and collaborative research. One can access Research News, Best of the Web, and *Scientific American*

Connection for cutting-edge research. Each learning activity has interactive self-assessment worksheets and notebooks. Results can be printed or e-mailed directly to instructors.

8. *American Botanical Society.* This site will be maintained on the Wadsworth Web page at http://www.thomson.com.

9. *BioLink 2.* With this presentation tool, instructors can easily assemble art and database files with lecture notes to create a fluid lecture that may help stimulate even the least-engaged students. It includes all illustrations in the book, animations and films from the student CD, and art from other Wadsworth biology textbooks. BioLink 2 also has a Kudo Browser with an easy drag-and-drop feature that allows file export into such presentation tools as Power Point. Upon its creation, a file or lecture with BioLink 2 can be posted to the Web, where students can access it for reference or for studying needs.

10. *Overheads and 35mm Slides.* All the micrographs and diagrams in the book are available as overheads and slides that are reproduced in vivid color with large, bold-lettered labels. All of the diagrams are on *CD-ROM*, Biolink 2.

11. *Cycles of Life: Exploring Biology.* Twenty-six programs of this telecourse feature compelling footage from around the world, original microvideography, and spectacular 3D animation. A student study guide, faculty manual, and laboratory manual are included. For information on course licensing and pricing, send queries to Coast Telecourses by fax 714-241-6286 or telephone 1-800-547-4748.

12. *Animations and Films from Cycles of Life Telecourse.* Tape One has animations for cell structure and function, and for principles of inheritance and evolution. Tape Two has more on diversity, plant structure and function, animal structure and function, and ecology and behavior.

13. *CNN Videos.* Produced by Turner Learning, nine videos can stimulate and engage students. They cover general biology, anatomy-physiology, and environmental science. New tapes are offered every year for these topics.

14. *Protecting Endangered Species* and *Science in the Rain Forest Videos.* Current treatment of these major issues.

15. *Life Science Video Library.* Films for the Humanities and CNN created this library.

16. *Wadsworth Biology Videodisc.* Carolina Biological Supply and Bill Surver (Clemson University) collaborated on this videodisc with new animations and films. Line art has large, boldface labels and often step-by-step or full-motion animation. It is available with fill-in-the-blank labels for tests. There are 3,500 still photographs, a correlation directory, bar code guide, and hypercard and toolbook software. All items are organized by book chapter.

17. *West's Biology Videodisc.* This videodisc provides thousands of additional images.

18. *Liquid Assets—The Ecology of Water.* Two double-sided level-3 videodiscs compare the ecologies of the Everglades and San Francisco Bay. A study guide is included.

19. *STELLA II.* Adopters can get a version of this software tool to develop critical thinking skills, and a workbook. This modeling software has 23 simulations. The textbook's critical thinking questions and 150 additional ones are arranged by chapter in *Critical Thinking Exercises.*

20. *Electronic Study Guide.* This has an average of 40 multiple-choice questions per book chapter that differ from those in the test-item booklet. Students respond to each question, and then an on-screen prompt allows them to review their answers and learn why they are correct or incorrect.

21. *West Nutrition CD.* This interactive learning tool has animations, video, hands-on exercises, and a glossary with pronunciation guides. In-depth sections allow students to learn more about the biochemistry of particular topics.

ADDITIONAL SUPPLEMENTS

Seven respected test writers created the *Test Items* booklet. The booklet has more than 4,500 questions in electronic form for IBM and Macintosh in a test-generating data manager. Questions also are available in Microsoft Word for Windows, DOS, and Mac.

For each book chapter, an *Instructor's Resource Manual* has an outline, objectives, list of boldface or italic terms, and detailed lecture outline, ideas for lectures, classroom and lab demonstrations, discussion questions, research paper topics, and annotations for filmstrips and videos. For those who wish to modify material, this resource manual is available electronically on the Wadsworth Web page.

An interactive *Study Guide and Workbook* lets students write answers to questions, which are arranged by chapter section with references to specific text pages. For those who wish to modify or select parts, *chapter objectives* are available on disk in the testing file.

A 100-page *Answer Booklet* answers the textbook's review questions. (The answers to the self-quizzes and genetics problems are given in appendixes to the book itself.)

Flashcards show 1,000 glossary items. A booklet, *Building Your Life Science Vocabulary,* helps students learn biological terms by explaining root words and their applications.

Study Skills for Science Students. Daniel Chiras's guide explains how to develop good study habits, sharpen memory and learning, prepare for tests, and produce term papers.

Strategies for Success: Learning Skills Booklets. Individual learning skills chapters from Gardner/Jewler's best-selling college success text can be customized together or singly and bundled with the text. Some chapters cover managing time, test taking, writing and speaking, note taking, reading and memory, computers and the Internet, critical thinking, and campus resources including the library.

Jim Perry and David Morton's *Laboratory Manual* has 38 experiments and exercises, with 600+ full-color, labeled photographs and diagrams. Many experiments are divided in parts for individual assignment, depending on available time. Each consists of objectives, discussion (introduction, background, and relevance), list of materials for each part, procedural steps, prelab questions, and post-lab questions. An *Instructor's Manual* for the lab manual lists quantities, procedure for preparing reagents, time requirements for each part of an exercise, hints to make the lab a success, and vendors of materials with item numbers. It has more investigative exercises that can be copied for lab use.

Customized Laboratory Manuals by Phillip Shelp and its accompanying instructor's manual can be tailored for individual courses. A new *Photo Atlas* has 700 full-color, labeled photographs and micrographs of the cells and organisms that students typically deal with in the lab.

Eight additional readings supplements are available:
• *Contemporary Readings in Biology* is a collection of articles on applications of interest. • *Current Perspectives in Biology* is another collection of articles. • *A Beginner's Guide to Scientific Method* is a supplement for those who wish to treat this topic in detail. • *The Game of Science* gives students a realistic view of what science is and what scientists do. • *Environment: Problems and Solutions* provides a brief 120-page introduction to environmental concerns. • *Green Lives, Green Campuses* is a hands-on workbook to help students evaluate the environmental impact of their own life-styles. • *Watersheds: Classic Cases in Environmental Ethics* is a collection of important case studies. • *Environmental Ethics: An Introduction to Environment Philosophy* surveys environmental ethics and recent philosophical positions.

A COMMUNITY EFFORT

One, two, or a smattering of authors can write accurately and often very well about their field of interest, but it takes more than this to deal with the full breadth of the biological sciences. For us, it takes an educational network that includes more than 2,000 teachers, researchers, and photographers in the United States, Canada, England, Germany, France, Australia, Sweden, and elsewhere. On the next two pages, we acknowledge reviewers whose contributions continue to shape our thinking. There simply is no way to describe the thoughtful effort these individuals and others before them gave to our books. We can only salute their commitment to quality in education.

In large part, *Biology: The Unity and Diversity of Life* is widely respected because it reflects the understandings of our general advisors and contributors and their abiding concern for students. Steve Wolfe, author of acclaimed books on cell biology, works out alternative phrasings with us, sometimes line by line over the phone. Daniel Fairbanks, scholar and gentleman, still finds the time to ferret out errors that creep into the genetics manuscripts. Katherine Denniston, another long-time advisor, wrote original critical thinking questions, itself no small feat. Aaron Bauer, Paul Hertz, Samuel Sweet, and Jerry Coyne helped chisel major parts of the evolution unit. Jerry also created our new computer simulation of genetic drift. The unit on plant structure and function is strong, thanks to the initial resource manuscripts from Cleon Ross.

And what would we do without Robert Lapen, who wrote the definitive manuscript on immunology and lived to tell about it? What would we do if our abiding friends Gene Kozloff, John Jackson, and Ron Hoham did not work diligently to stop us from inventing biology? Also for this edition, in collaboration with Linda Beidleman, Gene assembled the book's detailed Index.

We thank Rob Colwell, George Cox, and Tyler Miller for their contributions to the ecology unit. John Alcock, author of a respected book on behavior, provided resource manuscripts for our animal behavior chapter. Bruce Levin and Jim Bull provided prepublication data when we wrote a new essay on experimental testing. Lauralee Sherwood, author of a fine animal physiology text, helped refine the respiration chapter. As always, Jane Taylor and Larry Sellers were meticulous in their attention to detail.

Also over the years, Nancy Dengler, Bruce Holmes, David Morton, and Frank Salisbury have been guardians of accuracy and teachability. So has Tom Garrison, himself a seasoned author, who dispenses sympathy with wit better than just about anybody. This time, Tom and Don Collins generously opened their lecture hall and laboratory, and the revelations will shape our thinking for years to come.

Mary Douglas has now become the finest production manager in the business. We cannot imagine our complex revisions traveling through computerized production without guidance from talented, even-tempered, flexible Mary. She also falls in the category of author's shrink. Myrna Engler-Forkner, in charge of oversight and art coordination, has become as patient and superb as Mary. Gary Head, Quarkmeister, has the patience to teach even computer-challenged authors how to do page layouts. He designed the book and its cover. He went out on many photographic assignments and came back with such treats as Fred-and-Ginger, the streetwise snail in Figure 17.1.

Sandra Craig is a good sport and a fine shepherdess who keeps editorial production on its convoluted route. Kristin Milotich is in charge of a vast enterprise known as the book's supplements program. She is remarkable for her endurance, a big heart, and no attitude problem. Our developmental editor and good friend, Mary Arbogast, is in a class by herself. Pat Waldo, force of nature, outruns clocks and moves mountains to keep us at the forefront of multimedia for biology education. Thanks to Chris Evers, outstanding author, consultant, and Pat's secret weapon in digital publishing. Thanks also to the rest of the 0's and 1's Club, Stephen Rapley, Steve Bolinger, Jennie Redwitz, and Cooperative Media Group.

If authors could invent the most supportive president of a publishing house, they would come up with Susan Badger. If we could bottle Halee Dinsey's energy, we would be able to power all cities west of the Rockies. Gary Carlson, Michael Burgreen, Andrea Geanacopoulos, and John Walker add formidable talent to the biology team. Stephen Forsling signed on as photo researcher, and Roberta Broyer as permissions editor. Carol Lawson and Rebecca Linquist cheerfully kept us Quarking. Kathie Head, Pat Brewer, Peggy Meehan, and Stephen Rapley have made valuable contributions for many years.

Of the artists with whom we have worked, Raychel Ciemma remains the best. She works directly with us to turn rough sketches into works of art. We also are grateful for the dedication of Mary Roybal, Karen Stough, and the others listed on the copyright page.

Jim Gallagher, J.C. Morgan, and Jan Troutt at Precision Graphics are superb professionals. Tom Anderson, John Deady, Nancy Dean, and Rich Stanislawski at H&S Graphics cheerfully put up with our interminable pursuit of excellence.

Twenty-two years ago, Jack Carey convinced us to write this book. Ever since, he has remained close counseler and abiding friend. And nothing, in all that time, has shaken our shared belief in the intrinsic capacity of biology to enrich the lives of each new generation of students.

Current configurations of Earth's oceans and land masses —the geologic stage upon which life's drama continues to unfold. Thousands of separate images were pieced together to create this remarkable cloud-free view of our planet.

1 CONCEPTS AND METHODS IN BIOLOGY

Biology Revisited

Buried somewhere in that mass of tissue just above and behind your eyes are memories of first encounters with the living world. In that brain are early memories of discovering your hands and feet, your family, friends, the change of seasons, the smell of rain-drenched earth and grass. Still in residence are memories of early introductions to a great disorganized parade of spiders, flowers, frogs, and furred things, mostly living, sometimes dead. There, too, are memories of questions—"*What is life?*" and, inevitably, "*What is death?*" There are also memories of answers, some satisfying, others less so.

By making observations, asking questions, and gradually accumulating answers about the natural world, you have built up a store of knowledge about life. Education as well as experience has been refining your questions, and no doubt some answers are difficult to come by. Think of a young man, only twenty years old, whose motorcycle ran straight into a truck. Now he is in the hospital, with a brain that is functionally dead. If his breathing, heart rate, and other basic functions will proceed only as long as he stays hooked up to a respirator and other highly mechanized support systems, is he still "alive"? Or think of a recently fertilized egg

Figure 1.1 Think back on all you have ever known and seen. This is a foundation for your deeper probes into life.

inside a woman's body. At this early stage, a series of cell divisions has transformed the egg into a cluster of no more than a few dozen microscopically small cells. At what point in its development would you define the growing cell mass as a *human* life? If questions like this have ever crossed your mind, your thoughts about life obviously run deep.

The point is, the book you are starting to read is not your introduction to biology—the study of life—for you have been studying life ever since information started to penetrate your brain. The book simply is biology *revisited*, in ways that may help carry your thoughts to deeper, more organized levels of understanding.

Return to the question, *What is life?* Offhandedly, you might respond that you know it when you see it. To biologists, however, the question opens up a story that has been unfolding in countless directions for several billion years! "Life" is an outcome of ancient events by which nonliving materials became assembled into the first living cells. "Life" is a way of capturing and using energy and raw materials. "Life" is a way of sensing and responding to changes in the environment. "Life" is a capacity to reproduce, grow, and develop. And "life" evolves, meaning that details in the body plan and functions of organisms can change through successive generations.

Yet this short description only hints at the meaning of life. Deeper insight requires wide-ranging study of life's characteristics.

Throughout this book, you will come across many diverse examples of how organisms are constructed, how they function, where they live, and what they do. The examples support certain concepts which, taken together, will give you a sense of what "life" is. This chapter introduces you to the basic concepts. It also sets the stage for forthcoming descriptions of observations, experiments, and tests that help show how biologists develop, modify, and so refine their views of the world around them. As you continue reading the book, you may find it useful to return occasionally to this simple overview as a way of reinforcing your grasp of details.

KEY CONCEPTS

1. There is an underlying unity in the world of life, for all organisms are alike in key respects. They consist of the same kinds of substances, put together according to the same laws that govern matter and energy. Their activities depend on inputs of energy, which they must obtain from their environment. All organisms sense and respond to changing conditions in their environment. And they all grow and reproduce, based on instructions contained in their DNA.

2. There also is immense diversity in the world of life. Millions of different organisms inhabit Earth, and many millions more lived and became extinct over the past 3.8 billion years. And each kind of organism is unique in some of its traits—that is, in some aspects of its body plan, body functions, and behavior.

3. Theories of evolution, especially a theory of natural selection as first formulated by Charles Darwin, help explain the meaning of life's diversity.

4. Biology, like other branches of science, is based on systematic observations, hypotheses, predictions, and relentless observational and experimental tests. The external world, not internal conviction, is the testing ground for scientific theories.

Nothing Lives Without DNA

DNA AND THE MOLECULES OF LIFE   Picture a frog on a rock, busily croaking. Without even thinking about it, you know the frog is alive and the rock is not. But would you be able to explain why? At a fundamental level, both are no more than concentrations of the same units of matter, called protons, electrons, and neutrons. These units are the building blocks of atoms, which are building blocks of larger bits of matter called molecules. And it is at the molecular level that differences between living and nonliving things start to emerge.

You will never, ever find a rock made of nucleic acids, proteins, carbohydrates, and lipids. In the natural world, only cells build particular assortments of these complex molecules, in particular amounts. The **cell** is the smallest unit of matter having the capacity for life; all living things consist of one or more of them. And the cell's signature molecule is a nucleic acid popularly known as **DNA**. No chunk of granite or quartz has it.

Encoded in DNA's structure are the instructions for assembling a dazzling array of proteins from a limited number of smaller building blocks, the amino acids. By analogy, if you follow suitable instructions and invest some energy in the task, you can organize a heap of a few kinds of ceramic tiles (amino acids) into different patterns (diverse proteins), as shown in Figure 1.2.

Among the proteins are enzymes, a class of workers that build, juggle, and split *all* of the complex molecules of life when they get an energy boost. Some enzymes work as partners with another class of nucleic acids, the **RNAs**, in carrying out DNA's instructions. A simple way to think about this is to envision a flow from *DNA to RNA to protein*. As you will read in Chapter 14, this molecular trinity is central to our understanding of life.

THE HERITABILITY OF DNA   Although we humans tend to think we enter the world rather abruptly and leave it the same way, we are much more than this. *We and all other organisms are part of a journey that began almost 4 billion years ago, with the emergence of the first living cells.* Under present-day conditions on Earth, cells can arise only from cells that already exist. They do so by one of life's defining features, called **reproduction**. By

this process, parents transmit DNA instructions for duplicating their traits to offspring. Why do baby storks look like storks and not pelicans? They inherited stork DNA, which is not exactly the same as pelican DNA.

The process typically starts with a single cell that contains the DNA of one or two parents. Think of the first cell that forms from the fusion of a sperm (a single cell) with an egg (another single cell). That fertilized egg would not even exist if the sperm and egg had not formed earlier, according to DNA instructions passed on from cell to cell through countless generations.

For frogs and humans and other large organisms, DNA also encodes a developmental program by which single cells divide again and again, with most of their descendants becoming specialized in ways that form different tissues and organs. Think of a moth. Is it only a winged insect? Then what of the fertilized egg that a female moth deposits on a leaf (Figure 1.3)? Inside are DNA instructions for becoming an adult. They guide the egg's development into a caterpillar: an immature, larval stage adapted for rapid feeding and growth. The caterpillar eats and grows until an internal alarm clock goes off. Then tissues undergo drastic remodeling into a different developmental stage, called a pupa. Many cells die; others multiply and become organized in different patterns. In time, an adult emerges that is adapted for reproduction. It contains organs that produce sperm or eggs. And it flutters its distinctly colored and patterned wings at a frequency suitable for attracting a mate.

None of these stages is "the insect." The insect is a series of organized stages, from one fertilized egg to the next. Each stage is vital for the ultimate production of new moths. The instructions for each stage were written into moth DNA long before each individual moment of reproduction—and so the ancient moth story continues.

Nothing Lives Without Energy

ENERGY DEFINED   Everything in the entire universe has some amount of **energy**, which is most simply defined as a capacity to do work. And nothing—absolutely nothing—happens in the universe without a complete or partial *transfer* of energy. For example, a solitary atom does nothing except vibrate incessantly with its

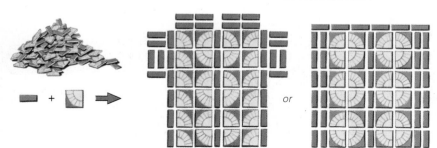

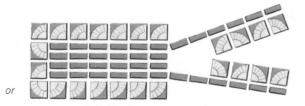

Figure 1.2   Examples of objects built from the same materials but with different assembly instructions.

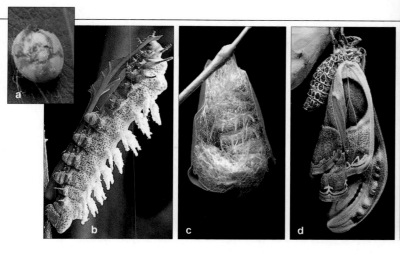

Figure 1.3 "The insect"—a continuous series of stages of development. Different adaptive properties emerge at each stage. Shown here, a silkworm moth, from the egg (**a**), to a larval stage (**b**), to a pupal stage (**c**), and on to the winged form of the adult (**d**,**e**).

own inherent energy. Suppose the atom absorbs extra energy from the sun's rays and starts vibrating faster. Now some energy is on the move, and it can do work by getting transferred elsewhere. By chance, the atom collides with a neighboring atom, and one may give up, grab, or share energy with the other. Molecules form, stay together, become rearranged, and may get split by such energy transfers. Without them, organisms cannot stay alive, grow, and reproduce.

METABOLISM DEFINED By now, you have an idea that a cell has the capacity to (1) obtain and convert energy from its surroundings and (2) use energy to maintain itself, grow, and make more cells. We call this capacity **metabolism**. Think of a food-producing cell in a leaf. By a process known as **photosynthesis**, it intercepts energy from the sun and uses it to form ATP, an energy carrier. ATP transfers energy to metabolic workers—in this case, enzymes that assemble sugars. Thus energy from the sun becomes stored as energy in sugar molecules. Later, by a process called **aerobic respiration**, some energy released from sugars and other molecules is intercepted to form more ATP, which helps drive hundreds of activities.

SENSING AND RESPONDING TO ENERGY It is often said that only organisms respond to the environment. Yet even a rock shows responsiveness, as when it yields to the force of gravity and tumbles down a hill or changes its shape slowly under the repeated battering of wind, rain, or tides. The difference is this: *Organisms sense changes in their surroundings, then they make controlled, compensatory responses to them*. How? Each organism has **receptors**, which are molecules and structures that can detect specific stimuli. A **stimulus** is any energy change in the environment, such as a variation in the amount of light or heat, that a receptor can detect.

Cells adjust metabolic activities in response to signals from receptors. Each cell (and organism) can withstand only so much heat or cold. It must rid itself of harmful substances. It requires certain foods, in certain amounts. Yet temperatures do shift, harmful substances might be encountered, and food is sometimes plentiful or scarce. Think of what happens after you finish a snack. Simple sugars leave the gut and enter your blood, which is part of your *internal* environment (the other part is tissue fluid that bathes cells). Over the long term, too much or too little blood sugar causes problems, such as diabetes. Normally, though, a glandular organ called the pancreas steps up its secretion of insulin as the sugar level rises. Most of your cells have receptors for this hormone, which stimulates cells to take up sugar. When they do, the sugar level in blood returns to normal.

What if you skip dinner and the sugar level declines? A different hormone stimulates liver cells to tap their stores of energy-rich molecules and degrade them to simple sugars. Sugars released from the cells enter the blood and help return the blood sugar level to normal.

Organisms respond so exquisitely to energy changes that their internal operating conditions remain within tolerable limits. We call this a state of **homeostasis**, and it is one of the key defining features of life.

All organisms consist of one or more cells, the smallest units of life. Under present-day conditions, new cells form only through the reproduction of cells that already exist.

DNA, the molecule of inheritance, encodes protein-building instructions, which RNAs help carry out. Many of the proteins are enzymes, and these metabolic workers are necessary to construct all of the complex molecules of life.

Cells live only as long as they engage in metabolism. They acquire and transfer energy about to assemble, break down, stockpile, and dispose of materials in ways that promote survival and reproduction.

Single cells and multicelled organisms sense and respond to specific energy changes in the environment in ways that help maintain their internal operating conditions.

1.2 ENERGY AND LIFE'S ORGANIZATION

Levels of Biological Organization

Taken as a whole, the metabolic activities of single cells and multicelled organisms maintain the great pattern of organization in nature, as sketched out in Figure 1.4. Consider this hierarchy. Life's properties emerge when DNA and other molecules become organized into cells. We can formally define the **cell** as the smallest unit of organization having a capacity to survive and reproduce on its own, given suitable conditions and building blocks, DNA instructions, and energy inputs. Amoebas, which are free-living single cells, clearly are like this. Does the definition still apply to **multicelled organisms**, which consist of specialized, interdependent cells that are typically organized in tissues and organs? Yes. You may find this a strange answer. After all, your own cells could never live alone in nature, because body fluids must continually bathe them. Yet even isolated human cells stay alive under controlled laboratory conditions. Investigators routinely maintain isolated human cells for use in important experiments, as in cancer studies.

BIOSPHERE
All those regions of Earth's waters, crust, and atmosphere in which organisms can exist

ECOSYSTEM
A community and its physical environment

COMMUNITY
The populations of all species occupying the same area

POPULATION
A group of individuals of the same kind (that is, the same species) occupying a given area

MULTICELLED ORGANISM
An individual composed of specialized, interdependent cells most often organized in tissues, organs, and organ systems

ORGAN SYSTEM
Two or more organs interacting chemically, physically, or both in ways that contribute to survival of the whole organism

ORGAN
A structural unit in which a number of tissues, combined in specific amounts and patterns, perform a common task

TISSUE
An organized group of cells and surrounding substances functioning together in a specialized activity

CELL
Smallest unit having the capacity to live and reproduce, independently or as part of a multicelled organism

ORGANELLE
Inside all cells except bacteria, a membrane-bound sac or compartment for a separate, specialized task

MOLECULE
A unit in which two or more atoms of the same element or different ones are bonded together

ATOM
Smallest unit of an element (a fundamental substance) that still retains the properties of that element

SUBATOMIC PARTICLE
An electron, proton, or neutron; one of the three major particles of which atoms are composed

Figure 1.4 Levels of organization in nature.

You typically find cells and multicelled organisms as part of a **population**, defined as a group of organisms of the same kind, such as a herd of zebras. The next level of organization is the **community**—all the populations of all species living in the same area (such as the African savanna's bacteria, grasses, zebras, lions, and so on). The next level, the **ecosystem**, is the community *and*

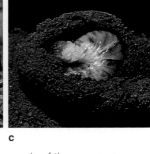

Producers trap, convert, and use or store some energy from the sun.

PRODUCERS

NUTRIENT CYCLING

CONSUMERS, DECOMPOSERS

ONE-WAY FLOW OF ENERGY

Energy is transferred from one organism to another; in time, all flows back to the environment.

a

b

c

Figure 1.5 An example of the one-way energy flow and the cycling of materials through the biosphere. (**a**) Plants of a warm, dry grassland called the African savanna capture energy from the sun and use it to build plant parts. Some of the energy ends up in plant-eating organisms, including this adult male elephant. He eats huge quantities of plants to maintain his eight-ton self and produces huge piles of solid wastes— dung—that still contain some unused nutrients. Thus, although most organisms would not recognize it as such, elephant dung is an exploitable food source.

(**b**) And so we next have little dung beetles rushing to the scene almost simultaneously with the uplifting of an elephant tail. Working rapidly, they carve fragments of moist dung into round balls, which they roll off and bury in burrows. In the balls the beetles lay eggs—a reproductive behavior that helps assure their forthcoming offspring (**c**) of a compact food supply. Thanks to beetles, dung does not pile up and dry out into rock-hard mounds in the intense heat of the day. Instead, the surface of the land is tidied up, beetle offspring get fed, and the leftover dung accumulates in beetle burrows—there to enrich the soil that nourishes the plants that sustain (among others) the elephants.

its physical and chemical environment. The **biosphere** encompasses all regions of Earth's atmosphere, waters, and crust in which organisms live. Astoundingly, *this globe-spanning organization begins with the convergence of energy, certain materials, and DNA in tiny, individual cells.*

Interdependencies Among Organisms

A great flow of energy into the world of life starts with the **producers**—plants and all other organisms that make their own food. Animals are **consumers**. Directly or indirectly, they depend on energy that became stored in the tissues of producers. For example, some energy is transferred to zebras after they browse on grasses. It is again transferred when lions devour a baby zebra that has wandered away from its herd. And it is transferred again when certain fungal and bacterial decomposers feed on tissues and remains of lions, elephants, or any other organism. **Decomposers** break down sugars and other biological molecules to simple materials—which

may be cycled back to producers. In time, all the energy that the producers initially captured from the sun's rays returns to the environment, but that's another story.

For now, keep in mind that organisms connect with one another by a one-way flow of energy through them and a cycling of materials among them, as in Figure 1.5. Their interconnectedness affects the structure, size, and composition of populations and communities. It affects ecosystems, even the biosphere. Understand the extent of their interactions and you will gain insight into acid rain, amplification of the greenhouse effect, and other modern-day problems.

Levels of organization exist in nature. The characteristics of life emerge at the level of single cells and extend through populations, communities, ecosystems, and the biosphere.

A one-way flow of energy through organisms and a cycling of materials among them organizes life in the biosphere. In nearly all cases, energy flow starts with energy from the sun.

So far, we have focused on life's unity, on the characteristics that all living things have in common. Think of it! They are put together from the same "lifeless" materials. They remain alive by metabolism —by ongoing energy transfers at the cellular level. They interact in their requirements for energy and for raw materials. They have the capacity to sense and respond to the environment in highly specific ways. They all have a capacity to reproduce, based on instructions encoded in DNA. And they all inherited their molecules of DNA from individuals of a preceding generation.

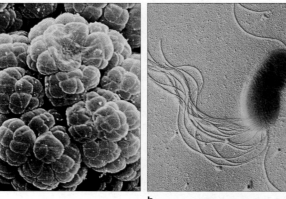
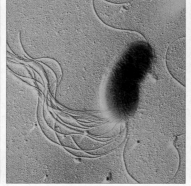
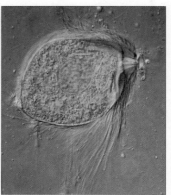

Superimposed on the common heritage is immense diversity. You share the planet with what may be many millions of different kinds of organisms, or **species**. Many millions more preceded you at some time during the past 3.8 billion years, but their lineages vanished; they became extinct. A few centuries ago, attempts to make sense of the confounding diversity resulted in a classification scheme that assigns each newly identified species a two-part name. The first part designates the **genus** (plural, genera). Each genus encompasses all of the species that seem to be related by way of descent from a common ancestor. The second part of the name designates a particular species within that genus.

For example, *Quercus alba* is the scientific name for the white oak, and *Q. rubra* is the name for the red oak. As this example suggests, once you spell out the name of the genus in a document, you may abbreviate the name wherever else it appears in that document.

Biologists also classify life's diversity by assigning species to groups at more encompassing levels. Among other things, they group genera that apparently share a common ancestor into the same *family*, related families into the same *order*, related orders into the same *class*, and related classes into the same *phylum* (plural, phyla). At a higher level, they assign related phyla to the same *kingdom*. They are still refining the groups. For instance, ancient scholars recognized only two kingdoms (animals and plants). Modern biologists have recognized as many as five kingdoms, but new evidence strongly suggests there should be six—the **Archaebacteria**, **Eubacteria**, **Protista**, **Fungi**, **Plantae**, and **Animalia** (Figure 1.6). We use a six-kingdom scheme throughout this book. At this point, it is enough simply to become familiar with a few of the defining features of their members.

All of the world's bacteria (singular, bacterium) are single cells, of a type called *prokaryotic*. The word means they do not have a nucleus, a membranous sac that otherwise would keep their DNA separated from the

d

Figure 1.6   A few representatives of life's diversity.

KINGDOM ARCHAEBACTERIA. (**a**) From the muck of an anaerobic (oxygen-free) habitat, a colony of methanogenic cells.

KINGDOM EUBACTERIA. (**b**) A eubacterium sporting a number of bacterial flagella, a structure that has uses in motility.

KINGDOM PROTISTA. (**c**) A trichomonad that lives in a termite's gut. Compared to bacteria, most protistans are much larger and have greater internal complexity. They range from microscopically small single cells to giant seaweeds.

KINGDOM FUNGI. (**d**) A stinkhorn fungus. Some species of fungi are parasites and some cause diseases, but the vast majority are decomposers. Without decomposers, communities would gradually become buried in their own garbage.

KINGDOM PLANTAE. (**e**) A grove of redwoods near the coast of California. Like nearly all plants, redwoods produce their own food by photosynthesis. (**f**) Flower of a plant called a composite. Its colors and patterning guide bees to nectar. The bees get food, and the plant gets help reproducing. Like many other organisms, they interact in a mutually beneficial way.

KINGDOM ANIMALIA. (**g,h**) Male bighorn sheep competing for females and so displaying a characteristic of the kingdom— they move actively through their environment.

e

f

g

h

rest of their interior. Different bacterial species are producers, consumers, and decomposers. And of all organisms, bacteria are the ones that show the greatest metabolic diversity.

Archaebacteria live only in extreme habitats, such as the ones that prevailed when life originated. The eubacteria are more successful in distribution; different species live nearly everywhere, including in or on other organisms. Those living in your gut and on your skin outnumber the trillions of cells that make up your body.

Protistans include single-celled species as well as some multicelled forms. Most are larger than bacteria, and they have far greater internal complexity. All are *eukaryotic*, meaning that their DNA is enclosed within a nucleus. A dazzling variety of microscopically small, single-celled producers and consumers are protistans. So are multicelled brown algae and other "seaweeds," some of which are giants of the underwater world.

Most fungi, including the common field mushrooms sold in grocery stores, are multicelled. These eukaryotic decomposers and consumers feed in a distinctive way. They secrete substances that digest food outside of the fungal body, then their cells absorb the digested bits.

Plants and animals include many familiar species and an astounding number of obscure ones. Members of both kingdoms are eukaryotic. Plants are multicelled producers; nearly all are photosynthetic. Animals are multicelled consumers that include diverse plant eaters, parasites, and meat eaters. Unlike plants, animals move about during at least some stage of their life.

Pulling this all together, you start to get a sense of what it means when someone says life shows unity *and* diversity.

Unity threads through the world of life, for all organisms are alike in several ways. They are composed of the same substances, which are assembled in the same basic ways. They engage in metabolism, and they sense and respond to their environment. They all have a capacity to reproduce, based on heritable instructions encoded in their DNA.

Immense diversity also threads through the world of life. As an outcome of differences in their DNA instructions, organisms differ enormously in body form, in the functions of their body parts, and in their behavior.

To make the study of life's diversity more manageable, we group organisms into six great kingdoms—archaebacteria, eubacteria, protistans, fungi, plants, and animals.

AN EVOLUTIONARY VIEW OF DIVERSITY

Given that organisms are so much alike, what could account for their great diversity? One key explanation is called evolution by means of natural selection. A few simple examples will be enough to introduce you to its assumptions, which build on the simple observation that variation in traits exists in all populations.

Mutation—Original Source of Variation

DNA has two striking qualities. Its instructions work to assure that offspring will resemble their parents, yet they also permit variations in the details of most traits. As an example, having five fingers on each hand is a human trait. Yet some humans are born with six fingers on each hand instead of five. This is an outcome of a **mutation**, a molecular change in the DNA. Mutations are the original source of variations in heritable traits.

Many mutations are harmful. A change in even a bit of DNA may be enough to sabotage the body's growth, development, or functioning. One such mutation causes *hemophilia A*. If a person affected by this blood-clotting disorder gets even a small cut or bruise, an abnormally lengthy time passes before a clot forms and stops the bleeding. Yet some variations are harmless or beneficial. A classic case is a mutation in light-colored moths that results in dark offspring. Moths fly at night and rest during the day, when the birds that eat them are active. A light-colored moth resting on a light-colored tree trunk is camouflaged (it "hides in the open," as in Figure 1.7), so birds tend not to see it. Suppose people build coal-burning factories nearby. Over time, soot-laden smoke darkens the tree trunks. Now the dark moths are less conspicuous to bird predators, so they have a better chance of living long enough to reproduce. Under the sooty conditions, the variant (dark) form of the trait is more adaptive. An **adaptive trait** simply is any form of a trait that helps an organism survive and reproduce under a given set of environmental conditions.

Evolution Defined

Think about a population of light-colored moths in a sooty forest. At some point, a DNA mutation arose in the population, and it resulted in a moth of a different color. When the mutated individual reproduced, some of its offspring inherited the trait. Birds saw and ate many light-colored moths, but most of the dark ones escaped detection and lived long enough to reproduce. So did their dark offspring, and so did *their* offspring. Over the generations, the frequency of the dark form of the trait increased and that of the light form decreased. As more time passes by, the dark form of the trait might even become the more common, and people might end up referring to "the population of dark-colored moths."

Figure 1.7 Example of different forms of the same trait (surface coloration), adaptive to two different environmental conditions. (**a**) On a light-colored tree trunk, light moths (*Biston betularia*) are hidden from predators, but dark ones stand out. (**b**) The dark color is more adaptive in places where tree trunks are darkened with soot.

Evolution is under way. In biology, the word refers to change occurring in a line of descent over time. As is true of moths, the individuals of most populations typically show different forms of many (or most) of their traits, and the frequencies of those different forms relative to one another can change over successive generations.

Natural Selection Defined

Long ago, the naturalist Charles Darwin used pigeons to explain the conceptual connection between evolution and variation in traits. Domesticated pigeons display splendid variation in size, feather color, and other traits (Figure 1.8). As Darwin knew, pigeon breeders select certain forms of traits. For example, if breeders prefer tail feathers that are black with curly edges, they will allow only those individual pigeons having the most black and the most curl in their tail feathers to mate and produce offspring. Over time, "black" and "curly" will become the most common forms of tail feathers in the captive population, and other forms of the two traits will become less common or will be eliminated.

Pigeon breeding is a case of **artificial selection**, for the selection among different forms of a trait is taking place in an artificial environment, under contrived, manipulated conditions. Yet Darwin saw the practice as a simple model for *natural* selection, a favoring of some forms of traits over others in nature. Whereas breeders are "selective agents" that promote reproduction of some individuals over others in captive populations, a pigeon-eating peregrine falcon is one of many selective agents that operate across the range of variation among pigeons in the wild. Generation after generation, the swifter or more effectively camouflaged pigeons have a better chance of living long enough to reproduce than the not-so-swift or too-conspicuous ones among them.

a

Figure 1.8 A few of the 300+ varieties of domesticated pigeons produced by artificial selection practices. Breeders began with variant forms of traits in captive populations of wild rock doves (**a**).

What Darwin identified as **natural selection** is no more than the outcome of differences in the survival and reproduction of a population's individuals that differ in one or more traits.

Unless you happen to be a pigeon breeder, Darwin's pigeon example may not be firing rockets through your imagination. So think about an example closer to home. Certain bacteria and fungi make *antibiotics*, metabolic products that can kill their bacterial competitors for nutrients in soil. Starting in the 1940s, we learned to use antibiotics to treat and control a variety of diseases that result when bacteria invade the human body and use it as a source of nutrients. Doctors routinely prescribed these so-called wonder drugs for mild as well as serious infections. The one called penicillin was even added to toothpaste, mouthwash, and chewing gum.

It turned out that antibiotics are powerful agents of natural selection. Over time, owing to mutations, some bacterial neighbors of the antibiotic producers have developed antibiotic resistance. Consider streptomycin, which binds with some essential bacterial proteins and inhibits their activity. In some variant strains of bacteria, mutations slightly changed the form of the proteins, so streptomycin is unable to bind with them. Such bacteria can escape streptomycin's effects.

In infected patients, an antibiotic acts against bacteria that are susceptible to its action—but it actually favors variant strains that have resistance to it! Presently, such antibiotic-resistant strains are making it difficult to treat typhoid, tuberculosis, gonorrhea, staph (*Staphylococcus*) infections, and some other bacterial diseases. In a few patients, the "superbugs" responsible for tuberculosis cannot be successfully eliminated.

As antibiotic resistance evolves by natural selection, so must antibiotics evolve if they are to overcome the defenses we unwittingly "selected." For example, drug companies have modified portions of the streptomycin molecule. Such molecular changes in the laboratory have produced more effective antibiotics—at least until

future generations of more resistant superbugs enter the deadly evolutionary competition for nutrients.

Later in the book, we will consider the mechanisms by which populations of moths, pigeons, bacteria, and all other organisms evolve. Meanwhile, keep in mind the following points about natural selection. They are central to biological inquiry, for they have consistently proved useful in explaining a great deal about nature.

1. Individuals of a population vary in form, function, and behavior, and much of the variation is heritable.

2. Some forms of heritable traits are more adaptive to environmental conditions than others. They improve an individual's chance of surviving and reproducing, as by helping it secure food, a mate, hiding places, and so on.

3. Natural selection is the outcome of differences in the survival and reproduction of individuals that show variation in one or more traits.

4. In time, natural selection results in a better fit with prevailing environmental conditions. The adaptive forms of traits tend to become more common and other forms less so. The population changes in its defining characteristics; it evolves.

In short, in the evolutionary view, *life's diversity is the sum total of variations in traits that have accumulated in different lines of descent over time, as by natural selection and other processes of change.*

Mutations in DNA introduce variations in heritable traits.

Although many mutations are harmful, some give rise to variations in form, function, or behavior that are adaptive under prevailing environmental conditions.

Natural selection is the outcome of differences in survival and reproduction among individuals of a population that show variation in one or more traits. The process helps explain evolution—changes in lines of descent over time.

The preceding sections sketched out the major concepts in biology. Now consider approaching this or any other collection of "facts" with a critical attitude. *"Why should I accept that they have merit?"* The answer requires insight into how biologists make inferences about observations and then test the predictive power of their inferences against actual experiences in nature or the laboratory.

Observations, Hypotheses, and Tests

To get a sense of "how to do science," start by following some practices that are pervasive in scientific research:

1. Observe some aspect of nature, carefully check what others have found out about it, and then frame a question or identify a problem related to your observation.

2. Develop **hypotheses**, or educated guesses, about possible answers to questions or solutions to problems.

3. Using hypotheses as a guide, make a **prediction**—that is, a statement of what you should observe in the natural world if you were to go looking for it. This is often called the "if-then" process. (*If* gravity does not pull objects toward Earth, *then* it should be possible to observe apples falling up, not down, from a tree.)

4. Devise ways to **test** the accuracy of your predictions, as by making systematic observations, developing models, and conducting experiments.

5. If the tests do not turn out as you expected, check to see what might have gone wrong. For example, maybe you overlooked a factor that influenced the test results. Or maybe the hypothesis is not a good one.

6. Repeat the tests or devise new ones—the more the better, for hypotheses that withstand many tests are likely to have a higher probability of being useful.

7. Objectively analyze and report the test results and the conclusions you have drawn from them.

You might hear someone refer to these practices as "the scientific method," as if all scientists march to the drumbeat of an absolute, fixed procedure. They do not. Many observe, describe, and report on some subject, then leave it to others to hypothesize about it. Some are lucky; they stumble onto information they are not even looking for, although chance does favor the prepared mind. It is not one single method they have in common. It is a critical attitude about being shown rather than told, and taking a logical approach to problem solving.

Logic encompasses thought patterns by which an individual draws a conclusion that does not contradict the evidence used to support it. Lick a lemon and you notice it's extremely sour. Lick ten more and you notice

the same thing each time, so you conclude all lemons are extremely sour. You have correlated one specific (lemon) with another (sour). By this pattern of thinking, called *inductive* logic, an individual derives a general statement from specific observations.

Express the generalization in "if-then" terms, and you have a hypothesis: "If you lick any lemon, then you will get an extremely sour taste in your mouth." By this pattern of thinking, called *deductive* logic, an individual makes inferences about specific consequences or specific predictions that must follow from a hypothesis.

You decide to test the hypothesis by tracking down and sampling all the varieties of lemons in the vicinity. One variety, the Meyer lemon, is actually mellow, for a lemon. You also discover that some people cannot taste anything. So you must modify the original hypothesis: "If most people lick any lemon *except* the Meyer lemon, they will get an extremely sour taste in their mouth." Suppose, after sampling all the known lemon varieties in the world, you conclude the modified hypothesis is a good one. You can never prove it beyond all shadow of a doubt, because there might be lemon trees growing in places people don't even know about. You *can* say the hypothesis has a high probability of not being wrong.

Comprehensive observations are a logical means to test the predictions that flow from hypotheses. So are **experiments**. These tests simplify observation in nature or the laboratory by manipulating and controlling the conditions under which observations are made. When suitably designed, observational and experimental tests allow you to predict that something will happen if a hypothesis isn't wrong (or won't happen if it *is* wrong).

For example, say you decided to use Darwin's view of natural selection to explain wing patterns of moths. By his reasoning, this trait or any other persists when it improves odds for reproductive success. You develop a hypothesis: *Moth wing patterning is a visual signal that helps males and females of a species identify each other and mate, and thereby produce offspring.* Next, you come up with a prediction based on it: *Moths mate at night, so if their wing pattern is a mating flag, then on moonless nights they will not see the pattern and I won't see moths mating.*

You test the prediction on a moonlit night, then on a moonless night. Moths mate both times. Do your direct observations mean the hypothesis is flawed? Maybe not; maybe your *prediction* is flawed. What if you overlooked a factor that affected the outcome? For example, what if moths, like cats, see better than you do in the dark?

You devise an experimental test to reveal whether wing pattern is not a mating flag. If moths having an *altered* wing pattern still attract mates, the prediction is probably flawed. So you capture moths, paint an altered pattern on their wings (Figure 1.9), and cage them with unaltered moths. You also set up a control group.

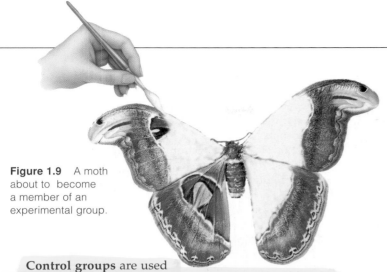

Figure 1.9 A moth about to become a member of an experimental group.

Control groups are used to identify side effects during a test that involves an experimental group. Ideally, a control group is identical to an experimental group in all respects *except* for the variable being studied. **Variables** are specific aspects of objects or events that may differ over time and among individuals. Experimenters directly manipulate a single variable that may support or disprove a prediction, and they also try to hold constant any other variables that may influence the results. In your case, a painted wing pattern is the key variable. But maybe paint fumes, not the painted pattern, will be repulsive to potential mates. Or maybe painting roughs up a moth in a way that will make it less attractive than moths of the control group. (For instance, wing fluttering attracts mates. What if the paint weighs enough to change the flutter frequency?)

To eliminate variables that are unrelated to testing your prediction, you paint moths of the control group with the same paint and brushes, and handle them the same way—but in their case you precisely *duplicate* the natural wing pattern. The point of the manipulations is to make the experimental and control groups identical *except* for the key variable. Now, if the moths having an altered wing pattern still turn out to be lucky in love, then the experiment will not support the prediction.

Have you been thinking that dabbing paint on moths is a fanciful idea of an experimental test? As reported in *Nature* in 1993, Karen Marchetti, a graduate student of the University of California at Davis, carefully used a paintbrush on birds in a forest in Kashmir, India. She found evidence for her hypothesis that bright feather color, not feather patterning, gives male yellow-browed leaf warblers a competitive edge in mating.

In early tests, Marchetti captured male birds and put a patternless dab of yellow paint on their head feathers. In later tests, she painted larger-than-normal yellow bands on the wings of an experimental group of male birds, then she painted out part of the bands of another group. She used transparent paint for a third group. (Can you guess why?) As Marchetti discovered, color-enhanced birds secured larger territories and produced more offspring, compared to the manipulatively toned-down birds in her experiments.

About the Word "Theory"

Suppose no one has disproved a hypothesis after years of rigorous tests. Suppose scientists use it to formulate *more* hypotheses that successfully explain a broad range of phenomena. When a hypothesis meets these criteria, it may become accepted as a **scientific theory**.

You may hear someone apply the word "theory" to a speculative idea, as in the expression "It's only a theory." However, a scientific theory differs from speculation for a simple reason: *Many researchers have tested its predictive power many times, in many different ways, and have found no evidence to disprove it*. That is why Darwin's view of natural selection is a highly regarded theory. Biologists use it successfully to explain such diverse issues as the origin of life, the relationship between plant toxins and plant-eating animals, the sexual advantages of brightly colored or patterned wings or feathers, the reason why certain cancers run in families, or why antibiotics that doctors often prescribe are no longer effective. By giving reasoned evidence that evolution occurred in the past, the theory even influenced views of Earth history.

An exhaustively tested theory might be as close to the truth as scientists can get with the evidence at hand. For example, Darwin's theory stands, with only minor modification, after more than a century of thousands of different tests. As biologists realize, we cannot show the theory holds under all possible conditions; an infinite number of tests would be required to do so. As for any theory, we can only say it has a *high or low probability* of being a good one. So far, biologists haven't found any evidence that calls Darwin's theory into question. Yet they still keep their eyes open for new information and new ways of testing that might disprove its premises.

And this point gets us back to the value of thinking critically. Scientists must keep asking themselves: *Will observations or experiments show that a hypothesis is false?* They expect one another to put aside pride or bias by testing ideas, even in ways that may prove them wrong. Even if an individual doesn't or won't do this, others will—for science proceeds as a community that is both cooperative and competitive. Ideally, its practitioners share their ideas, with the understanding that it is just as important to expose errors as it is to applaud insights. Individuals can and often do change their minds when presented with contradictory evidence. As you will see, this is a strength of science, not a weakness.

A scientific approach to studying nature is based on asking questions, formulating hypotheses, making predictions, devising tests, and objectively reporting the results.

A scientific theory is a testable explanation about the cause or causes of a broad range of related phenomena. It remains open to tests, revision, and tentative acceptance or rejection.

1.6

THE POWER AND PITFALLS OF EXPERIMENTAL TESTS

AN ASSUMPTION OF CAUSE AND EFFECT   Experiments start from a premise that any aspect of the natural world has one or more underlying causes, whether hidden or not. With this premise, science is distinct from faith in the supernatural (meaning "beyond nature"). Experiments must deal with *potentially falsifiable* hypotheses. Such hypotheses can be tested in the natural world in ways that might disprove them.

EXPERIMENTAL DESIGN   To get conclusive test results, experimenters rely on certain practices. They refine a test design by searching the literature for related information. They design their own experiments to test one prediction of a hypothesis at a time. Each time, they set up a control group as a standard for comparison with one or more experimental groups (Figure 1.10a).

Consider some experimental tests designed by Bruce Levin of Emory University and Jim Bull of the University of Texas. Both biologists were concerned with the ever increasing frequencies of antibiotic-resistant strains of pathogenic (disease-causing) bacteria. They had been studying literature on a possible alternative to antibiotics, a *biological therapy* that enlists the bacteriophages—the "bacteria-eaters." These viruses attack a narrow range of bacterial strains. When they contact a target, they typically inject a few enzymes and genetic material into it. What happens next is a hostile takeover of the cell's metabolic machinery. The cell itself builds many new viral particles, then it dies after its outer membrane ruptures. The virus particles released this way can infect more cells.

The biologists wondered, as others had decades ago, whether injections of bacteriophages could help people resist or fight off bacterial infections. The discovery of antibiotics in the 1940s had diverted attention away from the idea, at least in the West. Now, with so many lethal pathogens breaching the antibiotic arsenal, bacteria eaters were starting to look good again. Levin and Bull focused on some promising phage therapy experiments conducted in 1982 by H. Williams Smith and Michael Huggins, two British researchers. They started their research by testing whether Smith and Huggins's results could be duplicated.

They selected 018:K1:H, a strain of *Escherichia coli* originally isolated from a human patient with meningitis. Like the normally harmless *E. coli* strain that lives in the intestines of humans and other mammals, this one departs from the body in feces. The bacteriophage typically used in laboratory work ignored 018:K1:H7, so Levin and Bull went looking for some *E. coli* killers in samples from an Atlanta sewage treatment plant. They successfully isolated two kinds of bacteriophages and named them *H* (for *H*ero, the most effective killer) and *W* (for *W*imp).

With graduate student Terry DeRouin and laboratory technician Nina Moore Walker, the biologists grew a large population of 018:K1:H7 in a culture flask. They selected a specific strain of laboratory mice, all females of the same age. Then they ran the experiments outlined in Figure 1.10. Their results reinforced Smith and Huggins's conclusion that certain bacteriophages can be as good as or better than antibiotics at stopping specific bacterial infections.

Hypothesis: If bacteriophages specifically target and destroy cells of *E. coli* 018:K1:H7 in petri dishes, then they will do the same in laboratory mice that have been infected by that strain.

Prediction: Laboratory mice injected with a preparation that contains more than 10^7 particles of *H* bacteriophage will not die following an injection of *E. coli* strain 018:K1:H7.

Experimental test of the prediction:

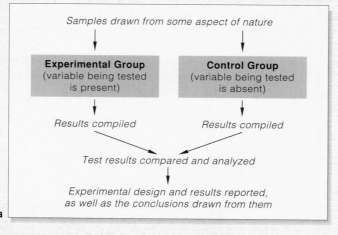

Researchers establish large populations of H bacteriophage and of E. coli 018:K1:H7 from which to draw their samples, and select a specific strain of laboratory mice.

EXPERIMENTAL GROUP	CONTROL GROUP
E. coli injected into right thigh of 15 of the mice; bacteriophage injected into their left thigh.	*E. coli* injected into 15 other mice; <u>no</u> bacteriophage injected into this sampling.

Test results:

All mice survive.	*All mice die within 32 hours.*

b

Figure 1.10 (**a**) Generalized sequence of steps involved in an experimental test of a prediction based on a hypothesis. (**b,c**) Two experiments comparing the effectiveness of bacteriophage injections and antibiotics in treating a bacterial infection.

Samples drawn from some aspect of nature

Experimental Group (variable being tested is present)	**Control Group** (variable being tested is absent)

Results compiled   *Results compiled*

Test results compared and analyzed

Experimental design and results reported, as well as the conclusions drawn from them

a

Each type of bacteriophage targets specific strains of one or at most a few bacterial species. At present, it takes too long for clinicians to discover which specific bacterium is infecting a patient, so the right bacteriophage might not be enlisted until it is too late to do the patient any good. On the bright side, procedures are now being developed that can dramatically accelerate the identification process.

IDENTIFYING IMPORTANT VARIABLES In nature, many factors can influence the outcome of a bacterial infection. For example, genetic differences among individuals lead to differences in how their immune system will respond to the invasion. So do the individual's age, nutrition, and health at the time of infection. Some strains of infectious bacteria are deadlier than others, and so on. That is why researchers try to simplify and control the variables in their experiments. Variables, recall, are specific aspects of objects or events that may differ over time and among individuals. All of the *E. coli* cells in Levin and Bull's experiments were descended from the same parent cell, raised on the same nutrients at the same temperature in the flask, and therefore could be expected to respond the same way to the bacteriophage attack. All of the mice were the same age and gender and were raised under identical

a Natalie, blindfolded, randomly plucks a single jellybean from a jar filled with 120 green and 280 black jumbo jellybeans. That is a ratio of 30 to 70 percent.

c Next, the still-blindfolded Natalie randomly plucks 50 jellybeans from the jar and ends up selecting 10 green and 40 black jellybeans.

b The jar is hidden before she removes her blindfold. She sees only a single green jellybean in her hand and assumes the jar contains green jellybeans only.

d Now, the larger sampling leads Natalie to assume that one-fifth of the jellybeans in the jar are green and four-fifths are black (a ratio of 20 to 80).

e Natalie's larger sampling more closely approximates the ratio of green to black jumbo jellybeans in the jar. The more times she repeats her sampling, the greater the likelihood that she will come close to determining the actual ratio.

Figure 1.11 A simple demonstration of sampling error.

laboratory conditions. Each mouse in every experimental group received the same amount of bacteriophages or streptomycin; every mouse in a given control group got the same injection of saline solution. Thus the focus was on *one variable*—a specific bacteriophage versus a specific antibiotic—in a simple, controlled, artificial situation.

SAMPLING ERROR As Levin and Bull did with their bacteriophage, *E. coli*, and mice, experimenters generally use samples (or subsets) of populations, events, and other aspects of nature. If they don't use large-enough samples, they run the risk of performing tests with groups that are not representative of the whole. In general, the larger the sample, the less likely it will be that differences among individuals will distort the results (Figure 1.11).

BIAS IN REPORTING THE RESULTS Whether intentional or not, experimenters run the risk of interpreting data in terms of what they want to prove or dismiss. A few have even been known to fake measurements or nudge findings in ways that reinforce their own bias. That is why science emphasizes presenting test results in *quantitative* terms—that is, with actual counts or some other precise form. Doing so allows other experimenters to check or test the results readily and systematically, as Levin and Bull did. At this writing, they are assembling a detailed report of their own experiments, for publication in a science journal.

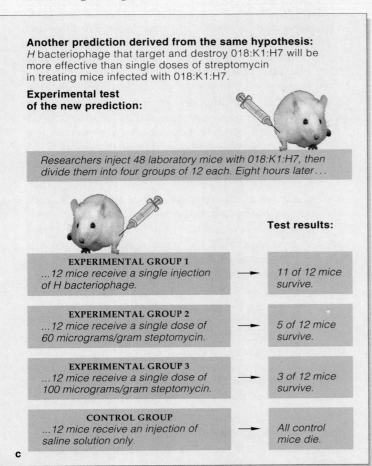

Another prediction derived from the same hypothesis:
H bacteriophage that target and destroy 018:K1:H7 will be more effective than single doses of streptomycin in treating mice infected with 018:K1:H7.

Experimental test of the new prediction:

Researchers inject 48 laboratory mice with 018:K1:H7, then divide them into four groups of 12 each. Eight hours later…

Test results:

EXPERIMENTAL GROUP 1
…12 mice receive a single injection of H bacteriophage. → 11 of 12 mice survive.

EXPERIMENTAL GROUP 2
…12 mice receive a single dose of 60 micrograms/gram steptomycin. → 5 of 12 mice survive.

EXPERIMENTAL GROUP 3
…12 mice receive a single dose of 100 micrograms/gram steptomycin. → 3 of 12 mice survive.

CONTROL GROUP
…12 mice receive an injection of saline solution only. → All control mice die.

c

1.7 THE LIMITS OF SCIENCE

The call for objective testing strengthens the theories that emerge from scientific studies. It also puts limits on the kinds of studies that can be carried out. Beyond the realm of science, some events remain unexplained. Why do we exist, for what purpose? Why does any one of us have to die at a particular moment? Such questions lead to *subjective* answers, which come from within, as an outcome of all the experiences and mental connections shaping our consciousness. Because people differ vastly in this regard, subjective answers do not readily lend themselves to scientific analysis and experiments.

This is not to say subjective answers are without value. No human society can function for long unless its members share a commitment to certain standards for making judgments, even subjective ones. The moral, aesthetic, philosophical, and economic standards vary from one society to the next. But they all guide their members in deciding what is important and good, and what is not. All attempt to give meaning to what we do.

Every so often, scientists stir up controversy when they happen to explain some part of the world that was considered to be beyond natural explanation—that is, as belonging to the "supernatural." This is often true when a society's moral codes have become interwoven with religious narratives. Exploring some longstanding view of the world from a scientific perspective may be misinterpreted as questioning morality, even though the two are not the same thing.

For example, centuries ago in Europe, Nicolaus Copernicus studied the planets and concluded the Earth circles the sun. Today this seems obvious enough. Back then it was heresy. The prevailing belief was that the Creator made the Earth (and, by extension, humankind) the immovable center of the universe. Later a respected scholar, Galileo Galilei, studied the Copernican model of the solar system, thought it was a good one, and said so. He was forced to retract his statement publicly, on his knees, and put the Earth back as the fixed center of things. (Word has it that when he stood up he muttered, "Even so, it *does* move.") Later still, Darwin's theory of evolution ran up against the same prevailing belief.

Today, as then, society has sets of standards. Those standards might be questioned when a new, natural explanation runs counter to supernatural beliefs. This doesn't mean that the scientists who raise questions are less moral, less lawful, less sensitive, or less caring than anyone else. It simply means one more standard guides their work: *The external world, not internal conviction, must be the testing ground for scientific beliefs.*

Systematic observations, hypotheses, predictions, tests— in all these ways, science differs from systems of belief that are based on faith, force, or simple consensus.

1.8 SUMMARY

1. There is unity in the living world, for all organisms have these characteristics in common:

   a. They are assembled from the same kinds of atoms and molecules, according to the same laws of energy.

   b. They survive by metabolism, and by sensing and responding to specific conditions in the environment.

   c. They have the capacity for growth, development, and reproduction, based on the heritable instructions encoded in the molecular structure of their DNA.

2. The characteristics of life extend from cells, through multicelled organisms, then populations, communities, ecosystems, and the biosphere.

3. Many millions of species (kinds of organisms) exist; many millions more lived in the past and became extinct. Classification schemes place species in ever more inclusive groupings, from genus on up through (for instance) family, order, class, phylum, and kingdom.

4. The diversity among organisms has arisen through mutations, which are changes in the structure of DNA molecules. The changes are the foundation for variation in heritable traits. These are traits that parents bestow on offspring, including most details of the body's form and functioning.

5. Darwin's theory of evolution by natural selection is a cornerstone of biological inquiry. Its key premises are:

   a. Individuals of a given population differ in their versions of the same heritable trait. Variant forms of traits may affect the ability to survive and reproduce.

   b. Natural selection is the outcome of differences in survival and reproduction among individuals that show variation in one or more traits. Adaptive forms of a given trait tend to become more common; less adaptive ones become less common or disappear. Thus the traits we use to define a population can change over time; the population can evolve.

6. There are many diverse methods of scientific inquiry. The following terms are important to all of them:

   a. Theory: An explanation of a broad range of related phenomena, supported by many tests. An example is Darwin's theory of evolution by natural selection.

   b. Hypothesis: A possible explanation of a specific phenomenon. Sometimes called an educated guess.

   c. Prediction: A claim about what can be expected in nature, based on the premises of a theory or hypothesis.

   d. Test: An attempt to produce actual observations that match predicted or expected observations.

   e. Conclusion: A statement about whether a theory or hypothesis should be accepted, modified, or rejected, based on tests of the predictions derived from it.

7. Systematic observations, hypotheses, predictions, and relentless tests are the foundation of scientific theories. The external world, rather than internal conviction, is the testing ground for those theories.

Review Questions

1. Why is it difficult to formulate a simple definition of life? *CI*. [For this and subsequent chapters, *italics* after the review questions identify section where you can find the answers to those questions. They include section numbers and *CI* (for *Chapter Introduction*).]

2. What is the molecule of inheritance? What is so important about the flow of events from DNA to RNA to protein? *1.1*

3. Write out simple definitions of the following terms: *1.1*
   - a. cell
   - b. energy
   - c. metabolism
   - d. photosynthesis
   - e. ATP
   - f. aerobic respiration

4. By what mechanisms do organisms sense changes in their surroundings? *1.1*

5. List the shared characteristics of life. *CI, 1.3*

6. Study Figure 1.4. Then, on your own, arrange and define the levels of biological organization. *1.2*

7. Study Figure 1.5. Then, on your own, make a sketch of the one-way flow of energy and the cycling of materials through the biosphere. To the side of the sketch, write out definitions of the producer, consumer, and decomposer organisms. *1.2*

8. Each kind of organism is called a separate species. What are the two components of its name? List the six kingdoms of species and name some of their general characteristics. *1.3*

9. Define mutation and adaptive trait. *1.4*

10. Explain the connection between mutations and the immense diversity of life. *1.4*

11. Write out simple definitions of evolution, artificial selection, and natural selection. *1.4*

12. Define and distinguish between: *1.5*
   - a. hypothesis or speculation and scientific theory
   - b. observational test and experimental test
   - c. inductive and deductive logic

13. With respect to experimental tests, define variable, control group, and experimental group. What is meant by the statement that scientific experiments are based on an assumption of cause and effect? *1.5, 1.6*

14. What is a sampling error? *1.6*

Self-Quiz *(Answers in Appendix IV)*

1. The _____ is the smallest unit of life.

2. _____ may be defined as the capacity of cells to extract and transform energy from their environment and use it to maintain themselves, grow, and reproduce.

3. _____ is a state in which the internal environment is being maintained within tolerable limits.

4. If a form of a trait improves chances for surviving and reproducing in a given environment, it is a(n) _____ trait.

5. The capacity to evolve is based on variations in heritable traits, which originally arise through _____ .

6. You have some number of traits that also were present in your great-great-great-great-grandmothers and -grandfathers. This is an example of _____ .
   - a. metabolism
   - b. homeostasis
   - c. a control group
   - d. inheritance

7. DNA molecules _____ .
   - a. contain instructions for traits
   - b. undergo mutation
   - c. are transmitted from parents to offspring
   - d. all of the above

8. For many years in a row, a dairy farmer allowed his best milk-producing cows but not the poor producers to mate. Over many generations, milk production increased. This outcome is an example of _____ .
   - a. natural selection
   - b. artificial selection
   - c. evolution
   - d. both b and c

9. Match the terms with the most suitable descriptions.
   - ____ adaptive trait
   - ____ natural selection
   - ____ theory
   - ____ hypothesis
   - ____ prediction

   a. statement of what you should observe in nature if you were to go looking for it
   b. educated guess
   c. improves chance of surviving and reproducing in environment
   d. related set of hypotheses that form a broad-reaching, testable explanation about nature
   e. outcome of differences in survival and reproduction that occurred among individuals that differ in one or more traits

Critical Thinking

1. A certain type of fishing spider (*Dolomedes*) that feeds on insects around ponds occasionally captures tadpoles and small fishes, as in Figure 1.12, and isn't that fun to think about. While they are immature, the female spiders confine themselves to a small area above the pond. When they are sexually mature, they mate and store sperm that fertilizes their eggs. Only then do they start to migrate through larger areas around the pond. Develop hypotheses to explain this observation, and then design an experiment to test each hypothesis.

Figure 1.12 A captured minnow, the unfortunate recipient of a paralyzing venom and digestive enzymes delivered by a fishing spider (*Dolomedes triton*), which is now sucking predigested juices from it.

2. Witnesses in a court of law are asked to "swear to tell the truth, the whole truth, and nothing but the truth." What are some of the problems inherent in the question? Can you think of a better alternative?

3. Design an experiment to support or refute the following hypothesis: Body fat appears yellow in certain rabbits—but only when those rabbits also eat leafy plants that contain an abundance of a yellow pigment molecule called xanthophyll.

4. A group of scientists devised the following experiments to shed light on whether different species of fishes of the same genus compete with one another in their natural habitat. They set up twelve ponds that were identical in chemical composition and physical characteristics. In each pond, they released the following:

Ponds 1,2,3:	species A	(300 individuals each pond)
Ponds 4,5,6:	species B	(300 individuals each pond)
Ponds 7,8,9:	species C	(300 individuals each pond)
Ponds 10,11,12:	species A,B,C	(300 of each in each pond)

Does this experimental design take into consideration all factors that can affect the outcome? If not, how would you change it?

5. Many popular magazines publish an astounding number of articles on diet, exercise, and other health-related topics. Often the authors recommend a specific diet or dietary supplement. What kinds of evidence do you think the articles should describe so that you can decide whether to accept their recommendations?

6. The males of many species of birds have brightly colored feathers, whereas the females are often rather drab. Why do you suppose drab coloration might be an adaptive trait for the females?

7. Surgical incisions in the skin of the African clawed frog (*Xenopus laevis*) do not become infected after the toad is placed in an aquarium with water that contains a variety of bacteria. Develop hypotheses to explain this observation. Then design an experiment to test each hypothesis.

8. An experimental pesticide is being sprayed in facilities where chickens are raised. The following year, an unusually large number of chicks with birth defects are hatched. Develop hypotheses to explain this observation. Then design an experiment to test each hypothesis.

9. The most recent manned mission to the planet Mars has returned. One of the astronauts recovered a gelatinous red substance from its surface. Terika, a NASA biologist, has been assigned the responsibility of determining whether the substance is alive or is a product of a living organism. What features should she look for?

Selected Key Terms

For this and subsequent chapters, these are the **boldface** terms that appear in the text, in the sections indicated here by *italic* numbers (or *CI*, short for Chapter Introduction). Make a list of the terms, write a definition next to each, and then check it against the one in the text. You will be using these terms later on. Becoming familiar with each one now will help give you a foundation for understanding the material in later chapters.

adaptive trait *1.4*	energy *1.1*	photosynthesis *1.1*
aerobic	Eubacteria *1.3*	Plantae *1.3*
respiration *1.1*	evolution *1.4*	population *1.2*
Animalia *1.3*	experiment *1.5*	prediction *1.5*
Archaebacteria *1.3*	Fungi *1.3*	producer *1.2*
artificial	genus *1.3*	Protista *1.3*
selection *1.4*	homeostasis *1.1*	receptor *1.1*
biosphere *1.2*	hypothesis *1.5*	reproduction *1.1*
cell *1.1*	logic *1.5*	RNA *1.1*
community *1.2*	metabolism *1.1*	species *1.3*
consumer *1.2*	multicelled	stimulus *1.1*
control group *1.5*	organism *1.2*	test, scientific *1.5*
decomposer *1.2*	mutation *1.4*	theory, scientific *1.5*
DNA *1.1*	natural selection *1.4*	variable *1.5*
ecosystem *1.2*		

Readings

Carey, S. 1994. *A Beginner's Guide to the Scientific Method*. Belmont, California: Wadsworth. Paperback.

Committee on the Conduct of Science. 1989. *On Being a Scientist*. Washington, D.C.: National Academy of Sciences. Paperback.

Dott, R., and R. Batten. 1988. *Evolution of the Earth*. Fourth edition. New York: McGraw-Hill.

Larkin, T. June 1985. "Evidence Versus Nonsense: A Guide to the Scientific Method." *FDA Consumer* 19: 26–29.

McCain, G., and E. Segal. 1988. *The Game of Science*. Fifth edition. Pacific Grove, California: Brooks/Cole. Paperback.

Moore, J. 1993. *Science as a Way of Knowing—The Foundations of Modern Biology*. Cambridge, Massachusetts: Harvard University Press.

Web Site See *http://www.wadsworth.com/biology* for practice quiz questions, hypercontents, BioUpdates, and critical thinking. The Wadsworth Biology Resource Center provides a wealth of information fully organized and integrated by chapter.

FACING PAGE: *Living cells of a green plant* (Elodea), *as seen with the aid of a microscope. Each rectangular cell contains efficient chemical factories called chloroplasts (the green spheres).*

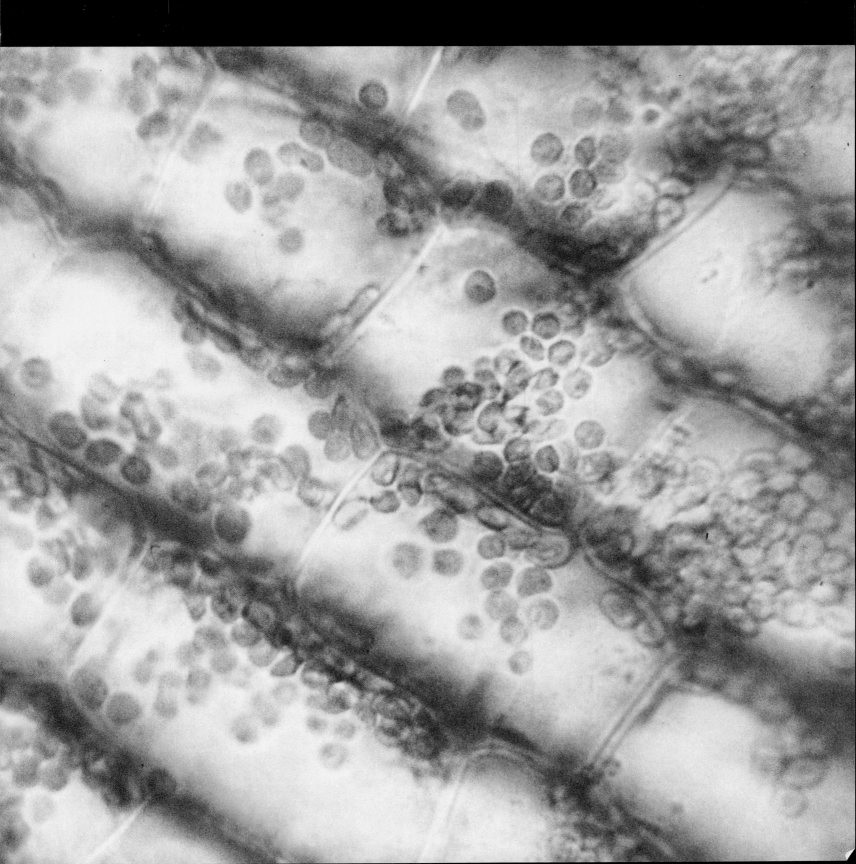

CHEMICAL FOUNDATIONS FOR CELLS

Leafy Clean-Up Crews

Right now you are breathing in oxygen. You would die without it. Two centuries ago, no one had a clue to what oxygen is, where it comes from, and how it helps keep people alive. Then researchers started unlocking the secrets of this chemical element and others. As their knowledge of chemistry deepened, they began to conjure up such amazing things as nuclear power, synthetic fertilizers, nylons, lipsticks, fabric cleaners, aspirin, antibiotics, and plastic parts of refrigerators, computers, television sets, jet planes, and cars.

Today, our chemical "magic" brings us benefits *and* problems. Think of the astounding scale of agriculture necessary to sustain the human population, which now surpasses 5.8 billion. If farmers did not use synthetic fertilizers to boost crop yields, more humans than you might ever imagine would starve to death. Fertilizers contain nitrogen, selenium, and other elements that serve as nutrients; they help plants grow. The nutrients dissolve in irrigation water, and a portion accumulates in runoff from the fields. Farmers commonly divert the runoff to ponds. There the water evaporates, and the leftover elements become even more concentrated. Especially troublesome is an ionized form of selenium

in irrigation water. At extremely high concentrations, it contaminates soil and kills wildlife.

In 1996, Norman Terry and his research colleagues at the University of California, Berkeley, started a large-scale experiment in Corcoran, California. They grew cattails, bulrushes, *Spartina*, and other grasses in ten experimental plots, each measuring a quarter of an acre, near some agricultural fields. As they knew, plants can incorporate selenium in their tissues and also have the metabolic means to convert some to dimethyl selenide —a gas about 600 times less toxic than selenium. Plants, they hypothesized, will be able to remove selenium from contaminated soil. They planned to measure how much selenium remains in mud of wetland plots, and how much gets incorporated in plant tissues and also escapes in gaseous form to the air. From earlier laboratory work, Terry knew that selenium levels can vary, depending on the soil type and plant species. He predicted that before runoff trickles on through to evaporation ponds,

Figure 2.1 At experimental wetlands in Corcoran, California, cattails, bulrushes, *Spartina*, and other grasses take up selenium from agricultural runoff and thereby help clean up the water.

a　　　　　**b**

Figure 2.2 (**a**) Young poplars taking up TCE-contaminated water. (**b**) Sunflowers, with their roots dangling in pondwater, take up radioactive elements near Chernobyl, in the Ukraine. As described in Section 50.8, the containment walls of Chernobyl's nuclear power plant exploded open in 1986, and dangerous radioactive elements entered the surrounding air, soil, and water.

the plants that he selected for his plots should be able to reduce selenium levels to less than 2 ppb. (That is, there should be less than two parts of selenium among every billion parts of other elements in the water.)

Terry got the idea for his experiment after gathering data on the selenium that became incorporated in plants growing in a wetland near an oil refinery in Richmond, California (Figure 2.1). As his measurements revealed, each day those plants remove 90 percent of the selenium present in 10 million liters of the refinery's wastewater!

Terry's research is an example of *phytoremediation*— the use of living plants to withdraw harmful substances from the environment. Similarly, Milton Gordon and his colleagues at the University of Washington utilize deep-rooted poplar plants to cleanse groundwater that became contaminated with TCE (Figure 2.2*a*). This chemical is now banned, but it was once common in dry cleaning and degreasing products. The poplars release TCE to the air, and sunlight breaks it down. As a final example, a phytoremediation company is using sunflowers to clean a pond that became contaminated with ^{90}strontium, ^{137}cesium, and other extremely nasty radioactive elements (Figure 2.2*b*).

As you might deduce from this story, safeguarding the environment, our food supplies, and our health depends on knowledge of chemistry. So do efforts to minimize side effects of its applications on biological systems. You owe it to yourself and others to gain insight into the structure and behavior of chemical substances. By demystifying chemistry's "magic," you will be better equipped to assess its benefits and risks.

KEY CONCEPTS

1. All substances consist of one or more elements, such as hydrogen, oxygen, and carbon. Each element is composed of atoms, which are the smallest units that still display the element's properties. In turn, its atoms are composed of protons, electrons, and (except for hydrogen) neutrons.

2. All of the atoms that make up each kind of element have the same number of protons and electrons, but they may vary slightly from one another in their number of neutrons. Variant forms of an element's atoms are called isotopes.

3. Atoms have no overall electric charge unless they become ionized—that is, unless they lose electrons or acquire more of them. An ion is an atom or molecule that has gained or lost one or more electrons and thereby has acquired an overall positive or negative charge.

4. Whether a given atom will interact with other atoms depends on how many electrons it has and how they are structurally arranged within the atom. When the electron structures of two or more atoms become united in some way, this is a chemical bond.

5. The molecular organization and the activities of living things arise largely from the interactions called ionic, covalent, and hydrogen bonds.

6. Life originated in water, and it is exquisitely adapted to its properties. Foremost among these properties are water's temperature-stabilizing effects, cohesiveness, and capacity to dissolve or repel a variety of substances.

7. Life depends on the controlled formation, use, and disposal of hydrogen ions (H^+). The pH scale is a measure of the concentration of these ions in different solutions.

People, pesticides, pumpkins, the solid earth beneath your feet, the very air you breathe—*everything in and around you is "chemistry."* Every solid, liquid, or gaseous substance you care to think about is a collection of one or more kinds of elements. Think of these **elements** as fundamental forms of matter that occupy space, have mass, and cannot be broken apart into something else, at least not by any ordinary means. If you could break a chunk of some element into ever smaller bits, the smallest bit you would end up with would be the same element. Ninety-two elements occur naturally on Earth, and many others have been produced under extreme conditions in laboratories (Appendix VI).

Each element has a unique symbol that is internationally recognized, regardless of its name in different countries. One such element is *nitrogen* in English, *azoto* in Italian, and *stickstoff* in German, but its symbol is always N. Similarly, sodium's symbol is always Na (from the Latin *natrium*). Table 2.1 lists the symbols for other elements.

Typically, more than 95 percent of the body weight of an organism is composed of only four elements: oxygen, carbon, hydrogen, and nitrogen. The remainder consists of calcium, phosphorus, potassium, sulfur, sodium, chlorine, magnesium, and more than a dozen trace elements. A **trace element** simply is one that represents less than 0.01 percent of body weight.

Figure 2.3 lists your own body's elements. The amounts may seem trivial, but normal functioning depends on them. For example, your daily meals usually include a trace of magnesium. Without it, muscles would weaken and your brain would not work properly. As another example, leaves droop and die without magnesium. Or think of a weed that absorbs 2,4D, a pesticide. It is stimulated to grow rapidly, but it cannot absorb trace elements fast enough to sustain the accelerated growth. It literally will grow itself to death.

Table 2.1	Atomic Number and Mass Number of Elements Common in Living Things		
Element	Symbol	Atomic Number	Most Common Mass Number
Hydrogen	H	1	1
Carbon	C	6	12
Nitrogen	N	7	14
Oxygen	O	8	16
Sodium	Na	11	23
Magnesium	Mg	12	24
Phosphorus	P	15	31
Sulfur	S	16	32
Chlorine	Cl	17	35
Potassium	K	19	39
Calcium	Ca	20	40
Iron	Fe	26	56
Iodine	I	53	127

Structure of Atoms

What are the smallest particles that retain the properties of an element? **Atoms**. A line of about a million of them would fit in the period ending this sentence. Small as atoms are, physicists have split them into more than a hundred kinds of smaller particles. The only subatomic particles you will need to consider in this book are the ones called **protons**, **electrons**, and **neutrons**.

All atoms have one or more protons, which carry a positive electric charge (p^+). Except for hydrogen atoms, they also have one or more neutrons, which carry no charge. Protons and neutrons make up the atom's core region, or atomic nucleus (Figure 2.4). Zipping about the nucleus and occupying most of the atom's volume are one or more electrons, which carry a negative charge (e^-). Each atom has just as many electrons as protons. This means atoms carry no net charge, overall.

Each element has a unique **atomic number**, which refers to the number of protons present in its atoms. To give examples, the atomic number is 1 for the hydrogen

EARTH'S CRUST		HUMAN		PUMPKIN	
Oxygen	46.6	Oxygen	65	Oxygen	85
Silicon	27.7	Carbon	18	Hydrogen	10.7
Aluminum	8.1	Hydrogen	10	Carbon	3.3
Iron	5.0	Nitrogen	3	Potassium	0.34
Calcium	3.6	Calcium	2	Nitrogen	0.16
Sodium	2.8	Phosphorus	1.1	Phosphorus	0.05
Potassium	2.6	Potassium	0.35	Calcium	0.02
Magnesium	2.1	Sulfur	0.25	Magnesium	0.01
Other elements:	1.5	Sodium	0.15	Iron	0.008
		Chlorine	0.15	Sodium	0.001
		Magnesium	0.05	Zinc	0.0002
		Iron	0.004	Copper	0.0001
		Iodine	0.0004	Other:	0.00005

Figure 2.3 Proportions of elements making up the human body and the fruit of pumpkin plants, compared to the proportions of elements in the materials of our planet's crust. In what respects are the proportions similar? In what respects do they differ?

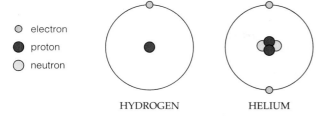

electron
proton
neutron

HYDROGEN   HELIUM

Figure 2.4 The hydrogen atom and helium atom according to one model of atomic structure. The model is highly simplified; the nucleus of these two representative atoms really would be an invisible speck at the scale employed here.

atom (which has a single proton) and 6 for the carbon atom (which has six protons).

Each element also has a **mass number**, which is the number of protons *and* neutrons in the atomic nucleus. A carbon atom with six protons and six neutrons has a mass number of 12. Some people call the relative mass of atoms "atomic weight." Mass is not quite the same thing as weight, but use of this imprecise term continues.

Why bother with atomic and mass numbers? They can give you an idea of whether and how substances will interact. *And this knowledge can help you predict how substances might behave in cells, in multicelled organisms, and in the environment, under a variety of conditions.*

Isotopes—Variant Forms of Atoms

All of the atoms of an element have the same number of protons and electrons, but they may not have the same number of neutrons. If an atom of a given element has more or fewer neutrons than those having the most common number, it is an **isotope**. As an example, "a carbon atom" might be carbon 12 (six protons and six neutrons, this being the most common form), carbon 13 (six protons and seven neutrons), or carbon 14 (with six protons and eight neutrons). Often you will see symbols for isotopes, as for the carbon isotopes ^{12}C, ^{13}C, and ^{14}C.

All isotopes of an element interact with other atoms in the same way. As you will see, this means that cells can use any isotope of an element during metabolism.

You have probably heard of radioactive isotopes, or **radioisotopes**. Henri Becquerel, a physicist, discovered them in 1896, after he put a heavily wrapped rock on an unexposed photographic plate in a desk drawer. The rock happened to contain uranium isotopes. A few days later, the plate bore a faint image of the rock, which must have been emitting energy. Becquerel's coworker, Marie Curie, named the phenomenon "radioactivity."

As we now know, radioisotopes are unstable atoms with dissimilar numbers of protons and neutrons, and they spontaneously lose subatomic particles and energy at a known rate. This process, called **radioactive decay**, transforms an unstable atom into an atom of a different

Figure 2.5 Chemical bookkeeping. We use symbols for elements when writing *formulas*, which identify the composition of compounds. For example, water has the formula H_2O. The subscript indicates two hydrogen (H) atoms are present for every oxygen (O) atom. We use such symbols and formulas when writing *chemical equations*, which are representations of the reactions among atoms and molecules. The substances entering a reaction (reactants) are to the left of the reaction arrow, and the products are to the right, as shown by the following chemical equation for photosynthesis:

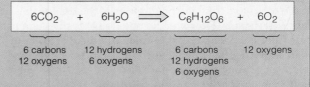

$$6CO_2 \quad + \quad 6H_2O \quad \Longrightarrow \quad C_6H_{12}O_6 \quad + \quad 6O_2$$

6 carbons	12 hydrogens	6 carbons	12 oxygens
12 oxygens	6 oxygens	12 hydrogens	
		6 oxygens	

You may read about reactions for which reactants and products are expressed in moles. A *mole* is a certain number of atoms or molecules of any substance, just as "a dozen" can refer to any twelve cats, roses, and so forth. Molar weight, in grams, equals the total atomic weight of all the atoms making up that substance.

For example, the atomic weight of carbon is 12, so one mole of carbon weighs 12 grams. A mole of oxygen (atomic weight 16) weighs 16 grams. Can you state why a mole of water (H_2O) weighs 18 grams, and why a mole of glucose ($C_6H_{12}O_6$) weighs 180 grams?

element. Carbon 14, for instance, decays to nitrogen 14. You will read about some uses of this radioisotope and others in the next section.

When Atom Bonds With Atom

Very shortly, we will consider **chemical bonds**, which are unions between the electron structures of atoms. Take a moment to review Figure 2.5, which summarizes a few conventions that are used to describe chemical reactions.

Elements are forms of matter that occupy space, have mass, and cannot be broken apart into something else by ordinary means.

Atoms, the smallest particles that are unique to each element, have one or more positively charged protons, negatively charged electrons, and (except for hydrogen) neutrons.

Isotopes are atoms of an element that differ in the number of neutrons. A radioisotope has uneven numbers of protons and neutrons. Such unstable atoms spontaneously emit particles and energy and so become transformed to a different atom.

USING RADIOISOTOPES TO DATE THE PAST, TRACK CHEMICALS, AND SAVE LIVES

RADIOMETRIC DATING Probably for as long as they have been digging up the earth, people have been turning up fossilized leaves, shells, skeletons, and other stone-hard evidence of past life, entombed in rocks. Figure 2.6*a* and *b* shows examples. Some time ago, geologists hypothesized that if newly formed rocks slowly accumulate on top of older rocks, then fossils in older rock layers must be more ancient than those in more recently deposited ones. They used this perception to count backward through great numbers of rock layers and thereby construct a chronology of Earth history—a geologic time scale. As described in Section 20.1, they used sequences of fossils and other clues in the rocks to define the boundaries of the time scale's successive spans. However, no one was able to assign firm dates along the length of the scale until the discovery of radioactive decay, which made it possible to convert the relative ages into absolute ones.

By a method called **radiometric dating**, researchers now measure the proportions of (1) a radioisotope in a mineral that became trapped long ago in a brand-new rock and (2) a daughter isotope that formed from it in the same rock. Remember, the atomic nucleus of each radioactive element

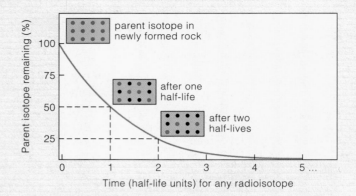

Figure 2.7 Generalized diagram of radioactive decay. Notice the proportion of the parent radioisotope in a rock sample over time, compared to the proportion of each decay product. During *each* half-life unit, the amount of the parent radioisotope will diminish by half.

consists of an *unstable* number of protons and neutrons. The nucleus spontaneously decays—it gives up energy and one or more particles of itself—until it reaches a more stable configuration. (Such nuclear emissions were the invisible culprits that exposed Becquerel's film.)

Half-life is the time it takes for half of a given quantity of any radioisotope to decay into a different, less unstable daughter isotope (Figure 2.7). The rate of decay is constant; changes in pressure, temperature, or chemical state cannot alter it. For example, ^{238}uranium occurs in zircon, which is a mineral in most volcanic rocks. Its nuclei decay into ^{234}thorium, a still-unstable daughter isotope that in turn decays into something else, and so on through a series of intermediate daughter isotopes to ^{206}lead—the final, most stable configuration for this series (Table 2.2). By using ^{238}uranium, with its half-life of 4.5 billion years, researchers realized the Earth formed more than 4.6 billion years ago.

Radiometric dating works for volcanic rocks or ashes. However, most fossil-containing rocks formed by the compaction of sand and other sediments. Thus the only way to date most fossil-containing rocks is to determine their position relative to volcanic rocks in the same area.

a   **b**

c

Figure 2.6 (**a**) Fossilized frond of a tree fern, one of many plant species that lived more than 250 million years ago. (**b**) Fossil of a worm that lived in the seas about 600 million years ago. (**c**) One of many glassy spherical particles, no bigger than sand grains, that are between 3.5 billion and 2.5 billion years old. The fact that it was buried in sediments indicates it was not volcanic in origin. It contains a large amount of iridium, an element that is rare on the Earth but abundant in meteorites. It may have formed early in Earth history, during the same bombardments from space that cratered the moon.

Table 2.2	Radioisotopes Commonly Used in Radiometric Dating		
Radioisotope (unstable)	More Stable Product	Half-Life (years)	Useful Range (years)
^{147}Samarium →	^{143}Neodymium	106 billion	>100 million*
^{87}Rubidium →	^{87}Strontium	48.8 billion	>100 million
^{232}Thorium →	^{208}Lead	14 billion	>200 million
^{238}Uranium →	^{206}Lead	4.5 billion	>100 million
^{40}Potassium →	^{40}Argon	1.25 billion	>100,000
^{235}Uranium →	^{207}Lead	700 million	>100 million
^{14}Carbon →	^{14}Nitrogen	5,730	0–60,000

* The symbol > means greater than.

You may have heard of dating methods based on "carbon 14." The greatest numbers of this radioisotope form continually in the upper atmosphere. There it combines with oxygen to form carbon dioxide. Along

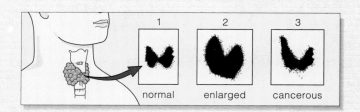

a

with the more stable isotopes of carbon, small amounts of ^{14}carbon enter the web of life by photosynthesis. Every organism incorporates it. When organisms die, the amount of ^{14}carbon starts to decrease as an outcome of radioactive decay. ^{14}Carbon has a half-life of 5,730 years, so amounts in bone, wood, shells, and other organic substances are too small to detect by direct analysis. Instead, researchers use scintillation counters to monitor emissions from a given sample. The older the sample, the fewer emissions (counts) will be recorded in a given period.

TRACKING TRACERS Radioisotopes also make fine tracers. A **tracer** is a substance with a radioisotope attached to it, rather like a shipping label, that researchers can track after they deliver it into a cell, a body, an ecosystem, or some other system. Scintillation counters and other devices can track a tracer's precise movement through some pathway or pinpoint its final destination.

Consider how Melvin Calvin and other botanists figured out the steps by which plants build carbohydrates during photosynthesis. They knew all isotopes of an element have the same number of electrons, so all must interact with other atoms the same way. Plant cells, they hypothesized, should be able to use any isotope of carbon when they build carbon compounds. By putting plant cells in a medium enriched with ^{14}carbon, these botanists were able to track the uptake of carbon through all the reaction steps leading to the formation of sugars and starches.

As another example, tracers are being used to identify how plants take up and use soil nutrients and synthetic fertilizers. The findings may help improve crop yields.

SAVING LIVES Radioisotopes have uses as diagnostic tools in medicine. For safety considerations, clinicians use only the kinds with extremely short half-lives. Consider the human thyroid, located in front of the windpipe. It is the only gland of ours that takes up iodine, a key building block for

Figure 2.9 (**a**) Patient being moved into a PET scanner. Inside (**b**), a ring of detectors intercepts the radioactive emissions from labeled molecules that were injected into the patient. Computers analyze and color-code the number of emissions from each location in the scanned body region. (**c**) Brain scan of a child who has a neurological disorder. Different colors in a brain scan signify differences in metabolic activity. The right half of this brain shows very little activity. By comparison, cells of the left half absorbed and used the labeled molecule at expected rates. ✵

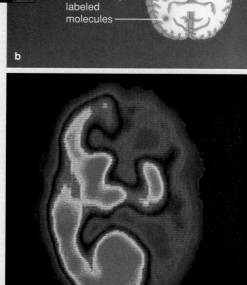

detector ring

body section within ring

ring intercepts any emissions from radioactively labeled molecules

b

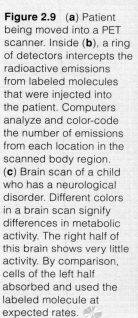

c

the synthesis of thyroid hormones that affect growth and metabolism. If a patient's symptoms point to a thyroid hormone imbalance, clinicians may inject a trace amount of ^{123}iodine into his or her blood, then use a photographic imaging device to scan the gland. Figure 2.8 shows three examples of the resulting image.

Also consider *PET* (for *Positron-Emission Tomography*). It yields images of metabolically active and inactive tissues in a patient. Suppose clinicians attach a radioisotope to glucose (or some other molecule) that cells require for metabolism. They inject the radioactively labeled glucose into the patient, who is moved into a PET scanner of the sort shown in Figure 2.9*a*. Cells in the patient's tissues will absorb the glucose. Detectors intercept emissions from it, and these are used to form an image of any variations or abnormalities in metabolic activity, as in Figure 2.9*b,c*.

Finally, radioisotopes have uses in some therapies. For example, the energetic emissions from ^{238}plutonium drive artificial pacemakers, which help correct irregular heartbeats. In such devices, the radioisotope is sealed inside a case so that its dangerous emissions won't damage tissues. As another example, clinicians may use radioisotopes for *radiation therapy*, which destroys or impairs living cells. In some therapies, they bombard localized cancers with the energetic emissions from a source of ^{226}radium or ^{60}cobalt.

1   2   3

normal   enlarged   cancerous

Figure 2.8 Scans of the thyroid gland from three patients.

THE NATURE OF CHEMICAL BONDS

In a given chemical reaction, atoms may acquire extra electrons, share them, or donate them to another atom. The atoms of certain elements do this rather easily, but others do not. What determines whether one atom will interact with another in such ways? *The outcome depends on the number and arrangement of their electrons.*

Electrons and Energy Levels

If you've ever tinkered with magnets, you have a sense of the attractive force between unlike charges (+ −) and the repulsive force between like charges (++ or −−). All electrons carry a negative charge. In an atom, they repel each other but are attracted to the positive charge of the protons. They spend as much time as possible near the protons and far away from each other by moving about in different orbitals. Think of **orbitals** as *volumes of space* around the atomic nucleus in which electrons are likely to be at any instant. (As a rough analogy, picture three preschoolers circling a cookie jar, not yet expert in the fine art of sharing. Each is drawn inexorably to the cookies but dreads being shoved away by the others. Two of them might duck and weave about on opposite sides of the jar in order to avoid a direct hit. But all three never, ever will occupy the same space at the same time.)

Just as the number of electrons differs among atoms, so does the number of occupied orbitals. One or at most two electrons can occupy any given orbital.

Consider hydrogen, the simplest atom of all. Its lone electron occupies a spherical orbital that is closest to the nucleus, which corresponds to the *lowest available energy level.* In all other atoms, two electrons fill this orbital.

electron orbital
nucleus

HYDROGEN ATOM

In other atoms, another two electrons occupy a second spherical orbital surrounding the first. Up to six more electrons may be distributed in three dumbbell-shaped orbitals, even more in other orbitals, and so on (Figure 2.10). These electrons spend more time farther away from the nucleus; they are at *higher energy levels.*

The **shell model** in Figure 2.11 is a simple though not quite accurate way to think about how electrons are distributed in a given atom. By this model, a series of shells encompasses all the orbitals that are available to electrons. The first shell encloses the first orbital, which is spherical. A second shell (at a higher energy level) encloses the first, and the next four available orbitals fit inside it. Additional orbitals fit inside a third shell, a fourth shell, and so on up to the large, complex atoms of heavier elements, as listed in Appendix VI.

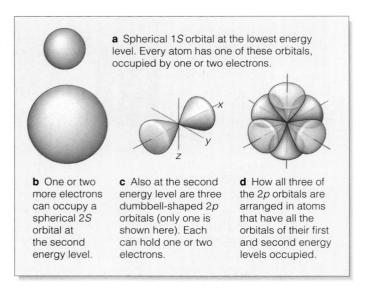

a Spherical 1S orbital at the lowest energy level. Every atom has one of these orbitals, occupied by one or two electrons.

b One or two more electrons can occupy a spherical 2S orbital at the second energy level.

c Also at the second energy level are three dumbbell-shaped 2p orbitals (only one is shown here). Each can hold one or two electrons.

d How all three of the 2p orbitals are arranged in atoms that have all the orbitals of their first and second energy levels occupied.

Figure 2.10 Arrangement of electrons in atoms. (**a**) One or at most two electrons occupy a ball-shaped volume of space—an orbital—close to the nucleus. When electrons occupy this orbital, they are at the atom's lowest energy level. (**b**–**d**) At the second energy level, there can be as many as eight more electrons; that is, up to two electrons in each of four more orbitals.

Orbital shapes get tricky, but for our purposes we can ignore them. Simply think of the total number of orbitals at a given energy level as being somewhere inside a "shell" that surrounds the nucleus of an atom. As you can deduce from Figure 2.11, higher energy levels correspond to shells farther from the nucleus.

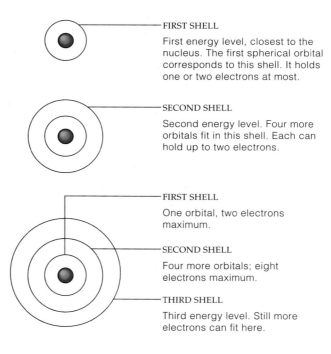

FIRST SHELL
First energy level, closest to the nucleus. The first spherical orbital corresponds to this shell. It holds one or two electrons at most.

SECOND SHELL
Second energy level. Four more orbitals fit in this shell. Each can hold up to two electrons.

FIRST SHELL
One orbital, two electrons maximum.

SECOND SHELL
Four more orbitals; eight electrons maximum.

THIRD SHELL
Third energy level. Still more electrons can fit here.

Figure 2.11 Shell model of electron distribution in atoms. Only three of the many possible energy levels (shells) are shown.

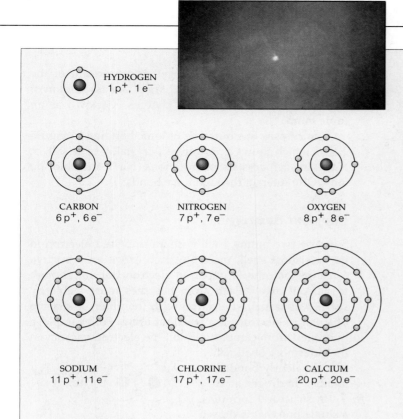

| | | | Distribution of Electrons | | | |
Element	Symbol	Atomic Number*	First Shell	Second Shell	Third Shell	Fourth Shell
Hydrogen	H	1	1	—	—	—
Helium	He	2	2	—	—	—
Carbon	C	6	2	4	—	—
Nitrogen	N	7	2	5	—	—
Oxygen	O	8	2	6	—	—
Neon	Ne	10	2	8	—	—
Sodium	Na	11	2	8	1	—
Magnesium	Mg	12	2	8	2	—
Phosphorus	P	15	2	8	5	—
Sulfur	S	16	2	8	6	—
Chlorine	Cl	17	2	8	7	—
Calcium	Ca	20	2	8	8	2

* The number of protons in the nucleus.

Figure 2.12 Examples of the shell model of the distribution of electrons in atoms. Hydrogen, carbon, and other atoms having electron vacancies (unfilled orbitals) in their outermost shell tend to give up, accept, or share electrons. Helium, neon, and other atoms with no electron vacancies in their outermost shell show no such tendency. Amazingly, physicists confined electrons in a bubble of liquid helium, then used fluctuating sound waves to pop the bubble. The photograph shows a white-centered red flash given off by an electron that escaped. It allows us to see how a single electron can make something happen.

Electrons and the Bonding Behavior of Atoms

How can you tell whether any two atoms will interact? Check for any electron vacancies in their outermost shell. Vacancies mean an atom might give up, gain, or share electrons under suitable conditions, which of course will change the distribution or number of its electrons. You can use the shell model to visualize what goes on here.

In Figure 2.12, which shows the electron distribution for several atoms, electrons are assigned to an energy level (a circle, or shell). Count the electron vacancies— that is, one or more unfilled orbitals in the outermost shell of each atom. As the table shows, helium and neon have no vacancies. They are among the *inert* atoms, which show little tendency to enter chemical reactions.

Now look again at Figure 2.3, which lists the most abundant of the elements making up a typical organism (a human). These elements include hydrogen, oxygen, carbon, and nitrogen. Atoms of all four elements have electron vacancies. Because of this, they tend to form *bonds* with other atoms—and thereby fill the vacancies.

From Atoms to Molecules

When two or more atoms bond together, the outcome is a **molecule**. Many molecules are composed of only one element. Molecular nitrogen (N_2), with its two nitrogen atoms, is like this.

The molecules of **compounds** are composed of two or more different elements in proportions that never vary. Water is a prime example of a compound. Each molecule of water has one oxygen atom chemically bonded to two hydrogen atoms. The molecules of water in rainclouds, the ocean, a Siberian lake, your bathtub, or anywhere else always have twice as many hydrogen as oxygen atoms.

By contrast, in a **mixture**, two or more elements are simply intermingling in proportions that can vary (and usually do). Consider the sugar sucrose, a compound of carbon, hydrogen, and oxygen. Swirl sucrose molecules and water molecules together, and you get a mixture.

In an atom, electrons occupy orbitals, which are volumes of space around the nucleus. By a simplified model, orbitals are arranged inside a series of shells that surround the nucleus. The shells correspond to levels of energy that become greater with distance from the nucleus.

One or two electrons at most occupy any orbital. Atoms with unfilled orbitals in their outermost shell tend to interact with other atoms; those with no vacancies do not.

In molecules of an element, all of the atoms are of the same kind. In molecules of a compound, atoms of two or more elements are bonded together, in unvarying proportions.

IMPORTANT BONDS IN BIOLOGICAL MOLECULES

Eat your peas! Drink your milk! Probably for longer than you care to remember, somebody has been telling you to eat foods that are rich in carbohydrates, proteins, and other "biological molecules." Only living organisms put together and use these molecules, which consist of a few kinds of atoms held together by only a few kinds of bonds. Foremost among the molecular interactions are the ionic, covalent, and hydrogen bonds.

Ion Formation and Ionic Bonding

An atom, recall, has just as many electrons as protons, so it carries no net charge. However, that balance can change for atoms with a vacancy—an unfilled orbital—in their outermost shell. For example, a chlorine atom can acquire an extra electron and thus fill the vacancy, as shown in Figure 2.13*a*. Similarly, the lone electron in a sodium atom's outermost shell can be knocked out of an orbital or pulled completely away from it.

Any atom that has either gained or lost one or more electrons is an **ion**. The balance between its protons and its electrons has shifted, so the atom is now ionized; it has become positively or negatively charged.

In living cells, neighboring atoms commonly accept or donate electrons among one another. When one atom loses an electron and one gains, both become ionized. Depending on cellular conditions, the two ions may not separate; they may remain together as a result of the mutual attraction of opposite charges. An association of two ions that have opposing charges is known as an **ionic bond**.

You can see one outcome of ionic bonding in Figure 2.13*b*, which shows a portion of a crystal of table salt, or NaCl. In such crystals, sodium ions (Na^+) and chloride ions (Cl^-) interact through ionic bonds.

Covalent Bonding

Suppose two atoms, each with an unpaired electron in its outermost shell, meet up. Each exerts an attractive force on the other's unpaired electron but not enough to yank it away. Each atom becomes more stable by *sharing* its unpaired electron with the other. A sharing of a pair of electrons is a **covalent bond**. For example, a hydrogen atom can partially fill the electron vacancy in its outermost shell when it is covalently bonded to another hydrogen atom.

 MOLECULAR HYDROGEN (H_2)

In structural formulas, a single line that is drawn between two atoms represents a *single* covalent bond. Molecular hydrogen has such a bond: H—H. In a *double* covalent bond, two atoms share two pairs of electrons. This is the case with molecular oxygen (O_2), or O=O.

In a *triple* covalent bond, two atoms share three pairs of electrons. This is true of molecular nitrogen (N_2), which may be written as N≡N. These three examples happen to be gaseous molecules. Each time you breathe in some air, you draw a great number of H_2, O_2, and N_2 molecules into your nose.

Covalent bonds are nonpolar or polar. In a *nonpolar* covalent bond,

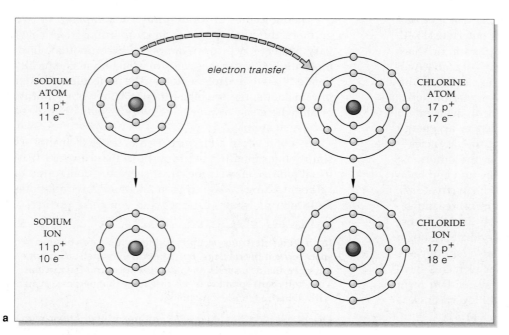

electron transfer

SODIUM ATOM
11 p$^+$
11 e$^-$

CHLORINE ATOM
17 p$^+$
17 e$^-$

SODIUM ION
11 p$^+$
10 e$^-$

CHLORIDE ION
17 p$^+$
18 e$^-$

a

Figure 2.13 (**a**) Ionization by way of an electron transfer. In this case, a sodium atom donates the lone electron in its outermost shell to a chlorine atom, which has an unfilled orbital in *its* outermost shell. A sodium ion (Na^+) and a chloride ion (Cl^-) are the outcome of this interaction.

(**b**) In each crystal of table salt, or NaCl, many sodium and chloride ions remain together because of the mutual attraction of their opposite charges. Their interaction is an example of ionic bonding.

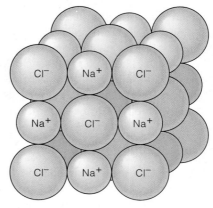

b A crystal of sodium chloride (NaCl)

participating atoms exert the same pull on the electrons and share them equally. The term "nonpolar" implies there is not any difference in charge at the ends of the bond (that is, at its two poles). Molecular hydrogen is a simple example of this. Its two H atoms, each with one proton, attract the shared electrons equally.

In a *polar* covalent bond, atoms of different elements (which have different numbers of protons) do not exert the same pull on shared electrons. The more attractive atom ends up with a slight negative charge; the atom is "electronegative." Its effect is balanced out by the other atom, which ends up with a slight positive charge. In other words, taken together, the atoms interacting in a polar covalent bond have no *net* charge—but the charge is distributed unevenly between the bond's two ends.

Consider the water molecule, which has two polar covalent bonds: H—O—H. In this molecule, electrons are less attracted to the hydrogens than to the oxygen, which has more protons. A water molecule carries no *net* charge, but you will see shortly that its polarity can weakly attract neighboring polar molecules and ions.

Hydrogen Bonding

The patterns of electron sharing in covalent bonds hold atoms together in specific arrangements in molecules. Some of the patterns also give rise to weak attractions and repulsions between charged functional groups of molecules, as well as between molecules and ions. Like interacting skydivers, such interactions break and form easily (Figure 2.14). Yet they have important roles in the structure and functioning of biological molecules.

For example, in a **hydrogen bond**, a small, highly electronegative atom of a molecule weakly interacts with a hydrogen atom that is already participating in a

Figure 2.14 Like skydivers who briefly clasp hands to form an orderly pattern, weak attractions within and between molecules, and between ions and molecules, can form and break easily.

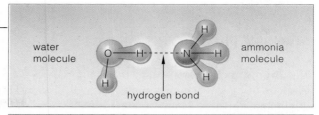

a

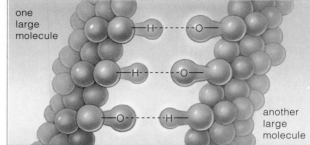

b

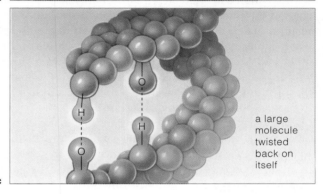

c

Figure 2.15 Examples of hydrogen bonds. Compared to covalent bonds, a hydrogen bond is easier to break. Collectively, however, such bonds are important in water and many other substances.

polar covalent bond. During this interaction, hydrogen shows a slight positive charge, and this attracts other atoms that carry a slight negative charge (Figure 2.15).

Hydrogen bonds may form between two or more molecules. They also may form in different parts of the *same* molecule, where it twists and folds back on itself. For example, many such bonds form between the two strands of a DNA molecule. Individually, the hydrogen bonds break easily. Collectively, they stabilize DNA's structure. Similarly, hydrogen bonds form between the molecules that make up water. As you will read next, they contribute to water's life-sustaining properties.

In an ionic bond, two ions of opposite charge attract each other and stay together. Ions form when atoms gain or lose electrons and so acquire a net positive or negative charge.

In a covalent bond, atoms share a pair of electrons. If the atoms share the electrons equally, the bond is nonpolar. If the sharing is not equal, the bond itself is polar—slightly positive at one end, slightly negative at the other.

In a hydrogen bond, a small, highly electronegative atom of a molecule interacts weakly with a neighboring hydrogen atom that is already taking part in a polar covalent bond.

PROPERTIES OF WATER

No sprint through basic chemistry is complete unless it leads us to the collection of molecules called water. Life originated in water. Many organisms still live in it. The ones that don't cart water around with them, in cells and tissue spaces. Many metabolic reactions require water as a reactant. Cell shape and internal structure depend on it. These topics will repeatedly occupy our attention in the book, so you may find it useful to become familiar with the following points about water's properties.

Polarity of the Water Molecule

A water molecule, remember, has no *net* charge, but the charges that it does carry are unevenly distributed. As a result of its electron arrangements and bond angles, the water molecule's oxygen "end" is a bit negative and its other end is a bit positive. Figure 2.16a shows a simple way to think about this charge distribution. Because of the resulting polarity, one water molecule attracts and hydrogen-bonds with others (Figure 2.16b).

The weak polarity of the water molecule also *attracts* other polar molecules, including sugars. We call these **hydrophilic** (water-loving) **substances**, for they readily hydrogen-bond with water. By contrast, water's polarity *repels* nonpolar molecules, including oils. We call these **hydrophobic** (water-dreading) **substances**. Observe this for yourself by shaking a bottle that contains water and salad oil, then setting it on a table. Soon, hydrogen bonds replace the ones that broke when you shook the bottle. As water molecules reunite, the oil molecules are pushed aside and forced to cluster as droplets or as a film at the water's surface. Such hydrophobic interactions are vital. For example, a thin, oily membrane is almost all that separates a cell's watery surroundings and its watery interior. The organization of that membrane starts with hydrophobic interactions, as described in Section 5.1.

Water's Temperature-Stabilizing Effects

Cells consist mostly of water, and they release a great deal of heat energy during their metabolic activities. If it were not for the hydrogen bonds in liquid water, cells might cook in their own juices. To see why this is so, start with these observations: Each molecule of water or any other substance is in constant motion, and when it absorbs heat, its motion increases. The **temperature** of a given substance is simply a measure of its molecular motion. Compared to most fluids, water can absorb more heat energy before its temperature increases measurably. Why? Much of the incoming energy disrupts hydrogen bonding *between* neighboring water molecules instead of directly causing the motion of individual molecules to increase. The stupendous number of hydrogen bonds in liquid water can buffer large swings in temperature

Figure 2.16 Water—a substance vital for life. (**a**) Polarity of the water molecule. (**b**) Hydrogen bonding between water molecules in liquid water. Dashed lines are hydrogen bonds. (**c**) Hydrogen bonding between water molecules in ice, which in vast quantities blankets the habitat of choice of polar bears. Below 0°C, each water molecule becomes locked, by four hydrogen bonds, into a crystal lattice. In the bonding pattern of ice, molecules are spaced farther apart than in liquid water at room temperature. When water is liquid, constant molecular motion usually prevents the maximum number of hydrogen bonds from forming.

inside cells. Similarly, hydrogen bonds help stabilize the temperature of aquatic habitats.

Even when the temperature of liquid water is not shifting much, hydrogen bonds are constantly breaking, but they also are forming again just as fast. By contrast, a large energy input can increase molecular motion so much that hydrogen bonds *stay* broken, and individual molecules at the water's surface escape into the air. By this process, called **evaporation**, heat energy converts liquid water to the gaseous state. As large numbers of molecules break free and depart, they carry away some energy and lower the water's surface temperature.

Evaporative water loss can help cool you and some other mammals when you work up a sweat on hot, dry days. Under such conditions, sweat—which is about 99 percent water—evaporates from your skin.

Below 0°C, hydrogen bonds resist breaking and lock water molecules in the latticelike bonding pattern of ice (Figure 2.16c). Ice is less dense than water. During winter freezes, ice sheets may form near the surface of ponds, lakes, and streams. Like a blanket, ice "insulates" the liquid water beneath it and helps protect many fishes, frogs, and other aquatic organisms against freezing.

c

Figure 2.17 A water strider (*Gerris*). This long-legged bug feeds on lightweight insects that land or fall onto the surface of water. Because of its fine, water-resistant leg hairs (and water's high surface tension), the bug scoots easily across the surface.

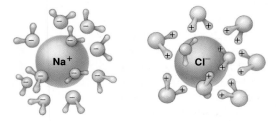

Figure 2.18 Spheres of hydration around two charged ions.

Water's Cohesion

Life also depends on water's cohesion. **Cohesion** means something has the capacity to resist rupturing when it is placed under tension—that is, stretched—as by the weight of a bug's legs (Figure 2.17).

Consider what goes on in a lake or swimming pool or some other large body of liquid water. Uncountable episodes of hydrogen bonding exert a continual inward pull on the water molecules at or near the surface. The ongoing, collective hydrogen bonding results in a high surface tension. It may be especially attention-grabbing if you swim in the lake or pool on a hot summer night, when far too many night-flying insects splat against the water and float on it.

Cohesion operates inside organisms as well as on the outside. For example, maple trees and other land plants require nutrient-laden water for their growth and metabolism. Largely because of the cohesion among water molecules, ribbonlike columns of liquid water move through vascular tissues that serve as pipelines from the roots to leaves and all other parts of the plant. On sunny days, water evaporates from the leaves. As individual molecules depart, hydrogen bonds "pull up" more water molecules into leaf cells as replacements, in ways that you will read about in Section 30.3.

Water's Solvent Properties

A final point to keep in mind is water's capacity as a solvent; ions and polar molecules readily dissolve in it. When dissolved in water, substances are called **solutes**. But what does "dissolved" mean? By way of example, pour some table salt (NaCl) into a glass of water. In time, the salt crystals separate into Na^+ and Cl^-. Each Na^+ attracts the negative end of water molecules; at the same time, each Cl^- attracts the positive end of others (Figure 2.18). Many water molecules cluster as "spheres of hydration" around ions and keep them dispersed in the fluid. In general, a substance is "dissolved" when such spheres have formed around its individual ions or molecules. This is what happens to solutes in the fluid portion of cells, in maple tree sap, in your own blood, and in all other fluids associated with life.

A water molecule has no net charge, yet it shows polarity. The polarity allows water molecules to hydrogen-bond with one another and with other polar (hydrophilic) substances. Water molecules tend to repel nonpolar (hydrophobic) substances.

Water has temperature-stabilizing effects, internal cohesion, and a capacity to dissolve many substances. These properties influence the structure and functioning of organisms.

ACIDS, BASES, AND BUFFERS

A great variety of ions dissolved in the fluids inside and outside cells influence cell structure and functioning. Among the most influential are **hydrogen ions**, or H^+. Hydrogen ions are the same thing as free (unbound) protons. They have far-reaching effects largely because they are chemically active and there are so many of them.

The pH Scale

At any instant in liquid water, some water molecules break apart into hydrogen ions and **hydroxide ions** (OH^-). This ionization of water is the basis of the **pH scale**, as shown in Figure 2.19. Biologists use this scale when measuring the H^+ concentration of seawater, tree sap, blood, and other fluids.

Pure water always contains just as many H^+ as OH^- ions. This condition also may occur in other fluids, and it signifies neutrality. We assign it a value of 7 at the midpoint of the pH scale, which ranges from 0 (the highest H^+ concentration) to 14 (the lowest). *The greater the H^+ concentration, the lower the pH.*

Starting at neutrality, each change by one unit of the pH scale corresponds to a tenfold increase or decrease in the H^+ concentration. An easy way to sense the logarithmic difference in scale is to dissolve a bit of baking soda (pH 9) on your tongue, then egg white (pH 8). Next, sip pure water (pH 7), then lemon juice (2.3).

Acids and Bases

As certain substances dissolve in water, they donate protons (H^+) to some water molecules. For a fleeting instant, the water molecules become H_3O^+, but these just as instantaneously donate the protons elsewhere, so for our purposes it simply is easier to think of all the water molecules as being H_2O. All substances that *donate* H^+ when dissolved in water are called **acids**. Certain other substances are **bases**. They *accept* H^+ when dissolved in water, and OH^- forms directly or indirectly after they do so. All *acidic* solutions, such as lemon juice, gastric fluid, and coffee, contain more H^+ than OH^-; their pH is below 7. *Basic* solutions, such as baking soda, seawater, and egg white, contain more OH^- than H^+. Such solutions, which are also referred to as "alkaline" fluids, have a pH above 7.

The solution inside most of your body's cells is neutral; the cell interior is about 7 on the pH scale. Most fluids bathing those cells are slightly higher, with pH values that range between 7.3 and 7.5. The same is true of the fluid portion of blood. By contrast, seawater is more alkaline than body fluids of organisms that live in it.

You can think of most acids as being either weak or strong. Weak ones such as carbonic acid (H_2CO_3) are reluctant H^+ donors. Depending on the pH, they can just as easily accept H^+ after giving it up, so they alternate between acting as an acid and a base. By contrast, strong acids totally give up H^+ when they dissociate in water. Hydrochloric acid (HCl), nitric acid (HNO_3), and sulfuric acid (H_2SO_4) are examples.

Think about what happens after someone sniffs and eats fried chicken, then sends it on its way to gastric fluid inside the saclike stomach. The meal stimulates cells in the lining of the stomach to secrete a strong acid, HCl, that dissociates into H^+ and Cl^-. These ions make the gastric fluid more acidic. The increased acidity activates enzymes, which digest chicken proteins and help kill most of the bacteria that lurked in or on the chicken. When people eat too much fried chicken, they may get an *acid stomach* and reach for an antacid such

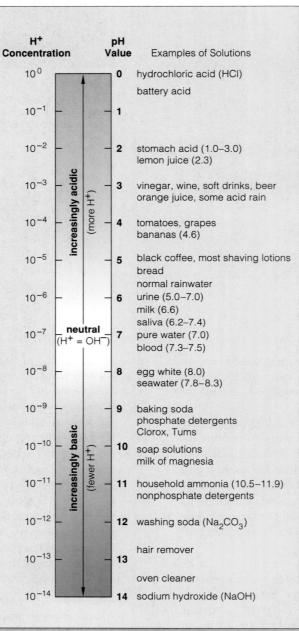

Figure 2.19 The pH scale, in which a liter of some solution is assigned a number according to the number of hydrogen ions that it contains. Also shown are approximate pH values for some common solutions. The pH scale ranges from 0 (most acidic) to 14 (most basic). A change of 1 on the scale means a tenfold change in H^+ concentration.

as milk of magnesia. When this strong base dissolves, it releases magnesium ions and OH⁻, which combines with some of the excess H⁺ in gastric fluid, raises the pH, and settles things down.

High concentrations of strong acids or bases can disrupt the environment and pose dangers to life. Read the labels on bottles of ammonia, drain cleaner, and other products that are often stored in homes. Many can cause severe *chemical burns*. So can the sulfuric acid in car batteries. Or think of strongly acidic industrial wastes. When airborne, they alter the pH of rain (Figure 2.20). In turn, the pH of acid rainwater can severely alter many freshwater and terrestrial habitats, as described in Section 50.1.

Buffers Against Shifts in pH

Metabolic reactions are sensitive to even slight shifts in pH, for H⁺ and OH⁻ can combine with many different molecules and alter their functions. Normally, control mechanisms minimize unsuitable shifts in pH, as they do when HCl enters the stomach in response to a meal. Many of the controls involve buffer systems.

A **buffer system** is a partnership between a weak acid and the base that forms when it dissolves in water. The two work as a pair to counter slight shifts in pH. Remember, when a strong base enters a fluid, the OH⁻ level rises. But a weak acid neutralizes part of the added OH⁻ by combining with it. By this interaction, its partner forms. Later, if a strong acid floods in, the base will accept H⁺ and so become its partner in the system.

Bear in mind, the action of a buffer system cannot make or eliminate hydrogen ions. It can only keep them locked up, out of the way, until other metabolic events restore the balance.

In all complex multicelled organisms, diverse buffer systems operate in the internal environment—that is, in blood and tissue fluids. For example, later in the book you will see how reactions in lungs and kidneys help control the acid-base balance of this environment, at levels suitable for life. For now, simply think of what happens when blood loses H⁺ and is not as acidic as it should be. Carbonic acid happens to be dissolved in the blood. At such times, it releases H⁺ and so becomes the partner base, bicarbonate:

$$H_2CO_3 \longrightarrow HCO_3^- + H^+$$
CARBONIC ACID   BICARBONATE

When blood becomes more acidic, more H⁺ becomes bound to the base, thus forming the partner acid:

$$HCO_3^- + H^+ \longrightarrow H_2CO_3$$
BICARBONATE   CARBONIC ACID

Figure 2.20 Sulfur dioxide emissions from a coal-burning power plant. Camera lens filters revealed the otherwise invisible emissions. Sulfur dioxide and other airborne pollutants dissolve in water vapor to form acidic solutions. They are a major component of acid rain.

Unless they are controlled, shifts in pH have drastic outcomes. If blood pH declines even to 7, an individual will enter into a *coma*, a sometimes irreversible state of unconsciousness. An increase to 7.8 can lead to *tetany*, a potentially lethal stage in which the body's skeletal muscles enter a state of uncontrollable contraction. In *acidosis*, carbon dioxide builds up in blood, too much carbonic acid forms, and blood pH severely decreases. *Alkalosis* is an uncorrected increase in blood pH. Both conditions weaken the body and can be lethal.

Salts

Salts are compounds that release ions *other than* H⁺ and OH⁻ in solutions. Salts and water often form when a strong acid and a strong base interact. Depending on a solution's pH value, salts can form and dissolve easily. Consider how sodium chloride forms, then dissolves:

$$HCl \text{ (acid)} + NaOH \text{ (base)} \longrightarrow NaCl \text{ (salt)} + H_2O$$
HYDROCHLORIC   SODIUM   SODIUM CHLORIDE
ACID   HYDROXIDE

$$Na^+ \quad Cl^- \text{ (ionization)}$$

Many salts dissolve into ions that serve key functions in cells. For example, nerve cell activity depends on ions of sodium, potassium, and calcium; plant cells take up water with the aid of potassium ions; and blood clots and muscles contract with the help of calcium ions.

Hydrogen ions (H⁺) and other ions dissolved in the fluids inside and outside cells affect cell structure and function.

When dissolved in water, acidic substances release H⁺ and basic (alkaline) substances accept them. Certain acid-base interactions, as in buffer systems, help maintain the pH value of a fluid—that is, its H⁺ concentration.

A buffer system counters slight shifts in pH by releasing hydrogen ions when their concentration is too low or by combining with them when the concentration is too high.

Salts are compounds that release ions other than H⁺ and OH⁻, and many of those ions have key roles in cell functions.

SUMMARY

1. Chemistry helps us understand the nature of all of the substances that make up cells, organisms, the Earth, its waters, and the atmosphere. Each substance consists of one or more elements. Of the ninety-two naturally occurring elements, oxygen, carbon, hydrogen, and nitrogen are the most common in organisms. Organisms have lesser and varying amounts of other elements, such as calcium, phosphorus, potassium, and sulfur.

2. Table 2.3 summarizes some key chemical terms that you will encounter throughout this book.

3. Elements consist of atoms. An atom has one or more positively charged protons, an equal number of negatively charged electrons, and (except for hydrogen) one or more uncharged neutrons. The protons and neutrons occupy a core region, the atomic nucleus. Atoms of each element may vary in the number of neutrons. Those that deviate from the most common number of neutrons are isotopes.

4. An atom has no *net* charge. But it may gain or lose one or more electrons and so become an ion, which has an overall positive or negative charge.

5. By one simplified model, electrons can occupy orbitals (volumes of space) inside a series of shells around the atomic nucleus. Whether one atom interacts with others depends on the number and arrangement of its electrons. Atoms with one or more unfilled orbitals in the outermost shell tend to bond with other elements.

6. A chemical bond is a union between the electron structures of atoms.

   a. In an ionic bond, a positive ion and negative ion remain together because of a mutual attraction of their opposite charges.

   b. Various atoms can share one or more pairs of electrons, in single, double, or triple covalent bonds. Electron sharing is equal in a nonpolar covalent bond and unequal in a polar covalent bond. The interacting atoms have no net charge, but the bond is slightly negative at one end and slightly positive at the other.

   c. In a hydrogen bond, a highly electronegative, small atom weakly interacts with a hydrogen atom that is already taking part in a polar covalent bond in the same molecule or in a different one.

7. By the pH scale, a solution is assigned a number that reflects its H^+ concentration. This ranges from 0 (highest concentration) to 14 (lowest). At pH 7, the H^+ and OH^- concentrations are equal. Acids release H^+ in water; bases combine with them. Buffer systems help maintain the pH values of blood, tissue fluids, and the fluid inside cells.

8. Polar covalent bonds join together the atoms of water molecules. The polarity promotes extensive hydrogen bonding between water molecules. Such bonding is the basis of liquid water's ability to resist temperature changes more than other fluids do, to show internal cohesion, and to easily dissolve polar or ionic substances. These properties profoundly influence the metabolic activity, shape, and internal organization of cells.

Table 2.3 Summary of Key Players in the Chemical Basis of Life

ELEMENT	Fundamental form of matter that occupies space, has mass, and cannot be broken apart into a different form of matter by ordinary physical or chemical means.
ATOM	Smallest unit of an element that still retains the characteristic properties of that element.
Proton (p^+)	Positively charged particle of the atomic nucleus. All atoms of an element have the same number of protons, which is the atomic number. A proton without an electron zipping around it is a hydrogen ion (H^+).
Electron (e^-)	Negatively charged particle that can occupy volumes of space (orbitals) around an atomic nucleus. All atoms of an element have the same number of electrons. Electrons can be shared or transferred among atoms.
Neutron	Uncharged particle of the nucleus of all atoms except hydrogen. For a given element, the mass number is the number of protons and neutrons in the nucleus.
MOLECULE	Unit of matter in which two or more atoms of the same element, or different ones, are bonded together.
Compound	Molecule composed of two or more different elements in unvarying proportions. Water is an example.
Mixture	Intermingling of two or more elements in proportions that can and usually do vary.
ISOTOPE	For a given element, an atom with more or fewer neutrons than atoms with the most common number.
Radioisotope	Unstable isotope, having an unbalanced number of protons and neutrons, that emits particles and energy.
Tracer	Molecule of a substance to which a radioisotope is attached. In conjunction with tracking devices, it can be used to follow the movement or destination of that substance in a metabolic pathway, the body, etc.
ION	Atom that has gained or lost one or more electrons, thus becoming positively or negatively charged.
SOLUTE	Any molecule or ion dissolved in some solvent.
Hydrophilic substance	Polar molecule or molecular region that can readily dissolve in water.
Hydrophobic substance	Nonpolar molecule or molecular region that strongly resists dissolving in water.
ACID	Substance that donates H^+ when dissolved in water.
BASE	Substance that accepts H^+ when dissolved in water; OH^- forms directly or indirectly afterward.
SALT	Compound that releases ions other than H^+ or OH^- when dissolved in water.

Review Questions

1. Name the four elements (and their symbols) that make up more than 95 percent of the body weight of all organisms. *2.1*

2. Define isotope, and describe how radioisotopes are used either in radiometric dating or as tracers. *2.1, 2.2*

3. How many electrons can occupy each orbital around an atomic nucleus? Using the shell model, explain how the orbitals available to electrons are distributed in an atom. *2.3*

4. Distinguish between:
   a. ionic and hydrogen bonds *2.4*
   b. polar and nonpolar covalent bonds *2.4*
   c. hydrophilic and hydrophobic interactions *2.5*

5. If a water molecule has no net charge, then why does it attract polar molecules and repel nonpolar ones? *2.5*

6. Define acid and base. Then describe the behavior of a weak acid in solutions having a high or low pH value. *2.6*

Self-Quiz *(Answers in Appendix IV)*

1. Electrons carry (a) _____ charge.
   a. positive   b. negative   c. zero

2. Atoms share electrons unequally in a(n) _____ bond.
   a. ionic   c. polar covalent
   b. nonpolar covalent   d. hydrogen

3. An individual water molecule shows _____ .
   a. polarity   d. solvency
   b. hydrogen-bonding capacity   e. a and b
   c. heat resistance   f. all of the above

4. In liquid water, spheres of hydration form around _____ .
   a. nonpolar molecules   d. solvents
   b. polar molecules   e. b and c
   c. ions   f. all of the above

5. Hydrogen ions (H^+) are _____ .
   a. the basis of pH values   d. dissolved in blood
   b. unbound protons   e. both a and b
   c. targets of certain buffers f. all of the above

6. When dissolved in water, a(n) _____ donates H^+; however, a(n) _____ accepts H^+ and leads to OH^- formation.

7. Match the terms with their most suitable descriptions.
   ____ trace element   a. weak acid and its partner base work
   ____ buffer system   as a pair to counter pH shifts
   ____ chemical bond   b. union between electron structures
   ____ temperature   of two atoms
                         c. less than 0.01% of body weight
                         d. measure of molecular motion in
                            some defined region

Critical Thinking

1. A hospital patient is administered a 3.2-milligram dose of ^{131}iodine, which has a half-life of 8.1 days. How much of this radioisotope will remain in the patient's body after twenty-four days, assuming none is lost except by way of radioactive decay?

2. An ionic compound forms when calcium combines with chlorine. Referring to Figure 2.12, give the compound's formula. (Hint: Be sure the outermost shell of each atom is filled.)

3. The molecular weight of hydrochloric acid (HCl), a strong acid, is 36 grams/mole, and that of sodium hydroxide (NaOH), a strong base, is 40 grams/mole. A one molar (1M) solution of any compound can be made by dissolving one mole in a liter of water.

Tara has 1 liter of a 1M solution of HCl. How many grams of NaOH must she add to neutralize the solution (adjust pH to 7)?

4. David, an inquisitive three-year-old, touched the water in a metal pan on the stove and found it was warm. Then he touched the pan and got a nasty burn. Devise a hypothesis to explain why water in a metal pan heats up far more slowly than the pan itself.

5. When molecules absorb microwaves, a form of electromagnetic radiation, they move more rapidly. Explain why a microwave oven can heat foods.

6. From what you know about cohesion, devise a hypothesis to explain why water forms droplets.

7. Edward is trying to study a chemical reaction that an enzyme catalyzes (speeds up). H^+ forms during the reaction, but the enzyme is destroyed at low pH. What can he include in his reaction mix to protect the enzyme while he studies the reaction? Explain how your suggestion might solve the problem.

8. Many metabolic reactions proceed on molecular regions of enzymes and other proteins. Cells must have access to those regions. By interactions with water and ions, a soluble protein stays dispersed in cellular fluid rather than settling against some cell structure. An electrically charged cushion around the protein makes this happen.
Using the diagram at right as a guide, explain the chemical interactions by which such a cushion forms, starting with major bonds within the protein itself.

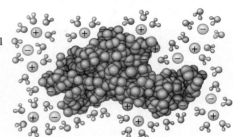

Selected Key Terms

acid *2.6*	hydrogen	orbital, atomic *2.3*
atom *2.1*	ion (H^+) *2.6*	pH scale *2.6*
atomic	hydrophilic	proton *2.1*
number *2.1*	substance *2.5*	radioactive
base *2.6*	hydrophobic	decay *2.1*
buffer system *2.6*	substance *2.5*	radioisotope *2.1*
chemical bond *2.1*	hydroxide	radiometric
cohesion *2.5*	ion (OH^-) *2.6*	dating *2.2*
compound *2.3*	ion *2.4*	salt *2.6*
covalent bond *2.4*	ionic bond *2.4*	shell model *2.3*
electron *2.1*	isotope *2.1*	solute *2.5*
element *2.1*	mass number *2.1*	temperature *2.5*
evaporation *2.5*	mixture *2.3*	trace element *2.1*
half-life *2.2*	molecule *2.3*	tracer *2.2*
hydrogen bond *2.4*	neutron *2.1*	

Readings

Pennisi, E. February 20, 1993. "Water, Water, Everywhere." *Science News.*

Ritter, P. 1996. *Biochemistry: A Foundation.* Pacific Grove, California: Brooks/Cole.

Web Site See *http://www.wadsworth.com/biology* for practice quiz questions, hypercontents, BioUpdates, and critical thinking. The Wadsworth Biology Resource Center provides a wealth of information fully organized and integrated by chapter.

CARBON COMPOUNDS IN CELLS

Carbon, Carbon, in the Sky—Are You Swinging Low and High?

High in the mountains of the Pacific Northwest, vast coniferous forests have endured one more murderously cold winter (Figure 3.1). As is true of other organisms, these evergreen, cone-bearing trees cannot grow or reproduce in the absence of liquid water. Yet through the winter months, water in the surroundings is locked away from them, in the form of snow and ice. The trees get by anyway. At the surface of their needle-shaped leaves, a layer of interconnected cells with thick, waxy walls forms a covering, an epidermis, that prevents water loss. The only way precious water can escape is through small gaps across the epidermis that close or open in response to changing conditions. During the cool, dry days of autumn, the trees enter dormancy. Metabolic activities idle and growth ceases, but water conserved inside them is enough to keep their cells alive.

With the arrival of spring, rising temperatures and water from melting snow stimulate renewed growth. Tree roots soak up mineral-laden water. And carbon dioxide, a gaseous molecule of one carbon atom and two oxygen atoms, moves in from the air, through gaps in the leaf epidermis. With their photosynthetic magic, the conifers turn these simple materials into sugars, starches, and other carbon-based compounds. They are premier producers of the northern forests. Producer organisms, recall, use the self-made compounds as structural materials and as packets of energy. So do other organisms. One way or another, every consumer and decomposer is nourished by the producer organisms of forests and all other ecosystems the world over.

Today, plants of the great prevailing forests and plains at northern latitudes

Figure 3.1 Conifers beneath the first snows of winter on Silver Star Mountain, Washington. As is true of all other organisms, the structure, activities, and very survival of these trees start with the carbon atom and its diverse molecular partners in organic compounds.

of the United States, Canada, and other parts of the Northern Hemisphere are breaking dormancy sooner than they did just two decades ago. Why? No one knows for sure, but the change in their life cycles might be one outcome of long-term change in the global climate.

Researchers in the Northern Hemisphere have been studying the concentration of carbon dioxide in the air around us since the early 1950s. Among other things, they found out that the concentration shifts with the seasons. It declines during spring and summer, when photosynthesizers take up stupendous amounts of the gas. It rises during other times of year, when huge populations of decomposers that release the gas as a metabolic by-product show rapid growth.

According to the researchers' measurements, the spring decline now starts a full week earlier than it did in the mid-1970s. Besides this, the seasonal swings are becoming more pronounced—by as much as 20 percent in Hawaii and a whopping 40 percent in Alaska.

Swings in the atmospheric concentration of carbon dioxide were greatest in 1981 and 1990. Intriguingly, temperatures of the lower atmosphere also have been rising—and in 1981 and 1990, they were uncommonly high. Are we in the midst of a long-term, worldwide rise in atmospheric temperature? And is this *global warming* promoting a longer growing season, hence the wider seasonal swings?

The picture gets more intricate. We humans burn great quantities of coal, gasoline, and other fossil fuels for energy. The fossil fuels are rich in carbon, which is released during the burning processes. The released carbon may be contributing to the global warming, in ways that you will read about in later chapters.

For now, the point to keep in mind is this: *Carbon permeates the entire world of life, from the energy-requiring activities and structural organization of individual cells, to physical and chemical conditions that span the globe and influence life everywhere.* With this chapter, we turn to the life-giving properties that emerge out of the molecular structure of carbon-rich compounds. Study the chapter well. It will serve as your foundation for understanding how different organisms put such compounds together, how they use them, and how the effects of these uses can ripple through the biosphere.

KEY CONCEPTS

1. Organic compounds have a backbone of one or more carbon atoms to which hydrogen, oxygen, nitrogen, and other atoms are attached. We define cells partly by their capacity to assemble the organic compounds known as carbohydrates, lipids, proteins, and nucleic acids.

2. Cells put together large biological molecules from their pools of smaller organic compounds, which include simple sugars, fatty acids, amino acids, and nucleotides.

3. Glucose and other simple sugars are carbohydrates. So are organic compounds composed of two or more sugar units, of one or more types, that are covalently bonded together. The most complex carbohydrates that cells assemble are polysaccharides, many of which consist of hundreds or thousands of sugar units.

4. Lipids are greasy or oily compounds that show little tendency to dissolve in water but that can dissolve in nonpolar compounds, including other lipids. They include neutral fats, phospholipids, waxes, and sterols.

5. Cells use carbohydrates and lipids as building blocks and as their major sources of energy.

6. Proteins have truly diverse roles. Many are structural materials. Many are enzymes, a type of molecule that enormously increases the rate of specific metabolic reactions. Other kinds transport cell substances, contribute to cell movements, trigger changes in cell activities, and defend the body against injury and disease.

7. For living organisms, ATP and other nucleotides are crucial players in metabolism. DNA and RNA, strandlike nucleic acids assembled from nucleotide subunits, are the basis of inheritance and reproduction.

The molecules of life are **organic compounds**, meaning they consist of one or more elements covalently bonded to carbon atoms. The term is a holdover from a time when chemists thought "organic" substances were the ones they got from animals and vegetables, as opposed to "inorganic" substances obtained from minerals. The term persists, even though researchers now synthesize organic compounds in laboratories. And it persists even though there are reasons to believe organic compounds were present on Earth *before* organisms were.

Effects of Carbon's Bonding Behavior

By far, organisms consist mainly of oxygen, hydrogen, and carbon (Figure 2.3). Much of the oxygen and the hydrogen is in the form of water. Remove the water, and carbon makes up more than half of what's left.

Carbon's importance in life arises from its versatile bonding behavior. *Each carbon atom can share pairs of electrons with as many as four other atoms.* Each covalent bond formed this way is quite stable. Such bonds link carbon atoms together in chains and rings. These serve as a backbone to which hydrogen, oxygen, and other elements become attached, as in Figure 3.2.

The flattened structural formulas shown in Figure 3.2 might lead you to believe that the molecules they represent are flat, like paper dolls. Yet the molecules have wonderful three-dimensional shapes, which begin with bonding arrangements in the carbon backbone.

When a carbon atom bonds with four other atoms, this three-dimensional shape emerges:

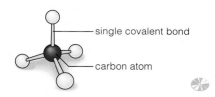

In a carbon backbone, some of the carbon atoms rotate freely around a *single* covalent bond:

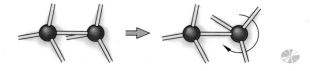

However, a *double* covalent bond uniting two carbon atoms restricts their rotation. Where such bonds occur in the backbone, the atoms rigidly hold their position in space:

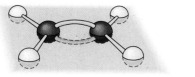

Now imagine a backbone with some atoms locked in position and others rotating at different angles in space. The resulting orientations of these atoms may promote

Figure 3.2 Carbon compounds. There is a Tinkertoy quality to carbon compounds, for a single atom can be the start of diverse molecules assembled from "straight-stick" covalent bonds. Start with the simplest hydrocarbon, methane (CH_4). Stripping one hydrogen from methane leaves a methyl group, which occurs in fats, oils, and waxes:

methane   methyl group

Imagine stripping a hydrogen atom from each of two methane molecules, then bonding the two molecules together. If the resulting structure were to lose a hydrogen atom, you would end up with an ethyl group:

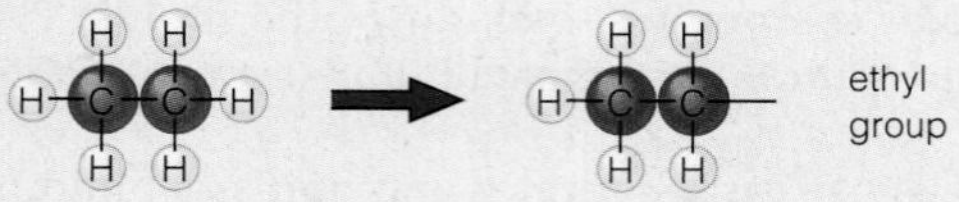

ethyl group

You could go on building a continuous chain:

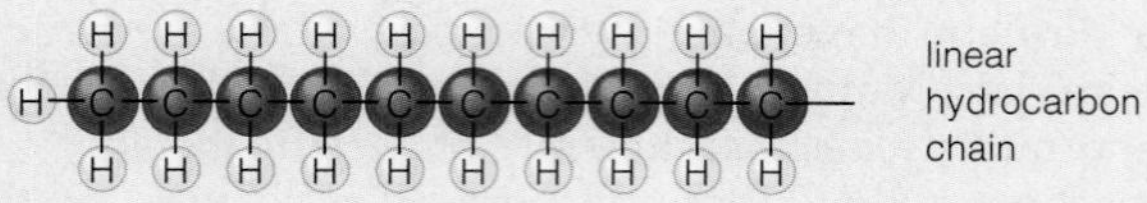

linear hydrocarbon chain

And you could add branches to the chain:

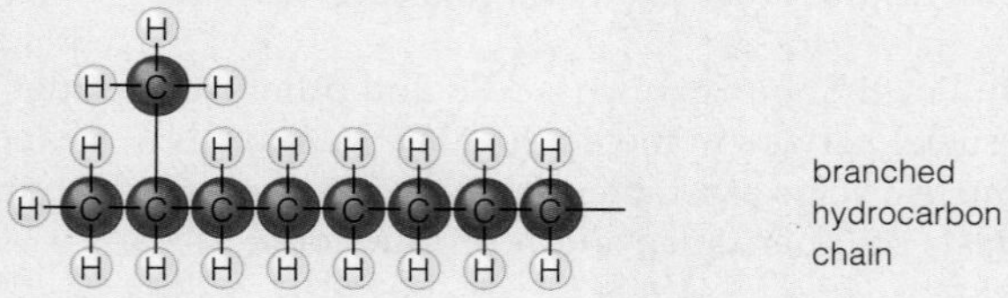

branched hydrocarbon chain

You might have the chain coil back on itself into a ring, which you could diagram in various ways:

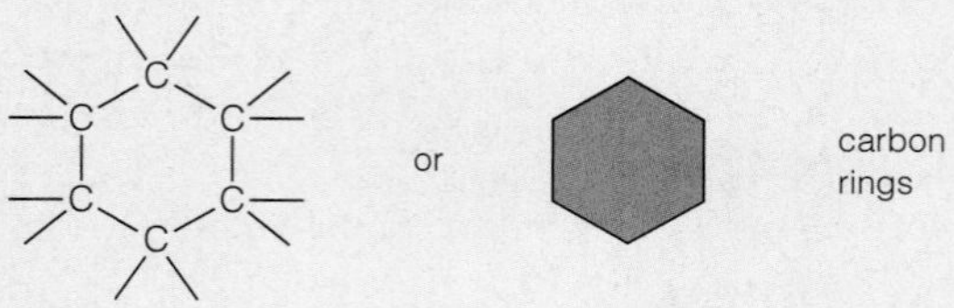

or   carbon rings

You might even join many carbon rings together and so produce larger molecules:

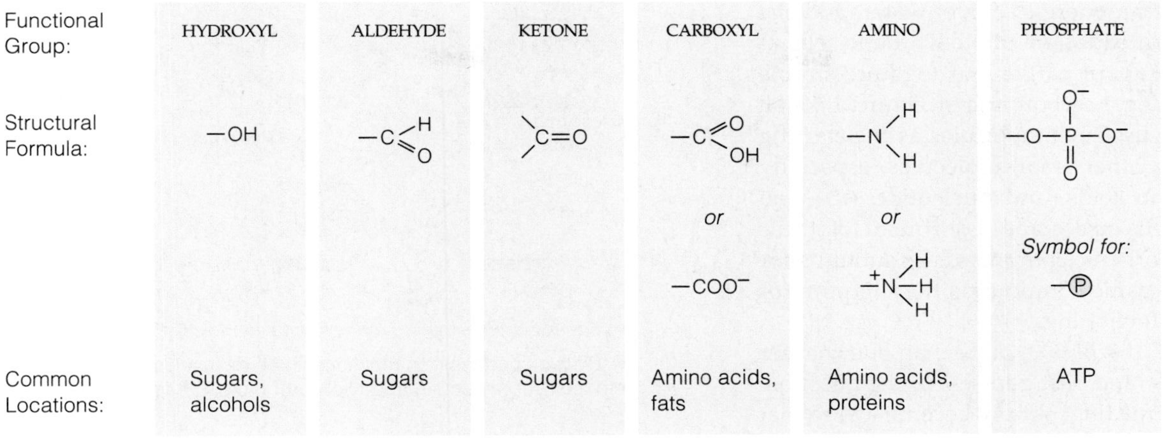

Functional Group:	HYDROXYL	ALDEHYDE	KETONE	CARBOXYL	AMINO	PHOSPHATE
Structural Formula:	—OH	$-C{\small\begin{matrix}H\\O\end{matrix}}$	$\begin{matrix}\\C=O\end{matrix}$	$-C{\small\begin{matrix}O\\OH\end{matrix}}$ *or* —COO⁻	$-N{\small\begin{matrix}H\\H\end{matrix}}$ *or* $\overset{+}{-}N{\small\begin{matrix}H\\-H\\H\end{matrix}}$	$-O-\overset{\overset{O^-}{\mid}}{\underset{\underset{O}{\parallel}}{P}}-O^-$ *Symbol for:* —Ⓟ
Common Locations:	Sugars, alcohols	Sugars	Sugars	Amino acids, fats	Amino acids, proteins	ATP

Figure 3.3 Examples of functional groups and some of their common locations in biological molecules. The carboxyl and amino groups also are shown in their ionized form, which is most common in cells and in the body fluids of multicelled organisms.

a   female wood duck   male wood duck

b   HO— AN ESTROGEN

c   O= TESTOSTERONE

Figure 3.4 Notable differences in traits between male and female wood ducks (*Aix sponsa*). Two different sex hormones have key roles in the development of feather color and other traits that help the two ducks recognize each other (and thereby influence reproductive success).

Both of the hormones have the same carbon ring structure. As you can see, however, the ring structures have different functional groups attached to them.

or discourage interactions with other substances in the vicinity. As you will see, *such interactions give rise to the shapes and functions of biological molecules.*

Hydrocarbons and Functional Groups

"Hydrocarbons" have only hydrogen atoms attached to a carbon backbone. They do not break apart very easily, and they form the stable portions of most biological molecules. Biological molecules also incorporate some number of **functional groups**, which are single atoms or clusters of atoms covalently bonded to the carbon backbone. To get a sense of their importance, consider a few of the groups shown in Figure 3.3. Sugars and other organic compounds classified as **alcohols** have one or more hydroxyl groups (—OH). Alcohols are quick to dissolve in water, for water molecules readily form

hydrogen bonds with —OH groups. Proteins have a backbone formed by reactions between amino groups and carboxyl groups. The unique three-dimensional structure of each protein starts with bonding patterns associated with this backbone. Also, amino groups can combine with H⁺ and act as buffers against decreases in pH. As a final example, Figure 3.4 shows two different functional groups that are a molecular starting point for differences between males and females of many species.

Organic compounds have diverse, three-dimensional shapes and functions. Their diversity begins with flexible and rigid bonding arrangements in their carbon backbones.

Functional groups covalently bonded to the carbon backbones of organic compounds add enormously to the structural and functional diversity.

HOW CELLS USE ORGANIC COMPOUNDS

By using sunlight as an energy source, water, and the carbon from carbon dioxide, the photosynthetic cells of those trees you read about earlier put together simple sugar molecules having a carbon ring structure. Like all living cells, they also use such molecules as the starting point for assembling other small molecules, especially the fatty acids, amino acids, and nucleotides. As you will read shortly, cells use some assortment of these four classes of small organic compounds as subunits for building all of the organic compounds they require for their structure and functioning.

How do they do it? It will take more than one chapter to sketch out answers (and best guesses) to the question. At this point in your reading, simply become aware that the reactions by which cells build organic compounds, and even rearrange them and break them apart, require more than an energy input. They also require the class of proteins called **enzymes**, which make specific metabolic reactions proceed faster than they would on their own.

Five Classes of Reactions

Different enzymes mediate different kinds of reactions, but most of the reactions fall into five categories:

1. **Functional-group transfer**. One molecule gives up a functional group, which another molecule accepts.

2. **Electron transfer**. One or more electrons stripped from one molecule are donated to another molecule.

3. **Rearrangement**. A juggling of internal bonds converts one type of organic compound into another.

4. **Condensation**. Through covalent bonding, two molecules combine to form a larger molecule.

5. **Cleavage**. A molecule splits into two smaller ones.

To get a sense of what goes on, consider two examples of these events. First, in many condensation reactions, enzymes remove a hydroxyl group from one molecule and an H atom from another, then speed the formation of a covalent bond between the two molecules at their exposed sites (Figure 3.5a). As a typical but incidental outcome of the reaction, the discarded atoms join to form a water molecule. A series of condensation reactions can produce starches and other polymers. A **polymer** is a large molecule with three to millions of subunits, which may or may not be identical. Often the subunits are called **monomers**, as in the sugar monomers of starch.

As the second example, a cleavage reaction called **hydrolysis** is a bit like condensation in reverse (Figure 3.5b). Enzymes that act on covalent bonds at functional groups split molecules into two or more parts, then attach —H and —OH derived from a water molecule

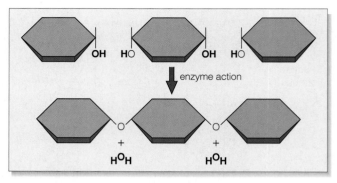

a By two condensation reactions, three molecules covalently bond into a larger molecule, and two water molecules typically form.

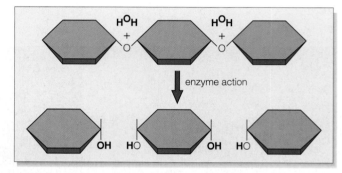

b Hydrolysis, a water-requiring cleavage reaction.

Figure 3.5 Examples of reactions by which most biological molecules are put together, rearranged, and broken apart.

(**a**) Two condensation reactions. (**b**) Two hydrolysis reactions.

to the exposed sites. Hydrolysis allows cells to cleave large polymers such as starch into smaller units when these are required for building blocks or energy.

The Molecules of Life

Under present-day conditions in nature, *only living cells synthesize carbohydrates, lipids, proteins, and nucleic acids.* These are the molecules characteristic of life. As you will see, different classes of biological molecules function as the cells' packets of instant energy, energy stores, structural materials, metabolic workers, and libraries of hereditary information.

Carbohydrates, lipids, proteins, and nucleic acids are the main biological molecules, the organic compounds that only living cells can assemble under conditions that now occur in nature.

Cells assemble, rearrange, and degrade organic compounds mainly through enzyme-mediated reactions involving the transfer of functional groups or electrons, rearrangement of internal bonds, and a combining or splitting of molecules.

3.3

FOOD PRODUCTION AND A CHEMICAL ARMS RACE

The next time you shop for groceries, reflect on what it takes to provide you with your daily supply of organic compounds. For example, those heads of lettuce typically grew in fertilized cropland. Possibly they competed with weeds, which didn't know the nutrients in the fertilizer were meant for the lettuce. Leaf-chewing insects didn't know the lettuce wasn't meant to be their salad bar. Each year, these food pirates and others ruin or gobble up nearly half of what people all over the world try to grow.

Most plants aren't entirely defenseless. They evolved under intense selection pressure from attacks by insects, fungi, and other organisms, and in natural settings they often can repel the attackers with toxins. A **toxin** is an organic compound, a normal metabolic product of one species, but its chemical effects can harm or kill individuals of a different species that come in contact with it. Humans, too, encounter traces of natural toxins in most of what they eat, even in such familiar edibles as hot peppers, potatoes, figs, celery, rhubarb, and alfalfa sprouts. Still, we do not die in droves from these natural toxins, so apparently our bodies have chemical defenses against them.

In 1945, we took a cue from plants and started using newly developed, synthetic toxins to protect crop yields, food stores, freshwater supplies, and even our health, pets, and ornamental plants. The *herbicides* kill weeds by disrupting metabolism and growth. Most *insecticides* clog a target insect's airways, jangle its nerve and muscle cells, or block its reproduction. *Fungicides* work against harmful fungi, including a mold that produces aflatoxin, one of the deadliest poisons known. By 1995, people in the United States alone were spraying or spreading more than 1.25 billion pounds of these toxins through fields, gardens, homes, offices, and industrial sites (Figure 3.6).

There are downsides to pesticide applications. Along with pests, some of the toxic organic compounds kill birds and other predators which, in natural settings, help control pest population sizes. Worse, the targeted pests have been developing resistance to the chemical arsenal, for reasons

Figure 3.6 One consumer of food crops, and a low-flying crop duster with its rain of pesticides.

that you read about in Chapter 1. Also, pesticides cannot be released haphazardly in the environment, for people might end up inhaling them, ingesting them with food, or absorbing them through the skin. Different types remain active for days, weeks, even years. Some trigger severe headaches, rashes, hives, asthma, and joint pain in millions of people. Some trigger life-threatening allergic reactions in abnormally sensitive individuals.

Presently, the long-lived pesticides are banned in the United States. Even rapidly degradable ones, including malathion and other organophosphates, are subjected to rigorous application standards and ongoing safety tests. However, with so many pests looming around us, the search for effective countermeasures goes on.

CARBOHYDRATES

Consider first the carbohydrates—the most abundant of all biological molecules, which cells use as structural materials and transportable or storage forms of energy. Most **carbohydrates** consist of carbon, hydrogen, and oxygen in a 1:2:1 ratio, but there are exceptions. Three classes are the **monosaccharides**, **oligosaccharides**, and **polysaccharides**.

The Simple Sugars

"Saccharide" comes from a Greek word meaning sugar. A *mono*saccharide, meaning "one monomer of sugar," is the simplest carbohydrate. It has an aldehyde or a ketone group and at least two —OH groups joined to the carbon backbone. Most monosaccharides are sweet tasting and readily dissolve in water. The most common have a backbone of five or six carbon atoms that tends to form a ring structure when dissolved in cells or body fluids. Ribose and deoxyribose, the sugar components of RNA and DNA, respectively, have five carbon atoms. Glucose has six (Figure 3.7a). Besides being the main energy source for most organisms, glucose is a precursor (parent molecule) of many compounds and a building block for larger carbohydrates. Vitamin C (a sugar acid) and glycerol (an alcohol with three —OH groups) are other examples of compounds based on sugar monomers.

Short-Chain Carbohydrates

Unlike the simple sugars, an *oligo*saccharide is a short chain of two or more sugar monomers that are bonded covalently. (*Oligo-* means a few.) The simplest kinds, the *di*saccharides, have only two sugar units. Lactose, sucrose, and maltose are examples. Lactose (a glucose and a galactose unit) is present in milk. Sucrose, the most plentiful sugar in nature, consists of a glucose and a fructose unit (Figure 3.7c). Plants continually convert carbohydrate stores to sucrose, which can be transported through leaves, stems, and roots. Table sugar is sucrose crystallized from sugarcane and sugar beets. Proteins and other large molecules often have oligosaccharides attached as side chains to the carbon backbone. Some chains take part in the body's defenses against disease, others in cell membrane functions.

Complex Carbohydrates

The "complex" carbohydrates, or *poly*saccharides, are straight or branched chains of *many* sugar monomers, often hundreds or thousands of the same or different kinds. Starch, cellulose, and glycogen, the most common polysaccharides, consist only of glucose. In plants, cells store sugars in the form of large molecules of starch, which enzymes can readily hydrolyze to glucose units. Plant cells use cellulose as a structural material in their walls. Fibers made of cellulose molecules are tough and insoluble. Like the steel rods in reinforced concrete, the fibers withstand considerable weight and stress.

If both cellulose and starch consist of glucose, why are their properties so different? The answer lies with differences in covalent bonding patterns between the monomers, which in both cases are strung in chains. In starch, the bonds angle each monomer with respect to the next in line, and the chain coils like a spiral staircase (Figure 3.8a,b). Such coils are not especially stable, and in starches that have branched chains, they are even less so. Many —OH groups project outward from the coiled chains, so they are readily accessible to enzymes.

By contrast, in cellulose, many glucose chains stretch out side by side and hydrogen-bond to one another at —OH groups (Figure 3.8c,d). This bonding arrangement

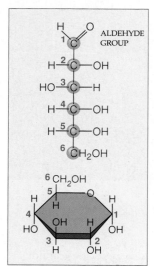

a Structure of glucose.

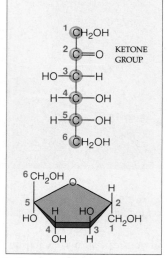

b Structure of fructose.

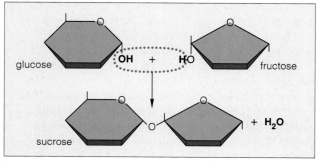

c Formation of sucrose by a condensation reaction.

Figure 3.7 Straight-chain and ring forms of (**a**) glucose and (**b**) fructose. For reference purposes, the carbon atoms of simple sugars are commonly numbered in sequence, starting at the end closest to the aldehyde or ketone group. (**c**) Condensation of two monosaccharides into a disaccharide.

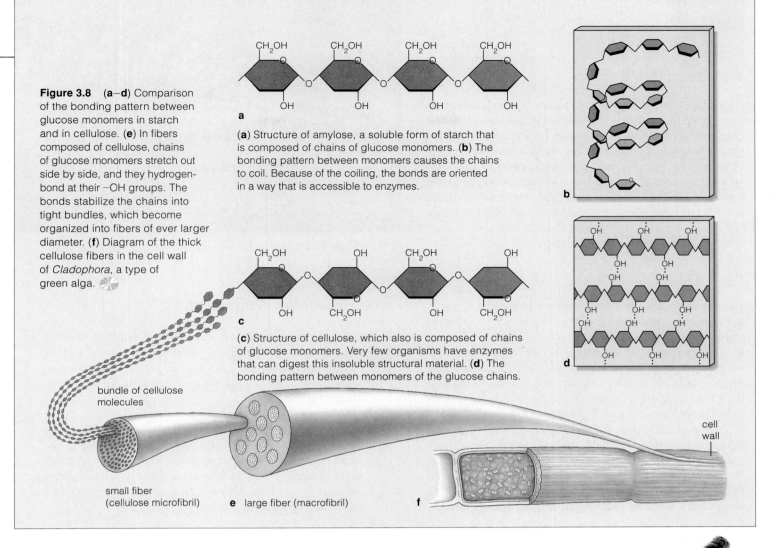

Figure 3.8 (**a–d**) Comparison of the bonding pattern between glucose monomers in starch and in cellulose. (**e**) In fibers composed of cellulose, chains of glucose monomers stretch out side by side, and they hydrogen-bond at their —OH groups. The bonds stabilize the chains into tight bundles, which become organized into fibers of ever larger diameter. (**f**) Diagram of the thick cellulose fibers in the cell wall of *Cladophora*, a type of green alga.

(**a**) Structure of amylose, a soluble form of starch that is composed of chains of glucose monomers. (**b**) The bonding pattern between monomers causes the chains to coil. Because of the coiling, the bonds are oriented in a way that is accessible to enzymes.

(**c**) Structure of cellulose, which also is composed of chains of glucose monomers. Very few organisms have enzymes that can digest this insoluble structural material. (**d**) The bonding pattern between monomers of the glucose chains.

bundle of cellulose molecules

small fiber (cellulose microfibril)

e large fiber (macrofibril)

f

cell wall

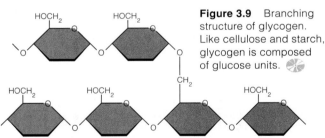

Figure 3.9 Branching structure of glycogen. Like cellulose and starch, glycogen is composed of glucose units.

Figure 3.10 Scanning electron micrograph of a tick. Its body covering is a protective, chitin-reinforced cuticle.

stabilizes the chains in a tightly bundled pattern that resists digestion, at least by most enzymes.

For long-term storage, plants squirrel away glucose in the form of amylose, the starch in Figure 3.8*a*. They also build amylopectin, a branched starch that invites rapid mobilization of glucose. In animals, especially in liver and muscle tissue, the sugar-storage equivalent of amylopectin is glycogen. When the sugar level in blood falls, liver cells degrade glycogen and release glucose to the blood. When you exercise briefly but intensely, muscle cells tap their glycogen stores for quick energy. Figure 3.9 shows many one of glycogen's branchings.

Chitin, another polysaccharide, has nitrogen atoms attached to the backbone. Chitin is the main structural material in the external skeletons and other hard body parts of numerous animals, such as crabs, earthworms, insects, and ticks (Figure 3.10). Chitin also is the main structural material in the cell walls of many fungi.

Carbohydrates are either simple sugars (such as glucose) or molecules composed of many sugar units (such as starch). Every cell requires carbohydrates as structural materials, stored forms of energy, and transportable packets of energy.

3.5 LIPIDS

As you may know, **lipids** are mainly hydrocarbons that show very little tendency to dissolve in water, although they readily dissolve in nonpolar substances. Lipids are greasy or oily to the touch. In nearly all organisms, certain lipids are the main reservoirs of stored energy. Others are structural materials in cells (such as membranes) and cell products (such as surface coatings). Here we consider neutral fats, phospholipids, and waxes, all of which have fatty acid components. We also consider the sterols, each with a backbone of four carbon rings.

Figure 3.12 Triglyceride-protected penguins taking the plunge.

Fatty Acids

The lipids called **fatty acids** have a backbone of up to thirty-six carbon atoms, a carboxyl group (—COOH) at one end, and hydrogen atoms occupying most or all of the remaining bonding sites. When combined with other molecules, fatty acids typically stretch out like flexible tails. Tails with one or more double bonds in their backbone are *unsaturated*. Those with single bonds only are *saturated*. Figure 3.11a shows examples.

When abundant in a substance, saturated fatty acid tails snuggle in parallel by mutual weak attractions, and this gives the substance a rather solid consistency. By contrast, the double and triple bonds in unsaturated fatty acids put rigid kinks in the tails. These packing arrays are less stable and impart fluidity to substances.

Triglycerides (Neutral Fats)

Butter, lard, and oils are examples of **triglycerides**, or neutral fats—the body's most abundant lipids and its richest energy source. These lipids have three fatty acid tails attached to a backbone of glycerol, as shown in Figure 3.11b.

Gram for gram, triglycerides yield more than twice as much energy as complex carbohydrates when they are degraded. In vertebrates, the cells of adipose tissue store quantities of triglycerides as fat droplets. A thick layer of triglycerides insulates penguins and some other animals against the near-freezing temperatures of their habitats (Figure 3.12).

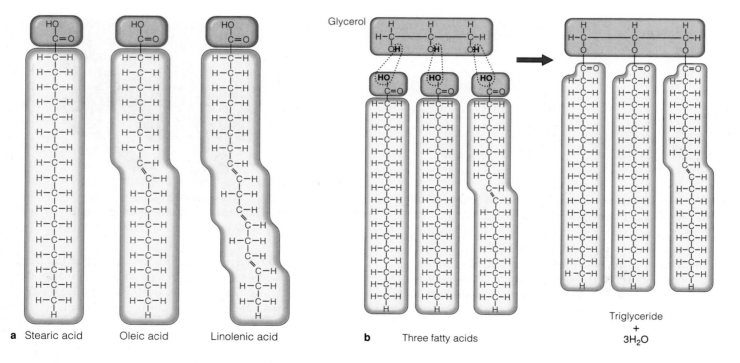

a Stearic acid   Oleic acid   Linolenic acid

b   Three fatty acids

Glycerol

Triglyceride
+
$3H_2O$

Figure 3.11 (**a**) Structural formulas for three fatty acids. Stearic acid's carbon backbone is fully saturated with hydrogen atoms. Oleic acid, with a double bond in its backbone, is unsaturated. Linolenic acid, with three double bonds, is a "polyunsaturated" fatty acid. (**b**) Condensation of fatty acids and glycerol into a triglyceride.

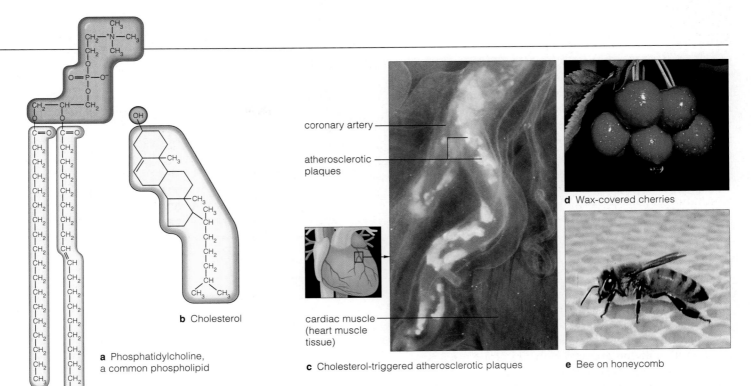

coronary artery

atherosclerotic plaques

cardiac muscle (heart muscle tissue)

d Wax-covered cherries

c Cholesterol-triggered atherosclerotic plaques

e Bee on honeycomb

b Cholesterol

a Phosphatidylcholine, a common phospholipid

Figure 3.13 (**a**) Structural formula of a typical phospholipid in animal and plant cell membranes. Its hydrophilic head is shaded *orange*. Are its hydrophobic tails (*yellow*) saturated or unsaturated? (**b**) Structural formula of cholesterol, the major sterol of animal tissues. (**c**) Your own liver synthesizes enough cholesterol for the body. A fat-rich diet may result in excessively high cholesterol levels in blood and the formation of abnormal masses of material in the blood vessels called arteries. Such *atherosclerotic plaques* may clog the arteries that deliver blood to the heart. (**d**) Demonstration of the water-repelling attribute of a cherry cuticle. (**e**) Honeycomb, a structural material constructed of a firm, water-repellent, waxy secretion called beeswax.

Phospholipids

A **phospholipid** has a glycerol backbone, two fatty acid tails, and a hydrophilic "head" with a phosphate group and another polar group (Figure 3.13*a*). Phospholipids are the main materials of cell membranes, which have two layers of lipids. Heads of one layer are dissolved in the cell's fluid interior, and heads of the other layer are dissolved in the surroundings. Sandwiched between the two are all the fatty acid tails, which are hydrophobic.

Sterols and Their Derivatives

Sterols are among the many lipids that have no fatty acid tails. Sterols differ in the number, position, and type of their functional groups, but they all have a rigid backbone of four fused-together carbon rings:

sterol backbone

Sterols occur in eukaryotic cell membranes. Cholesterol is the main type in animal tissues (Figure 3.13*b* and *c*). Cells also remodel cholesterol into compounds such as vitamin D (required for good bones and teeth), steroids (such as estrogen and testosterone, sex hormones that govern sexual traits and gamete formation), and bile salts (which assist fat digestion in the small intestine).

Waxes

The lipids called **waxes** have long-chain fatty acids tightly packed and linked to long-chain alcohols or to carbon rings. They have a firm consistency and repel water. Waxes and another lipid, cutin, make up most of the cuticles that cover aboveground plant parts. These coverings help plants conserve water and fend off some parasites. A waxy cherry cuticle is an example (Figure 3.13*d*). In many animals, waxy secretions from cells are incorporated into coatings that protect, lubricate, and impart pliability to skin or hair. Among waterfowl and other birds, wax secretions help keep feathers dry. As a final example, beeswax is the material of choice when bees construct their honeycombs (Figure 3.13*e*).

Being largely hydrocarbons, lipids can dissolve in polar substances but tend not to dissolve in water.

Neutral fats (triglycerides), which have a glycerol head and three fatty acid tails, are the body's main energy reservoirs. Phospholipids are the main components of cell membranes.

Sterols such as cholesterol are membrane components as well as precursors of steroid hormones and other compounds. Waxes are firm yet pliable components of water-repelling and lubricating substances.

AMINO ACIDS AND THE PRIMARY STRUCTURE OF PROTEINS

Of all the large biological molecules, **proteins** are the most diverse. The ones called enzymes make metabolic events proceed much faster than they otherwise would. Structural types are the stuff of feathers and webs, bone and cartilage, and a dizzying array of other body parts and products. Transport proteins move molecules and ions across cell membranes and through body fluids. Nutritious proteins abound in milk, eggs, and a variety of seeds. Protein hormones and other regulatory types are signals for change in cell activities. Many proteins are weapons against disease-causing bacteria and other invaders. Amazingly, cells build diverse proteins from their pools of only twenty kinds of amino acids.

Structure of Amino Acids

Every **amino acid** is a small organic compound that consists of an amino group, a carboxyl group (an acid), a hydrogen atom, and one or more atoms known as its R group. As you can see from the structural formula in Figure 3.14, these parts are covalently bonded to the same carbon atom. Figure 3.15 shows some specific amino acids that we will consider later in the book.

Primary Structure of Proteins

When a cell synthesizes a protein, amino acids become linked, one after the other, by peptide bonds. As Figure 3.16 shows, this is the type of covalent bond that forms between one amino acid's amino group (NH_3^+) and the carboxyl group ($—COO^-$) of the next amino acid.

When peptide bonds join two amino acids together, we have a **dipeptide**. When they join three or more, we have a **polypeptide chain**. In such chains, the carbon backbone has nitrogen atoms positioned in this regular pattern: —N—C—C—N—C—C—.

For each particular kind of protein, different amino acid units are selected one at a time from the twenty kinds available. Their orderly progression is prescribed by the cell's DNA. Overall, the resulting sequence of amino acids is unique for each kind of protein, and it represents the protein's *primary* structure (Figure 3.17).

Now consider this: different cells make thousands of different proteins. Many of the proteins are fibrous, with polypeptide chains organized as strands or sheets. Collectively, many such molecules contribute to the shape and internal organization of cells. Other kinds of proteins are globular, with one or more polypeptide

amino group
carboxyl group
R group

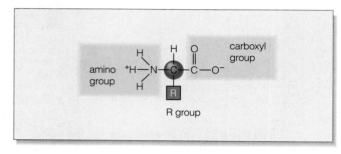

Figure 3.14 Generalized structural formula for the amino acids.

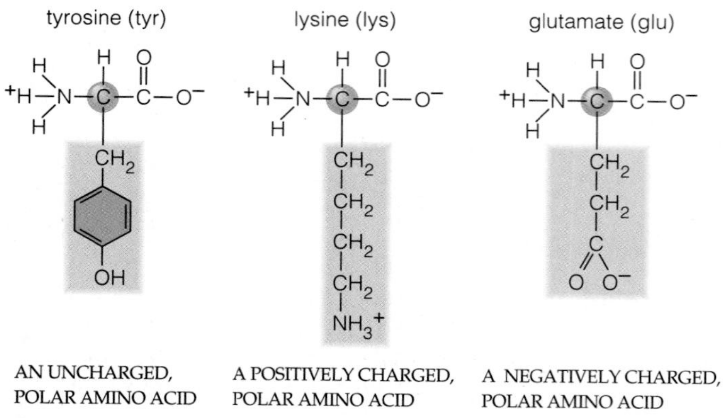

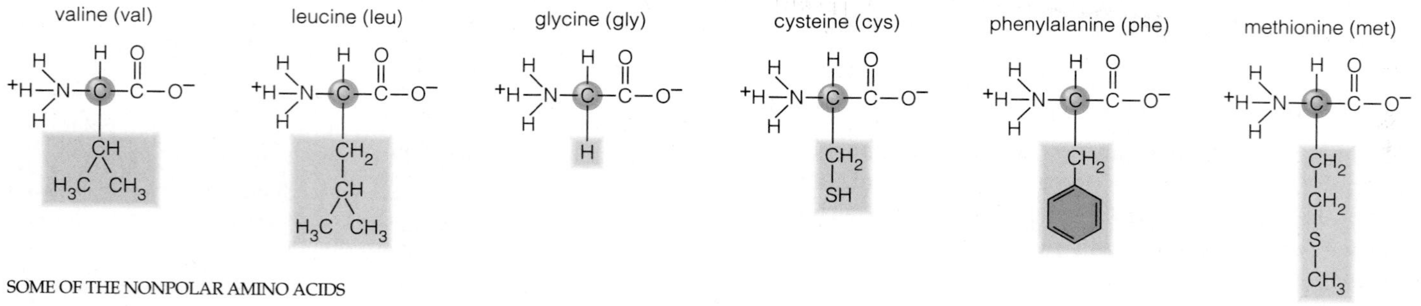

tyrosine (tyr)
lysine (lys)
glutamate (glu)
AN UNCHARGED, POLAR AMINO ACID
A POSITIVELY CHARGED, POLAR AMINO ACID
A NEGATIVELY CHARGED, POLAR AMINO ACID

valine (val)
leucine (leu)
glycine (gly)
cysteine (cys)
phenylalanine (phe)
methionine (met)
SOME OF THE NONPOLAR AMINO ACIDS

Figure 3.15 Structural formulas for nine of the twenty common amino acids. *Green* boxes highlight R groups (side chains that contain functional groups). The side chain gives each amino acid its distinctive properties.

Figure 3.16 Peptide bond formation during protein synthesis.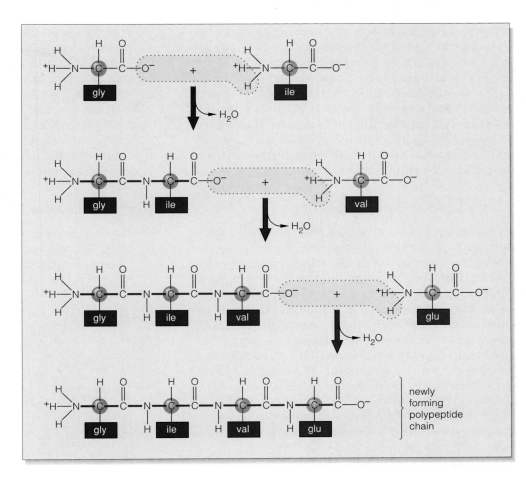

a The first two amino acids in this example are glycine and isoleucine. They are at the start of the sequence for one of two polypeptide chains that make up the protein insulin in cattle. Through a condensation reaction, the isoleucine becomes joined to glycine by a peptide bond. Water forms as a by-product of the reaction.

b A peptide bond forms between the isoleucine and valine, which is another amino acid, and water again forms.

c Remember, DNA specifies the order in which the different kinds of amino acids follow one another in a growing polypeptide chain. In this case, glutamate is the fourth amino acid specified. Chapter 14 provides more details on the steps that lead from DNA instructions to proteins.

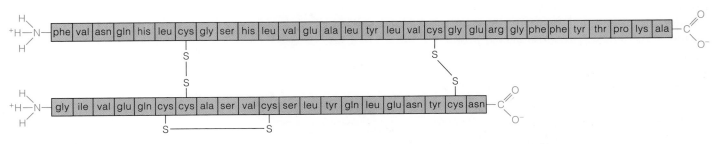

Figure 3.17 Simplified diagram of the amino acid sequence for the insulin molecule in cattle, as deduced by Frederick Sanger. This protein functions as a hormone, and it consists of two polypeptide chains. Disulfide bridges (—S—S—), each formed by a condensation reaction at two sulfhydryl groups, link the chains.

chains folded in compact, rather rounded shapes. Most enzymes are globular proteins. So are actin and most of the other proteins that contribute to cell movement.

Regardless of the type of protein, its shape and its function arise from the primary structure—that is, from information built into its amino acid sequence. As you will see next, that information dictates which parts of a polypeptide chain will coil, bend, or interact with other chains nearby. And the type and arrangement of atoms in the coiled, stretched-out, or folded regions determine whether that protein will function as, say, an enzyme, a transporter, a receptor, or sometimes as an inadvertent target for a bacterium or virus.

A protein consists of one or more polypeptide chains, in which amino acids are strung together. The amino acid sequence (which kind of amino acid follows another in the chain) is unique for each kind of protein and gives rise to its unique structure, chemical behavior, and function.

EMERGENCE OF THE THREE-DIMENSIONAL STRUCTURE OF PROTEINS

The preceding section gave you a sense of how amino acids are strung together in a polypeptide chain, which is a protein's primary structure. Now consider just a few examples of the protein shapes that emerge.

For the most part, the primary structure gives rise to a protein's shape in two ways. First, it allows hydrogen bonds to form between different amino acids along the length of a polypeptide chain. Second, it puts R groups into positions that force them to interact. Through these interactions, the chain is forced to bend and twist.

Second Level of Protein Structure

Hydrogen bonds form at regular, short intervals along a new polypeptide chain, and they give rise to a coiled or extended pattern known as the *secondary* structure of a protein. Think of a polypeptide chain as a set of rigid dominoes joined by links that can swivel a bit. Each "domino" is a peptide group (Figure 3.18a). Atoms on either side of it can rotate slightly around their covalent bonds and form bonds with neighboring atoms. For instance, in many chains, hydrogen bonds readily form between every third amino acid. The bonding pattern forces the peptide groups to coil helically, like a spiral staircase (Figure 3.18b). In other proteins, a hydrogen-bonding pattern holds two or more chains side by side, in an extended, sheetlike array (Figure 3.18c).

Third Level of Protein Structure

Most polypeptide chains that have a coiled secondary structure undergo more folding, owing to the number and the location of certain amino acids (such as proline) along their length. These amino acids bend a chain in certain directions and at certain angles. The chain loops out, and R groups far apart along its length interact to hold the loops in characteristic positions. A polypeptide chain that is folded as a result of its bend-producing amino acids and R-group interactions has a *tertiary* structure, which is the third level of protein structure.

Figure 3.19a shows the compact, tertiary structure of a globin molecule, an outcome of hydrogen bonds and disulfide bridges at functional groups along its length. Each disulfide bridge links the sulfur atom of a cysteine subunit in the polypeptide chain with that of another cysteine subunit in the chain (refer also to Figure 3.17).

Fourth Level of Protein Structure

Now imagine that bonds also form between *four* globin molecules and that an iron-containing functional group, a heme group, is positioned near the center of each one. You have hemoglobin, an oxygen-transporting protein

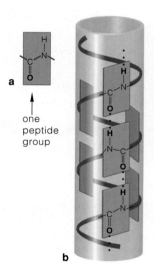
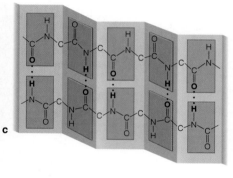

a
one peptide group

b

c

Figure 3.18 The major bonding patterns between the peptide groups (**a**) of polypeptide chains. Extensive hydrogen bonding (*dotted lines*) can bring about (**b**) coiling or (**c**) sheetlike formations.

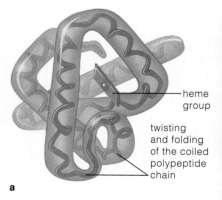

a

heme group

twisting and folding of the coiled polypeptide chain

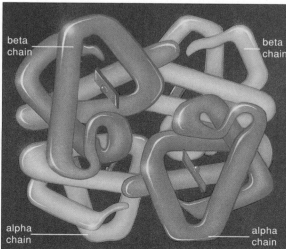

beta chain

beta chain

alpha chain

alpha chain

b

Figure 3.19 (**a**) A globin molecule. It is a single coiled polypeptide chain and a heme group associated with it. This iron-containing group is strongly attractive to oxygen. There is a whole family of globin molecules. They are identical in critical regions of their amino acid sequences, but they differ in other regions. In this diagram, the artist drew a transparent "green noodle" around the chain to help you visualize how it folds in three dimensions.

(**b**) Hemoglobin, a protein that transports oxygen in blood, is composed of four globin molecules. To keep this diagram from looking like a tangle of noodles, the two polypeptide chains in the foreground are tinted differently from the two in the background.

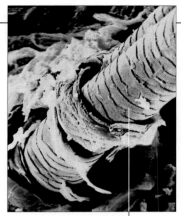

dead, flattened cells around a developing hair shaft

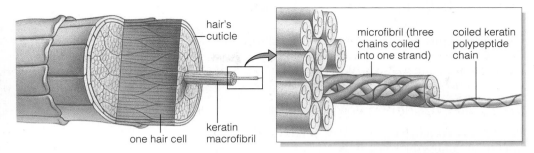

Figure 3.20 Structure of hair. Hair cells, which are derived from skin cells, synthesize polypeptide chains of keratin. Three chains become linked as fine fibers (microfibrils), which become bundled into larger, cablelike fibers (macrofibrils). These essentially fill the cells, which eventually die off. Dead, flattened cells form a tubelike cuticle around the developing hair shaft, shown in the photograph.

(Figure 3.19*b*). As you read this, each mature red blood cell in your body is transporting a billion or so oxygen molecules, bound to 250 million hemoglobin molecules.

Hemoglobin is a fine example of the fourth level of protein structure, or *quaternary* structure. In all proteins at this level of organization, two or more polypeptide chains have become joined together by numerous weak interactions (such as hydrogen bonds) and sometimes by covalent bonds between sulfur atoms of R groups.

Like most enzymes, hemoglobin is globular. Many other proteins having quaternary structure are fibrous. Keratin, a structural protein of hair and fur, is like this (Figure 3.20). So is collagen, the most common animal protein. Skin, bones, corneas, blood vessels, and other body parts depend on the strength inherent in collagen.

Glycoproteins and Lipoproteins

Some proteins have other organic compounds attached to their polypeptide chains. For example, **lipoproteins** form when certain proteins circulating in blood combine with cholesterol, triglycerides, and phospholipids that were absorbed from the gut after a meal. Lipoproteins differ in their particular protein and lipid components. Similarly, most **glycoproteins** have linear or branched oligosaccharides covalently bonded to them. Most form when new polypeptide chains are being modified into mature proteins. Nearly all the proteins at the surface of animal cells are glycoproteins. So are most protein secretions from cells and many proteins in blood.

Structural Changes by Denaturation

Breaking the weak bonds of a protein or any other large molecule disrupts its three-dimensional shape, an event called **denaturation**. For example, hydrogen bonds are weak, and they are sensitive to increases or decreases in temperature and pH. If the temperature or pH exceeds a protein's range of tolerance, its polypeptide chains will unwind or change shape, and the protein will no longer function.

Consider the protein albumin, concentrated in the "egg white" of uncooked chicken eggs. When you cook eggs, the heat doesn't disrupt the strong covalent bonds of albumin's primary structure. But it destroys weaker bonds contributing to the three-dimensional shape. For some proteins, denaturation might be reversed when normal conditions are restored—but albumin isn't one of them. There is no way to uncook a cooked egg.

We now leave this overview of protein structure, the levels of which are summarized in Table 3.1.

Proteins have a primary structure, which is the sequence of different kinds of amino acids along a polypeptide chain. An individual's DNA specifies that sequence.

Proteins have a secondary structure, a coiled or extended, sheetlike pattern that arises through hydrogen bonding at short, regular intervals along a polypeptide chain.

Many proteins have a folded, tertiary structure. It arises when certain bend-producing amino acids make a coiled chain loop at certain angles and in certain directions, and when interactions among certain R groups that are far apart in the chain hold the loops in characteristic positions.

Many proteins have a quaternary structure. They incorporate two or more polypeptide chains, joined together by numerous hydrogen bonds and other interactions.

Table 3.1 Levels of Protein Structure	
Primary structure	The sequence of amino acids in a polypeptide chain; unique for each type of protein
Secondary structure	Coiled or extended shape owing to hydrogen bonds at short intervals along the chain
Tertiary structure	Further folding of a coiled chain owing to bend-producing amino acids and interactions among R groups far apart on a looped-out chain
Quaternary structure	Two or more polypeptide chains linked tightly by many hydrogen bonds and other interactions

We conclude this chapter with a brief introduction to the nucleotides and the larger nucleic acids, which are assembled from nucleotide monomers.

Nucleotides With Roles in Metabolism

Nucleotides are small organic compounds with a sugar, at least one phosphate group, and a base. The sugar is either ribose or deoxyribose. Both sugars have a five-carbon ring structure. The only difference is that ribose has an oxygen atom attached to carbon 2 in the ring and deoxyribose does not. The bases have a single or double carbon ring structure that incorporates nitrogen.

One nucleotide, **ATP** (adenosine triphosphate), has a string of three phosphate groups attached to its sugar component in the manner shown in Figure 3.21. An ATP molecule can readily transfer a phosphate group to many other molecules inside the cell, whereupon the acceptor molecules become energized enough to enter into a reaction. Although ATP acquires its phosphate groups at certain reaction sites, it can deliver them to nearly all other reaction sites in a cell, so in this respect ATP is absolutely central to metabolism.

Other nucleotides are subunits of **coenzymes**. These enzyme helpers accept hydrogen atoms and electrons stripped from molecules at a reaction site and transfer them elsewhere. Two important examples are NAD^+ (nicotinamide adenine dinucleotide) and FAD (flavin adenine dinucleotide). Still other nucleotides function as chemical messengers within and between cells. You will be reading about one of these (cyclic adenosine monophosphate, or cAMP) later in the book.

Nucleic Acids—DNA and RNA

Different kinds of nucleotides, including the ones shown in Figure 3.22, are building blocks for the large, single- or double-stranded molecules known as **nucleic acids**. In the backbone of such strands, the sugar component of each nucleotide is covalently bonded to the phosphate

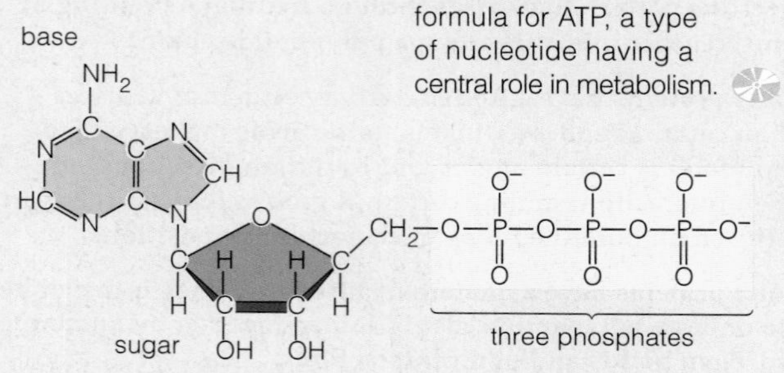

Figure 3.21   Structural formula for ATP, a type of nucleotide having a central role in metabolism.

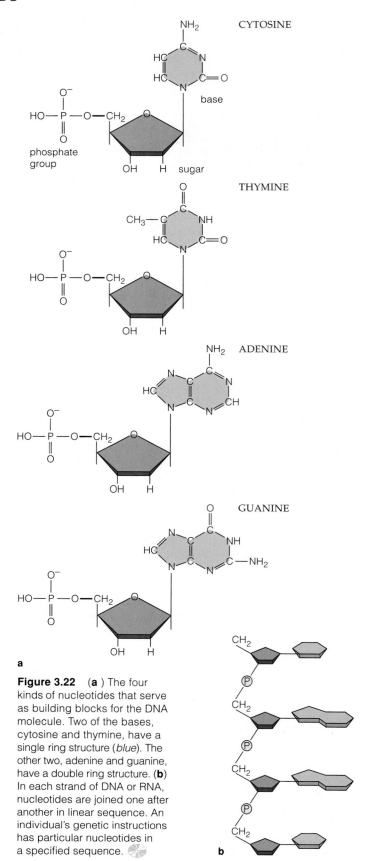

a

Figure 3.22   (**a**) The four kinds of nucleotides that serve as building blocks for the DNA molecule. Two of the bases, cytosine and thymine, have a single ring structure (*blue*). The other two, adenine and guanine, have a double ring structure. (**b**) In each strand of DNA or RNA, nucleotides are joined one after another in linear sequence. An individual's genetic instructions has particular nucleotides in a specified sequence.

b

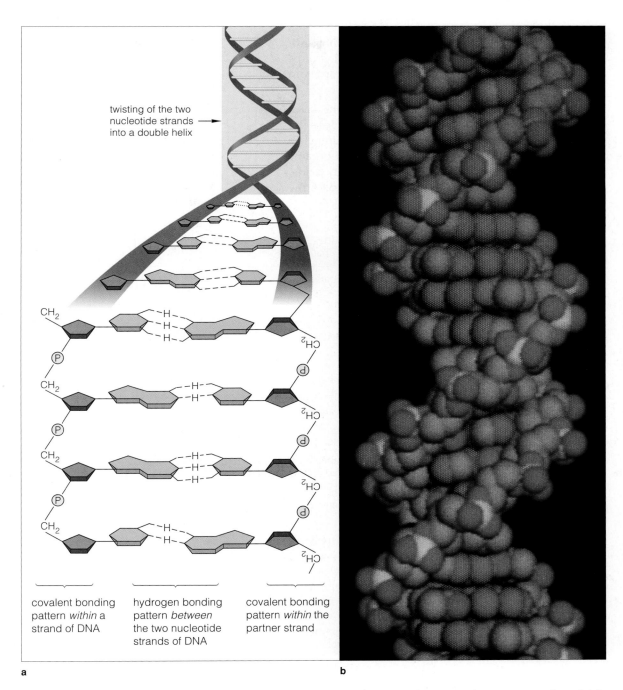

Figure 3.23 (**a**) Bonding patterns within the double-stranded DNA molecule. Notice how hydrogen bonds connect the bases (*blue*) of one strand with bases of the other strand. The base pairs are like the rungs of a ladder, and the sugar-phosphate backbones are like the ladder's posts. The "ladder" twists in the shape of a double helix.

(**b**) A computer-generated, three-dimensional model of DNA structure. Each ball signifies a single atom. The two nucleotide strands are shown twisted together. Base pairs of the sugar-phosphate backbones are shaded *blue*. The sugar components of the backbone are shaded *light purple*, and the phosphate groups *red* and *yellow*.

twisting of the two nucleotide strands into a double helix

covalent bonding pattern *within* a strand of DNA

hydrogen bonding pattern *between* the two nucleotide strands of DNA

covalent bonding pattern *within* the partner strand

a

b

group of adjacent nucleotides. Most likely, you already know something about **DNA** (deoxyribonucleic acid). To refresh your memory, DNA is a long, double strand of nucleotides twisted helically, as shown in Figure 3.23. Hydrogen bonding between the nucleotide bases holds the two strands of a DNA molecule together. Heritable (genetic) information is encoded in the base sequences. And in at least some regions along DNA's length, the sequence of particular bases is unique to each species.

Unlike DNA, the **RNAs** (ribonucleic acids) usually are single nucleotide strands. Ever since living cells first appeared, RNAs have functioned in processes by which genetic information is used to build proteins.

Nucleotides are small, nitrogen- and phosphorus-containing organic compounds. They are building blocks for DNA and RNA. Some, including ATP, have central roles in metabolism.

DNA, a double-stranded nucleic acid, has genetic information encoded in its sequence of nucleotide bases. The RNAs are single-stranded nucleic acids with roles in the processes by which DNA's genetic information is used to build proteins.

3.9 SUMMARY

1. Organic compounds consist of one or more elements covalently bonded to carbon atoms that commonly form linear and ring-shaped backbones. Functional groups attached to the backbone impart diverse properties to organic compounds.

2. In nature, only living cells can assemble the large organic compounds called biological molecules—the complex carbohydrates and lipids, proteins, and nucleic acids. To do so, they draw from small pools of organic compounds, which include simple sugars, fatty acids, amino acids, and nucleotides. Table 3.2 summarizes the major categories of these compounds.

3. Cells use the simple sugars (such as glucose) and oligosaccharides (such as sucrose) as building blocks and transportable forms of energy.

4. Cells use complex carbohydrates (including cellulose and starch) as energy storage forms and structural materials. They use lipids (including triglycerides) as energy storage forms, as structural components of cell membranes, and as precursors of other compounds.

5. Proteins, built of amino acids, are the most diverse biological molecules. They function in enzyme activity, structural support, cell transport, cell movements, changes in cell activities, and defenses against disease. The nucleic acids DNA and RNA, built of nucleotides, are the basis of inheritance and reproduction.

Table 3.2   Summary of the Main Organic Compounds in Living Things

Category	Main Subcategories	Some Examples and Their Functions	
CARBOHYDRATES . . . contain an aldehyde or a ketone group, and one or more hydroxyl groups	**Monosaccharides** (simple sugars)	Glucose	Energy source
	Oligosaccharides	Sucrose (a disaccharide)	Form of sugar transported in plants
	Polysaccharides (complex carbohydrates)	Starch Cellulose	Energy storage Structural roles
LIPIDS . . . are largely hydrocarbon; generally do not dissolve in water but dissolve in nonpolar substances	**Lipids with fatty acids:** *Glycerides:* one, two, or three fatty acid tails attached to glycerol backbone	Fats (e.g., butter) Oils (e.g., corn oil)	Energy storage
	Phospholipids: phosphate group, one other polar group, and (often) two fatty acids attached to glycerol backbone	Phosphatidylcholine	Key component of cell membranes
	Waxes: long-chain fatty acid tails attached to alcohol	Waxes in cutin	Water retention by plants
	Lipids with no fatty acids: *Sterols:* four carbon rings; the number, position, and type of functional groups differ among various types	Cholesterol	Component of animal cell membranes; precursor of many steroids and of vitamin D
PROTEINS . . . are polypeptides (up to several thousand amino acids, covalently linked)	**Fibrous proteins:** Individual polypeptide chains, often linked into tough, water-insoluble molecules	Keratin Collagen	Structural element of hair, nails Structural element of bone, cartilage
	Globular proteins: One or more polypeptide chains folded and linked into globular shapes; many roles in cell activities	Enzymes Hemoglobin Insulin Antibodies	Increase in rates of reactions Oxygen transport Control of glucose metabolism Tissue defense
NUCLEIC ACIDS (AND NUCLEOTIDES) . . . are chains of units (or individual units) that each consist of a five-carbon sugar, phosphate, and a nitrogen-containing base	**Adenosine phosphates**	ATP	Energy carrier
	Nucleotide coenzymes	NAD^+, $NADP^+$, FAD	Transport of protons (H^+), electrons from one reaction site to another
	Nucleic acids: Chains of thousands to millions of nucleotides	DNA, RNAs	Storage, transmission, translation of genetic information

Review Questions

1. Define organic compound. Name the type of chemical bond that predominates in the backbone of such a compound. *3.1*

2. Name the "molecules of life." Do they break apart most easily at their hydrocarbon portion or at functional groups? *3.1, 3.2*

3. Select one of the carbohydrates, lipids, proteins, or nucleic acids described in this chapter. Speculate on how its functional groups as the bonds between carbon atoms in its backbone contribute to its final shape and function. *3.4, 3.5, 3.6, 3.7, or 3.8*

4. Which category includes all of the other items listed? *3.5*
   a. triglyceride   c. wax   e. lipid
   b. fatty acid   d. sterol   f. phospholipid

5. Explain how a hemoglobin molecule's three-dimensional shape arises, starting with its primary structure. *3.6, 3.7*

Self-Quiz   (*Answers in Appendix IV*)

1. Each carbon atom can share pairs of electrons with as many as _____ other atoms.
   a. one   b. two   c. three   d. four

2. Hydrolysis is a _____ reaction.
   a. functional group transfer   d. condensation
   b. electron transfer   e. cleavage
   c. rearrangement   f. both b and d

3. _____ are simple sugars (monosaccharides).
   a. glucose   c. ribose   e. both a and b
   b. sucrose   d. chitin   f. both a and c

4. A saturated fat has fatty acid tails with one or more _____ .
   a. single covalent bonds   b. double covalent bonds

5. _____ are to proteins as _____ are to nucleic acids.
   a. sugars; lipids   c. amino acids; hydrogen bonds
   b. sugars; proteins   d. amino acids; nucleotides

6. Nucleotides include _____ .
   a. ATP   c. DNA and RNA subunits
   b. coenzyme subunits   d. all of the above

7. A denatured protein or DNA molecule has lost its _____ .
   a. hydrogen bonds   c. function
   b. shape   d. all of the above

8. Match each molecule with the most suitable description.
   ____ long sequence of amino acids   a. carbohydrate
   ____ the main energy carrier   b. phospholipid
   ____ glycerol, fatty acids, phosphate   c. protein
   ____ two strands of nucleotides   d. DNA
   ____ one or more sugar monomers   e. ATP

Critical Thinking

1. Jack decided to celebrate summer by making a crab salad and a peach pie for some friends. He had such a good time, he forgot to put the cap on the bottle of olive oil after making the salad. He also didn't tighten the lid on the can of shortening after making the pie crust. A few weeks later, he discovered the oil had turned rancid, but the shortening still smelled okay. Both substances are fats. Both were stored in a dark cupboard, at room temperature. Why did one spoil and the other stay fresh?

2. In the following list, identify which is the carbohydrate, fatty acid, amino acid, and polypeptide:
   a. $^+NH_3—CHR—COO^-$ c. (glycine)$_{20}$
   b. $C_6H_{12}O_6$ d. $CH_3(CH_2)_{16}COOH$

3. A clerk in a health-food store tells you that certain "natural" vitamin C tablets extracted from rose hips are better for you than synthetic vitamin C tablets. Given your understanding of the structure of organic compounds, what would be your response? Design an experiment to test whether these vitamins differ.

4. Rabbits that eat green, leafy vegetables containing xanthophyll accumulate this yellow pigment molecule in body fat but not in muscles. What chemical properties of the molecule might cause this selective accumulation?

5. Grasses can be digested in cows, but not in people. A cow's four-chambered stomach houses populations of certain microorganisms that are not normal inhabitants of the human stomach. What kind of metabolic reactions do you think the microorganisms carry out? What do you think might happen to a cow undergoing treatment with an antibiotic that killed the microorganisms?

6. A Gary Larsen cartoon states that sheep with steel wool have no natural enemies. You might laugh at the thought of such a preposterous animal—but what exactly *is* wool? It is the keratin-rich, soft undercoat of sheep, Angora goats, llamas, and other hairy mammals. Keratin, a protein, is a long polypeptide chain. Along its length, many hydrogen bonds hold it in a helical shape. Visualize three such chains coiled together. Many disulfide bridges in the end regions stabilize the trio in a tight, ropelike array. Each hair has linear aggregates of the ropelike fibers, which are highly resistant to stretching. Given keratin's structure, speculate why, when you wash a wool sweater or run it through the hot cycle of a dryer, it will shrink, pathetically and permanently.

Selected Key Terms

alcohol *3.1*	functional group *3.1*	phospholipid *3.5*
amino acid *3.6*	functional-group	polymer *3.2*
ATP *3.8*	transfer *3.2*	polypeptide chain
carbohydrate *3.4*	glycoprotein *3.7*	*3.6*
cleavage *3.2*	hydrolysis *3.2*	polysaccharide *3.4*
coenzyme *3.8*	lipid *3.5*	protein *3.6*
condensation *3.2*	lipoprotein *3.7*	rearrangement
denaturation *3.7*	monomer *3.2*	(of bonds) *3.2*
dipeptide *3.6*	monosaccharide *3.4*	RNAs *3.8*
DNA *3.8*	nucleic acid *3.8*	sterol *3.5*
electron transfer *3.2*	nucleotide *3.8*	toxin *3.3*
enzyme *3.2*	oligosaccharide *3.4*	triglyceride *3.5*
fatty acid *3.5*	organic compound *3.1*	wax *3.5*

Readings

Atkins, P. 1987. *Molecules.* New York: Scientific American Library. A molecular "glossary" includes the formula, three-dimensional stucture, and action of many organic compounds.

"The Molecules of Life." October 1985. *Scientific American* (253). The entire issue focuses on the structure and functioning of biological molecules.

Ritter, P. 1996. *Biochemistry: An Introduction.* Pacific Grove, California: Brooks/Cole. Chockful of human applications.

Wolfe, S. 1995. *Introduction to Molecular and Cellular Biology.* Belmont, California: Wadsworth.

Web Site See *http://www.wadsworth.com/biology* for practice quiz questions, hypercontents, BioUpdates, and critical thinking. The Wadsworth Biology Resource Center provides a wealth of information fully organized and integrated by chapter.

4

CELL STRUCTURE AND FUNCTION

Animalcules and Cells Fill'd With Juices

Early in the seventeenth century, a scholar by the name of Galileo Galilei arranged two glass lenses within a cylinder. With this instrument he happened to look at an insect, and later he described the stunning geometric patterns of its tiny eyes. Thus Galileo, who was not a biologist, was the first to record a biological observation made through a microscope. The study of the cellular basis of life was about to begin. First in Italy, then in France and England, scholars set out to explore a world whose existence had not even been suspected.

At midcentury Robert Hooke, Curator of Instruments for the Royal Society of England, was at the forefront of these studies. When Hooke first turned a microscope to thinly sliced cork from a mature tree, he observed tiny compartments (Figure 4.1*c*). He gave them the Latin name *cellulae*, meaning small rooms—hence the origin of the biological term "cell." They actually were the walls of dead cells, which is what cork is made of, but Hooke did not think of them as being dead because he did not know cells could be alive. In other plant tissues, he discovered cells "fill'd with juices" but could not imagine what they represented.

Given the simplicity of their instruments, it is amazing that the pioneers in microscopy observed as much as they did. Antony van Leeuwenhoek, a Dutch shopkeeper, had exceptional skill in constructing lenses and possibly the keenest vision (Figure 4.1*a*). By the late 1600s, he was observing wonders everywhere, including "many very small animalcules, the motions of which were very pleasing to behold," in scrapings of tartar from his own teeth. He discovered diverse protistans, sperm, even a bacterium—an organism so small that it would not be seen again for another two centuries!

In the 1820s, improvements in lenses brought cells into sharper focus. Robert Brown, a botanist, was noticing an opaque spot in

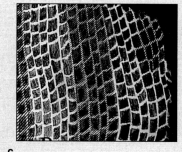

Figure 4.1 Early glimpses into the world of cells. (**a**) Antony van Leeuwenhoek, with microscope in hand. (**b**) Robert Hooke's compound microscope and (**c**) his drawing of cell walls from cork tissue. (**d**) One of van Leeuwenhoek's early sketches of sperm cells. (**e**) Some cartoon evidence of the startling impact of microscopic observations on nineteenth-century London.

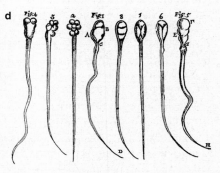

a variety of cells. He called it a nucleus. In 1838 still another botanist, Matthias Schleiden, wondered if the nucleus had something to do with a cell's development. As he hypothesized, each plant cell must develop as an independent unit even though it is an integral part of the plant.

By 1839, after years of studying animal tissues, the zoologist Theodor Schwann had this to say: Animals as well as plants consist of cells and cell products—and even though the cells are part of a whole organism, to some extent they have an individual life of their own.

A question remained: Where do cells come from? A decade later, Rudolf Virchow, a physiologist, completed his own studies of a cell's growth and reproduction— that is, its division into two daughter cells. Every cell, he decided, comes from a cell that already exists.

And so, by the middle of the nineteenth century, microscopic analysis had yielded three generalizations, which together constitute the **cell theory**. *First, every organism is composed of one or more cells. Second, the cell is the smallest unit having the properties of life. Third, the continuity of life arises directly from the growth and division of single cells.* All three insights still hold true.

This chapter provides an overview of our current understanding of the structure and function of cells. It also introduces some of the modern microscopes that transport us more deeply into the spectacular worlds of juice-fill'd cells and animalcules.

KEY CONCEPTS

1. All organisms are composed of one or more cells. The cell is the smallest unit that still retains the characteristics of life. And each new cell arises from a preexisting cell. These are the three generalizations of the cell theory.

2. All cells have an outermost, double-layered component, the plasma membrane, that helps their interior remain distinct from the surroundings. Cells contain cytoplasm, an internal region that is highly organized for energy conversions, protein synthesis, movements, and other activities necessary for survival. They also contain a nucleus or a comparable internal region in which their DNA is located.

3. The plasma membrane and internal cell membranes consist largely of phospholipid and protein molecules. The phospholipids form two adjacent layers that give the membrane its basic structure and prevent water-soluble substances from freely crossing it. Proteins embedded in those layers or positioned at its surfaces carry out most membrane functions.

4. The nucleus is one of many organelles. Organelles are membrane-bound compartments inside eukaryotic cells. They physically separate different metabolic reactions and allow them to proceed in orderly fashion. Prokaryotic cells (bacteria) do not have comparable organelles.

5. When cells grow, they increase faster in volume than in surface area. This physical constraint on increases in size influences cell size and shape.

6. Different microscopes modify light rays or accelerated beams of electrons in ways that allow us to form images of incredibly small specimens. They are the foundation for our current understanding of cell structure and function.

BASIC ASPECTS OF CELL STRUCTURE AND FUNCTION

Inside your body and at its moist surfaces, trillions of cells live in interdependency. In scummy pondwater, a single-celled amoeba moves about freely, thriving on its own. For humans, amoebas, and all other organisms, the **cell** is the smallest entity that retains the properties of life. A cell either can survive on its own or has the potential to do so. Its structure is highly organized, and it engages in metabolism. A cell senses and responds to changes in the environment. And it has the potential to reproduce, based on inherited instructions in its DNA.

Structural Organization of Cells

Cells differ enormously in size, shape, and activities, as you might gather by comparing a tiny bacterium with one of your relatively giant liver cells. Yet they are alike in three respects. All cells start out life with a plasma membrane, a region of DNA, and a region of cytoplasm:

1. **Plasma membrane**. This thin, outermost membrane maintains the cell as a distinct entity. By doing so, it allows metabolic events to proceed apart from random events in the environment. A plasma membrane does not *isolate* the cell interior. Substances and signals continually move across it, in highly controlled ways.

2. **DNA-containing region**. DNA occupies part of the cell interior, along with molecules that can copy or read its hereditary instructions.

3. **Cytoplasm**. The cytoplasm is everything in the cell interior *except* for the region of DNA. It includes a fluid portion, **ribosomes** (two-unit structures upon which proteins are synthesized), and other components.

This chapter introduces two fundamentally different kinds of cells. The cytoplasm of **eukaryotic cells** includes organelles, which are tiny sacs and other compartments bounded by membranes. One, the nucleus, houses the DNA. It is the defining feature of eukaryotic cells:

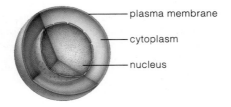

plasma membrane

cytoplasm

nucleus

By contrast, **prokaryotic cells** have no nucleus; they have no membranes intervening between their region of DNA and the cytoplasm that surrounds it. Bacteria are the only prokaryotic cells. Beyond the bacterial realm, all other organisms—from amoebas to peach trees and puffball mushrooms to zebras—are eukaryotes.

The Lipid Bilayer of Cell Membranes

All cell membranes have the same structural framework of two sheets of lipid molecules. Figure 4.2 shows this **lipid bilayer** arrangement for the plasma membrane, the continuous boundary that bars the free passage of water-soluble substances into and out of the cell. Within eukaryotic cells, membranes subdivide the cytoplasm into specific zones in which substances are synthesized, processed, stockpiled, or degraded.

Diverse proteins embedded in the lipid bilayer or positioned at or near one of its surfaces carry out most membrane functions. For example, some of the proteins serve as passive channels for water-soluble substances.

Figure 4.2 Organization of the lipid bilayer of cell membranes. (**a**) Symbol for phospholipid molecules, the most abundant membrane components. These and other lipids are arranged as two layers. (**b**) The hydrophobic tails of the molecules are sandwiched between their hydrophilic heads, which are dissolved in cytoplasm on one side of the bilayer and in extracellular fluid on the other (**c**).

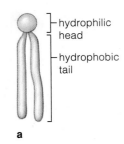

hydrophilic head

hydrophobic tail

a

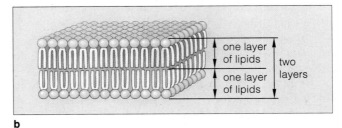

one layer of lipids

one layer of lipids

two layers

b

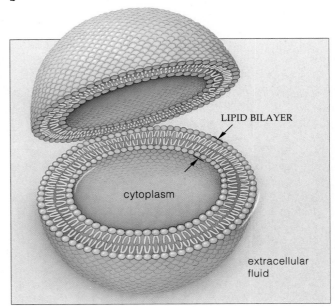

LIPID BILAYER

cytoplasm

extracellular fluid

c

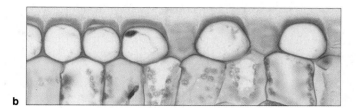

Diameter (cm):	0.5	1.0	1.5
Surface area (cm^2):	0.79	3.14	7.07
Volume (cm^3):	0.06	0.52	1.77
Surface-to-volume ratio:	13.17:1	6.04:1	3.99:1

a

Figure 4.3 (**a**) An example of the surface-to-volume ratio. This physical relationship between increases in volume and in surface area imposes restrictions on the size and shape of cells, including the ones near the surface of leaves (**b**) and in skeletal muscles (**c**).

Others transport electrons or pump substances across the lipid bilayer. Still others are receptors, which latch onto hormones and other types of signaling molecules that trigger alterations in cell activities. You will read more about membrane proteins in the chapter to follow.

Cell Size and Cell Shape

You may be wondering how small cells really are. Can any be observed with the unaided human eye? There are a few, including the "yolks" of bird eggs, cells in the red part of watermelons, and the fish eggs we call caviar. However, most cells cannot be observed without microscopes. To give you a sense of cell sizes, red blood cells are about 8 millionths of a meter across. You could fit a string of 2,000 of them across your thumbnail!

Why are most cells so small? A physical relationship called the **surface-to-volume ratio** constrains increases in a cell's size. By this relationship, an object's volume increases with the cube of the diameter, but its surface area increases only with the square. Figure 4.3*a* gives an example of this. Simply put, *if a cell expands in diameter during growth, its volume will increase more rapidly than its surface area will.*

Suppose you figure out a way to make a round cell grow four times wider than it normally would. Its volume increases 64 times (4^3), but its surface area only increases 16 times (4^2). As a result, each unit of the cell's plasma membrane must now serve four times as much cytoplasm as it did previously! Moreover, past a certain point, the inward flow of nutrients and outward flow of wastes will not be fast enough, and the cell will die.

A large, round cell also would have trouble moving materials *through* its cytoplasm. In small cells, such as those shown in Figure 4.3*b*, the random, tiny motions of molecules can easily distribute materials. If a cell is not small, you usually can expect it to be long and thin or to have outfoldings and infoldings that increase its surface relative to its volume. *The smaller or narrower or more frilly-surfaced the cell, the more efficiently materials cross its surface and become distributed through the interior.*

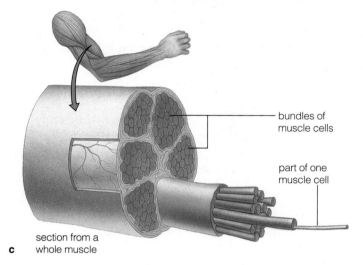

b

bundles of muscle cells

part of one muscle cell

section from a whole muscle

c

Multicelled body plans show evidence of surface-to-volume constraints, also. For example, cells attach end to end in strandlike algae, and each one interacts directly with its surroundings. Your skeletal muscle cells are very thin, but each one is as long as a biceps or some other muscle of which it is part (Figure 4.3*c*). Your circulatory system delivers materials to muscle cells and trillions of others, and it also sweeps away their metabolic wastes. Its many "highways" cut through the volume of tissue and so shrink the distance to and from individual cells.

All cells have an outermost plasma membrane, an internal region of cytoplasm, and an internal region of DNA.

Besides the plasma membrane, eukaryotic cells have internal, membrane-bound compartments, including a nucleus.

Each membrane has a bilayer structure, consisting largely of phospholipids. The hydrophobic parts of its lipid molecules are sandwiched between the hydrophilic parts, which are dissolved in the fluid surroundings.

A lipid bilayer imparts structure to the cell membrane and serves as a barrier to water-soluble substances.

Proteins embedded in the bilayer or positioned at its surfaces carry out most membrane functions.

During their growth, cells increase faster in volume than they do in surface area. This physical constraint on increases in size influences the size and shape of cells. It also influences the body plans of multicelled organisms.

4.2 MICROSCOPES—GATEWAYS TO THE CELL

Modern microscopes, including the three kinds sketched in Figure 4.4, are gateways to astounding worlds. Some even afford glimpses into the structure of molecules. Figure 4.5 gives an idea of the range of magnifications possible. The micrographs in Figures 4.5 and 4.6 hint at the details now being observed. A **micrograph** simply is a photograph of an image that was formed with the aid of a microscope.

LIGHT MICROSCOPES Light microscopes work by bending (refracting) light rays. Any rays directly entering a glass lens, except at the very center, are bent. The farther they are from the center of the lens, the more they will bend:

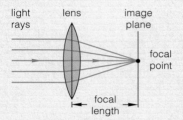

The angle at which rays of light enter a glass lens and the molecular structure of the glass dictate the extent to which the rays will bend.

In a *compound light microscope,* two or more sets of glass lenses bend incoming rays to form an enlarged image of a cell or some other specimen. A living cell must be small or thin enough for light to pass through. It would help if cell parts differed in color and density from the surroundings, but most are nearly colorless and appear uniformly dense. That is why microscopists stain cells (expose them to dyes that react with some parts but not others). Staining may alter and kill cells. Dead cells break down very fast, so they typically are pickled or preserved before being stained.

Suppose you are using the best glass lens system. When you magnify the diameter of a specimen by 2,000 times or more, you discover that cell parts appear larger but are not clearer. What limits the resolution of smaller details? The answer lies with a certain property of visible light.

Picture a train of waves moving across an ocean. Each **wavelength** is the distance from one wave's peak to the peak of the wave behind it. Light also travels as waves. In fact, different colors of light correspond to waves of certain, unvarying lengths. The wavelength is about 750 nanometers for red light and 400 nanometers for violet; all other colors fall in between. *And if a cell structure is less than one-half of a wavelength long, rays of light streaming by it will not be disturbed enough to make the object visible.*

ELECTRON MICROSCOPES Electrons are particles, but they, too, behave like waves. Electron microscopy is based on accelerating the flow of streams of electrons that have wavelengths of about 0.005 nanometer—about 100,000 times shorter than those of visible light. Electrons cannot pass through glass lenses, but they can be bent from their paths and focused by a magnetic field. In a *transmission electron microscope,* such a field is the "lens." Accelerated electrons are directed through a specimen, focused into an image, and magnified. With *scanning electron microscopes,* a narrow beam of electrons moves back and forth across a specimen thinly coated with metal. The metal responds by emitting some of its own electrons. A detector connected to electronic circuitry transforms the electron energy into an image of the specimen's surface on a television screen. Most of the images have fantastic depth.

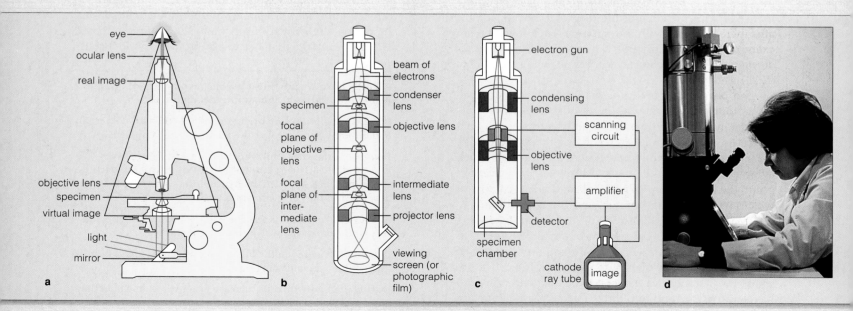

Figure 4.4 Diagrams of (**a**) a light microscope, (**b**) a transmission electron microscope, and (**c**) a scanning electron microscope. (**d**) Exterior view of an electron microscope.

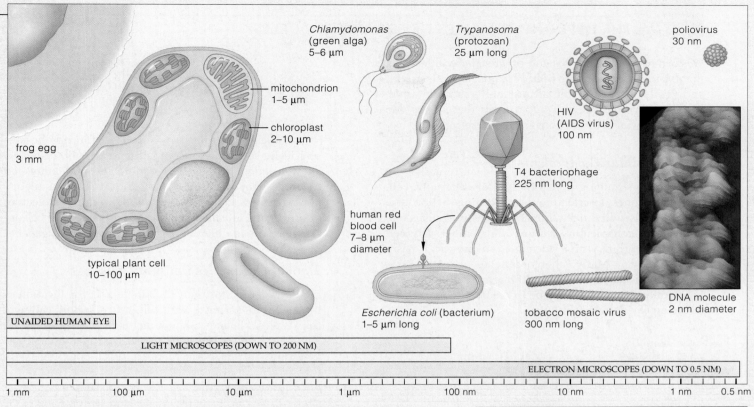

Figure 4.5 Units of measure used in microscopy. A *scanning tunneling microscope*, which can provide magnifications up to 100 million, gave us the photomicrograph of part of a DNA molecule. Its needlelike probe has a single atom at its tip. Voltage that is applied between the tip and an atom at a specimen's surface causes a detectable tunnel to form in the electron orbitals. As the tip moves over a specimen's contours, a computer analyzes the motion and creates a three-dimensional view of the surface atoms.

1 centimeter (cm)	= 1/100 meter, or 0.4 inch
1 millimeter (mm)	= 1/1,000 meter
1 micrometer (μm)	= 1/1,000,000 meter
1 nanometer (nm)	= 1/1,000,000,000 meter

1 meter = 10^2 cm = 10^3 mm = 10^6 μm = 10^9 nm

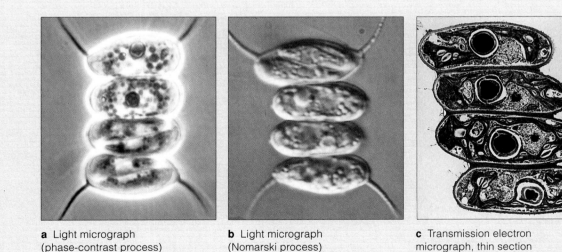

a Light micrograph (phase-contrast process)

b Light micrograph (Nomarski process)

c Transmission electron micrograph, thin section

d Scanning electron micrograph

10 μm

Figure 4.6 How different microscopes can reveal different aspects of the same organism—in this case, a green alga (*Scenedesmus*). The images of all four specimens are at the same magnification. The phase-contrast and Nomarski processes mentioned in (**a**) and (**b**) can create optical contrasts without staining the cells. Both processes enhance the usefulness of light micrographs.

As for other micrographs in the book, the short horizontal bar below the micrograph in (**d**) provides you with a visual reference for size. A micrometer (μm) is 1/1,000,000 of a meter.

THE DEFINING FEATURES OF EUKARYOTIC CELLS

We turn now to organelles and other structural features that typically occur in the cells of plants, animals, fungi, and protistans. We define an **organelle** as an internal, membrane-bound sac or compartment that serves one or more specialized functions inside these cells.

Major Cellular Components

Observe some micrographs of a typical eukaryotic cell, such as the ones described in the preceding section, and you probably will quickly notice that the nucleus is one of its most conspicuous features. Remember, any cell that starts out life with a nucleus is eukaryotic; it has a "true nucleus." Many other organelles and structures also are typical of these cells, although the numbers and kinds differ from one cell type to the next. Here is a list of the most common features:

Organelle or Structure	Main Function
Nucleus	*Localizing the cell's DNA*
Ribosomes	*Assembling polypeptide chains*
Endoplasmic reticulum	*Routing and modifying the newly formed polypeptide chains; also, synthesizing lipids*
Golgi body	*Modifying polypeptide chains into mature proteins; sorting and shipping proteins and lipids for secretion or for use inside cell*
Various vesicles	*Transporting or storing variety of substances; digesting substances and structures within the cell; other functions*
Mitochondria	*Producing many ATP molecules in highly efficient fashion*
Cytoskeleton	*Imparting shape and internal organization to cell; moving the cell and its internal structures*

Think about this list, and you might well find yourself asking: What is the advantage of partitioning the cell interior among many organelles? *Compartmentalization allows a large number of activities to proceed simultaneously in very limited space.* Consider a photosynthetic cell in a leaf. It can put together starch molecules by one set of reactions *and* break them apart by another set. Yet the cell would gain absolutely nothing if the synthesis and breakdown reactions proceeded at the same time on the same starch molecule. Without organelle membranes, the balance of diverse chemical activities that help keep eukaryotic cells alive would spiral out of control.

Figure 4.7 (*Facing page*) Generalized sketches showing some of the features that are typical of many plant cells (**a**) and animal cells (**b**).

Organelle membranes have another function besides physically separating incompatible reactions. They also allow reactions that are compatible and interconnected to proceed at different times. For instance, a plant's photosynthetic cells produce and store starch molecules in an organelle called a chloroplast, then later release starch for use in different reactions in the same organelle.

Typical Organelles in Plant Cells

Figure 4.7*a* can start you thinking about the location of organelles in a typical plant cell. Bear in mind, calling a cell "typical" is like calling a cactus or a water lily or an elm tree a "typical" plant. As is true of animal cells, variations on the basic plan are mind-boggling. With this qualification in mind, also take a close look at the micrograph in Figure 4.8, on the subsequent page. It shows the locations of organelles and structures you are likely to observe in many specialized plant cells.

Typical Organelles in Animal Cells

Next, start thinking about the organelles of a typical animal cell, such as the one shown in Figures 4.7*b* and 4.9. Like the plant cell, it contains a nucleus, numerous mitochondria, and the other components listed earlier. *These structural similarities point to basic functions that are necessary for survival, regardless of cell type.* We will return to this concept throughout the book.

Comparing Figures 4.7 through 4.9 also will give you an initial idea of how plant and animal cells differ in their structure. For example, you will never observe an animal cell surrounded by a cell wall. (You might see assorted fungal and protistan cells with one, however.) What other differences can you identify?

Eukaryotic cells contain a profusion of organelles, which are internal, membrane-bound sacs and compartments that serve specific metabolic functions.

Organelles physically separate chemical reactions, many of which are incompatible.

Organelles separate different reactions in time, as when certain molecules are put together, stored, and then used later in other reaction sequences.

All eukaryotic cells contain certain organelles (such as the nucleus) and structures (such as ribosomes) that perform functions essential for survival. Specialized cells also may incorporate additional kinds of organelles and structures.

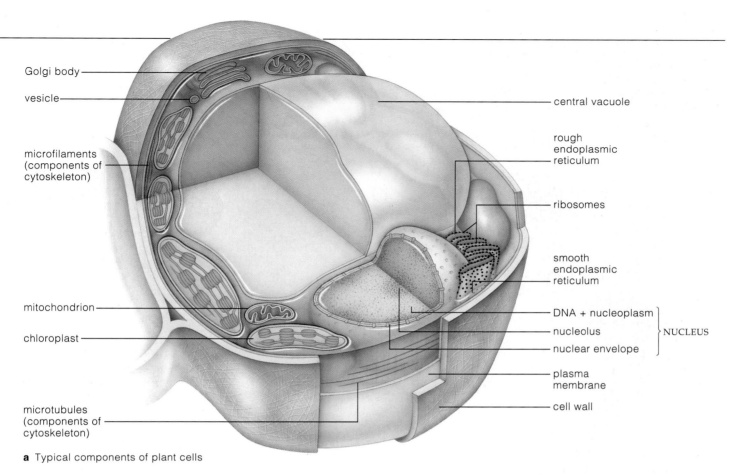

Golgi body

vesicle

microfilaments
(components of
cytoskeleton)

mitochondrion

chloroplast

microtubules
(components of
cytoskeleton)

central vacuole

rough
endoplasmic
reticulum

ribosomes

smooth
endoplasmic
reticulum

DNA + nucleoplasm

nucleolus   NUCLEUS

nuclear envelope

plasma
membrane

cell wall

a Typical components of plant cells

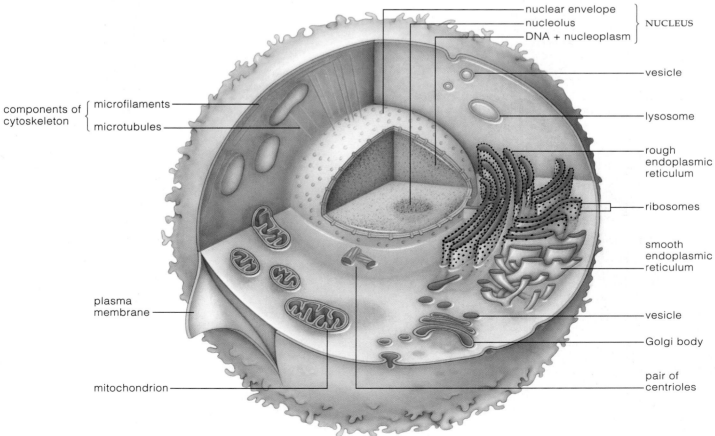

nuclear envelope

nucleolus   NUCLEUS

DNA + nucleoplasm

vesicle

lysosome

rough
endoplasmic
reticulum

ribosomes

smooth
endoplasmic
reticulum

vesicle

Golgi body

pair of
centrioles

components of
cytoskeleton
  microfilaments

  microtubules

plasma
membrane

mitochondrion

b Typical components of animal cells

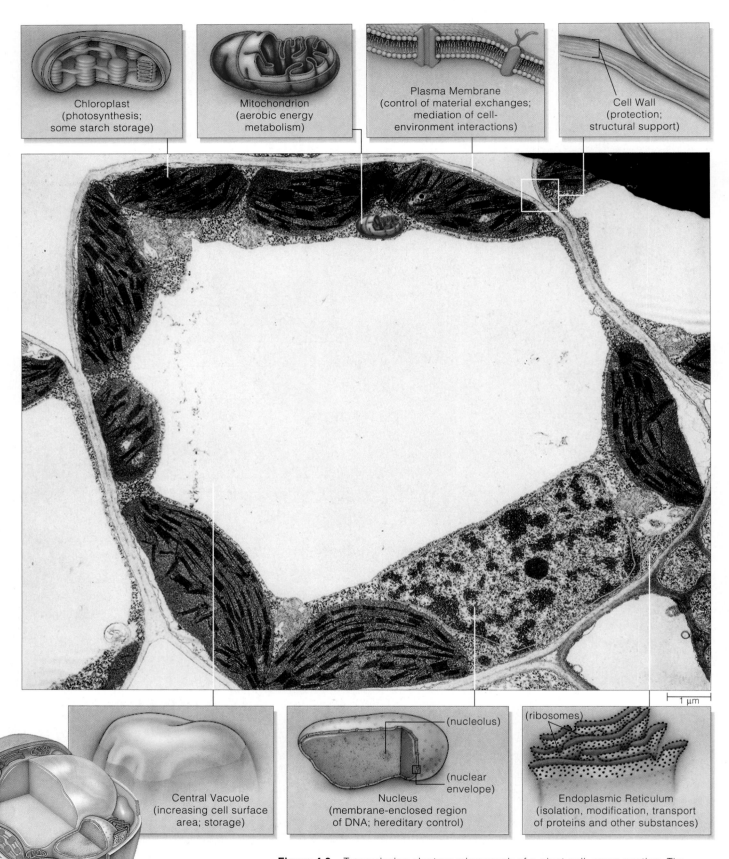

Chloroplast
(photosynthesis; some starch storage)

Mitochondrion
(aerobic energy metabolism)

Plasma Membrane
(control of material exchanges; mediation of cell-environment interactions)

Cell Wall
(protection; structural support)

1 μm

Central Vacuole
(increasing cell surface area; storage)

Nucleus
(membrane-enclosed region of DNA; hereditary control)

(nucleolus)

(nuclear envelope)

(ribosomes)

Endoplasmic Reticulum
(isolation, modification, transport of proteins and other substances)

Figure 4.8 Transmission electron micrograph of a plant cell, cross-section. The sketches highlight key organelles. The specimen is a photosynthetic cell from a blade of Timothy grass (*Phleum pratense*). This plant is one of the important forage grasses that ranchers grow on lands that are suitable for grazing but not for agriculture.

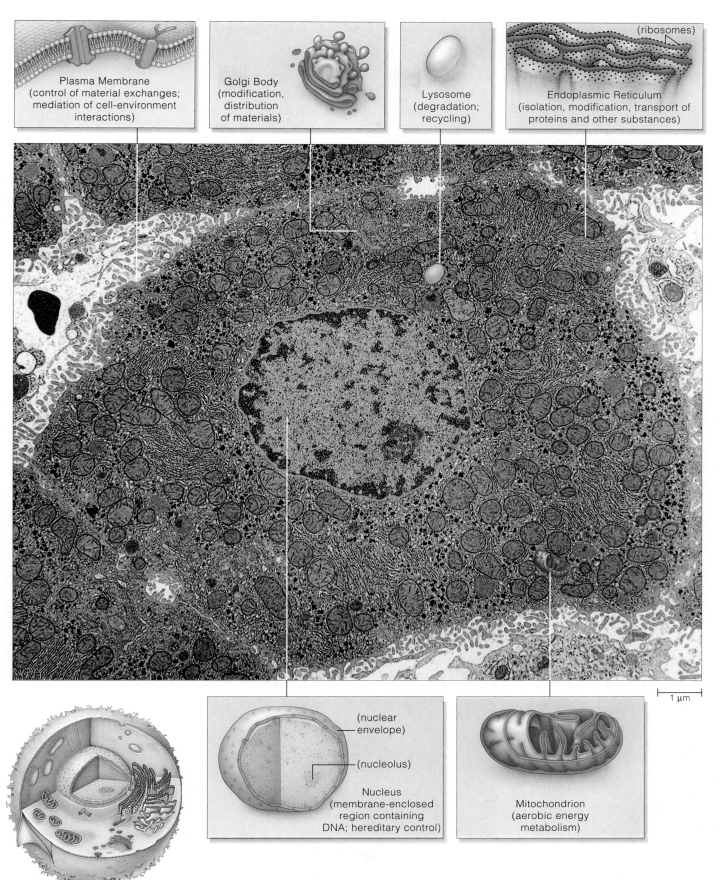

Plasma Membrane (control of material exchanges; mediation of cell-environment interactions)

Golgi Body (modification, distribution of materials)

Lysosome (degradation; recycling)

Endoplasmic Reticulum (isolation, modification, transport of proteins and other substances)

(ribosomes)

1 µm

(nuclear envelope)

(nucleolus)

Nucleus (membrane-enclosed region containing DNA; hereditary control)

Mitochondrion (aerobic energy metabolism)

Figure 4.9 Transmission electron micrograph of an animal cell, cross-section. The sketches highlight key organelles. The specimen is a cell from the liver of a rat.

THE NUCLEUS

Constructing, operating, and reproducing cells simply cannot be done without carbohydrates, lipids, proteins, and nucleic acids. It takes a class of proteins—enzymes—to build and use these molecules. Said another way, a cell's structure and function begin with proteins. *And instructions for building proteins are contained in DNA.*

Unlike bacteria, eukaryotic cells have their genetic instructions distributed through several to many DNA molecules of various lengths. For example, each of your body cells contains forty-six DNA molecules. Stretched end to end, they would be about one meter long. DNA in frog cells would be ten meters, end to end. Compared to the single molecule in bacteria, that's a lot of DNA!

Eukaryotic DNA resides in the **nucleus**. This type of organelle has a distinctive structure, as shown by the example in Figure 4.10, and it serves two key functions. *First*, the nucleus physically tucks away all of the DNA molecules, apart from the intricate metabolic machinery of the cytoplasm. This localization of DNA makes it easier to copy the genetic instructions before the time comes for a cell to divide. The DNA molecules can be assorted into parcels—one parcel for each new cell that is produced. *Second*, the membranous boundary of the nucleus helps control the exchange of substances and signals between the nucleus and the cytoplasm.

Nuclear Envelope

The outermost component of the nucleus is the **nuclear envelope**. This double-membrane system has *two* lipid bilayers, one wrapped over the other (Figure 4.11). It completely surrounds the fluid portion of the nucleus, or nucleoplasm. As is true of all other cell membranes, the lipid bilayers prevent water-soluble substances from moving across. However, pores composed of clusters of proteins span both bilayers. The nuclear pores allow ions and small, water-soluble molecules to move across the nuclear envelope. They also control the passage of ribosomal subunits, proteins, and other large molecules.

The innermost surface of the nuclear envelope has attachment sites for protein filaments. These anchor the DNA molecules to the membrane and help keep them organized. Peppering the outermost surface are many of the cell's ribosomes, which serve in protein synthesis.

Nucleolus

What is the dense, globular mass of material within the nucleus shown in Figure 4.10? As eukaryotic cells grow, one or more of the masses appear in the nucleus. Each is a **nucleolus** (plural, nucleoli). It is a site where the

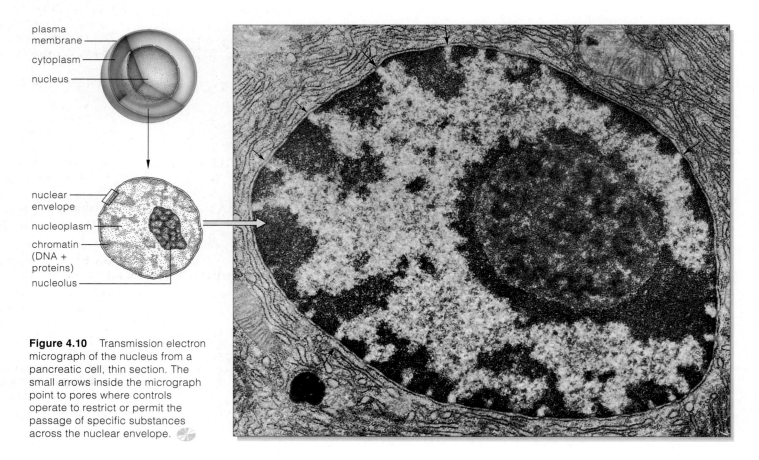

plasma membrane
cytoplasm
nucleus

nuclear envelope
nucleoplasm
chromatin (DNA + proteins)
nucleolus

Figure 4.10 Transmission electron micrograph of the nucleus from a pancreatic cell, thin section. The small arrows inside the micrograph point to pores where controls operate to restrict or permit the passage of specific substances across the nuclear envelope.

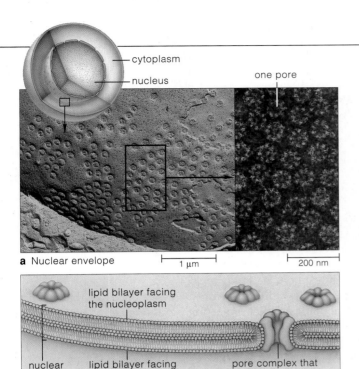

cytoplasm

nucleus

one pore

a Nuclear envelope

1 µm

200 nm

lipid bilayer facing
the nucleoplasm

nuclear
envelope

lipid bilayer facing
the cytoplasm

pore complex that
spans both bilayers

b

Figure 4.11 (**a**) Surface view of a nuclear envelope, which consists of two pore-studded lipid bilayers. *Left:* This specimen was deliberately fractured for microscopy, in the manner described in Section 5.2. *Right:* Closer view of the pores. Each pore across the envelope is an organized array of membrane proteins. It permits the selective transport of larger molecules into and out of the nucleus. (**b**) Sketch of the nuclear envelope.

protein and RNA subunits of ribosomes are assembled. Later, the subunits are shipped from the nucleus, into the cytoplasm. At the time when the polypeptide chains of proteins are synthesized, the subunits join together as intact, functional ribosomes.

Chromosomes

Between cell divisions, eukaryotic DNA is threadlike. Many enzymes and other proteins are attached to it, like beads on a string. Except at extreme magnification, the beaded threads look grainy, as in Figure 4.10. Before a cell divides, however, it duplicates its DNA molecules (so each new cell will get all of the required hereditary instructions). Besides this, the molecules get folded and twisted into condensed structures, proteins and all.

Table 4.1	Summary of the Components of the Nucleus
Nuclear envelope	Pore-riddled double-membrane system that selectively controls the passage of various substances into and out of the nucleus
Nucleolus	Dense cluster of RNA and proteins that will be assembled into subunits of ribosomes
Nucleoplasm	Fluid interior portion of the nucleus
Chromosome	One DNA molecule and many proteins that are intimately associated with it
Chromatin	Total collection of all DNA molecules and their associated proteins in the nucleus

Early microscopists bestowed the names *chromatin* on the seemingly grainy substance and *chromosomes* on the condensed structures. We now define **chromatin** as a cell's collection of DNA, together with all the proteins associated with it (Table 4.1). Each **chromosome** is one DNA molecule and its associated proteins, regardless of whether it is in threadlike or condensed form:

unduplicated,
not condensed
chromosome
(one DNA double
helix + proteins)

duplicated but
not condensed
chromosome
(two DNA double
helices + proteins)

duplicated and
now condensed
chromosome

In other words, the "chromosome" doesn't always look the same during the life of a eukaryotic cell. We will consider different aspects of chromosomes in chapters to come, so it will be helpful to keep that point in mind.

What Happens to the Proteins Specified by DNA?

Enzymes assemble polypeptide chains on ribosomes in the cytoplasm. What happens to the newly formed chains? Many are used or stockpiled in the cytoplasm. Many others pass through the cytomembrane system. As you will read in the next section, this system is a series of organelles, including endoplasmic reticulum, Golgi bodies, and vesicles.

Many proteins take on final form and get packaged in vesicles in the cytomembrane system. Also, lipids are assembled and packaged here. Some vesicles deliver proteins and lipids to regions of the cell where new membranes are to be built. Others store proteins or lipids for specific uses. Still others move to the plasma membrane and release their contents outside the cell.

The nucleus, a double-membrane organelle, sequesters the DNA molecules of eukaryotic cells away from the metabolic machinery of the cytoplasm.

The sequestration makes it easier to organize and copy the DNA, before a cell divides.

Pores across the nuclear envelope help control the passage of larger molecules between the nucleus and the cytoplasm.

THE CYTOMEMBRANE SYSTEM

The **cytomembrane system** is a series of organelles in which lipids are assembled and new polypeptide chains are modified into final proteins. Its products are sorted and shipped to different destinations. Figure 4.12 shows how its organelles—the ER, Golgi bodies, and various vesicles—are functionally related to one another.

Endoplasmic Reticulum

The functions of the cytomembrane system begin with **endoplasmic reticulum**, or **ER**. In animal cells, the ER membranes start at the nucleus and curve through the cytoplasm. Different regions of these membranes have a rough or smooth appearance. The difference arises largely from the presence or absence of ribosomes on the side of the membrane facing the cytoplasm.

Rough ER is often organized as stacked, flattened sacs that have many ribosomes attached (Figure 4.13a). All polypeptide chains are assembled on ribosomes. But only the newly forming chains having a built-in signal can enter the space within rough ER or get incorporated into ER membranes. (The signal is a string of fifteen to twenty specific amino acids.) Once the chains are inside rough ER, enzymes may attach oligosaccharides and other side chains to them. Many specialized cells secrete the final proteins. Rough ER is abundant in such cells. For example, ER-rich, glandular cells of the pancreas produce and secrete enzymes that end up in the small intestine, where they help digest your meals.

Smooth ER is free of ribosomes. It curves through the cytoplasm like connecting pipes (Figure 4.13b). In many cells, smooth ER is the main site of lipid synthesis. It is especially developed in seeds. In liver cells, it even inactivates certain drugs and harmful metabolic by-products. Sarcoplasmic reticulum, a type of smooth ER in skeletal muscle cells, functions in muscle contraction.

Golgi Bodies

In **Golgi bodies**, enzymes put the finishing touches on proteins and lipids, sort them out, and package them in vesicles for shipment to specific locations. For example, an enzyme in one Golgi region might attach a phosphate group to a new protein, thereby giving it a mailing tag to its proper destination.

Each Golgi body is composed of flattened membrane sacs that rather resemble a stack of pancakes (Figure 4.14). Vesicles form at the final region of a Golgi body (the uppermost pancakes) when parts of the membrane bulge, then break away. In animal cells, this is the region that is closest to the plasma membrane.

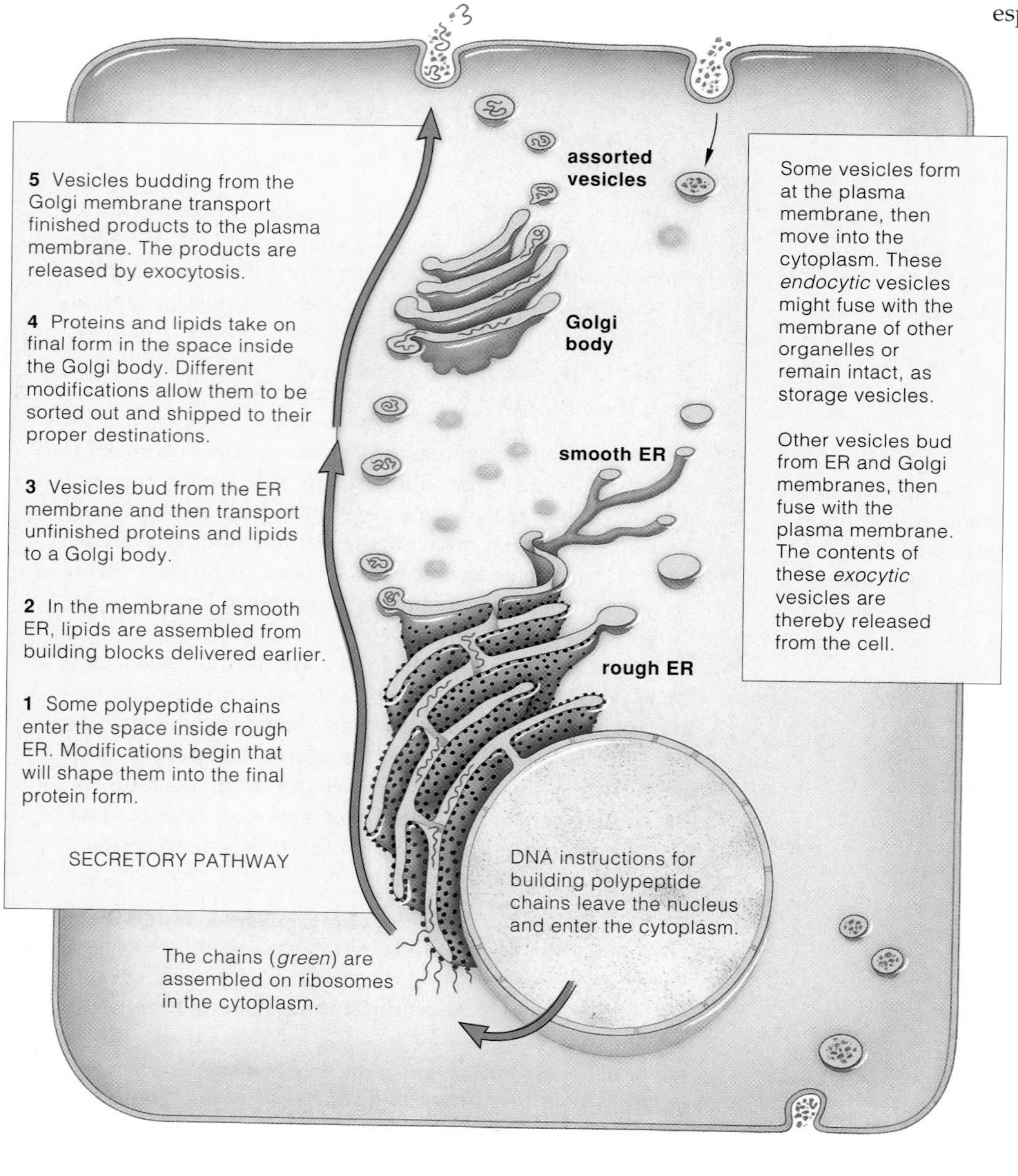

5 Vesicles budding from the Golgi membrane transport finished products to the plasma membrane. The products are released by exocytosis.

4 Proteins and lipids take on final form in the space inside the Golgi body. Different modifications allow them to be sorted out and shipped to their proper destinations.

3 Vesicles bud from the ER membrane and then transport unfinished proteins and lipids to a Golgi body.

2 In the membrane of smooth ER, lipids are assembled from building blocks delivered earlier.

1 Some polypeptide chains enter the space inside rough ER. Modifications begin that will shape them into the final protein form.

SECRETORY PATHWAY

assorted vesicles

Golgi body

smooth ER

rough ER

Some vesicles form at the plasma membrane, then move into the cytoplasm. These *endocytic* vesicles might fuse with the membrane of other organelles or remain intact, as storage vesicles.

Other vesicles bud from ER and Golgi membranes, then fuse with the plasma membrane. The contents of these *exocytic* vesicles are thereby released from the cell.

DNA instructions for building polypeptide chains leave the nucleus and enter the cytoplasm.

The chains (*green*) are assembled on ribosomes in the cytoplasm.

Figure 4.12 Cytomembrane system. This membrane system in the cytoplasm functions in the assembly, modification, packaging, and shipment of proteins and lipids. *Green* arrows highlight a secretory pathway by which certain proteins and lipids are packaged and then released from many types of cells. Among these are glandular cells that secrete mucus, sweat, and digestive enzymes.

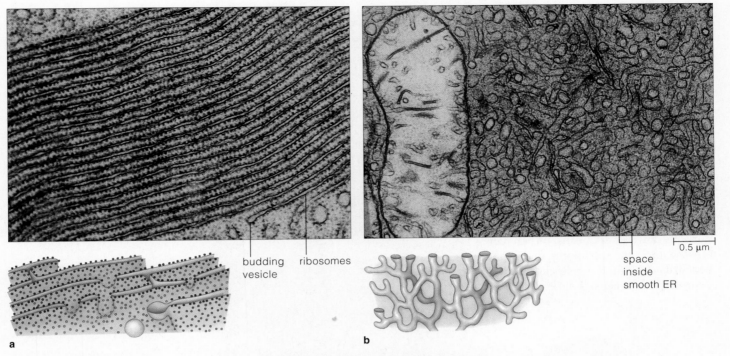

Figure 4.13 Transmission electron micrographs and sketches of endoplasmic reticulum. (**a**) Ribosomes pepper the flattened surfaces of rough ER that face the cytoplasm. (**b**) This section reveals the diameters of the many interconnected, pipelike regions of smooth ER. The large organelle at left is a mitochondrion.

budding vesicle — ribosomes

space inside smooth ER

0.5 μm

a

b

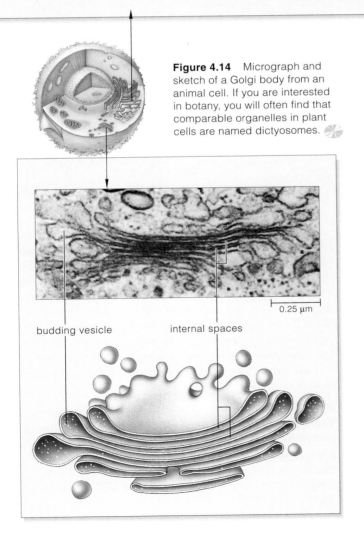

Figure 4.14 Micrograph and sketch of a Golgi body from an animal cell. If you are interested in botany, you will often find that comparable organelles in plant cells are named dictyosomes.

budding vesicle — internal spaces

0.25 μm

A Variety of Vesicles

Assorted vesicles move through the cytoplasm or take up positions inside it. Consider the **lysosome**, a type of vesicle that buds from the Golgi membranes of animal cells and some fungal cells. Lysosomes are organelles of intracellular digestion. They contain a potent brew, rich with enzymes that speed the breakdown of complex carbohydrates, proteins, nucleic acids, and some lipids. Often, lysosomes fuse with other vesicles that formed at the plasma membrane. As described in Section 5.6, such vesicles contain particles, bacteria, or other items that docked at the plasma membrane. Lysosomes may even digest whole cells or some of their parts. For example, when a tadpole is developing into an adult frog, its tail disappears. Lysosomal enzymes help destroy cells of the tail as part of a controlled program of development.

Or consider the **peroxisomes**. These sacs of enzymes break down fatty acids and amino acids. A potentially harmful product, hydrogen peroxide, forms during the reactions. But enzyme action converts it to water and oxygen or channels it into reactions that break down alcohol. After someone drinks alcohol, nearly half of it is degraded in peroxisomes of liver and kidney cells.

In the ER and Golgi bodies of the cytomembrane system, many proteins take on final form and lipids are synthesized.

Lipids, proteins (including enzymes), and other items become packaged in vesicles destined for export, storage, membrane building, intracellular digestion, and other cell activities.

MITOCHONDRIA

Recall, from Section 3.8, that ATP molecules are premier energy carriers. Energy associated with their phosphate groups can be delivered to nearly all reaction sites and drives nearly all cell activities. In eukaryotic cells, *many* ATP molecules can form when organic compounds are degraded in organelles called **mitochondria** (singular, mitochondrion). Figure 4.15 shows one of these organelles. The ATP-forming reactions require oxygen, and they can extract far more energy than can be done by any other means. When you breathe in, you are taking in oxygen primarily for all the mitochondria in your trillions of cells.

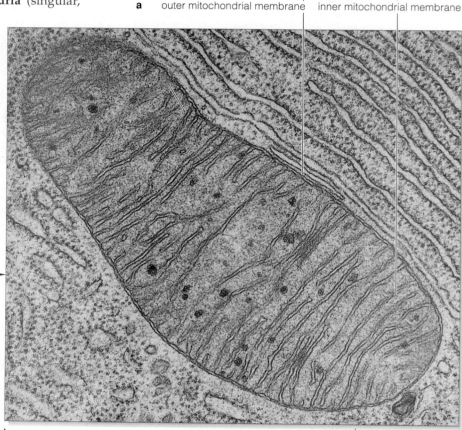

a outer mitochondrial membrane   inner mitochondrial membrane

Figure 4.15 (**a**) Sketch and (**b**) transmission electron micrograph of a thin slice through a typical mitochondrion. Reactions inside this organelle produce quantities of ATP, which is the major energy carrier between different reaction sites in cells. The mitochondrial reactions cannot proceed without oxygen.

b

0.5 μm

A mitochondrion has a double-membrane system. As you can tell from Figure 4.7, the cytoplasm bathes the outermost membrane. Most often, the inner membrane repeatedly folds back on itself. Each inner fold is called a crista (plural, cristae). The double-membrane system creates two distinct compartments in the mitochondrion. Enzymes and other types of proteins embedded in the inner mitochondrial membrane or positioned near its surface function as operating equipment, used for the formation of ATP. Free oxygen helps keep the machinery running. Chapter 8 gives details of the reactions.

All eukaryotic cells have one or more mitochondria. Maybe you will find only one in a single-celled yeast. But you might find a thousand or more in energy-demanding cells, such as those of muscles. Take a look at the profusion of mitochondria in Figure 4.9, which is a micrograph of merely one thin slice from a liver cell. It alone tells you that the liver is an exceptionally active, energy-demanding organ.

In terms of size and biochemistry, mitochondria resemble bacteria. They even have their own DNA and some ribosomes, and they divide on their own. Possibly they evolved from ancient bacteria that were engulfed by a predatory, amoebalike cell yet managed to escape digestion. Perhaps they were able to reproduce inside the cell and its descendants. If they became permanent, protected residents, they may have lost structures and functions required for independent life while becoming mitochondria. We return to this topic later in the book, in Section 21.4.

The organelles called mitochondria are the ATP-producing powerhouses of all eukaryotic cells.

Energy-releasing reactions proceed at the compartmented, internal membrane system of mitochondria. The reactions, which require oxygen, produce far more ATP than can be made by any other cellular reactions.

SPECIALIZED PLANT ORGANELLES

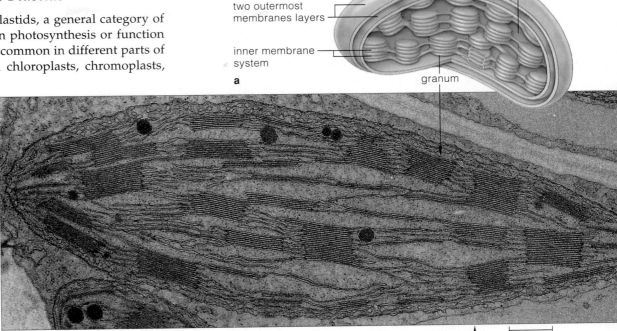

stroma (fluid interior)

two outermost membranes layers

inner membrane system

granum

a

Chloroplasts and Other Plastids

Many plant cells contain plastids, a general category of organelles that specialize in photosynthesis or function in storage. Three types are common in different parts of the plant. They are called chloroplasts, chromoplasts, and amyloplasts.

Chloroplasts occur only in photosynthetic, eukaryotic cells. Within these organelles, energy is harnessed from the sun's rays, ATP forms, and sugars and other organic compounds are synthesized from water and carbon dioxide. Chloroplasts commonly are oval or disk-shaped. They have two outer membrane layers, one wrapped over the other. The layers surround a semifluid interior, the stroma. Here, another membrane forms an inner system of interconnecting, disk-shaped compartments (Figure 4.16). The compartments are stacked together in many chloroplasts. Each stack is a granum (plural, grana).

The first stage of photosynthesis proceeds at light-trapping pigments, enzymes, and other proteins of the inner membrane system. That is where light energy is absorbed and ATP forms. The synthesis stage proceeds in the stroma. There, sugars and other organic compounds are put together. There also, starch grains (clusters of new starch molecules) may be temporarily stored.

Of the photosynthetic pigments, chlorophylls (green) are the most abundant, followed by carotenoids (yellow, orange, and red). As described in Section 7.3, the relative abundances of various pigments in plant parts influence their coloration. Some parts may be green, others golden brown, and so on.

In many ways, chloroplasts resemble photosynthetic bacteria. Like mitochondria, they might have evolved from bacteria that were engulfed by predatory cells, yet escaped digestion and became permanent residents in them. We return to this idea in Section 21.4.

Unlike chloroplasts, chromoplasts have an abundance of carotenoids but no chlorophylls. They are the source of red-to-yellow colors of many flowers, autumn leaves, ripening fruits, and carrots or other roots. The pigments also may visually attract animals that pollinate plants or that disperse seeds. Amyloplasts have no pigments. They often store starch grains and are abundant in cells of stems, potato tubers, and many seeds.

b

⊢ 0.5μm ⊣

Figure 4.16 (**a**) Generalized sketch of the chloroplast, the key defining feature of every photosynthetic eukaryotic cell. (**b**) This transmission electron micrograph shows a chloroplast, thin section.

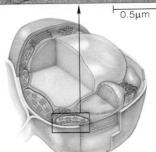

Central Vacuole

Many mature, living plant cells have a **central vacuole** (Figure 4.7). This fluid-filled organelle stores amino acids, sugars, ions, and toxic wastes. As it enlarges, it causes fluid pressure to build up inside the cell and so forces the cell's still-pliable cell wall to enlarge. The cell itself enlarges under the force, and its increased surface area enhances the rate at which substances can cross the plasma membrane. In most cases, the central vacuole increases so much in volume that it takes up 50 to 90 percent of the cell's interior. The cytoplasm ends up as a very narrow zone between the central vacuole and the plasma membrane.

Photosynthetic eukaryotic cells contain chloroplasts and other plastids that function in food production and storage.

Many plant cells have a central vacuole. When this storage vacuole enlarges during growth, cells are forced to enlarge, and this increases the surface area available for absorption.

COMPONENTS OF THE CYTOSKELETON

An interconnected system of fibers, threads, and lattices extends between the nucleus and plasma membrane of eukaryotic cells. This system, the **cytoskeleton**, gives cells their internal organization, shape, and capacity to move. Some elements reinforce the plasma membrane and nuclear envelope; others form scaffolds that hold protein clusters in specific membranes or cytoplasmic regions. Still others are like railroad tracks upon which organelles are shipped from one site to another!

The micrograph in Figure 4.17 shows an animal cell, isolated from its home tissue, as it was stretching out from left to right across a glass slide. The internal mesh of filaments only hints at the extent and intricacy of this cell's cytoskeleton. In this case, researchers were studying its **microtubules** and **microfilaments**. These two classes of cytoskeletal elements, acting singly or collectively, underlie nearly all movements of eukaryotic cells. Some animal cells also have **intermediate filaments**, a class of ropelike cytoskeletal elements that impart mechanical strength to cells and tissues. Many of the elements are permanent. Others appear only at certain times in a cell's life. Before a cell divides, for instance, many new microtubules form a "spindle" structure that moves chromosomes, then disassemble when the task is done.

Microtubules

Figure 4.18a shows part of a microtubule. This hollow cylinder, the cytoskeleton's largest structural element, is twenty-five nanometers wide. Many tubulin subunits make up the cylinder. Tubulin is a protein consisting of two chemically distinct polypeptide chains, each folded into the shape of a ball. While a microtubule is being assembled, all the subunits become oriented in the same direction. The uniform orientation puts slightly different chemical and electrical properties at opposite ends of the growing microtubule. The difference causes one end (the *plus* end) to elongate much faster than the other, *minus* end, which tends to lose subunits if not stabilized. The minus end typically gets anchored in loose material in the cytoplasm or at an **MTOC** (microtubule organizing center). MTOCs are sites of dense material that give rise to quantities of microtubules. You will be reading about various types, which go by such names as centrioles, centrosomes, and kinetochores. Once a newly forming microtubule is anchored, its plus end rapidly grows in a prescribed direction. For example, this is how many microtubules form an organized spindle apparatus.

Microtubules govern the division of cells and some aspects of their shape as well as many cell movements. Cells can't do much without them. As you might well imagine, microtubules have become prime targets in the chemical warfare between vulnerable species and their attackers. For example, autumn crocus and other plants of the genus *Colchicum* synthesize colchicine. This poison inhibits the assembly and promotes the disassembly of microtubules, with dire effects on browsing animals. The plant cells themselves are insensitive to colchicine, which cannot bind well to their tubulin subunits.

Taxol, a poison produced by the western yew (*Taxus brevifolia*), stabilizes microtubules that already exist in a cell and so prevents formation of new ones that the cell may require for other tasks. For example, like colchicine and some other microtubule poisons, taxol interferes with the formation of microtubular spindles, so it can inhibit cell division. Doctors have used it to reduce the uncontrolled cell divisions that underlie the growth of some benign and malignant tumors.

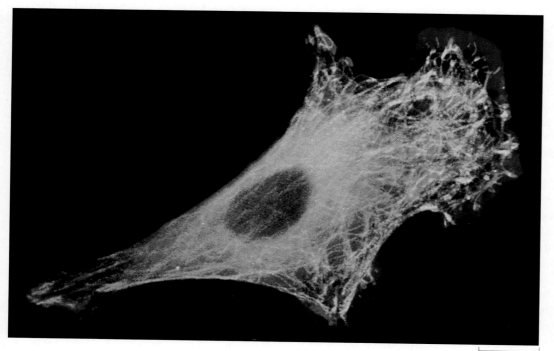

Figure 4.17 A view of many of the structural elements that make up the cytoskeleton of a fibroblast, a type of animal cell that gives rise to certain animal tissues. This composite of three micrograph images reveals the presence of microtubules (tinted *green* in this image). Two kinds of microfilaments are tinted *blue* and *red*.

10 μm

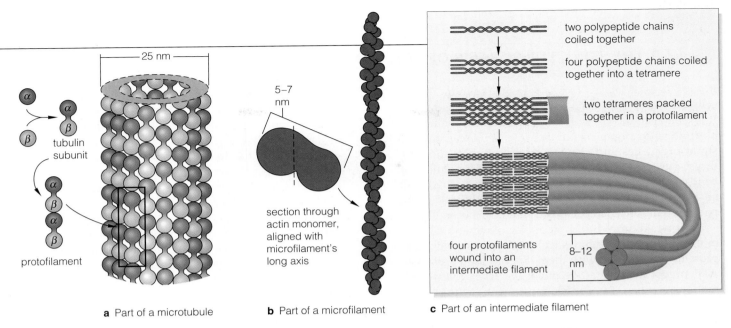

a Part of a microtubule   **b** Part of a microfilament   **c** Part of an intermediate filament

Figure 4.18 Structural organization of (**a**) microtubules, (**b**) microfilaments, and (**c**) an intermediate filament.

Microfilaments

Microfilaments, the thinnest of cytoskeletal elements, are only five to seven nanometers wide (Figure 4.18*b*). Each consists of two helically twisted, polypeptide chains that are assembled from U-shaped actin monomers. Like tubulins, actins are added preferentially to one end of a growing microfilament and removed from the other end.

As you will see in several chapters throughout this book, microfilaments take part in a great variety of movements, especially the kinds that proceed at the cell surface. They also contribute to the development and maintenance of animal cell shapes.

Myosin and Other Accessory Proteins

Both tubulin and actin have been highly conserved over evolutionary time; the microtubules and microfilaments of all eukaryotic species are assembled from them. In spite of the uniformity, monomers of a variety of other proteins can become attached to them. Thus embellished, microtubules and microfilaments that were assembled according to the same basic pattern can serve different functions in different regions of the cell.

For example, monomers of myosin, dynein, or some other "motor protein" typically extend from the surface of microtubules and microfilaments that have roles in cell movement. Myosin is abundant in muscle cells, and it is part of the machinery by which they contract. "Crosslinking proteins" splice adjacent microfilaments together. Certain kinds take part in the formation of the cell cortex, an extensive, three-dimensional network of microfilaments and other proteins just beneath the plasma membrane. The cortex reinforces the cell surface and facilitates movements as well as changes in shape. Other crosslinking proteins splice microfilaments in a

stable network having the properties of a gel. Spectrin, another accessory protein, attaches microfilaments to the plasma membrane. As a final example, the integrins span this outermost membrane, attach to microfilaments inside the cell, and connect them with proteins outside.

Intermediate Filaments

Intermediate filaments, the most stable elements of the cytoskeleton, are between eight and twelve nanometers wide (Figure 4.18*c*). The six known groups mechanically strengthen cells or cell parts and help maintain their shape. For example, the desmins and vimentins help support machinery by which muscle cells contract. The lamins help form a scaffold that reinforces the nucleus. Diverse cytokeratins structurally reinforce the cells that give rise to nails, claws, horns, and hairs.

Unlike the other two classes of cytoskeletal elements, the intermediate filaments occur only in animal cells of specific tissues. Moreover, because each cell usually has only one or sometimes two kinds, researchers can use intermediate filaments to identify cell type. Such typing has proved to be a useful tool in diagnosing the tissue origin of different forms of cancer.

Every eukaryotic cell has a cytoskeleton, the diverse elements of which are the basis of its shape, its internal structure, and its capacity for movements.

Microtubules are key organizers of the cytoskeleton and help move certain cell structures. Microfilaments take part in diverse movements and in the formation and maintenance of cell shape. Intermediate filaments structurally reinforce cells and internal cell structures.

THE STRUCTURAL BASIS OF CELL MOTILITY

If a cell lives, it moves. This is true of free-living single cells, such as the protistan and sperm cells in Figure 4.19. It is true also of cells with fixed positions in the tissues of multicelled organisms. Eukaryotic cells rearrange or shunt organelles and chromosomes, contract (shorten), thrust out long or fat lobes of cytoplasm, stir fluids, or bend a tail and propel the cell body forward. These and all other movements of cell structures, the cell itself, and entire multicelled organisms are forms of motility.

Section 4.8 introduced the major players in cellular motility—the microtubules, microfilaments, and other proteins that interact with them. Energy inputs, as from ATP, trigger motions at the molecular level. Many such motions combine to produce movement at the cellular level, as when a muscle cell contracts. The coordinated movements of many cells cause movements of tissues or organs, as when bundles of skeletal muscle cells interact with bones to make a thumb and finger turn a page.

Mechanisms of Cell Movements

Microfilaments, microtubules, or both take part in most aspects of motility. They do so by three mechanisms.

First, *the length of a microtubule or microfilament can grow or diminish by the controlled assembly or disassembly of its subunits.* When either one lengthens or shortens, a chromosome or some other structure attached to it is pushed or dragged through the cytoplasm.

Also, the rapid assembly of microfilaments produces dynamic surface extensions by which some cells crawl about. *Amoeba proteus*, a soft-bodied protistan, crawls on **pseudopods** ("false feet"; they actually are lobelike protrusions from the cell body). Rapid polymerization may distend the network of microfilaments beneath its plasma membrane. The animal cell in Figure 4.17 was migrating on sheetlike distensions of such a network.

Second, *parallel arrays of microfilaments or microtubules actively slide in controlled directions.* Consider a muscle cell. It contains a series of contractile units, each with parallel arrays of microfilaments and of motor proteins. Briefly, ATP activates the motor proteins (polymers of myosin), which have oarlike projections that bind and release the adjacent microfilaments in a series of power strokes. The repeated action causes the microfilaments to slide past them. Cumulatively, the directional sliding and shortening of all contractile units shortens the cell.

A similar sliding mechanism might be operating as cells crawl about. If the microfilament network beneath the plasma membrane at the cell's trailing end contracts, then some cytoplasmic gel would be squeezed into the leading end, which would bulge forward in response.

Third, *microtubules or microfilaments shunt organelles or portions of the cytoplasm from one location to another.* For example, chloroplasts in photosynthetic cells move to different light-intercepting positions in response to the changing angle of the sun's overhead position. Myosin monomers are "walking" over microfilaments and are carrying the attached chloroplasts with them. This shunt mechanism also causes cytoplasmic streaming: an active, directional, and often rapid flowing of components in cytoplasmic gel. Observe a living plant cell with a light microscope, and you may see pronounced streaming.

Case Study: Flagella and Cilia

To biologists, the **flagellum** (plural, flagella) and **cilium** (plural, cilia) are classic examples of cell motility. As Figure 4.20 shows, both motile structures contain a ring of nine pairs of microtubules around a central pair. A system of spokes and links holds the "9 + 2 array" together. A sliding mechanism operates in the outer ring of the array to bend the motile structure (Figure 4.21).

Each 9 + 2 array arises from a **centriole.** This barrel-shaped structure is one type of MTOC. The centriole remains at the base of the completed array, where it is often called a **basal body** (Figure 4.20).

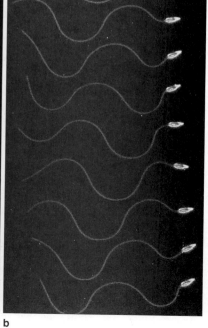

a   ├──────┤   b
              20 µm

Figure 4.19 Two free-living, motile cells. (**a**) Scanning electron micrograph of *Paramecium*, a heterotrophic protistan. Multiple rows of cilia at its surface beat in a synchronized way. The beating propels the cell through its aquatic habitat. It also sweeps food-laden water into the gullet, visible at left as a depression in the cell surface. (**b**) Time-lapse micrographs arranged to show the beating of a sperm's flagellum. Wavelike motions start at the flagellum's base and move to its tip.

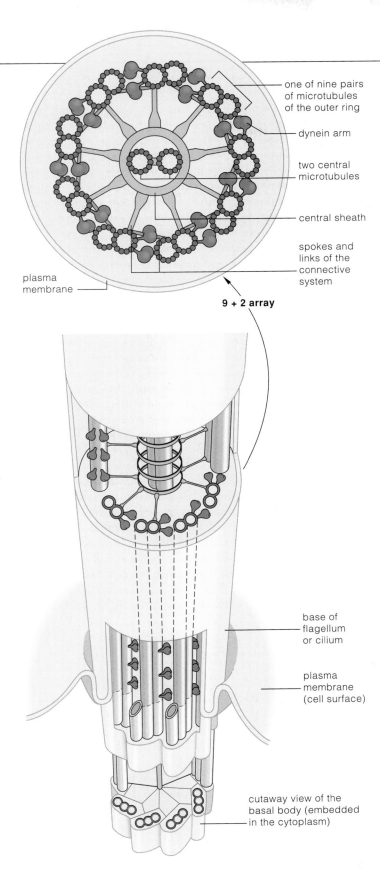

one of nine pairs of microtubules of the outer ring

dynein arm

two central microtubules

central sheath

spokes and links of the connective system

plasma membrane

9 + 2 array

base of flagellum or cilium

plasma membrane (cell surface)

cutaway view of the basal body (embedded in the cytoplasm)

Figure 4.20 Internal organization of flagella and cilia. Both motile structures have a system of microtubules and a connective system of spokes and linking elements. Nine pairs (doublets) of the microtubules are arranged as an outer ring around two central microtubules. This is called a 9 + 2 array.

Figure 4.21 Model of a sliding mechanism responsible for the beating of flagella and cilia, as proposed by cell biologists K. Summers and R. Gibbons. Inside an unbent flagellum (or cilium), all microtubule doublets extend the same distance into the tip. When the flagellum bends, the doublets on the side that is bending the most are being displaced farthest from the tip, as shown here:

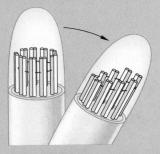

A sliding mechanism operates between the microtubule doublets of the outer ring. A series of motor proteins (dynein) form short arms that extend from each doublet toward the next doublet in the ring. Inputs of ATP energy cause the arms of one doublet to attach to the doublet in front of them, tilt in a short power stroke that pulls on the attached doublet, then release their hold. Repeated power strokes and cross-bridgings force the doublet to slide in a direction toward the base of the flagellum. As the attached doublet moves, *its* arms force the *next* doublet in line to slide down a bit, which forces the *next* doublet to slide down a bit also, and so on in sequence.

The system of interconnecting spokes and links extends through the length of the flagellum and prevents doublets from sliding totally out of the 9 + 2 array. This restriction forces the entire flagellum to bend to accommodate the internal displacement of the sliding doublets.

In short, in the 9 + 2 array of flagella and cilia, energized motor proteins make the nine doublets slide in sequence, and restrictions imposed by a system of spokes and links convert the doublet sliding into a bending motion.

The protistan cell in Figure 4.19*a* bears cilia, which typically are shorter and far more profuse than flagella. Sperm and many other free-living cells use flagella as whiplike tails to move in fluid environments (Figure 4.19*b*). In multicelled organisms, some ciliated epithelial cells stir their surroundings. For example, thousands line airways of your respiratory tract. Their beating cilia direct mucus-trapped particles away from the lungs.

Cell contractions and migrations, chromosome movements, and all other forms of cell motility arise at organized arrays of microtubules, microfilaments, and accessory proteins.

Different mechanisms cause these cytoskeletal elements to assemble or disassemble, slide past one another, and shunt structures to new locations.

This survey of eukaryotic cells concludes with a look at cell walls and some other specialized surface structures. Many of these architectural marvels are constructed of various secretions from the cells themselves. Others are cytoplasmic bridges or sets of membrane proteins that connect neighboring cells and allow them to interact.

Eukaryotic Cell Walls

Single-celled eukaryotic species are directly exposed to their surroundings. Many have a **cell wall**, an outermost structure that wraps continuously around the plasma membrane. The protistan shown in Figure 4.22 is an example. Cell walls help support and protect their owners. Because walls are porous, water and solutes can easily move to and from the plasma membrane. Walls also occur on various cells of plants and fungi.

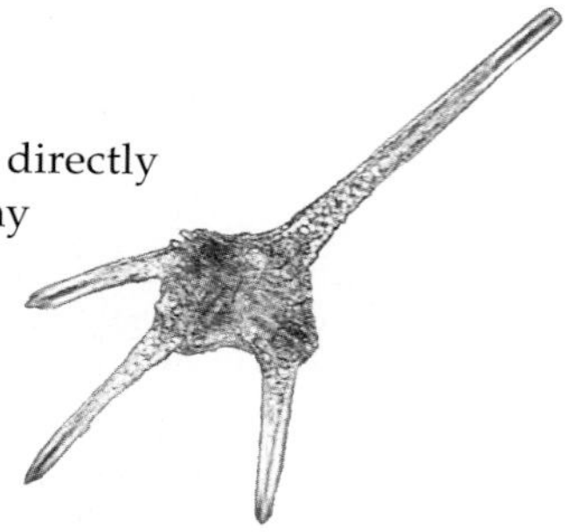

Figure 4.22   From a freshwater habitat, a single-celled, walled protistan (*Ceratium hirundinella*, one of the dinoflagellates).

For example, young plant cells in actively growing regions secrete pectin and hemicellulose (two gluelike polysaccharides), glycoproteins, and cellulose. The cellulose molecules join into ropelike strands that become embedded in the gluey matrix. Together, the secretions form a **primary wall** (Figure 4.23a). Primary walls are sticky, and they cement the walls of adjacent cells together. They also are thin and pliable, so the cell surface area continues to enlarge under the pressure of incoming water. Many cells develop only this thin wall; they are the ones that retain the capacity to divide or change shape as the plant grows and develops. At the cell surfaces exposed to the surrounding air, waxes and other deposits build up as a semitransparent protective covering, a cuticle, that helps reduce evaporative water loss (Figure 4.24a).

When other plant cells mature, they stop enlarging and start secreting more material on the *inside* of the primary wall. These deposits combine to form a rigid, **secondary wall** that reinforces cell shape (Figure 4.23c). Whereas cellulose made up less than 25 percent of the primary wall, the added deposits now contribute more to structural support. In woody plants, up to 25 percent of the secondary cell wall consists of lignin. Lignin is a complex phenolic (a three-carbon chain and an oxygen atom are attached to its six-carbon ring structure). It makes plant parts stronger, more waterproof, and less inviting to insects and other plant-attacking organisms.

Matrixes Between Animal Cells

Although animal cells have no walls, diverse matrixes composed of cell secretions and even materials drawn from the surroundings intervene between many of them. Think of cartilage at the knobby ends of your leg bones.

Cartilage consists of scattered cells and collagen or elastin fibers of their own secretion, all embedded within a "ground substance" of secreted, modified polysaccharides. Similarly, mature bone is mostly an extensive intercellular matrix (Figure 4.24b).

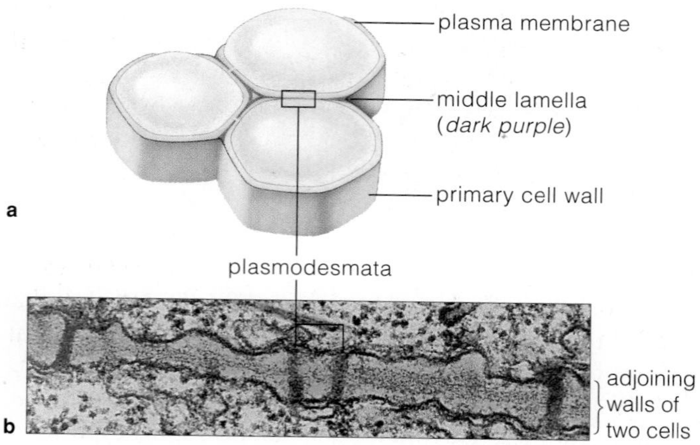

plasma membrane

middle lamella (*dark purple*)

primary cell wall

a

plasmodesmata

adjoining walls of two cells

b

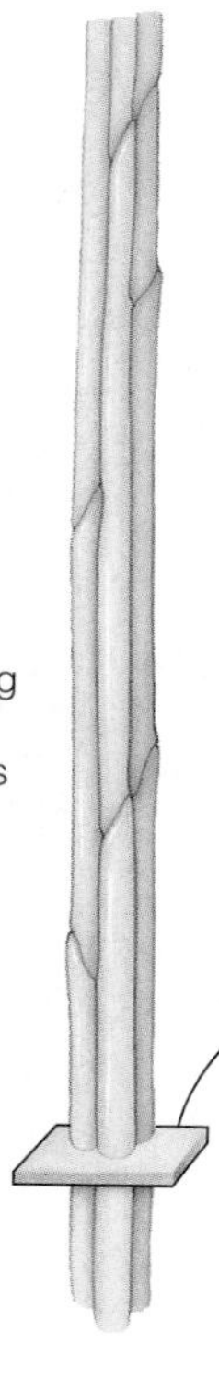

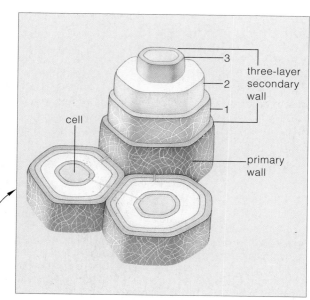

3

2

1

three-layer secondary wall

cell

primary wall

c

Figure 4.23   (**a**) Primary cell wall of young cells in plant tissues. Cell secretions form the middle lamella, a layer between the walls of adjoining cells. The layer is thickest in adjoining corners. (**b**) Plasmodesmata, which are membrane-lined channels across the adjacent walls, connect the cytoplasm of neighboring cells. (**c**) In many types of cells, such as the long fiber cells shown at right, more layers become deposited inside the primary wall. These stiffen the wall and help maintain its shape. Later, the cell may die, leaving the stiffened walls behind. This also is true of the water-conducting pipelines that thread through most plant tissues. They are "tubes" of many interconnected, stiffened cell walls.

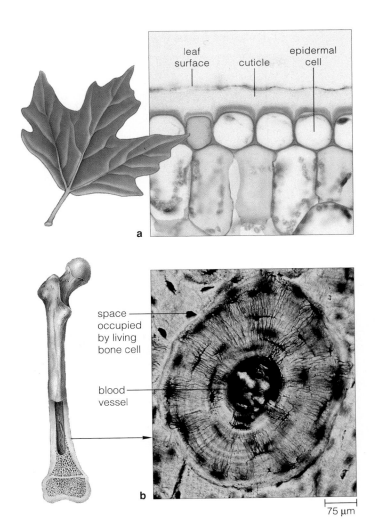

Figure 4.24 (a) Section through a plant cuticle, a surface layer composed of cell secretions. (b) Section through compact bone tissue, stained for microscopy.

labels for (a): leaf surface, cuticle, epidermal cell

labels for (b): space occupied by living bone cell, blood vessel, 75 μm

Cell-to-Cell Junctions

Even when a wall or some other structure imprisons a cell in its own secretions, the only contact that cell has with the outside world is *through* its plasma membrane. Certain components of that membrane project outward, into adjacent cells as well as the surrounding medium. Among those components are junctions where the cell sends and receives diverse signals and materials, where it recognizes and cements itself to kindred cells.

In plants, for instance, numerous tiny channels cross the adjacent primary walls of living cells and connect their cytoplasm. Figure 4.23b shows a few. Each channel is a **plasmodesma** (plural, plasmodesmata). The plasma membranes of the adjoining cells have merged to fully line the channels, so there can be an uninterrupted flow of substances between cells. Thus, all living cells in the plant body have the potential to exchange substances.

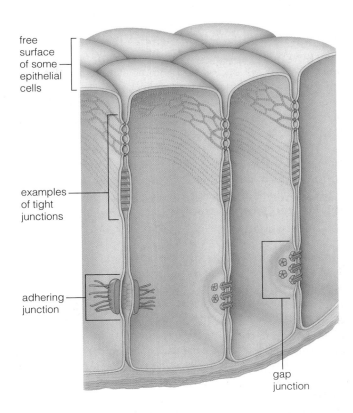

labels: free surface of some epithelial cells, examples of tight junctions, adhering junction, gap junction

Figure 4.25 Composite of types of cell junctions in animals.

In most animal tissues, three categories of cell-to-cell junctions are common (Figure 4.25). *Tight* junctions link the cells of epithelial tissues, which line the body's outer surface and internal cavities and organs. They seal adjoining cells together so water-soluble substances can't leak between them. Thus gastric fluid cannot leak across the stomach's lining and damage surrounding tissues. *Adhering* junctions join cells in tissues of the skin, heart, and other organs subject to stretching. *Gap* junctions, which link the cytoplasm of neighboring cells, are open channels for the rapid flow of signals and substances.

We will be returning to the cell walls, intercellular substances, and cell-to-cell interactions in later chapters. For now, these are the points to remember:

A variety of protistan, plant, and fungal cells have a porous but protective wall that surrounds the plasma membrane. The cells themselves secrete the wall-forming materials.

Secretions from the surface of certain cells form the cuticles of plants and some animals, the extracellular matrixes in animal tissues, and other specialized structures.

In multicelled organisms, coordinated cell activities depend on cell-to-cell junctions, which are protein complexes or cytoplasmic bridges that serve as physical links and forms of communication between cells.

PROKARYOTIC CELLS—THE BACTERIA

We turn now to the bacteria. Unlike the cells we have considered so far, all species of bacteria are prokaryotic; their DNA is *not* enclosed within a nucleus. *Prokaryotic* means "before the nucleus." Microbiologists selected this word as a reminder that bacteria existed on Earth before the nucleus appeared in the forerunners of all eukaryotic cells.

With only one recently discovered exception, bacteria are the smallest of all cells. They usually are not much more than one micrometer wide; even rod-shaped species are only about a few micrometers long. In structural terms, they

are the simplest kinds of cells to think about. Most species have a semirigid or rigid cell wall that surrounds the plasma membrane, physically supports the cell, and imparts shape to it (Figure 4.26*a*). Dissolved substances can still move freely to and from the plasma membrane, for the wall is porous. Sticky polysaccharides cover the cell wall of many species. They help a bacterium attach to rocks, teeth, the vagina, and many other interesting surfaces. In many disease-causing (pathogenic) species, the polysaccharides form a thick, jellylike capsule that surrounds and helps protect the wall.

As is true of eukaryotic cells, bacteria have a plasma membrane that controls the movements of substances to and from the cytoplasm. Their plasma membrane, too, contains proteins that function as channels, transporters, and receptors. It incorporates built-in machinery for metabolic reactions, such as the breakdown of energy-rich compounds. In the photosynthetic types, clusters of some membrane proteins harness light energy and convert it to the chemical energy of ATP.

Bacterial cells are too small to contain more than a small volume of cytoplasm, but they have many ribosomes upon which polypeptide chains are assembled. Apparently, these cells are small enough and so internally simple that they do not require a cytoskeleton.

The cytoplasm of a bacterial cell is distinct from that of eukaryotic cells in being continuous with an irregularly shaped region of DNA. Membranes do not surround this region, which is named a **nucleoid**. As Figure 4.26*c* indicates, a single, circular molecule of DNA occupies this region.

Extending from the surface of many bacterial cells are one or more threadlike motile structures known as **bacterial flagella** (singular, flagellum). These are not the same as eukaryotic flagella, for the 9+2 array of microtubules is absent. Bacterial flagella help a cell

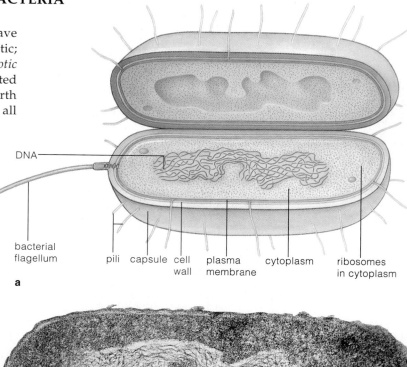

DNA

bacterial flagellum — pili — capsule — cell wall — plasma membrane — cytoplasm — ribosomes in cytoplasm

a

b

0.5 μm

Figure 4.26 (**a**) Generalized prokaryotic body plan. (**b**) Micrograph of *Escherichia coli*.

Your own gut is home to a large population of a normally harmless strain of *E. coli*. In 1993, a harmful strain of contaminated meat that was sold to some fast-food restaurants. The same strain also contaminated hard apple cider sold at a few roadside stands. Cooking the meat thoroughly or boiling the cider would have killed the bacterial cells. Where this was not done, people who ate the meat or drank the cider became quite sick. Some died.

Facing page: (**c**) Researchers manipulated this *E. coli* cell to release its single, circular molecule of DNA. (**d**) Cells of different bacterial species are shaped like balls, rods, or corkscrews. The ball-shaped cells of *Nostoc*, a photosynthetic bacterium, stick together inside a thick, gelatinlike sheath of their own secretions. Chapter 22 gives other splendid examples. (**e**) Like this *Pseudomonas marginalis* cell, many species have one or more bacterial flagella, motile structures that propel the cell through fluid environments.

move rapidly through the fluid surroundings. Other surface projections include pili (singular, pilus), which are protein filaments that help many kinds of bacterial cells attach to various surfaces, even to one another.

Taken as a whole, the bacteria are certainly the most metabolically diverse organisms. They have managed to exploit energy and raw materials in just about every kind of environment. Besides this, ancient prokaryotes gave rise to all the protistans, plants, fungi, and animals

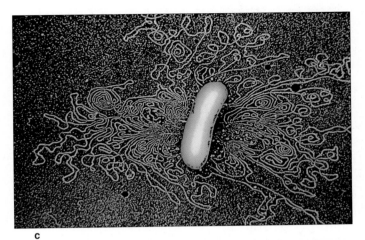

c

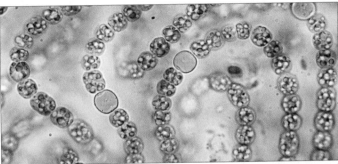

d

1 µm

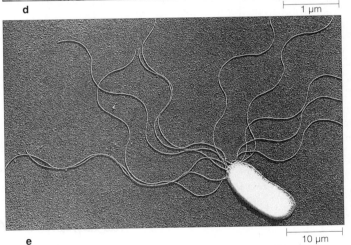

e

10 µm

ever to appear on Earth. The evolution, structure, and functioning of these remarkable cells are topics of later chapters.

Bacteria alone are prokaryotic cells; their DNA is not housed inside a nucleus. Most species have a cell wall around the plasma membrane. Generally, their cytoplasm does not have any organelles comparable to those of eukaryotic cells.

Bacteria are the simplest cells but as a group show the most metabolic diversity. Their metabolic activities proceed at the plasma membrane and at many ribosomes in the cytoplasm.

SUMMARY

1. Three generalizations constitute the cell theory:

a. All living things are composed of one or more cells.

b. The cell is the smallest entity that retains the properties of life. That is, it either lives independently or has a built-in, genetic capacity to do so.

c. New cells arise only from cells that already exist.

2. At the minimum, a newly formed cell has a plasma membrane, a region of cytoplasm, and a region of DNA.

a. The plasma membrane (a thin, outer membrane) maintains the cell as a distinct, separate entity. It allows metabolic events to proceed apart from random events in the environment. Many substances and signals are continually moving across it, in highly controlled ways.

b. The cytoplasm is all the fluids, ribosomes, structural elements, and (in eukaryotic cells) organelles between the plasma membrane and the region of DNA.

3. Membranes are crucial to cell structure and function. They consist of lipids (phospholipids, for the most part) and proteins. The lipids are arrayed as two layers, with all the hydrophobic tails of both layers sandwiched in between all the hydrophilic heads. This lipid bilayer imparts structure to the membrane and bars passage of water-soluble substances across it. Diverse proteins are embedded in the bilayer or attached to its surfaces.

4. Proteins carry out most cell membrane functions. For example, many serve as channels or pumps that allow or promote passage of water-soluble substances across the lipid bilayer. Others are receptors for extracellular substances that trigger changes in cell activities.

5. Cell membranes divide the cytoplasm of eukaryotic cells into functional compartments called organelles. Prokaryotic cells do not have comparable organelles.

6. Organelle membranes separate metabolic reactions in the space of the cytoplasm and allow different kinds to proceed in orderly fashion. (In bacteria, many similar reactions proceed at the plasma membrane.)

a. The nuclear envelope functionally separates the DNA from the metabolic machinery of the cytoplasm.

b. The cytomembrane system includes the ER, Golgi bodies, and vesicles. Many new proteins are modified into final form and lipids are assembled in this system. Finished products are packaged and then shipped off to destinations inside or outside the cell.

c. Mitochondria are specialists in oxygen-requiring reactions that produce many ATP molecules.

d. Chloroplasts trap sunlight energy and produce organic compounds in photosynthetic eukaryotic cells.

7. The cytoskeleton of a eukaryotic cell functions in cell shape, internal organization, and movements.

8. Table 4.2 on the next page summarizes the defining features of both prokaryotic and eukaryotic cells.

Table 4.2 Summary of Typical Components of Prokaryotic and Eukaryotic Cells

Cell Component	Function	PROKARYOTIC Archaebacteria, Eubacteria	EUKARYOTIC Protistans	Fungi	Plants	Animals
Cell wall	Protection, structural support	✓*	✓*	✓	✓	None
Plasma membrane	Control of substances moving into and out of cell	✓	✓	✓	✓	✓
Nucleus	Physical separation and organization of DNA	None	✓	✓	✓	✓
DNA	Encoding of hereditary information	✓	✓	✓	✓	✓
RNA	Transcription, translation of DNA messages into polypeptide chains of specific proteins	✓	✓	✓	✓	✓
Nucleolus	Assembly of subunits of ribosomes	None	✓	✓	✓	✓
Ribosome	Protein synthesis	✓	✓	✓	✓	✓
Endoplasmic reticulum (ER)	Initial modification of many of the newly forming polypeptide chains of proteins; lipid synthesis	None	✓	✓	✓	✓
Golgi body	Final modification of proteins, lipids; sorting and packaging them for use inside cell or for export	None	✓	✓	✓	✓
Lysosome	Intracellular digestion	None	✓	✓*	✓*	✓
Mitochondrion	ATP formation	**	✓	✓	✓	✓
Photosynthetic pigment	Light–energy conversion	✓*	✓*	None	✓	None
Chloroplast	Photosynthesis; some starch storage	None	✓*	None	✓	None
Central vacuole	Increasing cell surface area; storage	None	None	✓*	✓	None
Bacterial flagellum	Locomotion through fluid surroundings	✓*	None	None	None	None
Flagellum or cilium with 9 + 2 microtubular array	Locomotion through or motion within fluid surroundings	None	✓*	✓*	✓*	✓
Cytoskeleton	Cell shape; internal organization; basis of cell movement and, in many cells, locomotion	None	✓*	✓*	✓*	✓

* Known to occur in at least some groups.

** Oxygen-requiring (aerobic) pathways of ATP formation do occur in many groups, but mitochondria are not involved.

Review Questions

1. Label the organelles in this diagram of a plant cell. 4.3

2. Label the organelles in this diagram of an animal cell. 4.3

3. State the three key points of the cell theory. *CI*

4. Describe three features that all cells have in common. After reviewing Table 4.2, write a paragraph on the key differences between prokaryotic and eukaryotic cells. *4.1, 4.3, 4.11*

5. Suppose you want to observe the three-dimensional surface of an insect's eye. Would you benefit most by using a compound light microscope, transmission electron microscope, or scanning electron microscope? *4.2*

6. Briefly characterize the structure and function of the nucleus, the nuclear envelope, and the nucleolus. *4.4*

7. Define chromosome and chromatin. Do chromosomes always have the same appearance during a cell's life? *4.4*

8. Which organelles are part of the cytomembrane system? *4.5*

9. Is this statement true or false: Plant cells have chloroplasts, but not mitochondria. Explain your answer. *4.6, 4.7*

10. What are the functions of the central vacuole in mature, living plant cells? *4.7*

11. Define cytoskeleton. How does it aid in cell functioning? *4.8*

12. Are all components of the cytoskeleton permanent? *4.8, 4.9*

13. What gives rise to the 9 + 2 array of cilia and flagella? *4.9*

14. Cell walls are typical of which organisms: animals, plants, fungi, protistans, or bacteria? Are the walls solid or porous? *4.10*

15. In certain plant cells, is a secondary wall deposited inside or outside the surface of the primary wall? *4.10*

16. In multicelled organisms, coordinated interactions depend on linkages and communications between cells. What types of junctions occur between adjacent animal cells? Plant cells? *4.10*

Self-Quiz *(Answers in Appendix IV)*

1. Cell membranes consist mainly of a _____ .
   a. carbohydrate bilayer and proteins
   b. protein bilayer and phospholipids
   c. lipid bilayer and proteins
   d. none of the above

2. Organelles _____ .
   a. are membrane-bound compartments
   b. are typical of eukaryotic cells, not prokaryotic cells
   c. separate chemical reactions in time and space
   d. all of the above are features of the organelles

3. Cells of many protistans, plants, and fungi, but not animals, commonly have _____ .
   a. mitochondria   c. ribosomes
   b. a plasma membrane   d. a cell wall

4. Is this statement true or false: The plasma membrane is the outermost component of all cells. Explain your answer.

5. Unlike eukaryotic cells, prokaryotic cells _____ .
   a. lack a plasma membrane   c. do not have a nucleus
   b. have RNA, not DNA   d. all of the above

6. Match each cell component with its function.
   ____ mitochondrion   a. synthesis of polypeptide chains
   ____ chloroplast   b. initial modification of new
   ____ ribosome   polypeptide chains
   ____ rough ER   c. final modification of proteins; lipid
   ____ Golgi body   synthesis; sorting and shipping tasks
                         d. photosynthesis
                         e. site of oxygen-requiring pathway
                            of ATP formation

Critical Thinking

1. Why is it likely that you will never encounter a predatory two-ton living cell on the sidewalk?

2. Many compound light microscopes have blue filters. Think of the spectrum of visible light (as in Figure 7.3) and explain why blue light is efficient when viewing objects at high magnification.

3. Your biology professor shows you an electron micrograph of a cell that contains very large numbers of mitochondria and Golgi bodies. You notice that this particular cell also has a great deal of rough endoplasmic reticulum. What kinds of cellular activities would require such an abundance of the three kinds of organelles?

4. Do the events that proceed in the nucleus and cytoplasm of eukaryotic cells seem too remote from topics of human interest? Then try starting with the human body and moving down to the cellular level. For example, *cystic fibrosis* is a disabling and eventually fatal genetic disorder. Glands of an affected person secrete far more than they should, with far-reaching effects. In time, digestive enzymes clog a duct between the pancreas and small intestine, food cannot be digested properly, and even if food intake increases, malnutrition results. Cysts form in the pancreas, which degenerates and becomes fibrous (hence the name of the disorder). Thick mucus builds up in the respiratory tract, so affected individuals have difficulty expelling airborne bacteria and particles that enter the lungs. Also, too much sweat forms. Midwives used to lick the forehead of newborns, and if they tasted far too much salt in the sweat, they predicted the new individual would develop lung congestion.

A defective form of a protein is implicated in the disorder. The protein is a plasma membrane component of cells of glands that secrete mucus, digestive enzymes, and sweat. After reviewing Sections 4.4 and 4.5, track the disorder back from the affected protein to its source at the cellular level.

Selected Key Terms

bacterial flagellum *4.11*	ER (endoplasmic reticulum) *4.5*	nucleolus *4.4*
basal body *4.9*	eukaryotic cell *4.1*	nucleus *4.4*
cell *4.1*	flagellum *4.10*	organelle *4.3*
cell theory *CI*	Golgi body *4.5*	peroxisome *4.5*
cell wall *4.10*	intermediate filament *4.8*	plasma membrane *4.1*
central vacuole *4.7*	lipid bilayer *4.1*	plasmodesma *4.10*
centriole *4.9*	lysosome *4.5*	primary wall *4.10*
chloroplast *4.7*	microfilament *4.8*	prokaryotic cell *4.1*
chromatin *4.4*	micrograph *4.2*	pseudopod *4.9*
chromosome *4.4*	microtubule *4.8*	ribosome *4.1*
cilium *4.9*	mitochondrion *4.6*	secondary wall *4.10*
cytomembrane system *4.5*	MTOC *4.8*	surface-to-volume ratio *4.1*
cytoplasm *4.1*	nuclear envelope *4.4*	wavelength, of light *4.2*
cytoskeleton *4.8*	nucleoid *4.11*	

Readings

deDuve, C. 1985. *A Guided Tour of the Living Cell*. New York: Freeman. Beautiful introduction to the cell; two short volumes.

Web Site See *http://www.wadsworth.com/biology* for practice quiz questions, hypercontents, BioUpdates, and critical thinking. The Wadsworth Biology Resource Center provides a wealth of information fully organized and integrated by chapter.

5

A CLOSER LOOK AT CELL MEMBRANES

It Isn't Easy Being Single

As small as it may be, a cell is a living thing engaged in the risky business of survival. Consider how something as ordinary as water can challenge its very existence. Water bathes cells inside and out, donates its individual molecules to many metabolic reactions, and dissolves ions that are necessary for cell functioning. If all goes well, the cell holds onto enough water and dissolved ions—not too little, not too much—to survive. But who is to say that life consistently goes well?

Think of a goose barnacle attached to the submerged side of a log that is drifting offshore, at the mercy of ocean currents. At feeding time, the barnacle opens its hinged shell and extends featherlike appendages, which trap bacteria and other bits of food suspended in the water (Figure 5.1a). The fluid bathing each living cell in the barnacle's body is salty, rather like the salt

composition of seawater. And seawater normally is in balance with the salty fluid inside the barnacle's cells.

However, suppose the log drifts into a part of the ocean where meltwater from a glacier has diluted the water (Figure 5.1b). For reasons you will explore in this chapter, salts inevitably move out of the barnacle's body—and out of its cells. The previously exquisite salt–water balance gradually spirals out of control, and the cells die. So, in time, does the barnacle.

The same thing happens to burrowing worms and other soft-bodied organisms that live between the high and low tide marks along a rocky shore. For instance, after an unpredictably fierce storm along the coast, seawater becomes highly dilute with runoff from the land. The dilution upsets the balance between body fluids that bathe cells of those organisms and the water

a

Figure 5.1 (a) Goose barnacles, which live attached to logs and other floating objects in the seas. As is true of every organism, drastic changes in salt concentration can threaten the cells of these marine animals. Such changes would occur if the barnacles were to accidentally end up in glacial meltwaters. (b) The dark-blue seawater in this photograph is quite salty. The lighter blue water is diluted by meltwater from the glacier in the background; it is much, much lower in salts than seawater.

moving in with the tides. At such times, the resulting death toll in the intertidal zone can be catastrophic.

With these examples we begin to see the cell for what it is: a tiny, organized bit of life in a world that is, by comparison, unorganized and sometimes harsh. How finely adapted a living cell must be to its environment! The cell must be built in such a way that it can bring in certain substances, release or keep out other substances, and conduct its internal activities with great precision.

For this bit of life, precision begins at the plasma membrane—a flimsy bilayer of lipids, dotted with diverse proteins, that surrounds the cytoplasm. Across the membrane, the cell exchanges substances with its surroundings in highly selective ways. For eukaryotic cells, precision continues at the membranes of internal compartments called organelles. Aerobic respiration, photosynthesis, and many other metabolic processes depend on the selective movement of substances across the membranes of organelles. Gain insight into the structure and function of cell membranes, and you will gain insight as well into survival at life's most fundamental level.

b

KEY CONCEPTS

1. A cell membrane consists largely of phospholipids and proteins. Its phospholipid molecules are organized as a double layer. This "bilayer" gives the cell membrane its basic structure and prevents the haphazard movement of water-soluble substances across it. Proteins embedded in the bilayer or associated with one of its surfaces carry out most membrane functions.

2. Many transport proteins span the bilayer of all cell membranes. Different kinds of transport proteins are open channels, gated channels, carriers, or pumps for specific water-soluble substances.

3. At the outer surface of a plasma membrane, diverse receptor proteins receive chemical signals that can trigger alterations in cell activities. Also present are recognition proteins, which are like molecular fingerprints; they identify a cell as being of a certain type.

4. Unless something prevents them from doing so, ions or molecules of a substance tend to move from one region into an adjacent region where they are less concentrated. This behavior is called diffusion.

5. In a form of molecular behavior known as osmosis, water diffuses across a membrane to a region where its concentration is lower. Osmosis only occurs at selectively permeable membranes that permit water to cross freely but that prevent the free passage of ions and large polar molecules.

6. Some membrane proteins function in passive transport. By this process, a solute enters or leaves a cell by diffusing through the protein's interior, in the direction that its concentration gradient takes it.

7. Other membrane proteins function in active transport. By this process, an energized transport protein actively pumps a solute across a cell membrane, against the concentration gradient.

MEMBRANE STRUCTURE AND FUNCTION

Earlier chapters provided you with a brief look at the structure of cell membranes and the general functions of their component parts. Here, we incorporate some of the background information in a more detailed picture.

The Lipid Bilayer of Cell Membranes

Fluid bathes the two surfaces of a cell membrane and is vital for its functioning. The membrane, too, has a fluid quality; it is not a solid, static wall between cytoplasmic and extracellular fluids. For instance, puncture a cell with a fine needle, and its cytoplasm will not ooze out. The membrane will flow over the puncture site and seal it!

How does a fluid membrane remain distinct from its fluid surroundings? To arrive at the answer, start by reviewing what we have already learned about its most abundant components, the phospholipids. Recall that a **phospholipid** has a phosphate-containing head and two fatty acid tails attached to a glycerol backbone (Figure 5.2a). The head is hydrophilic; it easily dissolves in water. Its tails are hydrophobic; water repels them. Immerse a number of phospholipid molecules in water, and they will interact with water molecules and with one another until they spontaneously cluster in a sheet or film at the water's surface. Their jostlings may even force them to become organized in two layers, with all fatty acid tails sandwiched between all hydrophilic heads. This **lipid bilayer** arrangement, remember, is the structural basis of cell membranes (Section 4.1 and Figure 5.2c).

The organization of each lipid bilayer minimizes the total number of hydrophobic

groups exposed to water, so the fatty acid tails do not have to spend a lot of energy fighting water molecules, so to speak. A "punctured" membrane exhibits sealing behavior precisely because a puncture is energetically unfavorable. It leaves far too many hydrophobic groups exposed to the surrounding fluid.

Ordinarily, few cells get jabbed by fine needles. But the self-sealing behavior of membrane phospholipids is good for more than damage control. Among other things, it functions in vesicle formation. For example, as vesicles bud away from ER or Golgi membranes, phospholipids interact hydrophobically with cytoplasmic water. They get pushed together, and the rupture seals. You will read more about vesicle formation later in the chapter.

Fluid Mosaic Model of Membrane Structure

Figure 5.3 shows a bit of membrane that corresponds to the **fluid mosaic model**. By this model, cell membranes are a mixed composition—a "mosaic"—of phospholipids, glycolipids, sterols, and proteins. The phospholipid heads as well as the length and saturation of the tails are not all the same. (Recall that unsaturated fatty acids have one or more double bonds in their backbone and fully saturated ones have none.) The glycolipids are structurally similar to phospholipids, but their head incorporates one or more sugar monomers. In animal cell membranes, cholesterol is the most abundant sterol (Figure 5.2b). Phytosterols are their equivalent in plant cell membranes.

Also by this model, the membrane is "fluid" owing to the motions and interactions of its component parts.

Figure 5.2
(**a**) Structural formula of phosphatidylcholine, a phospholipid that is one of the most common components of the membranes of animal cells. *Orange* indicates its hydrophilic head; *yellow* indicates its hydrophobic tails.

(**b**) Structural formula of cholesterol, the major sterol in animal tissues.

(**c**) Diagram showing how lipids that are placed in liquid water may spontaneously organize themselves into a bilayer structure.

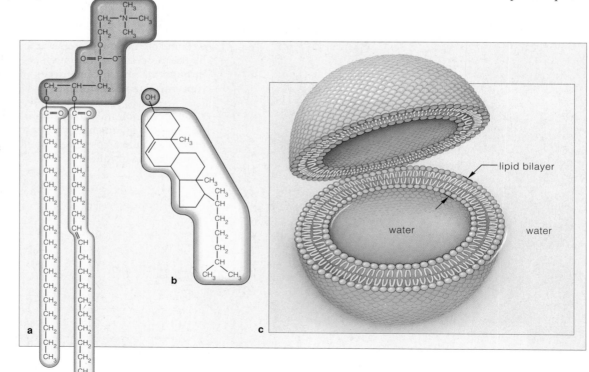

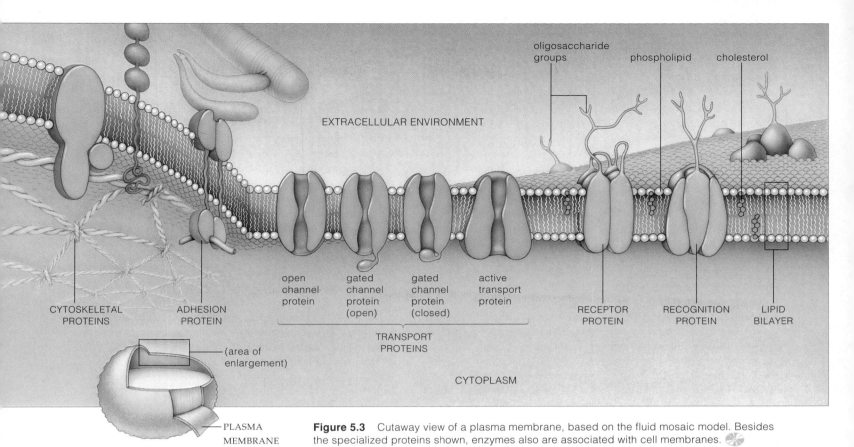

EXTRACELLULAR ENVIRONMENT

oligosaccharide groups   phospholipid   cholesterol

CYTOSKELETAL PROTEINS

ADHESION PROTEIN

open channel protein

gated channel protein (open)

gated channel protein (closed)

active transport protein

RECEPTOR PROTEIN

RECOGNITION PROTEIN

LIPID BILAYER

TRANSPORT PROTEINS

(area of enlargement)

CYTOPLASM

PLASMA MEMBRANE

Figure 5.3 Cutaway view of a plasma membrane, based on the fluid mosaic model. Besides the specialized proteins shown, enzymes also are associated with cell membranes.

The hydrophobic interactions that give rise to most of a membrane's structure are weaker than covalent bonds. This means most phospholipids and some proteins are free to drift sideways. Also, the phospholipids can spin about their long axis and flex their tails, which keeps neighboring molecules from packing together in a solid layer. Short or kinked (unsaturated) fatty acid tails also contribute to membrane fluidity.

The fluid mosaic model is a good starting point for exploring cell membranes. But bear in mind, membranes differ in the details of their molecular composition and arrangements, and they are not even the same on both surfaces of their bilayer. For example, oligosaccharides and other carbohydrates are covalently bonded to protein and lipid components of a plasma membrane, but only on its outward-facing surface (Figure 5.3). Moreover, they differ in number and kind from one species to the next, even among the different cells of the same individual.

Overview of Membrane Proteins

The proteins embedded in a lipid bilayer or attached to one of its surfaces carry out most membrane functions. Many are enzyme components of metabolic machinery. Others are **transport proteins** that allow water-soluble substances to move through their interior, which spans the bilayer. They bind molecules or ions on one side of the membrane, then release them on the other side.

The **receptor proteins** bind extracellular substances, such as hormones, that trigger changes in cell activities.

For example, certain enzymes that crank up machinery for cell growth and division become switched on when somatotropin, a hormone, binds with receptors for it. Different cells have different combinations of receptors.

Diverse **recognition proteins** at the cell surface are like molecular fingerprints; their oligosaccharide chains identify a cell as being of a specific type. For example, "self" proteins pepper the plasma membrane of your cells. Certain white blood cells chemically recognize the proteins and leave your own cells alone, but they attack invading bacterial cells having "nonself" proteins at their surface. Finally, **adhesion proteins** of multicelled organisms help cells of the same type locate and stick to one another and stay positioned in the proper tissues. They are glycoproteins with oligosaccharides attached. After tissues form, the sites of adhesion may become a type of cell junction, as described earlier in Section 4.10.

A cell membrane has two layers composed mainly of lipids, phospholipids especially. This lipid bilayer is the structural foundation for the membrane and also serves as a barrier to water-soluble substances.

Hydrophilic heads of the phospholipids are dissolved in fluids that bathe the two outer surfaces of the bilayer. Their hydrophobic tails are sandwiched between the heads.

Proteins associated with the bilayer carry out most membrane functions. Many are enzymes, transporters of substances across the bilayer, or receptors for extracellular substances. Other types function in cell-to-cell recognition or adhesion.

5.2 TESTING IDEAS ABOUT CELL MEMBRANES

TESTING MEMBRANE MODELS Recall, from Chapter 1, that scientific methods are used to reveal nature's secrets. To gain insight into how researchers discovered some structural details of cell membranes, imagine yourself duplicating some of their test methods. You can start by attempting to identify the molecular components of the plasma membrane.

Your first challenge is to secure a membrane sample that is large enough to study and not contaminated with organelle membranes. Using red blood cells will simplify the task. These cells are abundant, easy to collect, and structurally simple. As they are maturing, their nucleus disintegrates. Ribosomes, hemoglobin, and just a few other components remain, but these are enough to keep the cells functioning for their brief, four-month life span.

By placing a sample of red blood cells in a test tube filled with distilled water, you can separate the plasma membranes from the other cell components. Cells have more solutes and fewer water molecules than distilled water does. For reasons that you will read about shortly, the difference in solute concentrations between the two regions causes water to move across a plasma membrane, into the cells. These particular cells have no mechanisms for actively expelling the excess water, so they swell. In time they burst, and their contents spill out.

Now you have a mixture of membranes, hemoglobin, and other cell components in water. How do you separate all the bits of membrane? You place a tube that holds a solution of the cell parts in a **centrifuge**, which is a motor-driven rotary device that can spin test tubes at very high speed (Figure 5.4). Structures and molecules move in response to the force of centrifugation. How far they move depends on their mass, density, and shape—as well as on the mass, density, and fluidity of the solution in the tubes. If the bits of membrane have greater mass and density than the solution, they will move downward in the tube. If they have lesser mass or density, they will stay near the top of the tube.

Each cell component has a molecular composition that gives it a characteristic density. As the centrifuge spins at an appropriate speed, the components having the greatest density move toward the bottom of the tube. Other cell components take up positions in layers above them, according to their relative densities.

You perform centrifugation properly, so one layer in the solution consists only of membrane shreds. You draw off this layer carefully and examine it with a microscope to check whether the membrane sample is contaminated.

Afterward, by using standard procedures of chemical analysis, you discover that the plasma membrane of red blood cells consists of lipids and proteins.

In the past, two competing structural models guided research into membranes. According to the "protein coat" model, a cell membrane consists of a lipid bilayer that has a layer of proteins coating both of its outward-facing surfaces. According to the "fluid mosaic" model, proteins are largely embedded within the bilayer.

You decide to test the protein coat model. First you calculate how much protein it would take to coat the inner and outer surfaces of a known number of red blood cells. Then you separate a membrane sample into its lipid and protein fractions and measure the amount of each. In this way, you can compare the *observed* ratio of proteins to lipids against the ratio *predicted* on the basis of the model.

Such calculations were performed with membrane samples. There are indeed enough lipids for a bilayer arrangement. And there *might* have been enough proteins to cover both of its surfaces *if* all the membrane proteins had a stretched-out (not globular) structure. However, that possibility was discarded for two reasons. First, it became clear that proteins stretched out in a thin layer over lipids would be a most unfavorable arrangement, in terms of the energy cost to maintain it. Second, biochemical analysis showed that the proteins are globular. Such evidence does not favor the protein coat model.

Now you do an observational test of the fluid mosaic model. You prepare cell samples for microscopy by *freeze-fracturing* and *freeze-etching* them. By these research methods, you immerse the cell sample in an extremely cold fluid: liquid nitrogen. The cells freeze instantly. Then you strike them with an exceedingly fine blade, as in Figure 5.5. A suitably directed blow can fracture cells in such a way that one lipid layer of the plasma membrane will separate from

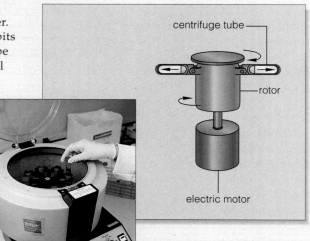

Figure 5.4 One example of a centrifuge. The diagram shows the arrangement of the tubes and the rotor of this high-speed spinning device.

centrifuge tube

rotor

electric motor

the other. Such preparations can be observed under extremely high magnifications.

If the protein coat model were correct, then you would observe a perfectly smooth, pure lipid layer. However, the freeze-fractured cell membranes you observe reveal many bumps and other irregularities in the separated layers of lipids. The bumps are proteins, incorporated directly in the bilayer—just as the fluid mosaic model predicts.

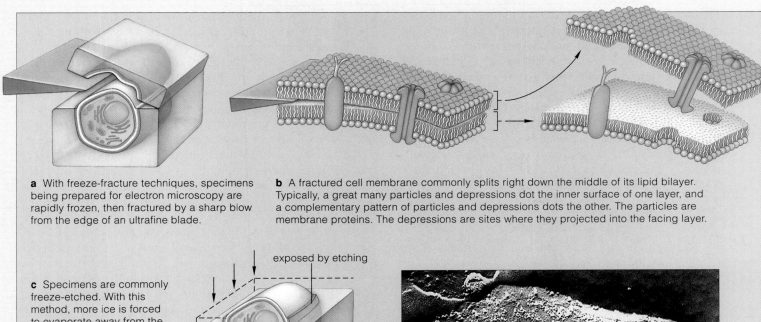

a With freeze-fracture techniques, specimens being prepared for electron microscopy are rapidly frozen, then fractured by a sharp blow from the edge of an ultrafine blade.

b A fractured cell membrane commonly splits right down the middle of its lipid bilayer. Typically, a great many particles and depressions dot the inner surface of one layer, and a complementary pattern of particles and depressions dots the other. The particles are membrane proteins. The depressions are sites where they projected into the facing layer.

c Specimens are commonly freeze-etched. With this method, more ice is forced to evaporate away from the fracture face, so that a portion of the outward-facing surface of the cell membrane is also exposed.

exposed by etching

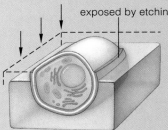

d By metal-shadowing methods, the fractured surface is coated with a layer of carbon and heavy metal, such as platinum. The coat is thin enough to replicate details of the exposed specimen surface. The metal replica, not the specimen, is used to make micrographs of the details.

(deposition of carbon and metal)

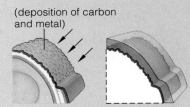

outer membrane layer exposed by etching   fracture edge

e

Figure 5.5 Freeze-fracturing and freeze-etching methods. The micrograph in (**e**) shows part of a replica of a red blood cell that was prepared by the techniques described in the text.

OBSERVING MEMBRANE FLUIDITY Consider now an experiment that yielded evidence of the fluid motion of membrane proteins. Researchers induced an isolated human cell and an isolated mouse cell to fuse. The plasma membranes of the cells from the two species merged to form a single, continuous membrane in the new, hybrid cell. Analysis showed that, in less than an hour, most of the membrane proteins were mixed together; they were free to drift laterally through the hybrid membrane:

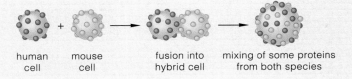

human cell   mouse cell   fusion into hybrid cell   mixing of some proteins from both species

Through observational tests, we also know that the fluidity of a membrane is subject to change, depending on environmental temperatures—and that at least some cells can respond to the danger.

For example, after some researchers decreased the temperature of a solution that contained live bacterial or yeast cells, the membranes of those cells started to stiffen. Such stiffening can disrupt the functioning of membrane proteins. However, the cells responded by rapidly synthesizing unsaturated fatty acids, and the infusion of those kinked lipids into the cell membranes countered the stiffening effect. However, past a certain low temperature, the membranes did solidify. The temperature at which this occurs varies, depending on the lipid composition of a given membrane.

HOW SUBSTANCES CROSS CELL MEMBRANES

Picture a cell membrane, with water bathing both sides of its bilayer. Plenty of substances are dissolved in the water, but the kinds and amounts are not the same on the two sides. The membrane itself helps establish and maintain those differences, which are essential for cell functioning. How does it accomplish this feat? Like every cell membrane, it shows **selective permeability**. *Because of its molecular structure, it allows some substances but not others to cross it in certain ways, at certain times.*

Carbon dioxide, molecular oxygen, and other small, nonpolar molecules can readily cross a cell membrane's lipid bilayer, and so can water molecules:

Although water molecules do show polarity, they might move through transient gaps that open up in the bilayer when hydrocarbon chains of the lipids flex and bend. Glucose and other large, polar molecules almost never move freely across the bilayer. Neither do ions:

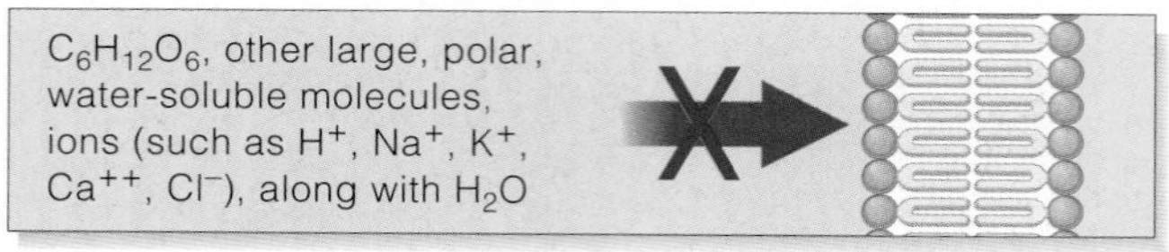

Such water-soluble substances must move across the bilayer through the interior of transport proteins. And the water they are dissolved in moves with them.

The question becomes this: Why does a substance move one way or another during any given interval? The answer starts with concentration gradients.

Concentration Gradients and Diffusion

"Concentration" refers to the number of molecules or ions of a substance in a specified region, as in volume of fluid or air. "Gradient" means the number in one region is not the same as it is in another. Thus, a **concentration gradient** is a difference in the number of molecules or ions of a given substance in two adjoining regions.

In the absence of other forces, a substance moves from a region where it is more concentrated to a region where it is less concentrated. *The energy inherent in its individual molecules, which keeps them in constant motion, drives the directional movement.* Although the molecules collide randomly and career back and forth, millions of times a second, the *net* movement is away from the place of greater concentration (and the most collisions).

Diffusion is the name for the net movement of like molecules or ions down a concentration gradient. It is a key factor in the movement of substances across cell membranes and through fluid parts of the cytoplasm.

In multicelled organisms, it figures in the movement of substances to and from cells, the fluids bathing them, and the environment. For example, when oxygen builds up in photosynthetic leaf cells, it diffuses across their plasma membrane, through water in leaves, and out to the air, where its concentration is lower. Carbon dioxide diffuses inward from the air, the region where it is most concentrated. Similarly, after you breathe in, oxygen in your lungs diffuses across a thin, moist lining, then into blood vessels, then into any tissue fluid having a lower oxygen concentration. From there, it diffuses into cells engaged in aerobic respiration. Carbon dioxide levels are highest in such cells but lower in tissue fluids, still lower in blood, and finally lowest in air in the lungs, so it diffuses in the opposite direction and is breathed out.

Like all other substances, oxygen or carbon dioxide diffuses in the direction set by its *own* concentration gradient. It makes no difference if other substances are dissolved in the same fluid. You can see the outcome of this tendency by squeezing a drop of dye into one part of a bowl of water. Dye molecules diffuse to the region where they are less concentrated, and water molecules move in the opposite direction, to the region where *they* are less concentrated (Figure 5.6).

Factors Influencing the Rate and Direction of Diffusion

Several factors influence the rate at which molecules or ions move down a concentration gradient. They include that gradient's steepness and molecular size, temperature, and electric or pressure gradients that may be present.

Diffusion is faster when the gradient is steep. In the region of greatest concentration, far more molecules are moving outward, compared to the number moving in. As the gradient decreases, the difference in the number of molecules moving one way or the other becomes less pronounced, and diffusion slows. When the gradient is gone, individual molecules are still in

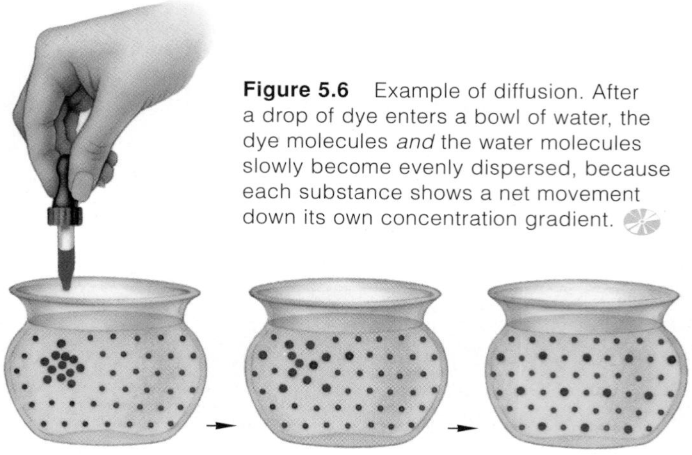

Figure 5.6 Example of diffusion. After a drop of dye enters a bowl of water, the dye molecules *and* the water molecules slowly become evenly dispersed, because each substance shows a net movement down its own concentration gradient.

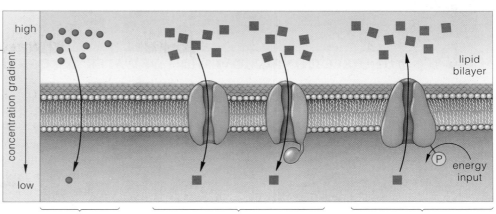

Figure 5.7 Overview of the major mechanisms by which solutes cross cell membranes. (Exocytosis and endocytosis proceed only at the plasma membrane.)

DIFFUSION ACROSS LIPID BILAYER
Lipid-soluble substances as well as water diffuse across.

PASSIVE TRANSPORT
Water-soluble substances, and water, diffuse through interior of transport proteins. No energy boost required. Also called facilitated diffusion.

ACTIVE TRANSPORT
Specific solutes are pumped through interior of transport proteins. Requires energy boost.

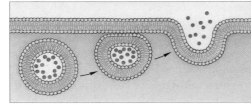

EXOCYTOSIS
Vesicle in cytoplasm moves to plasma membrane, fuses with it; contents released to the outside.

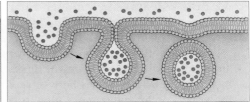

ENDOCYTOSIS
Vesicle forms from a patch of inward-sinking plasma membrane, enters cytoplasm.

motion. However, the total number moving one way or the other during a specified interval is approximately the same. When the net distribution of molecules becomes nearly uniform throughout two adjoining regions, we call this dynamic equilibrium.

Rates of diffusion are faster at higher temperatures. With more heat energy, molecules move faster and thus tend to collide more frequently. Diffusion rates also are influenced by molecular size. Small molecules tend to move faster than large ones.

In addition, the rate and direction of diffusion may be modified by an **electric gradient**: a difference in the electric charges of adjoining regions. For example, the fluid bathing each surface of a cell membrane has many kinds of dissolved ions, and each one contributes to its overall electric charge. Opposite charges attract, so the fluid with a more *negative* charge, overall, tends to exert the greatest pull on a *positively* charged substance—say, sodium ions. In the absence of other factors, the pull can enhance the flow of sodium across the membrane. Many processes, such as information flow in the nervous system, depend upon the combined force of electric and concentration gradients to attract substances across cell membranes.

Finally, as you will see shortly, the rate and direction of diffusion also may be modified by the presence of a **pressure gradient**, which is a difference in the pressure being exerted in two adjoining regions.

Mechanisms By Which Solutes Cross Cell Membranes

Figure 5.7 is an overview of the mechanisms that move solutes across all cell membranes. Through combinations of these mechanisms, cells and organelles are supplied with raw materials and are rid of wastes, at controlled rates. The mechanisms also help maintain the volume and pH of a cell or organelle within functional ranges.

Again, small nonpolar molecules (such as oxygen) and water diffuse across the lipid bilayer. By contrast, in **passive transport**, a polar substance diffuses down its concentration gradient through the interior of transport proteins spanning the bilayer. This directional movement is also called *facilitated* diffusion. In **active transport**, a solute also cross the membrane through the interior of transport proteins, but in this case the net movement is against the concentration gradient. This mechanism can only operate when transport proteins are activated, as by an energy boost from ATP.

With exocytosis or endocytosis, substances move in bulk across the plasma membrane. As you will read in Section 5.6, **exocytosis** involves fusion of a vesicle (that formed earlier in the cytoplasm) with the cell's plasma membrane. **Endocytosis** involves an inward sinking of a small patch of plasma membrane, which then seals back on itself to form a cytoplasmic vesicle.

Molecules or ions of a substance constantly collide because of their inherent energy of motion. The collisions result in diffusion, a net outward movement of a substance from one region into an adjoining region where it is less concentrated.

A concentration gradient is a form of energy that can drive the directional movement of a substance across membranes.

The steepness of the concentration gradient affects diffusion rates. So do temperature, molecular size, and the presence of electric or pressure gradients.

Nonpolar substances and water can diffuse across the lipid bilayer of cell membranes. Polar substances and water cross by passive transport (facilitated diffusion) and by active transport. Substances also move in bulk across the plasma membrane by exocytosis and endocytosis.

THE DIRECTIONAL MOVEMENT OF WATER ACROSS MEMBRANES

By far, more water diffuses across cell membranes than any other substance, so the key factors that influence its directional movement deserve some attention before we get into the details of transport mechanisms.

Water Movement by Osmosis

Turn on a faucet or watch a waterfall, and the moving water provides a demonstration of bulk flow. **Bulk flow** is the mass movement of one or more substances in response to pressure, gravity, or some other external force. It accounts for some movement of water through complex plants and animals. With each beat, your heart creates fluid pressure that drives a volume of blood, which is mainly water, through interconnected blood vessels. Sap runs inside conducting tissues that thread through maple trees, and this, too, is a case of bulk flow.

For cells, the diffusion of water from one region to another is a different movement. It occurs *only* when a membrane that intervenes between two regions allows the small, polar water molecules to cross but restricts the passage of ions and large polar molecules. **Osmosis** is the name for the diffusion of water in response to a water concentration gradient between two regions that are separated by a selectively permeable membrane.

Osmotic movement depends on the concentrations of solutes in the water on both sides of a membrane. *The side with more solute particles has a lower concentration of water.* To see why this is so, dissolve a small amount of glucose in water. That glucose solution has fewer water molecules than an equivalent volume of water, for each glucose molecule now occupies some of the space formerly occupied by a water molecule. Also, some water molecules are no longer free to contribute to the water concentration. They have become part of spheres of hydration around polar groups at the surface of each glucose molecule (Section 2.5).

It is mainly the *total number* of molecules or ions, not the type of solute, that dictates the concentration of water. Dissolve one mole of an amino acid or urea in 1 liter of water, and the water concentration decreases about as much as it did in the glucose solution. Add one mole of sodium chloride (NaCl) to 1 liter of water, and it dissociates into equal numbers of sodium ions and chloride ions. There are now two moles of solute particles—twice as many as in the glucose solution—so the water concentration has decreased proportionately.

Effects of Tonicity

Given that water molecules tend to move osmotically to a region where water is less concentrated, the direction of movement will be toward a region where solutes are more concentrated. Figure 5.8 illustrates this tendency.

Suppose you decide to make a simple observational test of this statement. You construct three sacs out of a membrane that water but not sucrose can cross, and you fill each with a 2M sucrose solution. (*M* stands for Molarity, the number of moles of a specified solute in 1 liter of fluid.) Then you immerse one of the sacs in 1 liter of distilled water (which has no solutes), one in a 10M sucrose solution, and the third in a 2M sucrose solution. In each case, the extent and direction of water movement are dictated by tonicity (Figure 5.9*a*).

Tonicity refers to the relative solute concentrations of two fluids. When two fluids on opposing sides of a membrane differ in solute concentration, the one having fewer solutes is called the **hypotonic solution**, and the one with more is the **hypertonic solution**. Water tends to diffuse from hypotonic to hypertonic fluids. **Isotonic solutions** have the same solute concentrations, so water shows no net osmotic movement from one to the other.

Normally, the fluid inside your cells and the tissue fluid bathing them are isotonic. If a tissue fluid became drastically hypotonic, so much water would diffuse into your cells that they would burst. If it became far too hypertonic, the outward diffusion of water would make the cells shrivel. Most cells have built-in mechanisms for adjusting to shifts in tonicity. Red blood cells do not; Figure 5.9*b* shows what happened to such cells during a demonstration of the effects of tonicity differences. That is why patients who are severely dehydrated are given infusions of a solution that is isotonic with blood. Such solutions move by bulk flow from a bottle mounted above the patient, through a flexible tube, and directly into the blood of an incised vein.

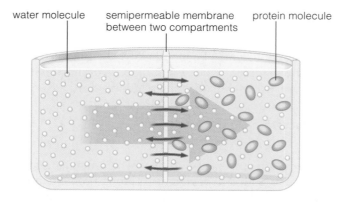

Figure 5.8 Effect of a solute concentration gradient on osmotic movement. Start with a container divided by a membrane that water but not proteins can cross. Pour water into the left half. Pour the same volume of a protein-rich solution into the right half. There, proteins occupy some of the space. And the polar groups of each protein are sites where spheres of hydration form and tie up some water molecules, which thereby cannot contribute to the concentration of free water. The net diffusion of water in this case is from left to right (*large blue arrow*).

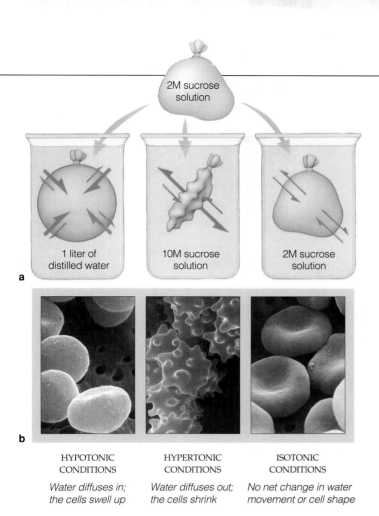

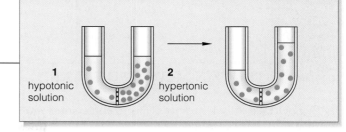

Figure 5.10 Increase in fluid volume as a result of osmosis. When the net diffusion across a membrane separating two compartments is equal, the fluid volume in compartment 2 is greater because the membrane is impermeable to solutes.

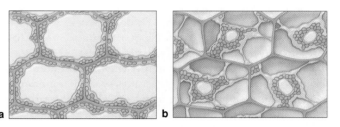

Figure 5.11 Plasmolysis, an osmotically induced loss of internal fluid pressure, in young plant cells. The cytoplasm and central vacuole shrink; the plasma membrane moves away from the wall.

HYPOTONIC CONDITIONS

Water diffuses in; the cells swell up

HYPERTONIC CONDITIONS

Water diffuses out; the cells shrink

ISOTONIC CONDITIONS

No net change in water movement or cell shape

Figure 5.9 Effect of tonicity on water movement, as described in the text. (**a**) The direction and relative amounts of water movement are indicated by arrow widths. The micrographs (**b**) correspond to the sketches. They show the kinds of shapes that human red blood cells will assume when placed in fluids of higher, lower, and equal solute concentrations. Normally, the solutions inside and outside of red blood cells are in balance. This type of cell does not have built-in mechanisms for adjusting to drastic changes in solute levels in its fluid surroundings.

Effects of Fluid Pressure

Animal cells generally can avoid bursting by engaging in the ongoing selective transport of solutes across the plasma membrane. Cells of plants and many protistans, fungi, and bacteria also avoid that unpleasant prospect with the help of pressure exerted on their cell walls.

Pressure differences as well as solute concentrations influence the osmotic movement of water. Take a look at Figure 5.10. It shows how water continues to diffuse from a hypotonic solution to a hypertonic one until its concentration becomes the same on both sides of the membrane between them. As you can see, the *volume* of the formerly hypertonic solution has increased (because its solutes can't diffuse out). Any volume of fluid exerts **hydrostatic pressure**, a force directed against a wall, membrane, or some other structure that encloses the fluid. The greater the solute concentration of the fluid, the greater will be the hydrostatic pressure it exerts.

As you know, living cells cannot increase in volume indefinitely (Section 4.1). At some point, the hydrostatic pressure that develops inside a cell counters the inward diffusion of water. That point is the **osmotic pressure**, the amount of force which prevents any further increase in the volume of a solution.

Consider young plant cells, with a pliable, primary wall. While they grow, water diffuses in and hydrostatic pressure increases against the wall. The wall expands, and the cell volume increases. The thin walls are strong enough for the cell's internal fluid pressure to develop to the point where it counterbalances water uptake.

Plant cells are more vulnerable to water loss, which can happen when soil dries or becomes too salty. Water stops diffusing in, the cells lose water, and internal fluid pressure drops. Such osmotically induced shrinkage of cytoplasm is called **plasmolysis** (Figure 5.11). Plants can adjust somewhat to the loss of pressure, as when they actively take up potassium ions against a concentration gradient by mechanisms outlined in the next section.

As you will read in Chapters 39 and 43, hydrostatic and osmotic pressure also influence the distribution of water in the blood, tissue fluid, and cells of animals.

Osmosis is the net diffusion of water between two solutions that differ in water concentration and that are separated by a selectively permeable membrane. The greater the number of molecules and ions dissolved in a solution, the lower its water concentration will be.

Water tends to move osmotically to regions of greater solute concentration (from hypotonic to hypertonic solutions). There is no net diffusion between isotonic solutions.

The fluid pressure that a solution exerts against a membrane or wall also influences the osmotic movement of water.

PROTEIN-MEDIATED TRANSPORT

Let's now take a look at how water-soluble substances diffuse into and out of cells or organelles. Whether by passive or active transport mechanisms, a great variety of transport proteins have roles in moving molecules or ions of such substances across cell membranes. These proteins span the lipid bilayer, and their interior is able to open on both sides of it (Figure 5.12).

To understand how these proteins work, you have to know they are not rigid blobs of atoms. When the protein interacts with a particular solute, it changes from one shape to another shape, then back again. The changes start when a solute becomes weakly bound at a specific site on the protein surface. Part of the protein closes in behind the bound solute and part opens up to the other side of the membrane. The solute dissociates (separates) from the site on that side. Think of the solute as hopping onto the transport protein on one side of the membrane, then hopping off on the other side.

Passive Transport

Transport proteins permit solutes to move both ways across a cell membrane. In cases of passive transport, the *net* direction of movement during a given interval depends on how many molecules or ions of the solute are making random contact with vacant binding sites in the interior of the proteins (Figure 5.13). The binding and transport simply proceed more often on the side of the membrane where the solute is more concentrated. Because there are more molecules around, the random encounters with the binding sites are more frequent than they are on the other side of the membrane

By itself, the passive two-way transport of a solute would continue until its concentrations became equal on both sides of the membrane. In most cases, however, other processes affect the outcome. For example, the bloodstream delivers glucose molecules to all of your body's tissues. Nearly all cells use this simple sugar as an energy source and as a building block for larger

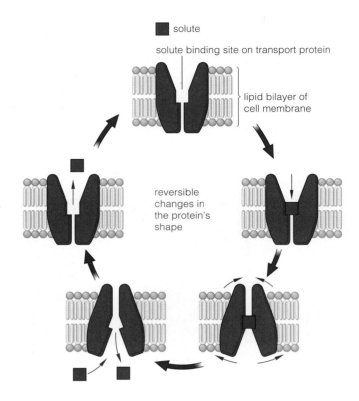

Figure 5.13 Passive transport across a cell membrane. A solute can move in both directions through transport proteins. In passive transport, however, the *net* movement will be down the solute's concentration gradient (from higher to lower concentration). This will continue until the concentrations become the same on both sides of the membrane.

compounds. When the glucose concentration in blood and tissues is high, cellular uptake is rapid. But as fast as some glucose molecules diffuse into a cell, others usually are leaving the cytoplasm and entering various metabolic reactions. Thus, when cells are rapidly using glucose, they are actually maintaining a concentration gradient that favors the uptake of *more* glucose.

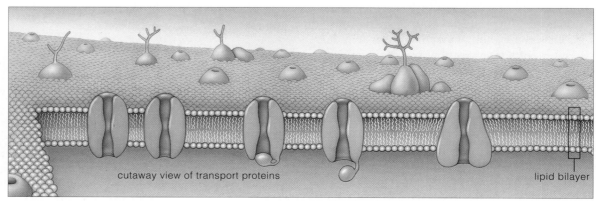

Figure 5.12 Sketch of some transport proteins that span the lipid bilayer of a plasma membrane. Transport proteins passively allow or actively assist ions as well as glucose and other large polar molecules to move through their interior, from one side of the membrane to the other.

cutaway view of transport proteins

lipid bilayer

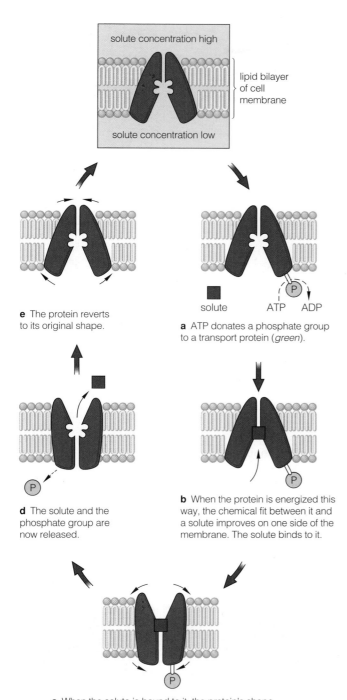

e The protein reverts to its original shape.

a ATP donates a phosphate group to a transport protein (*green*).

solute   ATP ADP

d The solute and the phosphate group are now released.

b When the protein is energized this way, the chemical fit between it and a solute improves on one side of the membrane. The solute binds to it.

c When the solute is bound to it, the protein's shape changes, and the bound solute ends up facing the opposite side of the membrane. There, the binding site reverts to the less attractive configuration.

solute concentration high

lipid bilayer of cell membrane

solute concentration low

Figure 5.14 Active transport across a cell membrane. As with passive transport, a solute can diffuse in both directions through the interior of the transport protein. However, with active transport, ATP transfers a phosphate group to the protein. The transfer sets in motion reversible changes in the protein's shape that result in a greater net movement of solute particles against the concentration gradient.

Active Transport

Transport proteins with roles in active transport across cell membranes move a solute against its concentration gradient. Unlike passive transport, this mechanism can continue until the solute becomes *more* concentrated on the side of the membrane to which it is being pumped.

Active transport does not proceed spontaneously. It requires an energy boost, most often from ATP. Briefly, ATP donates a phosphate group to a transport protein. When it does, the chemical fit between the binding site and the solute improves on one side of the membrane (Figure 5.14*a* and *b*). After a solute particle binds at the site, the protein's folded shape changes in such a way that the bound solute becomes exposed to fluid bathing the opposite side of the membrane (Figure 5.14*c*). Now the binding site reverts to its less attractive state and the solute is released. A less attractive site means fewer molecules or ions of the solute make the return trip. So the *net* movement is to the side of the membrane where the solute is more concentrated. By analogy, picture hordes of skiers rapidly hopping onto the chairs of a ski lift that moves them up a mountain. At the top, the chairs tilt in a way that is not very inviting for skiers who want to hop on quickly for a downhill ride. Thus, more skiers will be transported uphill than downhill.

One type of active transport system, the **calcium pump**, helps keep the calcium concentration in a cell at least a thousand times lower than outside. Another, the **sodium-potassium pump**, moves potassium ions (K^+) across the plasma membrane. Activation of this protein facilitates binding of a sodium ion (Na^+) on one side of a cell membrane. After the Na^+ makes the crossing, it is released. Its release facilitates the binding of K^+ at a different binding site on the protein, which reverts to its original shape after K^+ is delivered to the other side.

Through operation of such active transport systems, concentration and electric gradients are maintained across membranes. The gradients are vital to many cell activities and physiological processes, including muscle contraction and information flow in nervous systems. We will return to the mechanisms of active transport in later chapters. For now, keep these points in mind:

Transport proteins span the lipid bilayer of cell membranes. When they bind a solute on one side of a membrane, they undergo a reversible change in shape that shunts the solute through their interior, to the other side.

With passive transport, a solute simply diffuses through the protein; its net movement is down its concentration gradient.

With active transport, the net diffusion of a solute is uphill, against its concentration gradient. The transporting protein must be activated, as by ATP energy, to counter the energy inherent in the gradient.

EXOCYTOSIS AND ENDOCYTOSIS

Transport proteins can move ions and small molecules into or out of cells. However, when it comes to taking in or expelling large molecules or particles, cells rely on vesicles that form through exocytosis and endocytosis.

Transport To the Plasma Membrane

By exocytosis, a cytoplasmic vesicle moves to the cell surface, and the protein-studded lipid bilayer of its own membrane fuses with the plasma membrane. While this exocytic vesicle is losing its identity, its contents are released to the surroundings (Figure 5.15a).

Transport From the Plasma Membrane

By three pathways of endocytosis, a cell internalizes substances next to its surface. In all three cases, a small indentation forms at the plasma membrane, balloons inward, and pinches off. The resulting endocytic vesicle transports its contents or stores them in the cytoplasm (Figure 5.15b). By *receptor-mediated* endocytosis, the first pathway, membrane receptors chemically recognize and bind specific substances, such as lipoproteins, vitamins, iron, peptide hormones, growth factors, and antibodies. The receptors become concentrated in tiny depressions, or pits, in the plasma membrane (Figure 5.16). Each pit looks like a loosely woven basket on its cytoplasmic side. The basket consists of protein filaments (clathrin), interlocked in stable, geometric patterns. When the pit sinks into the cytoplasm, the basket closes back on itself and becomes the structural framework for the vesicle.

Cells are less finicky with *bulk-phase* endocytosis, the second pathway. An endocytic vesicle forms around a

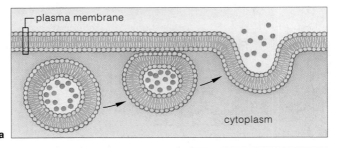

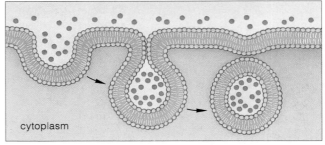

Figure 5.15 (**a**) Exocytosis. Substances are released outside the cell when an exocytic vesicle's membrane fuses with the plasma membrane. (**b**) Endocytosis. Substances are brought into the cell when a bit of plasma membrane balloons inward beneath them, self-seals, and so forms an endocytic vesicle.

small volume of extracellular fluid regardless of which substances happen to be dissolved in it. The bulk-phase pathway operates at a fairly constant rate in nearly all eukaryotic cells. By steadily sending patches of plasma membrane into the cytoplasm, the pathway compensates for membrane that steadily departs from the cytoplasm in the form of exocytic vesicles.

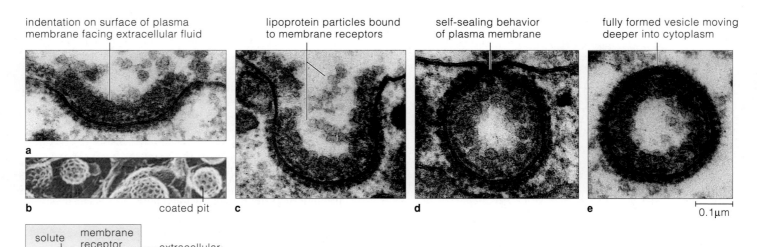

Figure 5.16 Example of receptor-mediated endocytosis, at the plasma membrane of an immature chicken egg. (**a**) This shallow indentation is a coated pit. (**b**) The cytoplasmic surface of each pit has a basketlike array of clathrin filaments. (**c**) Receptor proteins at the pit's outer surface preferentially bind lipoprotein particles. (**d**) The pit deepens and rounds out. (**e**) The formed endocytic vesicle encases lipoproteins that the cell will use or store.

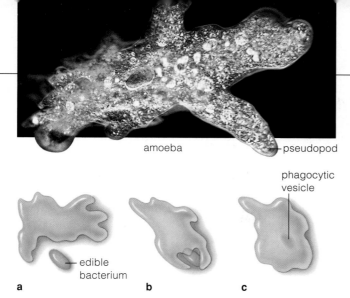

amoeba   pseudopod

phagocytic
vesicle

edible
bacterium

a   b   c

Figure 5.17 Phagocytosis in amoebas (including the *Amoeba proteus* cell in the photograph), certain white blood cells such as macrophages, and some other cells. Lobes of cytoplasm extend outward and surround the target. The plasma membrane of the extensions fuses together, thereby forming a phagocytic vesicle. Such vesicles move deeper into the cytoplasm and then fuse with lysosomes. Their contents are digested and, along with the vesicle's membrane components, are recycled elsewhere.

The third pathway, **phagocytosis**, is an active form of endocytosis by which the cell engulfs microorganisms, large edible particles, and cellular debris. (Phagocytosis literally means "cell eating.") Amoebas and some other protistans get food this way. In multicelled organisms, certain white blood cells engage in phagocytosis during responses that defend the body against harmful viruses, bacteria, and other threats to health.

As you read earlier in Section 4.9, a phagocytic cell shows amoeboid motion after a target binds to certain receptors that bristle from its plasma membrane. The binding sends signals into the cell. The signals trigger a directional assembly and crosslinking of microfilaments into a dynamic, ATP-requiring network just beneath the plasma membrane. The network contracts in ways that squeeze some cytoplasm toward the cell margins, thus forming the lobes called pseudopods (Figure 5.17). The pseudopods flow over the target and fuse at their tips. The result is a phagocytic vesicle, which sinks into the cytoplasm. There it fuses with lysosomes, the organelles of intracellular digestion in which trapped items are digested to fragments and smaller, reusable molecules.

Membrane Cycling

As long as a cell stays alive, exocytosis and endocytosis continually replace and withdraw patches of its plasma membrane. And they apparently do so at rates that can maintain the plasma membrane's total surface area.

As an example, consider the bursts of exocytosis by which neurons (nerve cells) release neurotransmitters, a type of signaling molecule that acts on neighboring cells. Two cell biologists, John Heuser and T. S. Reese,

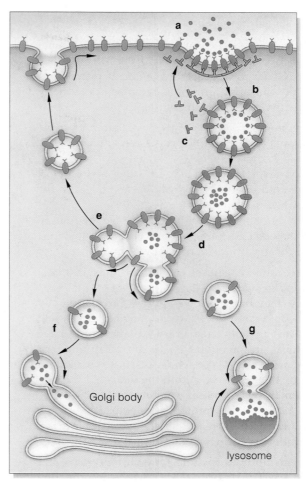

Golgi body

lysosome

a Molecules get concentrated inside coated pits of plasma membrane.

b Endocytic vesicles form from the pits.

c Vesicles lose molecules of clathrin, which return to the plasma membrane.

d Enclosed molecules are sorted and often released from receptors.

e Many sorted molecules are cycled back to the plasma membrane.

f, g Many other sorted molecules travel to Golgi bodies and stay there. Others are routed to spaces inside nuclear envelope and ER membranes, and still others to lysosomes.

Figure 5.18 The cycling of membrane lipids and proteins. This example starts with receptor-mediated endocytosis. The plasma membrane gives up small patches of itself to endocytic vesicles that form from coated pits. It gets membrane back from exocytic vesicles that budded from ER membranes and from Golgi bodies. The membrane that was initially appropriated by endocytic vesicles will cycle receptor proteins and lipids back to the plasma membrane.

documented a notably intense burst of endocytosis that immediately followed and counterbalanced an intense episode of exocytosis in neurons. Figure 5.18 shows some ways in which cells cycle their bits of membrane.

Whereas transport proteins in a cell membrane deal only with ions and small molecules, exocytosis and endocytosis move larger packets of material across the plasma membrane.

By exocytosis, a cytoplasmic vesicle fuses with the plasma membrane, so that its contents are released outside the cell. By endocytosis, a small patch of the plasma membrane sinks inward and seals back on itself, forming a vesicle inside the cytoplasm. Membrane receptors often mediate this process.

Membrane Structure and Function

1. The plasma membrane is a structural and functional boundary between the cytoplasm and the surroundings of all cells. In eukaryotic cells only, organelle membranes subdivide the fluid portion of the cytoplasm into many functionally diverse compartments.

2. A cell membrane consists of two water-impermeable layers of lipids (phospholipids especially) and proteins associated with those layers.

   a. In the bilayer, fatty acid tails and other hydrophobic parts of the lipid molecules are sandwiched between the hydrophilic heads.

   b. Many different kinds of proteins are embedded in the lipid bilayer or positioned at one of its two surfaces. The proteins carry out most membrane functions.

3. These are the key features of the fluid mosaic model of membrane structure:

   a. A cell membrane shows fluid behavior, mainly because its lipid components twist, move laterally, and flex hydrocarbon tails. Also, some of the lipids have ring structures, and many have kinked (unsaturated) or short fatty acid tails, all of which disrupt what might otherwise be tight packing within the bilayer.

   b. A membrane is a mosaic, or composite, of diverse lipids and proteins. The proteins are embedded in the bilayer and are at its surface. Its two layers differ in the number, kind, and arrangement of lipids and proteins.

4. All cell membranes include transport proteins and proteins that structurally reinforce the membrane. The plasma membrane also has proteins that serve in signal reception, cell recognition, and adhesion. Differences in the number and types of proteins among cells affect metabolism, cell volume, pH, and responsiveness to substances that make contact with the membrane.

   a. Transport proteins allow water-soluble substances to pass through their interior, which opens to both sides of the membrane.

   b. Receptor proteins bind extracellular substances that trigger alterations in metabolic activities.

   c. Recognition proteins are molecular fingerprints at the surface of each cell type. Infection-fighting white blood cells chemically recognize self proteins (of the body's own cells) and nonself proteins of foreign cells.

   d. In multicelled organisms, adhesion proteins help cells adhere to one another and form cell junctions.

Movement of Substances Into and Out of Cells

1. Molecules or ions of a substance tend to move from a region of higher to lower concentration. Movement in response to a concentration gradient is called diffusion.

   a. Diffusion rates are influenced by the steepness of the concentration gradient, temperature, and molecular size, as well as by gradients in electrical charge and pressure that may occur between two regions.

   b. Cells have built-in mechanisms that work with or against gradients to move solutes across membranes.

2. Osmosis is the technical name for the diffusion of water across any selectively permeable membrane in response to a concentration gradient.

   a. The direction of osmotic movement between two solutions (such as cytoplasmic fluid and tissue fluid) can be predicted by comparing their tonicity. Tonicity is a measure of the solute concentration of one solution *relative to* the solute concentration of another solution.

   b. Water tends to move from a hypotonic solution (with the lower solute concentration) to a hypertonic solution (with the higher solute concentration).

   c. If fluids are isotonic (equal solute concentrations), water will show no net movement in either direction.

3. Oxygen, carbon dioxide, and other small nonpolar molecules diffuse across a membrane's lipid bilayer. Ions, glucose, and other large, polar molecules cross it with the passive or active help of transport proteins. Water can move through the bilayer and the proteins.

4. Transport proteins bind specific solutes on one side of a membrane, and they shunt solutes to the other side through reversible changes in their shape.

   a. By passive transport, the protein allows a solute simply to diffuse through its interior, in the direction of its concentration gradient. The protein's shape changes without an input of energy.

   b. By active transport, the protein pumps a solute across the membrane against its concentration gradient. The changes in protein shape that trigger the process require an energy boost, as from ATP.

5. By exocytosis, vesicles in the cytoplasm move to the plasma membrane. When their membrane fuses with it, their contents are automatically released to the outside.

6. By endocytosis, a small patch of plasma membrane sinks into the cytoplasm and seals back on itself to form a vesicle. The three pathways are receptor-mediated endocytosis (requires recognition of specific solutes), bulk-phase endocytosis (indiscriminate uptake of some extracellular fluid), and phagocytosis (active uptake of large particles, cell parts, or whole cells).

Review Questions

1. Describe the fluid mosaic model of cell membranes. What imparts fluidity to the membrane? What makes it a mosaic? *5.1*

2. Structurally, what do all cell membranes have in common? In what ways do the membranes of different cell types vary? *5.1*

3. Distinguish among transport proteins, receptor proteins, recognition proteins, and adhesion proteins. *5.1*

4. Define diffusion. Does diffusion occur in response to a solute concentration gradient, an electric gradient, a pressure gradient, or some combination of these? *5.3*

5. Define osmosis. How do the solute concentrations of two solutions on either side of a membrane influence the osmotic movement of water down the water concentration gradient? *5.4*

6. Define hypertonic, hypotonic, and isotonic solutions. Does each term refer to a property inherent in a given type of solution? Or are they used only when comparing one solution to another? *5.4*

7. If *all* transport proteins shunt substances across cell membranes by changing shape, then how do the passive transporters differ from the active transporters? *5.5*

8. Define exocytosis and endocytosis. Describe the main features of the three pathways of endocytosis. *5.3, 5.6*

Self-Quiz *(Answers in Appendix IV)*

1. Cell membranes consist mainly of a _____ .
   a. carbohydrate bilayer and proteins
   b. protein bilayer and phospholipids
   c. lipid bilayer and proteins

2. In a lipid bilayer, _____ of lipid molecules are sandwiched between _____ .
   a. hydrophilic tails; hydrophobic heads
   b. hydrophilic heads; hydrophilic tails
   c. hydrophobic tails; hydrophilic heads
   d. hydrophobic heads; hydrophilic tails

3. Most membrane functions are carried out by _____ .
   a. proteins   c. nucleic acids
   b. phospholipids   d. hormones

4. All cell membranes of multicelled organisms incorporate _____ .
   a. transport proteins   d. recognition proteins
   b. receptor proteins   e. all of the above
   c. adhesion proteins

5. Immerse a living cell in a hypotonic solution, and water will tend to _____ .
   a. move into the cell   c. show no net movement
   b. move out of the cell   d. move in by endocytosis

6. A _____ is a device that spins test tubes and separates cell components according to their relative densities.

7. _____ can readily diffuse across a lipid bilayer.
   a. Glucose   c. Carbon dioxide
   b. Oxygen   d. b and c

8. Sodium ions cross a membrane through transport proteins that receive an energy boost. This is an example of _____ .
   a. passive transport   c. facilitated diffusion
   b. active transport   d. a and c

Critical Thinking

1. Water moves osmotically into *Paramecium*, a single-celled protistan of aquatic habitats. If unchecked, the influx would bloat the cell and rupture its plasma membrane. An energy-requiring mechanism involving contractile vacuoles expels the excess (Figure 5.19). Water enters tubelike extensions of this organelle and collects in a central space in the vacuole. When full, the vacuole contracts and squirts excess water through a pore that opens to the surroundings. Are the fluid surroundings hypotonic, hypertonic, or isotonic relative to *Paramecium*'s cytoplasm?

2. The bacterium *Vibrio cholerae* causes the disease cholera. Infected people have severe diarrhea and may lose up to twenty liters of fluid in a day. The bacterium enters the body if someone drinks contaminated water, then it adheres to the intestinal lining. It secretes a metabolic product that is toxic to cells of the lining,

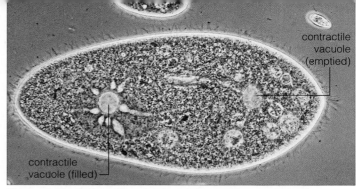

Figure 5.19 Contractile vacuoles of *Paramecium*, a protistan.

and they start secreting chloride ions (Cl⁻). Sodium ions (Na⁺) follow the chloride ions into the fluid in the intestines. Explain how this sequence of events causes the massive fluid loss.

3. Many cultivated fields in California require heavy irrigation. Over the years, most of the water has evaporated from the soil, leaving behind all of the irrigation water's solutes. What kinds of problems might the altered soil conditions cause for plants?

4. Certain species of bacteria thrive in environments where the temperatures approach the boiling point of water—for example, in the steam-venting fissures of volcanoes and in hot springs of Yellowstone National Park. Assume that the lipid bilayer of the bacterial cell membranes consists mainly of phospholipids. What features might the fatty acid tails of the phospholipids have that help stabilize the membranes at such extreme temperatures?

5. In hospitals, solutions of glucose having a concentration of 0.3M can be infused directly into the bloodstream of patients. The same is true of solutions of sodium chloride (NaCl), but these have a concentration of 0.15M. Only isotonic solutions can be infused safely into the blood. If that is so, then how can you explain the difference in the Molarity of the two solutions?

Selected Key Terms

active transport *5.3*	isotonic solution *5.4*
adhesion protein *5.1*	lipid bilayer *5.1*
bulk flow *5.4*	osmosis *5.4*
calcium pump *5.5*	osmotic pressure *5.4*
centrifuge *5.2*	passive transport *5.3*
concentration gradient *5.3*	phagocytosis *5.6*
diffusion *5.3*	phospholipid *5.1*
electric gradient *5.3*	plasmolysis *5.4*
endocytosis *5.3*	pressure gradient *5.3*
exocytosis *5.3*	receptor protein *5.1*
fluid mosaic model *5.1*	recognition protein *5.1*
hydrostatic pressure *5.4*	selective permeability *5.3*
hypertonic solution *5.4*	sodium-potassium pump *5.5*
hypotonic solution *5.4*	transport protein *5.1*

Readings

Bretscher, M. October 1985. "Molecules of the Cell Membrane." *Scientific American* 253(4): 100–108.

Dautry-Varsat, A., and H. Lodish. May 1984. "How Receptors Bring Proteins and Particles Into Cells." *Scientific American* 250(5): 52–58. Describes receptor-mediated endocytosis.

Singer, S., and G. Nicolson. 1972. "The Fluid Mosaic Model of the Structure of Cell Membranes." *Science* 175: 720–731.

Web Site See *http://www.wadsworth.com/biology* for practice quiz questions, hypercontents, BioUpdates, and critical thinking. The Wadsworth Biology Resource Center provides a wealth of information fully organized and integrated by chapter.

6

GROUND RULES OF METABOLISM

Growing Old With Molecular Mayhem

Somewhere in those slender strands of DNA in your cells are snippets of instructions for constructing two wonderful proteins. The proteins are called superoxide dismutase and catalase (Figure 6.1a). Both are enzymes, a type of molecule that makes metabolic reactions run much, much faster than they would spontaneously, on their own. And both kinds of enzymes help keep you from growing old before your time.

The two enzymes help your cells clean house, so to speak. Together, they produce and then disassemble hydrogen peroxide (H_2O_2), a normal but potentially toxic by-product of certain oxygen-requiring reactions. Oxygen (O_2) is supposed to pick up electrons from the reactions. Sometimes it picks up only one—which is not enough to complete the reaction but is enough to give the oxygen a negative charge (O_2^-).

Like other unbound, molecular fragments with the wrong number of electrons, O_2^- is a **free radical**. Free radicals are *so* reactive, they even attach to molecules that usually will not take part in just any reaction— molecules such as DNA. An attack by free radicals can disrupt DNA's structure and destroy its function.

Enter superoxide dismutase. Under its chemical prodding, two of the rogue oxygen molecules combine with hydrogen ions. H_2O_2 and O_2 are the outcome.

Enter catalase. Under *its* prodding, two molecules of hydrogen peroxide react and split into ordinary water and ordinary oxygen: $2H_2O_2 \longrightarrow 2H_2O + O_2$.

As people age, their capacity to produce functional proteins, including enzymes, begins to falter. Among those enzymes are superoxide dismutase and catalase.

Figure 6.1 (**a**) Adult owner of skin with a spattering of age spots—visible evidence of free radicals on the loose. At one time he, like the boy shown in (**b**), had a good supply of smoothly functioning catalase (**c**) and (**d**) superoxide dismutase molecules. Both of these enzymes help keep free radicals inside the body in check.

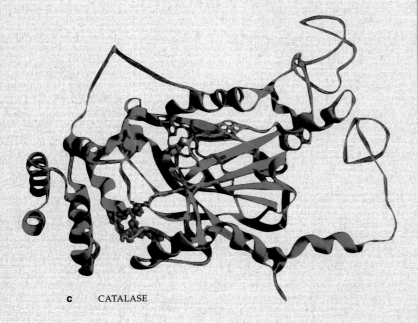

c CATALASE

d SUPEROXIDE DISMUTASE

Cells synthesize copies of the two enzymes in ever diminishing numbers, in crippled form, or both. When that happens, free radicals and hydrogen peroxide accumulate. Like loose cannons, they career through cells as tiny blasts at the structural integrity of proteins, DNA, membranes, and other vital components.

Cells under attack suffer or die outright. Those brown "age spots" you may have noticed on an older person's skin are evidence of assaults by free radicals. The irregular dark spots on the skin of the adult male in Figure 6.1*a* are examples. Each age spot is a mass of brownish-black pigment molecules that build up in cells when free radicals take over—all for the want of two enzymes.

With this chapter, we will start to examine the kinds of activities that keep all cells alive and functioning smoothly. At times the topics may seem remote from the world of your interests. But they help define who *you* are and who you will become, age spots and all.

KEY CONCEPTS

1. Cells engage in metabolism—that is, they use energy to build, stockpile, break apart, and eliminate substances in ways that help them survive and reproduce.

2. With each metabolic reaction, energy escapes into the environment. To stay alive, cells must balance their energy losses with energy gains. Yet they cannot create energy from scratch. They can only draw upon existing sources, such as light energy from the sun and chemical energy that has become tucked away in glucose and other substances.

3. Compared to their environment, cells maintain greater or lesser amounts of certain substances, as required for metabolism. They do so even though the molecules or ions of any substance have a natural tendency to diffuse into regions where they are less concentrated. Membrane transport proteins work with or against this tendency.

4. Metabolic pathways maintain, increase, or decrease the relative amounts of various substances in cells. Typically, the pathways couple reactions that release usable energy from substances to other reactions that require energy.

5. Chemical reactions proceed far too slowly on their own to sustain life. In living organisms, the action of specific enzymes greatly increases the rate of specific reactions.

6. ATP transports usable energy, in chemical form, from one reaction site to another. When it donates a phosphate group to glucose or some other substance, the substance becomes activated—that is, primed for chemical change. Metabolic pathways depend on such phosphate-group transfers.

Find a microscope and watch a living cell, suspended in a water droplet. The image might jar you: something that small pulsates with movement. Even as you watch, the cell takes in energy-rich solutes, builds membranes, stores things, replenishes enzymes, and checks out its DNA. It is alive; it is growing; it may divide in two. Multiply such activities by *trillions* of cells and you get an idea of what goes on in your own body, even when you do nothing more than sit quietly and watch a cell!

These dynamic activities are signs of **metabolism**— of a cell's capacity to acquire energy and use it to build, break apart, store, and release substances in controlled ways. Metabolism is the means by which cells survive and reproduce, and it all begins with energy.

Defining Energy

If you have ever watched a house cat stalking a mouse, you know it can "freeze" its position to avoid detection before springing at its unsuspecting prey. Like anything else in the whole universe that is stationary, the cat has a store of **potential energy**—a capacity to do work, simply

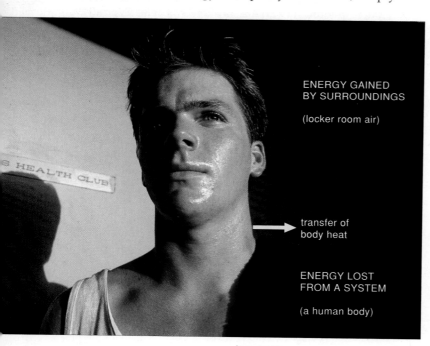

ENERGY GAINED
BY SURROUNDINGS

(locker room air)

transfer of
body heat

ENERGY LOST
FROM A SYSTEM

(a human body)

net energy change = 0

Figure 6.2 Example of how the total energy content of any system *together with its surroundings* remains constant.

"System" means all matter in a specific region, such as a human body, a plant, a DNA molecule, or a galaxy. "Surroundings" can be a small region in contact with the system or as vast as the entire universe. The system shown (a human male) is giving off heat to the surroundings (a locker room) by evaporative water loss from sweat. What one region loses, the other region gains, so the total energy content of both does not change.

owing to its position in space and the arrangement of its parts. When a cat springs, some of its potential energy is transformed into **kinetic energy**, the energy of motion. Energy on the move can do work by imparting motion to other things. In the cat's skeletal muscle cells, ATP gave up some potential energy to molecules of contractile units and set them in motion, which resulted in muscle movements. The transfer also resulted in the release of another form of kinetic energy—**heat**, or *thermal* energy.

The potential energy of molecules has its own name: **chemical energy**. It is measurable, as in kilocalories. A **kilocalorie** is the same thing as 1,000 calories, which is the amount of energy it takes to heat 1,000 grams of water from 14.5°C to 15.5°C at standard pressure.

How Much Energy Is Available?

Like single cells, you can't create energy from scratch; you must get it from someplace else. Why? According to the **first law of thermodynamics**, the total amount of energy in the universe remains constant. That is, more energy cannot be created, and existing energy cannot be destroyed. It can only be converted from one form to some other form.

Think about what the law means. The universe has only so much energy, distributed in a variety of forms. One form can be converted to another, as when corn plants absorb energy from the sun and convert it to the chemical energy of starch. After you eat corn, your cells extract energy from starch and convert it to other forms, such as kinetic energy for your movements. During each metabolic conversion, a bit of energy escapes to the surroundings, as heat. Even when you are "doing nothing," your body gives off about as much heat as a 100-watt light bulb because of conversions proceeding in your cells. The energy being released is transferred to atoms and molecules that make up the air and so "heats up" the surroundings (Figure 6.2). There, kinetic energy increases the number of ongoing, random collisions among molecules. And with each collision, a bit more energy is released as heat. However, none of the energy ever vanishes.

The One-Way Flow of Energy

Most of the energy available for conversions in a cell resides in covalent bonds. Glucose, glycogen, starches, fatty acids, and other organic compounds have complex arrangements of many of these bonds and are said to have a high energy content. When the compounds enter metabolic reactions, certain bonds break or become rearranged. During the molecular commotion, some heat energy is lost to the surroundings. In general, cells cannot recapture energy lost as heat.

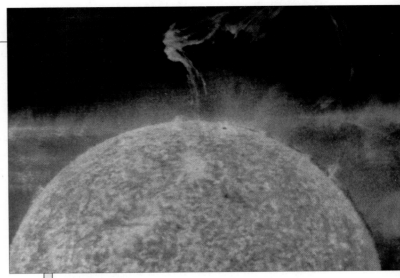

Figure 6.3 An example of the one-way flow of energy into the world of life that compensates for the one-way flow of energy out of it. The sun continuously loses energy, much of it in the form of wavelengths of light (Section 7.2). Living cells intercept some of the energy and convert it to useful forms of energy, stored in bonds of organic compounds. Each time a metabolic reaction proceeds in cells, stored energy is released—and some inevitably is lost to the surroundings, mostly as heat.

In the lower photograph, *green* "dots" are water-dwelling, photosynthetic cells (*Volvox*) living in tiny, spherical colonies. *Orange* dots are cells set aside for reproduction. They form new colonies inside the parent sphere.

ENERGY LOST
One-way flow of energy away from the sun

ENERGY GAINED
One-way flow of energy from the sun into organisms

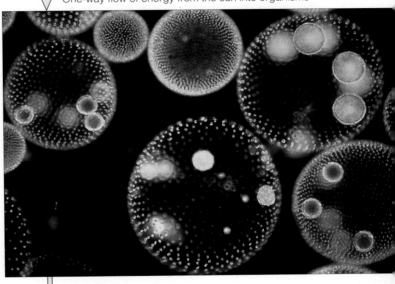

ENERGY LOST
One-way flow of energy away from organisms
to the surroundings

Think of what happens when your cells break all of the covalent bonds in glucose that they can until, after many steps, six molecules of carbon dioxide and six of water remain. They do so to release usable energy, some of which is conserved in ATP. Compared to glucose, the leftovers have more stable arrangements of atoms, but the chemical energy in all of their bonds is much less than the total chemical energy of glucose. Why? *Some energy was lost at each step leading to their formation*. Said another way, as a source of energy that cells can use, carbon dioxide is lower in quality than glucose.

What about the small amount of heat energy that was transferred to the surroundings when the carbon dioxide formed? It is very low quality; it does not lend itself to conversions in cells and so cannot be used to do work.

Bad news for cells of the remote future: The amount of "low-quality" energy in the universe is increasing. Because no energy conversion can ever be 100 percent efficient, the total amount of energy in the universe is spontaneously flowing from forms of higher to lower quality. That, basically, is the point to remember about the **second law of thermodynamics**.

Without energy inputs to maintain it, any organized system tends to get disorganized over time. **Entropy** is a measure of the degree of a system's disorder. Think of the Egyptian pyramids—originally organized, presently crumbling, and many thousands of years from now, dust. The ultimate destination of those pyramids and everything else in the universe is a state of maximum entropy. Billions of years from now, all of the energy available for conversions will be dissipated.

Can it be that life is one glorious pocket of resistance to the depressing flow toward maximum entropy? After all, every time a new organism grows, new bonds form and hold atoms together in precise arrays, so molecules become more organized and have a far richer store of energy, not poorer! Yet a simple example will show that the second law does indeed apply to life on Earth.

The primary energy source for life on Earth is the sun, which has been losing energy since it first formed. Plants capture sunlight energy, convert it in various ways, then lose energy to other organisms that feed, directly or indirectly, on plants. At each energy transfer

along the way, some energy is lost, usually as heat that joins the universal pool. *Overall, energy still flows in one direction*. The world of life maintains a high degree of organization only because it is being resupplied with energy that is being lost from someplace else (Figure 6.3).

The amount of energy in the universe remains constant. Energy can undergo conversions from one form to another, but it cannot be created out of nothing or destroyed.

The total amount of energy in the universe is spontaneously flowing from forms of higher to lower quality.

A steady flow of sunlight energy into the interconnected web of life compensates for the steady flow of energy leaving it.

Cells never stop building and tearing down molecules and moving them in specific directions until they die. What does the incessant juggling act accomplish? Think it through. Nearly all cells are microscopically small, they can take up and use only so many molecules at a given time, and they have only so much internal space. If they produce more of a substance than they are able to use, store, or secrete, they have big problems.

Consider *phenylketonuria*, or PKU. People affected by this heritable disorder produce a defective enzyme that prevents cells from using phenylalanine, which is one of the amino acids. When the amino acid accumulates, the excess enters reactions that produce phenylketones. An accumulation of *these* molecules damages the brain in less than a few months. In most developed countries, routine screening programs identify affected newborns, who can grow up symptom-free if they are placed on a phenylalanine-restricted diet.

Which Way Will a Reaction Run?

From the preceding chapter, you have an idea that cells control their internal concentrations of substances with respect to the surroundings, and that eukaryotic cells further control the concentrations of substances on both sides of organelle membranes. At any time, thousands of concentration gradients across cell membranes are helping to drive specific molecules and ions in specific directions. Even on their own, molecules or ions are in constant random motion, which puts them on collision courses. But the more concentrated they are, the more often they collide. Energy associated with the collisions might be enough to cause a **chemical reaction**—that is, to make a molecule combine with something else, split into smaller parts, or change its shape.

Nearly all chemical reactions in cells are reversible. In other words, they might start out in the "forward" direction, from starting substances to products. But they also can run in "reverse," with the products being converted back to starting substances. When you see a chemical equation with opposing arrows, this signifies the reaction is reversible. Each arrow means *yields*:

$$A + B \rightleftharpoons C$$

STARTING SUBSTANCES   PRODUCT

Which way such a reaction runs depends partly on the ratio of reactant to product. When the concentration of reactant molecules is high, this is an energetically favorable state, so the reaction runs spontaneously and strongly in the forward direction. When the product concentration is high enough, more molecules or ions of the product are available to revert spontaneously to reactants.

Any reversible reaction tends to run spontaneously toward **chemical equilibrium**, the time at which it will be running at about the same pace in both directions (Figure 6.4). Almost always, the *amounts* of the reactant and product molecules are not the same at that time. Picture a party with just as many people drifting in as drifting out of two rooms. The number in each room stays the same overall—say, thirty in one and ten in the other—even as the mix of people in each room changes.

Each type of reaction has a certain ratio of reactant to product molecules at chemical equilibrium. Consider the conversion of glucose-1-phosphate into glucose-6-phosphate (Figure 6.5). Both are glucose molecules with a phosphate group attached to one of their six carbon

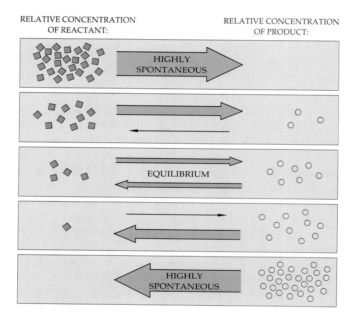

RELATIVE CONCENTRATION OF REACTANT: RELATIVE CONCENTRATION OF PRODUCT:

Figure 6.4 Chemical equilibrium. When the concentration of reactant molecules is high, a reaction runs most strongly in the forward direction (to products). When the concentration of product molecules is high, it runs most strongly in reverse. At equilibrium, the rates of the forward and reverse reactions are the same.

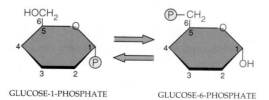

GLUCOSE-1-PHOSPHATE   GLUCOSE-6-PHOSPHATE

Figure 6.5 A reversible reaction. Glucose is primed to enter reactions when a phosphate group becomes attached to it. With a high concentration of glucose-1-phosphate, the reaction tends to run in the forward direction. With a high concentration of glucose-6-phosphate, it runs in reverse. (The 1 and 6 of these names simply identify which particular carbon atom of the glucose ring has a phosphate group attached to it.)

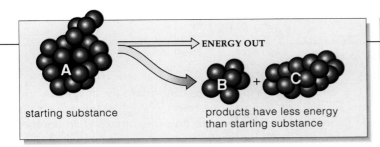

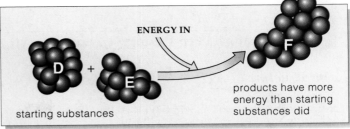

ENERGY OUT

starting substance

products have less energy
than starting substance

ENERGY IN

starting substances

products have more
energy than starting
substances did

a In exergonic reactions, starting substances have more energy than the products. In some reactions, including some steps of aerobic respiration, usable energy is released. (A bit of energy also is lost, as heat.)

b During endergonic reactions, starting substances have less energy than the products. Energy inputs drive such reactions, as when energy from the sun is harnessed to drive photosynthesis.

atoms. Depending on the ratio of reactants to products at the outset, the reaction runs in the forward or reverse direction. In time the forward and reverse reactions will proceed at the same rate, but only when there are about nineteen glucose-6-phosphate molecules for every one of glucose-1-phosphate. For this particular reaction, the ratio at equilibrium is 19:1.

No Vanishing Atoms at the End of the Run

When metabolic reactions run in either the forward or reverse direction, they rearrange atoms, but they never destroy them. By the **law of conservation of mass**, the total mass of all substances entering a reaction equals the total mass of all the products. When you study any chemical equation, count up the individual atoms of each reactant and product molecule. There should be as many atoms of each element to the right of the arrow as there are to the left, even though they are combined in different forms. When you write out equations for metabolic reactions, they must balance this way.

Energy Inputs Coupled With Outputs

Many metabolic reactions can release usable energy, as when cells degrade glucose to carbon dioxide and water. The products have more stable structures; it takes more energy to break them apart. Their total bond energies are lower than they were in glucose, however, for the released energy is channeled elsewhere or lost as heat.

As Figure 6.6a suggests, an **exergonic reaction** is one that ends with a net *loss* in energy. (*Exergonic* means "energy outward.") Such reactions proceed continually in living cells. In a well-fed adult human, for example, the cumulative daily breakdown of glucose and other organic compounds by exergonic reactions releases an average of 1,200 to 2,800 kilocalories of usable energy.

An **endergonic reaction** ends with a net *increase* in energy. (*Endergonic* means "energy in.") Such reactions are energetically unfavorable. They tend not to run toward equilibrium without prodding, as with an input of energy from an outside source (Figure 6.6b).

Think of an exergonic reaction as a downhill run, with equilibrium being the bottom of the hill. Getting it

c Cells conserve energy by coupling exergonic with endergonic reactions. For instance, some usable energy released by glucose breakdown is sent elsewhere, as to the energy-requiring steps of biosynthesis.

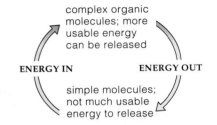

complex organic molecules; more usable energy can be released

ENERGY IN   ENERGY OUT

simple molecules; not much usable energy to release

Figure 6.6 Energy changes during metabolic reactions.

to run in the reverse direction is an uphill battle. For example, that is what happens when photosynthetic cells assemble glucose molecules from carbon dioxide and water. The synthesis steps are a steep climb, and they depend on inputs of energy from the sun.

Are steep energy hills worth the battle? Absolutely. The farther endergonic reactions can be made to run uphill, the more potential energy can be stored in the products. The trick is to drive these energy-requiring reactions with dependable sources of energy.

In cells, usable energy that is released during certain reactions becomes channeled into the formation of ATP. From earlier chapters, you are already familiar with the idea that this molecule delivers energy to thousands of reaction sites in cells. Now you know its deliveries can make reactions run in directions that are energetically unfavorable, that would not happen on their own. This is how cells build starch, fatty acids, and other "energy-rich" molecules from glucose, which they build from still smaller molecules (carbon dioxide and water) of lower energy content. With this thought in mind, turn now to the remarkable role of ATP in metabolism.

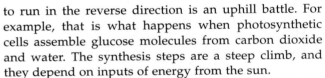

A cell can simultaneously increase, decrease, and maintain the concentrations of thousands of different substances by coordinating thousands of metabolic reactions.

Certain metabolic reactions release energy when they run toward equilibrium; others require energy inputs that drive them away from equilibrium. The molecules participating in the reaction show a net loss or net gain in energy, but the total number of atoms does not change.

The coupling of energy-releasing reactions with energy-requiring reactions, as by ATP, is central to metabolism.

The Structure of ATP

Recall, from Section 3.8, that **ATP** is the abbreviation for adenosine triphosphate, which is one of the small organic compounds known as nucleotides. An ATP molecule is composed of a five-carbon sugar (ribose) to which adenine (a nucleotide base) and a string of three phosphate groups are attached, as in Figure 6.7a and b. As you will see, the triphosphate tail of the molecule is where the action is, so to speak.

Phosphate-Group Transfers

In itself, a covalent bond is a stable interaction between atoms, and this is the type of bond that holds all the component parts of an ATP molecule together. Even so, not all bonds within the molecule are equally stable. Many hundreds of

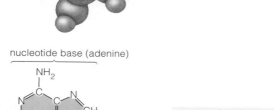

a

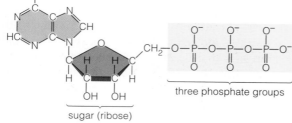

b

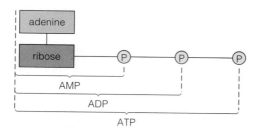

c

Figure 6.7 ATP—adenosine triphosphate, the main energy carrier in all cells. (**a**) Three-dimensional model showing ATP's component atoms. (**b**) Structural formula for ATP. (**c**) Adenosine diphosphate (ADP) forms when ATP gives up one phosphate group to another molecule. Two phosphate-group transfers leave adenosine monophosphate (AMP).

different enzymes can readily cleave the covalent bond that joins the two outermost phosphate groups of the molecule's triphosphate tail, then attach it to another substance.

The transfer of a phosphate group to a molecule of any sort is known as **phosphorylation**. A great deal of usable energy is transferred during this event. The energy is sufficient to activate hundreds of different molecules and drive hundreds of cellular activities. For example, phosphorylation is the energy transfer that drives the synthesis and breakdown of molecules, the active transport of substances across cell membranes against concentration gradients, and the contraction of muscle cells. It's been said that ATP molecules are like the coins of a nation; their phosphate groups are the common energy currency in all cells. That is why you often see a cartoon "coin" used to symbolize ATP, as shown by the sketch to the right.

Cells can renew their ATP supplies. At certain steps of many metabolic processes, such as aerobic respiration, an unbound phosphate atom (P_i) or a phosphate group that an enzyme cleaves from some substance becomes attached to adenosine diphosphate, or **ADP**. The result is an ATP molecule. When ATP transfers a phosphate group elsewhere, it reverts to ADP, thereby completing the steps of the **ATP/ADP cycle**:

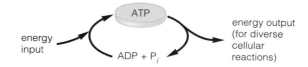

Another enzyme can remove and transfer the two outermost phosphate groups of an ATP molecule. The nucleotide that forms this way is cyclic adenosine monophosphate, or cAMP (Figure 6.7c). It is a player in pathways by which cells respond to outside signals.

In sum, phosphate-group transfers from ATP are a renewable, rapid, and near-universal mechanism for delivering energy where and when it is required to do work. Through such controlled transfers, energy drives transport proteins to pump solutes across membranes against concentration gradients. Transferred energy drives exergonic reactions uphill. It drives contractions and other movements of cell structures and organelles, of the cell itself, and of the multicelled organism. Figure 6.8 gives two representative examples.

ATP Output and Metabolic Pathways

At any instant, phosphate-group transfers are driving reactions that involve thousands of substances within the confines of a cell. All of the reactions don't proceed

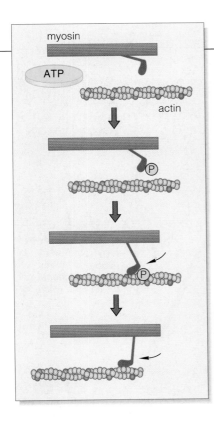

myosin

ATP

actin

a In a contractile unit of a muscle cell, ATP transfers a phosphate group to a filament of myosin, which binds a neighboring actin filament to itself and drags it along during a short power stroke, in a direction that helps shorten the contractile unit.

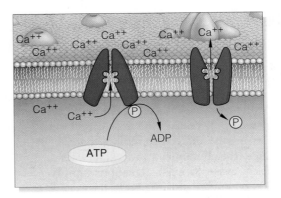

b ATP transfers a phosphate group to an active transport protein, which changes shape and pumps calcium ions out of the cell, against calcium's concentration gradient.

Figure 6.8 Two examples of the type of cellular work that can be triggered by phosphate-group transfers from ATP to enzymes and other participants in metabolic reactions.

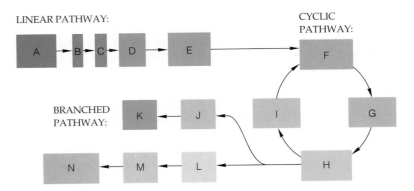

Figure 6.9 Types of reaction sequences of metabolic pathways.

willy-nilly. Most proceed in precisely ordered, enzyme-mediated sequences called **metabolic pathways**.

The main metabolic pathways are biosynthetic or degradative, overall. In a *biosynthetic* pathway, small molecules are assembled into complex carbohydrates, proteins, and other large molecules of higher energy content. In a *degradative* pathway, large molecules are broken down to products of lower energy content. Later in the book, you will consider examples of both kinds of pathways. You may find it useful to start thinking about their participants, as outlined here:

1. **Substrates** (also called reactants or precursors). The substances that enter a reaction.

2. **Intermediates**. Any of the substances that form between the start and end of a pathway.

3. **End products**. Substances remaining at the end of a metabolic reaction or pathway.

4. **Energy carriers**. ATP and a few other compounds that activate substances by transferring energy (via functional groups) to them.

5. **Enzymes**. Primarily proteins that speed specific reactions. (Some RNAs also show enzyme activity.)

6. **Cofactors**. Certain organic compounds or metal ions that assist enzymes or that transport electrons, atoms, or functional groups released during a reaction.

7. **Transport proteins**. Membrane-bound proteins by which concentration gradients are adjusted in ways that influence the direction of metabolic reactions.

The steps of many metabolic pathways proceed in linear sequence, from substrate to end products. Many others proceed in a circle, with the final end product serving as the beginning reactant if its concentration becomes high enough. Intermediates or end products of one pathway very commonly become substrates that enter different metabolic pathways. In other words, two pathways (or more) often become coupled, in linear or branching fashion. Figure 6.9 shows some of the many possibilities.

ATP is the main, renewable energy carrier between sites of metabolic reactions in cells. Its deliveries couple energy-releasing reactions with energy-requiring reactions.

When an organic compound becomes phosphorylated, as by a phosphate-group transfer from ATP, its store of energy increases and it becomes primed to enter a reaction.

Metabolic reactions and pathways proceed from substrates to the end products, often with intermediates forming in between. The participants include enzymes, cofactors, and transport proteins as well as energy carriers.

ENZYME STRUCTURE AND FUNCTION

Even when you are about to fall asleep, when you think your body is shutting down for the night, its metabolic machinery is synthesizing and degrading uncountable numbers of molecules. Much of the glucose from your latest meal was dismantled earlier for a quick energy fix in cells of your brain, heart, and skeletal muscles. More glucose is being converted to fat or something else. Hemoglobin molecules built sixty days ago are being digested to bits even as new molecules replace them.

Without enzymes, the dynamic, steady state called "you" would quickly cease to exist. Reactions simply would not proceed fast enough for the body to process food, build and tear down hemoglobin and other vital molecules, send signals to and from brain cells, make muscles contract, and do everything else to stay alive.

Four Features of Enzymes

Enzymes are *catalytic* molecules; they speed the rate at which reactions approach equilibrium. Nearly all of them are proteins, although some RNA molecules also show catalytic activity. Enzymes have four features in common. *First*, they do not make anything happen that could not happen on its own. But they usually make it happen hundreds to millions of times faster. *Second*, enzymes are not permanently altered or used up in a reaction; the same one may act over and over again. *Third*, the same enzyme usually works for the forward and the reverse directions of a reaction. *Fourth*, each type of enzyme is highly selective about its substrates.

An enzyme's substrates are specific molecules that it can chemically recognize, bind, and modify in certain ways. For example, thrombin, an enzyme involved in blood clotting, recognizes a side-by-side arrangement of two specific amino acids (arginine and glycine) and cleaves a peptide bond between them:

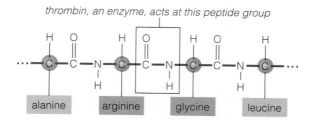

thrombin, an enzyme, acts at this peptide group

Enzyme-Substrate Interactions

A reaction occurs when participating molecules collide with some minimum amount of energy, known as the **activation energy**. This is true regardless of whether the reaction occurs spontaneously or with enzymes. Think of activation energy as an "energy hill" that must be surmounted before a reaction will proceed (Figure 6.10).

Enzymes make the energy hill smaller, so to speak.

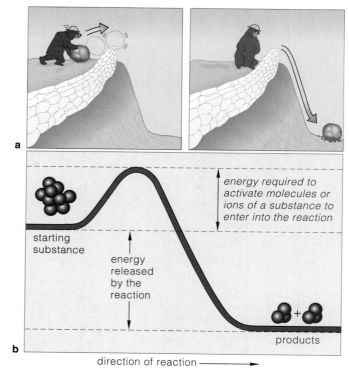

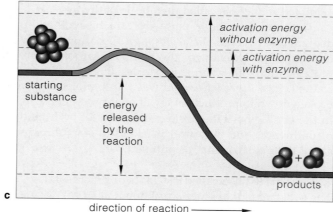

Figure 6.10 Activation energy. (**a,b**) Activation energy is like a barrier at the crest of a hill. It takes an input of energy to move a rock over the barrier before it can roll down spontaneously. (**c**) An enzyme enhances the rate at which a given number of molecules complete a reaction. It lowers the amount of energy required to boost reactants to the crest (transition state) of an energy hill.

How do they do this? Every enzyme has one or more **active sites**, or crevices in its surface where substrates interact with the enzyme and where a specific reaction is catalyzed. Figure 6.11*a* and *b* shows two views of the active site of an enzyme that phosphorylates glucose.

According to Daniel Koshland's **induced-fit model**, each substrate has a surface region that almost but not quite matches chemical groups in an active site. When substrates first settle in the site, the contact strains some of their bonds. Strained bonds are easier to break, which

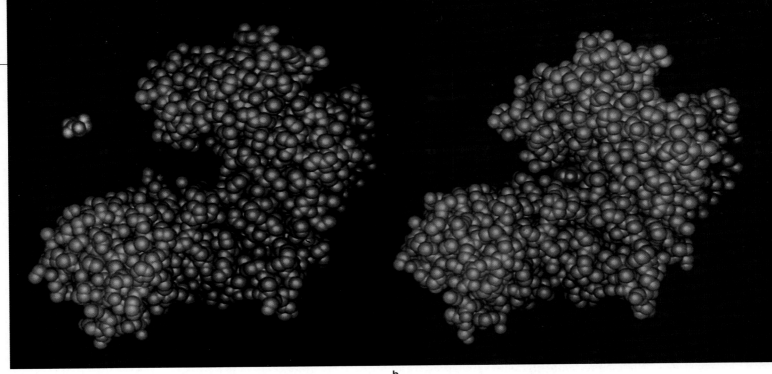

a

b

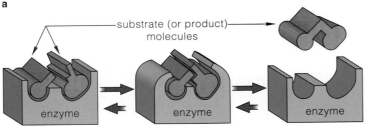

substrate (or product) molecules

enzyme → enzyme → enzyme

transition state
(tightest binding but least stable)

c

Figure 6.11 Model of an enzyme, hexokinase, at work. (**a**) The substrate is a glucose molecule (color-coded *red*). It is heading toward the active site, a cleft in the enzyme (*green*). (**b**) When it makes contact with the site, parts of the enzyme temporarily close in around it and prod the molecule to enter a reaction.

(**c**) Induced-fit model of enzyme-substrate interactions. Only when the substrate is bound in place is an enzyme's active site complementary to it. The fit is most precise during the transition state of a reaction. An enzyme-substrate complex is short-lived, for the attractive forces holding it together are usually weak.

promotes formation of new bonds (in products). Also, interactions among charged or polar groups in the site often favor a redistribution of electric charge in substrates that primes them for conversion to an activated state.

When substrates fit most precisely in the active site of an enzyme, they reach an activated, *transition* state (Figure 6.11*c*). Substrates in the transition state react spontaneously, just as a rock that has been pushed up and over a barrier at the crest of a hill rolls down on its own (Figure 6.10*a*).

What induces the transition state or gets substrates over the energy barrier once the state is reached? Several mechanisms are involved, including the following:

1. *Helping substrates get together.* Substrate molecules rarely collide if their concentrations are low. By binding substrates, the active site effectively boosts their local concentrations, which may increase reaction rates by 10,000 to a billion times.

2. *Orienting substrates in positions favoring reaction.* On their own, substrates collide from random directions. By contrast, weak but extensive bonding at an active site puts their mutually attractive chemical groups on precise collision courses much more frequently.

3. *Promoting acid-base reactions.* In many active sites, acidic or basic side groups of amino acids are poised to donate or accept hydrogen atoms from substrates. Such H^+ transfers destabilize electron sharing in covalent bonds, which become easier to break. Hydrolysis and stepwise electron transfers work this way.

4. *Shutting out water.* Some active sites bind substrates so tightly that they exclude some or all water molecules. The nonpolar environment can reduce the activation energy for certain reactions, such as the attachment of a carboxyl group ($-COO^-$) to a molecule, by as much as 500,000 times.

Depending on the enzyme, such mechanisms work alone or in combination to bring about the straining and warping that convert substrates to the transition state.

Enzymes catalyze (speed) the rate at which specific reactions reach equilibrium. They do so by lowering the amount of activation energy necessary to make substrates react.

Enzymes change the rate, not the outcome, of a reaction. They only act on specific substrates. And they may catalyze the same reaction repeatedly, as long as substrates are available.

FACTORS INFLUENCING ENZYME ACTIVITY

You probably don't get much done when you feel too hot or cold, or out of sorts because you ate too many sour plums or salty potato chips. Probably when the cupboard is bare, you focus on food. Maybe you call a friend to go shopping with you, and if you drive too fast to the grocery store, police tend to slow you down. In such respects, you have a lot in common with enzymes. They, too, respond to shifts in temperature, pH, and salinity, and to the relative abundances of particular substances. Many even engage helpers for specific tasks. And all normal enzymes respond to metabolic police.

Enzymes and Environmental Conditions

Temperature, recall, is a measure of molecular motion. You may think increases in temperature must improve the rate of enzyme-mediated reactions by making the substrates collide more frequently with active sites. This is true, but only until some point on the temperature scale. Past that point—which differs among enzymes— the increased molecular motion disrupts weak bonds holding the enzyme in its three-dimensional shape. The substrates can no longer bind to the active site, and so the reaction rate declines sharply, as in Figure 6.12a.

Expose an organism to temperatures that are far higher than it normally faces and its enzymes will unravel, thereby throwing its metabolic activities into turmoil. This can happen to sick people who develop extremely high fevers. They usually will die if their internal temperature reaches 44°C (112°F). Yet enzymes of a certain bacterium (*Thermoanaeobacter*), which lives in hot springs of Yellowstone National Park, perk along until the temperature passes 78°C!

Similarly, pH values that rise or sink beyond each enzyme's range of tolerance disrupt enzyme structure and functioning (Figure 6.13). Nearly all enzymes work best in the pH range between 6 and 8. For example, trypsin is active in the mammalian intestine, where pH is around pH 8. Pepsin, a protein-digesting enzyme, is one of the exceptions. It functions in gastric fluid, even though this highly acidic fluid denatures most enzymes.

Enzyme activity also will suffer if the environment gets far saltier than is normally encountered. Extremely high ion concentrations disrupt interactions that help hold most enzyme molecules in their three-dimensional shape. Here again we find some exceptions. Enzymes of *Halobacterium* and some algae function in such salty places as Utah's Great Salt Lake.

Control of Enzyme Function

Built into each living cell are controls over its enzyme activity. By coordinating its control mechanisms, the cell maintains, increases, and decreases concentrations of substances. Certain controls adjust the rate at which some enzymes are synthesized. They can limit or beef up the number of enzyme molecules for a key step in a metabolic pathway. Other controls switch on or inhibit enzymes that are already synthesized. For example, by **allosteric control**, enzymes that have already been formed are activated or inhibited when a specific substance combines with them at a binding site *other than* the active site (*allo-* means

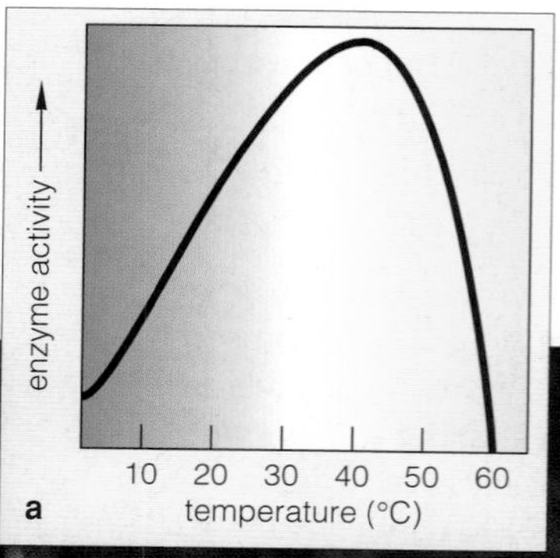

Figure 6.12 Example of how temperatures that fall outside the range of tolerance for an enzyme influence its activity. (**a**) Changes in the activity of one type of enzyme that had been subjected to changing temperatures. (**b**) Siamese cats show observable effects of such changes. Fur on the ears and paws contains more of a dark brown pigment, melanin, than the rest of the body does. A heat-sensitive enzyme controlling melanin production is less active in warmer body regions, and this results in lighter fur in those regions.

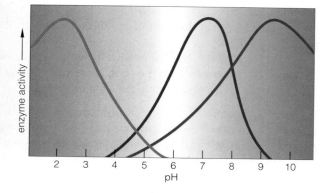

Figure 6.13 Diagram showing how the activity of three different enzymes is influenced by pH. One enzyme (the *brown* line) functions best in neutral solutions. Another enzyme (*red* line) functions best in basic solutions, and another (*purple* line) in acidic solutions.

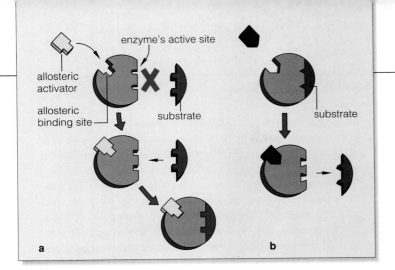

Figure 6.14 Allosteric control. (**a**) Activation of an allosteric enzyme. (**b**) Inhibition of an allosteric enzyme.

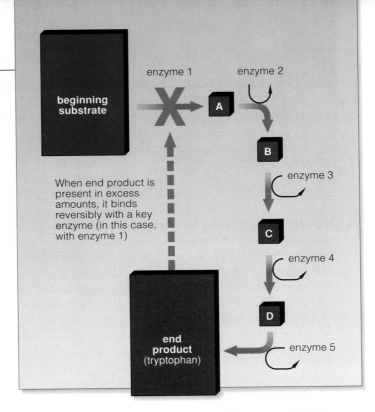

Figure 6.15 Example of feedback inhibition of a metabolic pathway. Five kinds of enzymes act in sequence to convert a substrate to an end product (tryptophan). When end product accumulates, some of the excess molecules bind to molecules of the first enzyme and thereby block the entire pathway.

different; *steric* means structure, or state). Figure 6.14 shows models of the binding, which is reversible.

Picture a bacterium, busily synthesizing tryptophan and other amino acids necessary to construct its proteins. After a bit, protein synthesis slows, so tryptophan is no longer required. But the tryptophan pathway is in full swing, so the cellular concentration of its end product continues to rise. Now **feedback inhibition** kicks in: A cellular change, caused by a specific activity, *shuts down the activity that brought it about*. In this case, the feedback loop starts and ends with a key allosteric enzyme in the pathway. Unused tryptophan molecules bind with the allosteric site. This inactivates the enzyme and blocks the pathway. By contrast, if few tryptophan molecules are around when the demand for them increases, the enzyme remains free of inhibition, and so tryptophan production rises. Such feedback loops can quickly adjust the concentrations of many substances (Figure 6.15).

In humans and other multicelled organisms, control of enzyme activity is just amazing. Cells not only work to keep themselves alive, they work with other cells in ways that benefit the whole body! Consider hormones, which are key signaling agents in this vast enterprise. Specialized cells release hormones into the blood. Any cell having receptors that are specific for a particular hormone will take it up, then its program for building a particular protein or some other activity will be altered. The hormone trips internal control agents into action—and the activities of specific enzymes change.

Enzyme Helpers

During many reactions, enzymes speed the transfer of one or more electrons, atoms, or functional groups from one substrate to another. Cofactors assist the reactions or briefly act as transfer agents. They include complex organic compounds called **coenzymes** as well as metal ions that associate with the enzyme molecule.

NAD$^+$ (for nicotinamide adenine dinucleotide) and FAD (for flavin adenine dinucleotide) are examples of coenzymes that are derived from vitamins. Both accept electrons and hydrogen atoms released during many reactions, such as glucose breakdown, and then transfer the electrons to other reaction sites. Being attracted to electrons, the unbound protons (H$^+$) go along for the ride. When loaded with electrons and hydrogen, NAD$^+$ and FAD are, respectively, abbreviated NADH and FADH$_2$. The coenzyme NADP$^+$ (nicotinamide adenine dinucleotide phosphate) functions in photosynthesis. Like NAD$^+$, it is a derivative of the vitamin niacin. When loaded down, it is abbreviated NADPH.

Ferrous iron (Fe^{++}), one of the metal ions that act as cofactors, is a component of cytochrome molecules. Cytochromes are proteins that show enzyme activity. They are embedded in cell membranes, including the membranes of chloroplasts and mitochondria.

Enzymes function best when the cellular environment stays within limited ranges of temperature, pH, and salinity. The actual ranges depend on the type of enzyme.

Control mechanisms govern the synthesis of new enzymes and stimulate or inhibit the activity of existing enzymes. By controlling enzymes, cells control the concentrations and kinds of substances available to them.

Some enzymes require cofactors (coenzymes and metal ions), which help catalyze reactions or transfer electrons, atoms, and functional groups from one substrate to another.

ELECTRON TRANSFERS THROUGH TRANSPORT SYSTEMS

Electrons, too, are transferred from one molecule to another in many metabolic pathways. Occasionally you may hear someone refer to an electron transfer as an **oxidation-reduction reaction**, but the name simply is a technical way of saying the same thing. The *donor* molecule that gives up electrons is said to be "oxidized." A molecule that *accepts* electrons is the one "reduced."

Such transfers often proceed after some atom of a molecule absorbs enough energy to boost one or more electrons farther from the nucleus or even completely out of the atom. An electron excited this way gives off energy as it returns to the lowest energy level available to it, either in the same atom or in a new one. Section 6.7 describes an example of this.

of atoms. Breaking the bonds in glucose puts electrons up for grabs, so to speak. In this metabolic pathway and many others in the cell, the released electrons are made to flow in ways that do useful work.

For example, consider an **electron transport system**, an organized array of enzymes and coenzymes that transfer electrons in sequence. One molecule donates electrons, and the next molecule in line accepts them. You find these systems in cell membranes, such as the ones inside mitochondria and chloroplasts (Figure 6.17).

An electron transport system operates by accepting electrons at higher energy levels, then releasing them at lower energy levels. Think of it as a staircase (Figure 6.16*b*). Excited electrons at the top step have the most

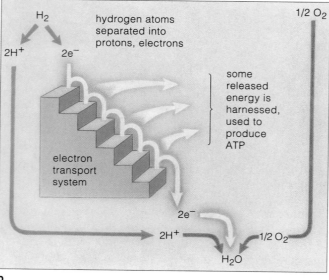
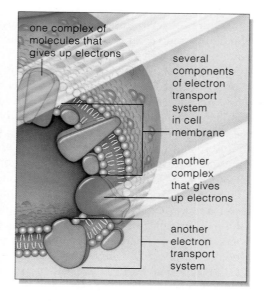

a   b

Figure 6.16 Examples of uncontrolled and controlled release of energy. (**a**) Hydrogen and oxygen exposed to an electric spark will react and release energy all at once. (**b**) In a cell, the same type of reaction proceeds in many small, enzyme-mediated steps that allow some released energy to be harnessed. These particular steps involve electron transfers, often between molecules that operate together as an electron transport system.

Figure 6.17 Example of electron transport systems embedded in the internal membrane system of chloroplasts. Each is positioned to accept electrons that sunlight energy knocks out of a neighboring complex of pigment molecules.

Cells conserve usable energy during certain electron transfers. Imagine tossing a bit of glucose into a wood fire. Its atoms would quickly let go of one another and combine with oxygen in the air, thereby forming CO_2 and H_2O. In that case, all of the released energy would be lost as heat to the surroundings (Figure 6.16*a*). By contrast, cells do not "burn" glucose all at once, so they do not waste much energy. Instead, enzymes cleave atoms from intermediates that form at different steps in a pathway by which glucose molecules are dismantled. Each step releases only some of the potential energy that has been stored in chemical bonds. Such bonds, remember, are unions between the electron structures

energy. They drop down, one step at a time, and release a bit of energy as they do. At certain steps, some energy is harnessed to do work. For instance, energy can move ions into a membrane-bound compartment and thereby set up concentration and electric gradients across the compartment's membrane. Such gradients are central to ATP formation. And that metabolic event, an absolutely central aspect of a cell's life, will occupy our attention in the next two chapters.

In many metabolic pathways, electrons as well as energy are transferred among molecules, as by electron transport systems.

YOU LIGHT UP MY LIFE—VISIBLE SIGNS OF METABOLIC ACTIVITY

BIOLUMINESCENCE   Fireflies and some other organisms show **bioluminescence**: they give off flashes of light when enzymes called luciferases convert chemical energy to light energy. The reactions start when ATP phosphorylates luciferin, a protein that, when activated, emits fluorescent light. Luciferases, with the help of oxygen, convert the activated luciferin to a different chemical form. The action excites electrons in the luciferin to a higher energy level. The excited electrons quickly return to a lower energy level. As they do, they release energy in the form of light.

In Jamaica's tropical forests, click beetles known locally as kittyboos are renowned for their communal flashings (Figure 6.18*a* and *b*). Different varieties emit green, yellow-green, yellow, or orange flashes. Amazingly, biologists now make *bioluminescent gene transfers*. They insert copies of the genes that specify bioluminescence into bacteria, plants, and other organisms (Figure 6.18*c* and *d*).

PUTTING BIOLUMINESCENCE TO USE   Besides being fun to think about, the transfers have practical applications. Each year, for example, 3 million people die from a lung disease caused by *Mycobacterium tuberculosis*. Different strains of this bacterium are resistant to different kinds of antibiotics, so effective treatment depends on identifying which strain is causing an infection. Bacterial cells taken from a patient can be cultured and exposed to luciferase genes. Some bacterial cells take up the genes, which, in a few cases, get inserted into their DNA. Clinicians identify the genetically modified cells, then expose colonies of their descendants to different antibiotics. If an antibiotic has no effect, the bacterial cells churn out gene products, including luciferase, and the colony glows. If it doesn't glow, this signifies the antibiotic destroyed the strain.

Christopher Contag and Pamela Contag, postdoctoral students at Stanford University, lit up bacteria that cause *Salmonella* infections in live laboratory mice! Researchers of viral or bacterial diseases infect dozens to hundreds of laboratory mice for their experiments. Before the Contags' discovery, the only option was to kill infected mice and examine their tissues to find out whether infection had occurred—a costly, tediously painstaking practice that also happens to offend animal rights activists.

The Contags approached David Benaron, a medical imaging researcher at Stanford, with a novel idea: If the infecting bacteria were bioluminescent, would the light

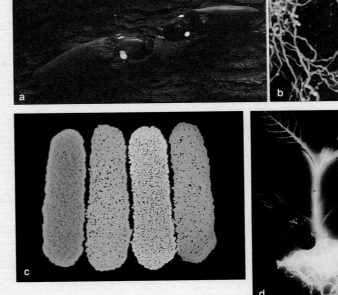

Figure 6.18   (**a**) Two kittyboos (*Pyrophorus noctilucus*). (**b**) Time-lapse photograph that reveals random paths of kittyboos on the wing. These beetles light up the night sky with bioluminescent flashes, which help potential mates find each other in the dark. (**c**) Micrograph of four colonies of bacterial cells. Each colony started with a parent bacterial cell that had taken up a kittyboo gene for a different glowing color. (**d**) A genetically modified tobacco plant that glows in the dark.

shine through tissues of *living* animals? In an experimental test of this idea, the researchers put glowing *Salmonella* cells in a thawed chicken breast. The glow showed through.

And so, in a series of experimental tests, the researchers transferred genes for bioluminescence into three strains of *Salmonella* and inoculated mice in three experimental groups. They used a digital imaging camera to track the course of infection in each group. The first strain was weakest; mice in the first group successfully fought off the infection within six days and did not glow. The second remained localized; it did not spread through the body. The third strain was dangerous; it spread rapidly through the mouse gut, and the entire gut glowed.

Bioluminescent gene transfer, in combination with the imaging of luciferase activity, has now proved its potential for verifying the effectiveness of new drug treatments (Figure 6.19). It may have application as well in *gene therapy*, whereby clinicians insert copies of functional genes into patients as replacements for defective or cancer-causing genes of their own. This is a fine story in itself, one to which we will return in Chapter 16.

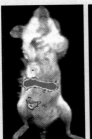
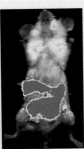
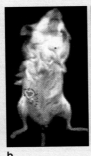

a   b

Figure 6.19   Using bioluminescent bacterial cells to chart the location and spread of infectious bacteria inside living laboratory mice. (**a**) False-color images in this pair of photographs show how infection spread in a control group that had not been dosed with antibiotics. (**b**) This pair shows how antibiotics had killed most of the infectious bacterial cells.

SUMMARY

1. Cells store, break down, and dispose of substances by acquiring and using energy from outside sources. The sum total of these energy-driven activities, called metabolism, underlies the survival of each living thing.

2. Two laws of thermodynamics affect life. First, energy undergoes conversion from one form to another, but its total amount never increases or decreases as a result of any conversion. Thus the total amount of energy in the universe holds constant. Second, energy spontaneously flows in one direction, from forms of higher to lower quality.

   a. All matter has some amount of potential energy (as measured by the capacity to do work) by virtue of its position in space and the arrangement of its parts.

   b. Potential energy may be transformed into kinetic energy, the energy of motion. Mechanical movements and heat, which corresponds to the degree of molecular motion, are two common forms of kinetic energy.

   c. Chemical energy (the potential energy inherent in molecular bonds) is often measured in kilocalories.

3. Like all organized systems, the cell tends to become disorganized without energy. It inevitably loses some of its chemical potential energy during every metabolic reaction, mainly in the form of heat. It stays organized and alive as long as it counters its energy expenditures (outputs) with energy replacements (inputs).

4. The sun is life's primary energy source. In plants and other photosynthesizers, cells trap energy from the sun's rays and convert it to the chemical energy of organic compounds. The plants, and then organisms that feed on plants and one another, use some of the energy that was stored in the organic compounds to do cellular work.

5. Most metabolic reactions proceed strongly in the forward direction (to products) when concentrations of starting substances are high. They also run strongly in reverse (to starting substances) when the concentrations of products are high.

   a. When left to themselves, reversible reactions run toward chemical equilibrium, at which time they are proceeding at about the same rate in both directions.

   b. But cellular mechanisms work with and against this tendency. At any time, reversible reactions help to increase, decrease, and maintain the concentrations of thousands of different substances required for specific reactions.

   c. Regardless of the direction of a metabolic reaction or the extent of the molecular cleavages, combinations, and rearrangements, the total number of atoms that end up in all the products will equal the total number of atoms that were present in the starting substances.

6. Metabolic reactions may release energy when they run toward equilibrium, or they may require energy inputs to drive them away from equilibrium.

7. An exergonic reaction (energy out) ends with a net loss in energy. An endergonic reaction (energy in) ends with a net gain in energy. Cells conserve a great deal of energy by coupling energy-releasing reactions with energy-requiring ones.

8. ATP is the main energy carrier in cells. It is the key coupling agent between energy-releasing and energy-requiring pathways.

9. ATP forms when energy released during specific reactions drives the phosphorylation (attachment of a phosphate molecule) to ADP. ATP later donates energy to other reactions by phosphorylating either starting substances or intermediates, which thereby become primed to enter into specific reactions.

10. Metabolic pathways are orderly, stepwise sequences of enzyme-mediated reactions.

   a. A substrate (or reactant) is a substance that enters a reaction or pathway. Intermediates are the substances that form between the reactants and end products.

   b. By the end of a biosynthetic pathway, energy-rich organic compounds have been assembled from smaller molecules of lower energy content.

   c. By the end of a degradative pathway, energy-rich molecules have been broken down to smaller ones of lower energy content.

11. Enzymes are catalysts; they greatly enhance the rate of a reaction involving specific substrates but do not alter the outcome. Nearly all enzymes are proteins, although some RNAs also show catalytic activity.

   a. Enzymes lower the activation energy required to start a reaction. They bind substrates at an active site, strain its bonds, and make the bonds easier to break.

   b. Each kind of enzyme acts best within limited ranges of temperature, pH, and salinity.

   c. Cofactors assist enzymes in speeding a reaction, or they carry electrons, hydrogen, or functional groups stripped from substrates to other sites. They include coenzymes (such as NAD^+) and metal ions.

   d. Controls stimulate or inhibit enzyme activity at key steps in metabolic pathways. They help coordinate the kinds and amounts of substances available.

12. Many energy conversions in cells involve a flow of electrons, as through electron transport systems.

Review Questions

1. Define free radical. How does it relate to aging? *CI*

2. State the first and second laws of thermodynamics. Does life violate the second law? *6.1*

3. Define and give examples of potential and kinetic energy. *6.1*

4. What can a chemical reaction do to a substrate molecule? *6.2*

5. Make a simple diagram of the ATP molecule; then highlight which part of it can be transferred to another molecule and later replaced. Why is such a transfer possible at so many different sites in the cell, and what can it accomplish? *6.3*

6. Define activation energy, then briefly describe four ways in which enzymes may lower it. *6.4*

7. Define feedback inhibition as it relates to the activity of an allosteric enzyme. *6.5*

8. What is an oxidation-reduction reaction? Explain how such reactions proceed in an electron transport system. *6.6*

Self-Quiz *(Answers in Appendix IV)*

1. _____ is the primary source of energy for life on Earth.
   a. Food
   b. Water
   c. The sun
   d. ATP

2. Which is *not* true of chemical equilibrium?
   a. Product and reactant concentrations are always equal.
   b. The rates of the forward and reverse reactions are the same.
   c. There is no further net change in product and reactant concentrations.

3. If we liken chemical equilibrium to the bottom of an energy hill, then an _____ reaction is an uphill run.
   a. endergonic
   b. exergonic
   c. ATP-assisted
   d. both a and c

4. Phosphate-group transfers from ATP to another molecule are a _____ mechanism for delivering energy.
   a. rapid
   b. renewable
   c. near-universal
   d. all of the above

5. Which statement is *not* correct? A metabolic pathway _____ .
   a. has an orderly sequence of reaction steps
   b. is mediated by one enzyme, which initiates the reactions
   c. may be biosynthetic or degradative, overall
   d. all of the above

6. Enzymes are _____ .
   a. enhancers of reaction rates
   b. influenced by temperature
   c. influenced by pH
   d. not influenced by salinity
   e. a through c
   f. all of the above

7. All enzymes incorporate a(n) _____ .
   a. active site
   b. coenzyme
   c. metal ion
   d. all of the above

8. The coenzymes NAD^+, FAD, and $NADP^+$ are _____ .
   a. cofactors
   b. allosteric enzymes
   c. metal ions
   d. both a and b

9. Electron transport systems involve _____ .
   a. enzymes and cofactors
   b. electron transfers
   c. cell membranes
   d. all of the above

10. Match each substance with the most suitable description.
    ____ coenzyme or metal ion
    ____ adjusts gradients at membrane
    ____ substance entering a reaction
    ____ substance formed while a reaction is proceeding
    ____ substance at end of reaction
    ____ enhances reaction rate
    ____ mainly ATP
    a. reactant or substrate
    b. enzyme
    c. cofactor
    d. intermediate
    e. product
    f. energy carrier
    g. transport protein

Critical Thinking

1. When cyanide, a toxic organic compound, binds to an enzyme that is a component of electron transport systems, the outcome is *cyanide poisoning*. Binding prevents the enzyme from donating electrons to a nearby acceptor molecule in the system. What effect will this have on ATP production? From what you know of ATP function, what effect will this have on a person's health?

2. AZT, or azidothymidine, is a drug used to alleviate symptoms of *AIDS* (acquired immunodeficiency syndrome). AZT is very similar in molecular structure to thymidine, one of the DNA nucleotides. AZT also can fit into the active site of an enzyme produced by the virus that causes AIDS. Infection puts the viral enzyme and viral genetic material (RNA) into a cell. There, the RNA is used as a template (structural pattern) for joining nucleotides to form a strand of DNA. The viral enzyme takes part in the assembly reactions. Propose a model to explain how AZT might inhibit replication of the virus inside cells.

3. The bacterium *Clostridium botulinum* is an obligate anaerobe, meaning it dies quickly in the presence of oxygen. It lives in oxygen-free pockets in soil, and it can enter a metabolically inactive, resting state by forming spores. The spores may end up on the surfaces of garden vegetables. When the picked vegetables are being canned, the spores must be destroyed. If they are not destroyed, *C. botulinum* may grow and produce botulinum, a toxin. When tainted canned foods are not cooked properly, the toxin remains active and causes a type of food poisoning called *botulism*. This bacterium is not able to produce either superoxide dismutase or catalase. Develop a hypothesis to explain how the inability to make the enzymes is related to its anaerobic life-style.

4. *Pyrococcus furiosus* thrives at 100°C, the boiling point of water. This species of bacterium was discovered growing in a volcanic vent in Italy. Enzymes isolated from *P. furiosus* cells do not function well below 100°C. What is it about the structure of these enzymes that allows them to remain stable and active at such high temperatures? (Hint: Review Section 3.7, which summarizes the interactions that maintain protein structure.)

Selected Key Terms

activation energy *6.4*	exergonic reaction *6.2*
active site *6.4*	feedback inhibition *6.5*
ADP *6.3*	first law of thermodynamics *6.1*
allosteric control *6.5*	free radical *CI*
ATP *6.3*	heat *6.1*
ATP/ADP cycle *6.3*	induced-fit model *6.4*
bioluminescence *6.7*	intermediate *6.3*
chemical energy *6.1*	kilocalorie *6.1*
chemical equilibrium *6.2*	kinetic energy *6.1*
chemical reaction *6.2*	law of conservation of mass *6.2*
coenzyme *6.5*	metabolic pathway *6.3*
cofactor *6.3*	metabolism *6.1*
electron transport system *6.6*	oxidation-reduction reaction *6.6*
end product *6.3*	phosphorylation *6.3*
endergonic reaction *6.2*	potential energy *6.1*
energy carrier *6.3*	second law of thermodynamics *6.1*
entropy *6.1*	substrate (reactant) *6.3*
enzyme *6.3*	transport protein *6.3*

Readings

Fenn, J. *Engines, Energy, and Entropy*. 1982. New York: Freeman. Deceptively simple paperback.

Rusting, R. December 1992. "Why Do We Age?" *Scientific American* 267(6): 131–141.

Web Site See *http://www.wadsworth.com/biology* for practice quiz questions, hypercontents, BioUpdates, and critical thinking. The Wadsworth Biology Resource Center provides a wealth of information fully organized and integrated by chapter.

7

ENERGY-ACQUIRING PATHWAYS

Sun, Rain, and Survival

Just before dawn in the American Midwest, the air is dry and motionless. Heat has scorched the land for weeks; it still rises from the earth and hangs in the air of a new day. There are no clouds in sight. There is no promise of rain. For hundreds of miles, crops stretch out, withered and nearly dead. All the marvels of modern agriculture can't save them. In the absence of one vital resource—water—life in each cell of those hundreds of thousands of plants has ceased.

In Los Angeles, a student wonders if the Midwest drought will bump up food prices. In Washington, D.C., economists analyze crop failures in terms of the tonnage available for domestic consumption and for export.

Thousands of kilometers away, in the vast Sahel Desert of Africa, grasses and cattle are dying after a similarly unrelenting drought. Children with bloated bellies and spindly legs wait passively for death. Deprived of nourishment for too long, living cells of their bodies will never function normally again.

You are about to explore key pathways by which cells secure and use energy. At first,

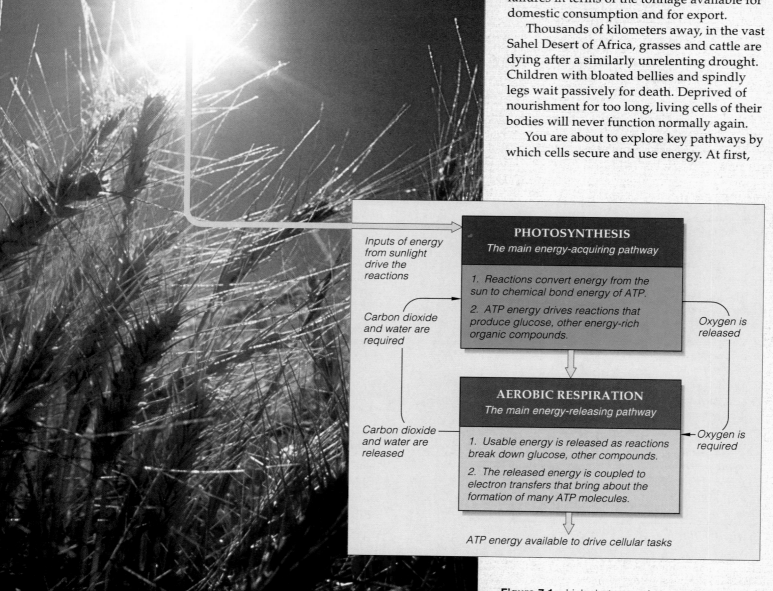

Inputs of energy from sunlight drive the reactions

Carbon dioxide and water are required

PHOTOSYNTHESIS
The main energy-acquiring pathway

1. Reactions convert energy from the sun to chemical bond energy of ATP.

2. ATP energy drives reactions that produce glucose, other energy-rich organic compounds.

Oxygen is released

Carbon dioxide and water are released

AEROBIC RESPIRATION
The main energy-releasing pathway

1. Usable energy is released as reactions break down glucose, other compounds.

2. The released energy is coupled to electron transfers that bring about the formation of many ATP molecules.

Oxygen is required

ATP energy available to drive cellular tasks

Figure 7.1 Links between photosynthesis and aerobic respiration, the main energy-acquiring and energy-releasing pathways in the world of life.

these pathways may seem far removed from your daily interests. *Yet the food that nourishes you and nearly all other organisms cannot be produced or used without them.*

We will return to this point in later chapters, when we address the important issues of human population growth, nutrition, limits on agriculture, and genetic engineering, as well as the impact of pollution on crops. Here, our point of departure is the *source* of our food—which isn't a farm or a supermarket or a refrigerator. What we call "food" was put together somewhere in the world by living cells from glucose and other organic compounds. Such compounds are built on a framework of carbon atoms, so the questions become these:

1. *Where does the carbon come from in the first place?*

2. *Where does the energy come from to drive the synthesis of carbon-based compounds?*

Answers to these questions depend on an organism's mode of nutrition.

Nutritionally, many organisms are **autotrophs**. They are "self-nourishing," which is what autotroph means. They latch onto carbon dioxide (CO_2) as their source of carbon. The air all around us has an abundance of this gaseous compound, which also is dissolved in aquatic habitats. What about energy? Plants, many protistans, and some bacteria are *photo*autotrophs; they trap energy from the sun. A few bacteria are *chemo*autotrophs; they use energy that is released when they break apart a variety of inorganic substances, such as sulfur.

Many other organisms are **heterotrophs**. They can't nourish themselves with inorganic substances. To secure carbon and energy, they feed on autotrophs, each other, and organic wastes. (*Hetero-* means other, as in "being nourished by other organisms.") That is how animals, fungi, many protistans, and most bacteria stay alive.

When you think about this grand pattern of who feeds on whom, one thing becomes clear. *Survival of nearly all organisms ultimately depends on photosynthesis, the main pathway by which carbon and energy enter the web of life.* Once glucose and other organic compounds are assembled, cells put them in storage or use them as building blocks. When cells require energy, they break such compounds apart by several pathways. *Of all the energy-releasing pathways, the most common is aerobic respiration.* Figure 7.1 provides you with a preview of the chemical links between photosynthesis and aerobic respiration, the focus of this chapter and the next.

KEY CONCEPTS

1. Organic compounds, with their backbone of carbon atoms, are the structural materials and energy stores of life. Plants, like other photoautotrophs, produce their own organic compounds by photosynthesis.

2. In the first stage of photosynthesis, sunlight energy is trapped and converted to chemical bond energy in ATP molecules. In the second stage, ATP transfers energy to reaction sites where glucose is synthesized from carbon dioxide and water. The glucose monomers become the building blocks for starch and other molecules.

3. Photosynthesis is the biosynthetic pathway by which most carbon and energy enter the web of life.

4. In plant cells, photosynthesis proceeds in organelles called chloroplasts. The pathway starts at a membrane system inside the chloroplast. The required metabolic machinery includes organized arrays of light-absorbing pigments, enzymes, and the coenzyme $NADP^+$, which delivers hydrogen and electrons to the synthesis reactions.

5. Photosynthesis is often summarized this way:

$$12H_2O + 6CO_2 \xrightarrow{\text{LIGHT ENERGY}} 6O_2 + C_6H_{12}O_6 + 6H_2O$$

PHOTOSYNTHESIS—AN OVERVIEW

Photosynthesis is an ancient pathway, and it evolved in distinct ways in different photoautotrophic organisms. To keep things simple, we can focus on what goes on in weeds, lettuces, and other leafy plants (Figure 7.2).

Energy and Materials for the Reactions

The pathway has two stages, each with its own set of reactions. In the *light-dependent* reactions, energy from the sun is absorbed and converted to ATP energy. Water molecules are split and a coenzyme, $NADP^+$, picks up the liberated hydrogen and electrons to form NADPH. In the *light-independent* reactions, ATP donates energy to sites where glucose ($C_6H_{12}O_6$) is put together from carbon, hydrogen, and oxygen atoms. NADPH delivers the hydrogen, and carbon dioxide (CO_2) that diffuses

Figure 7.2 Diagrams and photomicrographs showing where photosynthesis takes place inside the leaves of sow thistle (*Sonchus*), a common weed in many parts of the world. This plant is growing next to a country road in Germany.

a

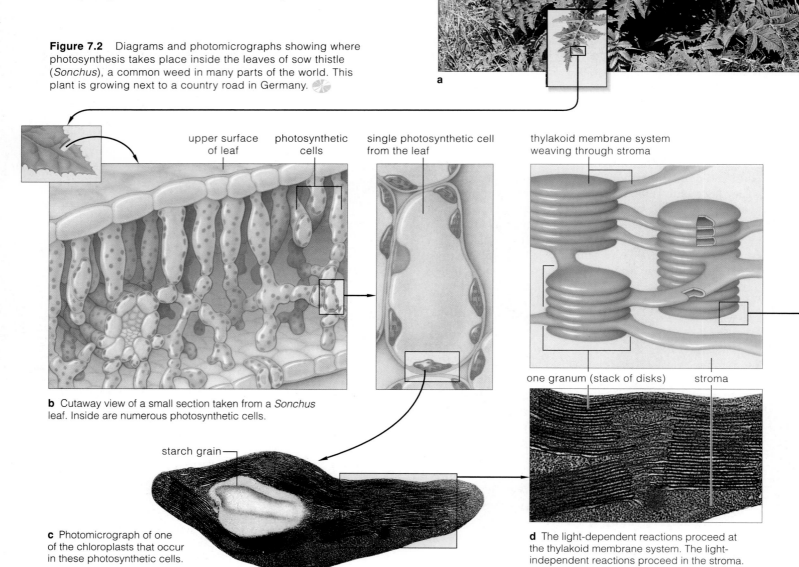

upper surface of leaf   photosynthetic cells

single photosynthetic cell from the leaf

thylakoid membrane system weaving through stroma

one granum (stack of disks)   stroma

b Cutaway view of a small section taken from a *Sonchus* leaf. Inside are numerous photosynthetic cells.

starch grain

c Photomicrograph of one of the chloroplasts that occur in these photosynthetic cells.

d The light-dependent reactions proceed at the thylakoid membrane system. The light-independent reactions proceed in the stroma.

into the leaf donates the carbon and oxygen. Overall, the reactions are often summarized in this manner:

$$12H_2O + 6CO_2 \xrightarrow{\text{LIGHT ENERGY}} 6O_2 + C_6H_{12}O_6 + 6H_2O$$

If you were to track the destinations of all the atoms of the reactants, you might end up with this picture:

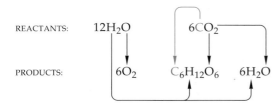

REACTANTS: $12H_2O$ $6CO_2$

PRODUCTS: $6O_2$ $C_6H_{12}O_6$ $6H_2O$

Notice how the summary equation shows glucose as the pathway's carbon-rich end product. Doing so does keep the chemical bookkeeping simple. However, the pathway does not really end with glucose molecules. Glucose and other simple sugars combine at once to form sucrose, starch, and other carbohydrates—the true end products of photosynthesis.

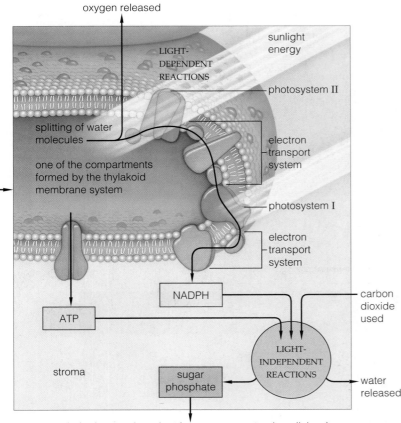

e Overview of the sites where the key steps of both stages of reactions proceed. The sections to follow will fill in the details.

Where the Reactions Take Place

The two stages of photosynthesis proceed at different sites inside each **chloroplast**. Only the photosynthetic cells of plants and some protistans contain this type of organelle (Section 4.7). Each chloroplast, recall, has two outer membranes that are wrapped around a largely fluid interior, called the **stroma**. Another membrane extends through the stroma. Often this membrane is repeatedly folded into numerous channels and stacked disks. Each stack of disks is a granum (plural, grana).

The first stage of photosynthesis proceeds at this much-folded inner membrane, which is the **thylakoid membrane system** (Figure 7.2e). The spaces within its channels and disks interconnect, as a compartment that collects hydrogen ions. The flow of these ions across the membrane is required for ATP production. The second stage of photosynthesis—the set of reactions by which sugars are assembled—takes place in the stroma.

The diagram below is a simplified version of Figure 7.2. Where you see it repeated in sections to follow, use it as a reminder to refer back to this figure to reinforce your grasp of how the details fit into the big picture:

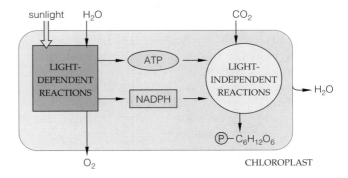

Before turning to the details, think about this: Two thousand chloroplasts, lined up single file, would be no wider than a dime. Imagine all the chloroplasts in just one weed or one lettuce leaf—each a tiny factory for producing sugars and starch—and you get an idea of the magnitude of metabolic events required to feed you and all other organisms on this planet.

Chloroplasts are organelles that are present only in plants and some protistans. They specialize in photosynthesis.

In the first stage of photosynthesis, energy from the sun's rays is absorbed and converted to ATP energy. This stage takes place at a membrane system inside the chloroplast.

In the second stage, hydrogen, carbon, and oxygen combine to form carbohydrates. This second stage takes place in the stroma, which is located between the chloroplast's outer and inner membrane systems.

SUNLIGHT AS AN ENERGY SOURCE

Each second, more than 2 million metric tons of the sun's mass enter thermonuclear reactions that release stupendous amounts of energy. It takes a mere eight minutes for a fraction of the released energy to reach the Earth's atmosphere, 160 million kilometers away. Each day, that fraction averages 7,000 kilocalories per square meter of the Earth as a whole. Almost a third is reflected back into space. Of the amount that does reach the Earth's surface, only about 1 percent is intercepted by photoautotrophs, the entry point for a one-way flow of energy through nearly all webs of life.

Properties of Light

Energy released from the sun radiates through space in an undulating motion, a bit like waves crossing the ocean. These are different forms of energy in motion. In both cases, however, the horizontal distance between the crests of every two successive waves is known as a **wavelength** (Figure 7.3a).

Wavelengths of radiant energy range from gamma rays, which are less than 10^{-5} nanometers long, to radio waves, which are more than 10 kilometers long. The entire range of all of these wavelengths represents the **electromagnetic spectrum** (Figure 7.3b).

Photoautotrophs can only harness the wavelengths between 400 and 750 nanometers. This is the range of **visible light**, which we can perceive as different colors. Wavelengths shorter than this have enough energy to break bonds of organic compounds and destroy cells. Ultraviolet (UV) radiation falls in this range. It can even force atoms to give up electrons, which is why it also is known as ionizing radiation.

Hundreds of millions of years ago, an ozone layer formed in the atmosphere and shielded the Earth from incoming ultraviolet radiation. Before this happened, the lethal bombardment kept photosynthetic bacteria

and other organisms confined below the surface of the seas. We return to this topic in Section 21.3.

Besides having wavelike properties, visible light also behaves as if its energy is contained in distinct *packets*. We call such a packet of light energy a **photon**. Different photons have different but fixed amounts of energy. The most energetic photons travel as very short wavelengths that correspond to blue-violet light. The least energetic photons travel as long wavelengths that correspond to red light (Figure 7.3b).

400-nanometer wavelength

750-nanometer wavelength

a

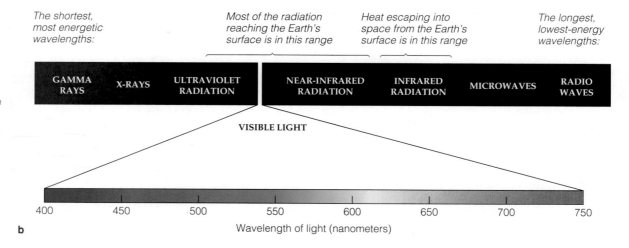

The shortest, most energetic wavelengths:

Most of the radiation reaching the Earth's surface is in this range

Heat escaping into space from the Earth's surface is in this range

The longest, lowest-energy wavelengths:

Figure 7.3
(**a**) Examples of wavelengths, the horizontal distance between crests of successive waves. (**b**) The entire electromagnetic spectrum. Visible light is only one of many forms of electromagnetic radiation in the spectrum.

| GAMMA RAYS | X-RAYS | ULTRAVIOLET RADIATION | NEAR-INFRARED RADIATION | INFRARED RADIATION | MICROWAVES | RADIO WAVES |

VISIBLE LIGHT

400   450   500   550   600   650   700   750

Wavelength of light (nanometers)

b

Figure 7.4 T. Englemann's 1882 observational test that correlated photosynthetic activity of *Spirogyra*, a strandlike green alga, with certain wavelengths of light. Oxygen is a by-product of photosynthesis. Like many organisms, certain free-living bacterial cells use oxygen for aerobic respiration. These cells move toward places where conditions favor their activities and away from less favorable conditions.

As Englemann reasoned, oxygen-requiring bacteria living in the same habitats as the green alga would congregate where oxygen was being produced. He put a strand of the alga in a drop of water containing the bacteria and then mounted it on a microscope slide. He positioned a crystal prism to cast a spectrum of colors across the slide. The bacterial cells congregated mostly where violet and red wavelengths crossed the algal strand. Englemann concluded these wavelengths are most effective for photosynthesis (and oxygen production).

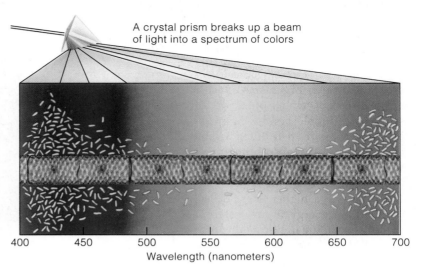

A crystal prism breaks up a beam of light into a spectrum of colors

Figure 7.5 (**a**) Two absorption spectra showing the efficiency with which two kinds of chlorophylls absorb certain wavelengths of visible light. (**b**) Similar absorption spectra for two kinds of accessory pigments. Peaks and valleys in the ranges of absorption correspond to measured amounts of energy absorbed from wavelengths and used in photosynthesis.

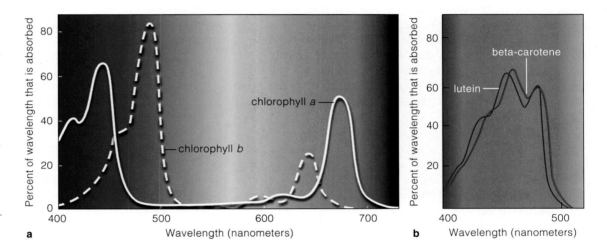

Pigments—The Molecular Bridge From Sunlight to Photosynthesis

The sun's rays contain all wavelengths of light, which collectively look white to us. A pyramid-shaped crystal prism that intercepts white light can sort it out into its component colors by bending different wavelengths by different degrees. As early as the 1800s, T. Englemann, a botanist, used prisms to find out which wavelengths were driving photosynthesis in a green alga (Figure 7.4). However, molecular biology was far into the future, so he could not identify the bridge between sunlight and photosynthetic activity.

Pigments are the molecular bridge. These molecules absorb wavelengths of light, and organisms put them to many uses. Most absorb only some wavelengths and transmit the rest. A few, such as the melanins in animals, absorb so many wavelengths they appear dark or black.

An **absorption spectrum** is a diagram that shows how effectively a given pigment molecule absorbs different wavelengths in the spectrum of visible light. Consider chlorophylls, the main pigments of photosynthesis. As Figure 7.5a shows, chlorophylls absorb all wavelengths except very little of the green and yellow-green ones, which they mainly transmit. (That is why plant parts that are chockful of chlorophylls appear green to us.) Other, *accessory* pigments that impart different colors also are present in different photoautotrophs. Figure 7.5b and the next section show how their assistance extends the range of wavelengths that can drive photosynthesis.

Radiation from the sun travels in waves that differ in length and energy content. Short wavelengths pack the most energy.

We perceive wavelengths of visible light as different colors and measure their energy content in packets called photons.

Chlorophylls and certain other pigments absorb specific wavelengths of visible light. They are the molecular bridge between the sun's energy and photosynthetic activity.

THE RAINBOW CATCHERS

The Chemical Basis of Color

How is it that pigment molecules can impart color to plants and other organisms? Each has a light-catching array of atoms, which often are joined by alternating single and double bonds. Figure 7.6 shows examples. Electrons that are distributed around the atomic nuclei of such arrays can absorb photons of specific energies, which correspond to specific colors of light.

Remember, when an atom's electrons absorb energy, they move to a higher energy level (Section 2.3). In a pigment, an energy input destabilizes the distribution of electrons in the light-catching array. Within 10^{-15} of a second, excited electrons return to a lower energy level, the electron distribution stabilizes, and energy is emitted in the form of light. When any destabilized molecule emits light as it reverts to a stable configuration, this is called a **fluorescence**.

Excitation occurs *only* if the quantity of energy of an incoming photon matches the amount of energy required to boost an electron to a higher energy level. Suppose a sunbeam strikes a pigment. Photons of its red or orange wavelengths match the amount of energy

required for the boost, so they are absorbed. Photons of its blue or violet wavelengths are a mismatch; they can't be absorbed. They are transmitted, reflected, and so impart a blue or violet color to the molecule.

On the Variety of Photosynthetic Pigments

Most pigments respond to only part of the rainbow of visible light. If acquiring energy is so vital for life, why doesn't each photosynthetic pigment go after the whole rainbow? In other words, why isn't each one black?

If the first photoautotrophs evolved in the seas, then so did their pigments. Ultraviolet and blue wavelengths do not penetrate water as deeply as green and red wavelengths do. Possibly natural selection favored the evolution of diverse pigments at different depths. Many existing red algae *are* nearly black, and they live in deep water. Green algae live in shallow water, and their main pigments respond to red wavelengths. Their accessory pigments help harvest light, and some even function as shields against ultraviolet radiation.

Today, the **chlorophylls** are the main pigments in all but one marginal group of photoautotrophs. Remember, photosynthetic pigments respond best to red and blue-to-violet light. The grass-green chlorophyll *a* is the main pigment inside chloroplasts (Figure 7.6). Chlorophyll *b*, a bluish-green pigment, occurs in plants, green algae, as well as a few photoautotrophic bacteria.

All photoautotrophs also produce **carotenoids**. These accessory pigments absorb blue-violet wavelengths that chlorophylls miss. Usually their backbone has a carbon ring at each end. Many types, including beta-carotene, are pure hydrocarbons (Figure 7.6). The xanthophylls also incorporate oxygen. Carotenoids impart yellow, orange, and red colors to many flowers, fruits, and vegetables. Often they have proteins attached, in which case they appear purple, violet, blue, green, brown, or black.

Carotenoids are less abundant than the chlorophylls in green leaves, but in many plants they become visible in autumn (Figure 7.7). Three weeks a year, tourists

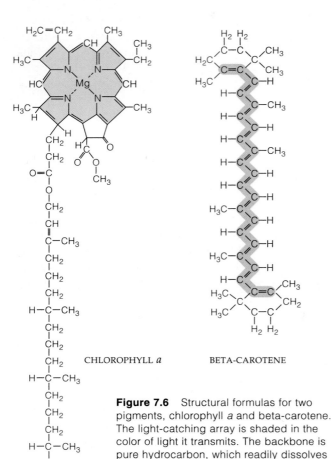

CHLOROPHYLL *a*   BETA-CAROTENE

Figure 7.6 Structural formulas for two pigments, chlorophyll *a* and beta-carotene. The light-catching array is shaded in the color of light it transmits. The backbone is pure hydrocarbon, which readily dissolves in the lipid bilayer of cell membranes.

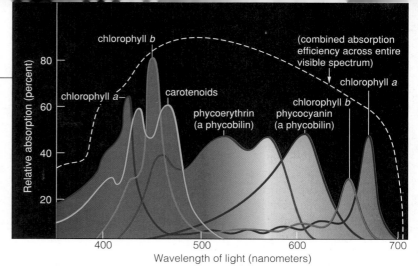

Figure 7.7 *Changes in leaf color in autumn. In intensely green leaves, the abundance of chlorophyll masks the presence of carotenoids and other pigments. Chlorophyll molecules are synthesized and degraded on a regular basis, but in many species, the gradual decline of daylength in autumn and other seasonal factors cause their synthesis to lag behind. Accessory pigments are unmasked, and more colors show through.*

Also in autumn, water-soluble anthocyanins accumulate in central vacuoles of leaf cells. These pigments appear red if fluids being transported through plants are slightly acidic, blue if the fluids are basic (alkaline), or colors in between if they are at intermediate pH levels. Soil conditions contribute to the pH differences. Birch, aspen, and other deciduous trees always have the same color in autumn. The leaf color of maple, ash, sumac, and other species varies around the country. It even varies from one leaf to the next, depending on the pigment combinations.

Figure 7.8 *Combined energy absorption efficiency, indicated by the dashed line, of certain chlorophylls (green lines), carotenoids (yellow-orange lines), and phycobilins (purple and blue lines).*

spend about a billion dollars just to watch the demise of chlorophyll in the deciduous trees of New England.

Other accessory pigments include the red and blue **anthocyanins** and **phycobilins**. The anthocyanins are pigments in many flowers, and the phycobilins are the signature pigments of red algae and cyanobacteria.

Collectively, the chlorophylls and accessory pigments can absorb nearly all wavelengths of visible light, as the absorption spectra in Figure 7.8 suggest.

What Happens to the Absorbed Energy?

Let's now consider what happens after electrons of a pigment's light-catching array have absorbed photons. Again, when electrons return to a lower energy level, a pigment molecule quickly gives up energy. If nothing else were around to intercept it, all the released energy would escape as fluorescent light and heat. However, pigments are not isolated. They have neighbors.

For example, peppering the thylakoid membrane of chloroplasts are **photosystems**, which are clusters of

proteins and 200 to 300 pigments. Most of the pigments only harvest photon energy. Then, instead of releasing energy as a fluorescent afterglow, a harvester directly transfers energy to a neighbor. As it reverts to a stable state, its neighbor enters a state of excitation, and so on. In this way, excitation energy randomly "walks" among the light-harvesting pigment molecules (Figure 7.9).

With each energy transfer, a bit of energy is lost as heat. Within 10^{-15} of a second, though, the energy still left corresponds to a wavelength that only a specialized chlorophyll a molecule can trap. By virtue of its position, that molecule is at the photosystem's **reaction center**. It accepts excitation energy but doesn't pass it on. Rather, the reaction center *gives up electrons* for photosynthesis. As you will see, a primary acceptor molecule, poised at the start of a transport system, accepts the gift.

About Those Roving Pigments

There are only about eight classes of pigments, but this limited group gets around in the world. For example, animals synthesize some pigments (such as melanin) but not carotenoids. These originate with photoautotrophs and move up through food webs, as when tiny aquatic snails feed on green algae and flamingos eat the snails. Flamingos modify the ingested carotenoids in diverse ways. For instance, beta-carotene molecules are split to form two molecules of vitamin A, the precursor of a visual pigment (retinol) that transduces light to electric signals in the flamingo's eyes. Beta-carotene molecules also are dissolved in fat reservoirs under the flamingo's skin, where they are taken up by skin cells that differentiate and give rise to glorious pink feathers.

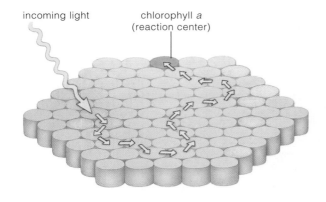

Figure 7.9 *Random walk of energy from photons among the pigments of a photosystem. Excitation energy quickly reaches a specialized chlorophyll a molecule, at the system's reaction center. Energy that reaches it cannot escape. When sufficiently activated, a reaction center gives up electrons to a nearby transport system.*

Chlorophylls and accessory pigments interact in light-harvesting photosystems. Photon absorption initiates a random walk of excitation energy among them and ends at a specialized chlorophyll a at the system's reaction center, which gives up electrons to ATP-producing machinery.

THE LIGHT-DEPENDENT REACTIONS

We are now ready to look more closely at the first stage of photosynthesis, the **light-dependent reactions**, as it proceeds in chloroplasts. Three events unfold during this stage. *First*, the pigments of photosystems absorb photon energy and give up excited electrons. *Second*, transfers of electrons and hydrogen through electron transport systems lead to ATP and NADPH formation. *Third*, the pigments that gave up electrons in the first place get electron replacements.

The ATP-Producing Machinery

The chloroplast's thylakoid membrane incorporates the light-harvesting photosystems, many thousands of them in some plants (Figure 7.10). Excited electrons from the reaction center of each photosystem are picked up by a primary acceptor molecule, which transfers them to an adjacent electron transport system in the membrane.

Electron transport systems, remember, are organized arrays of enzymes, coenzymes, and other proteins in certain cell membranes. As described earlier in Section 6.6, one component transfers electrons to another in orderly sequence. Some energy escapes at each transfer, but much of it is harnessed to drive specific reactions. We will focus on some of these reactions in Section 7.5. For now, simply note that photosystems and electron transport systems work together as the machinery that produces ATP, NADPH, or both. Chloroplasts happen to contain two different models of this machinery. They allow plants to produce ATP by two different pathways—one cyclic and the other noncyclic.

Cyclic Pathway of ATP Formation

The simplest pathway starts when a primary acceptor molecule picks up excited electrons from the reaction center, designated *P700*, of a *type I* photosystem. In the **cyclic pathway of ATP formation**, the electrons "cycle"

from the reaction center, through an adjacent electron transport system, then back to P700. Energy associated with the electron flow drives the formation of ATP from the chloroplast's pool of ADP and unbound phosphate:

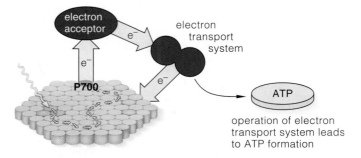

operation of electron transport system leads to ATP formation

The cyclic pathway is probably the oldest means of ATP production. The first cells to use it were as tiny as existing bacteria, so their body-building programs were scarcely enormous. ATP alone would have provided enough energy to build enough organic compounds. Building larger organisms requires far greater amounts of organic compounds—and vast amounts of hydrogen atoms and electrons. Long ago, in the forerunners of plants, the machinery of the cyclic pathway apparently underwent expansion and became the basis of a more efficient ATP-forming pathway. Amazingly, this bit of modified machinery changed the course of evolution.

Noncyclic Pathway of ATP Formation

The cyclic pathway still operates in the chloroplasts of trees, weeds, lettuces, and other members of the plant kingdom. But the **noncyclic pathway of ATP formation** now dominates. Electrons are not cycled through this pathway. They depart with NADPH, *and electrons from water molecules replace them*.

The pathway goes into operation when rays from the sun bombard a *type II* photosystem. Photon energy

Figure 7.10 Location of photosystems and electron transport systems relative to the lipid bilayer of the chloroplast's inner, thylakoid membrane system. All the components of electron transport systems are dark green. (The components of one transport system extend from P680 of photosystem II to P700 of photosystem I. The components of the second transport system extend from P700 and deliver electrons and hydrogen to NADP+.)

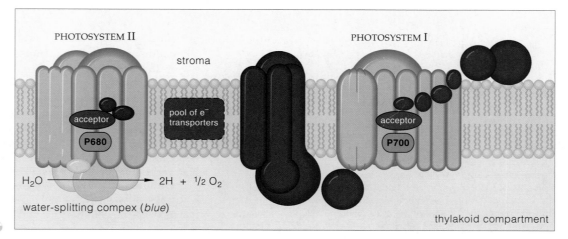

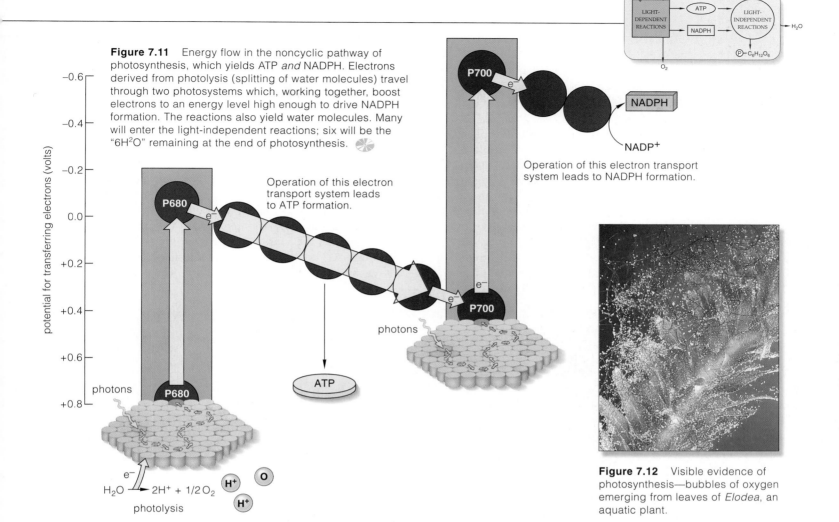

Figure 7.11 Energy flow in the noncyclic pathway of photosynthesis, which yields ATP *and* NADPH. Electrons derived from photolysis (splitting of water molecules) travel through two photosystems which, working together, boost electrons to an energy level high enough to drive NADPH formation. The reactions also yield water molecules. Many will enter the light-independent reactions; six will be the "$6H_2O$" remaining at the end of photosynthesis.

Operation of this electron transport system leads to ATP formation.

Operation of this electron transport system leads to NADPH formation.

$$H_2O \longrightarrow 2H^+ + 1/2\,O_2$$

photolysis

Figure 7.12 Visible evidence of photosynthesis—bubbles of oxygen emerging from leaves of *Elodea*, an aquatic plant.

makes this photosystem's reaction center (chlorophyll P680) give up electrons. The energy input also triggers **photolysis**, a reaction sequence in which molecules of water split into oxygen, hydrogen ions, and electrons. P680 attracts the less excited electrons as replacements for the excited ones that got away (Figure 7.11).

Meanwhile, excited electrons are moving through a transport system, then to P700 of *photosystem I*. They still have some extra energy left, and they get even more upon their arrival—because photons also happen to be bombarding P700. The energy boost in photosystem I puts electrons at a higher energy level that allows them to enter a second transport system. There a coenzyme, $NADP^+$, assists one of the enzyme components of the second electron transport system. It becomes NADPH when it accepts two of the electrons and a hydrogen ion from the enzyme, *then transfers them to the sites where organic compounds are built.*

The Legacy—A New Atmosphere

On sunny days, on the surfaces of aquatic plants, you can see bubbles of oxygen (Figure 7.12). The oxygen is a by-product of the noncyclic pathway of photosynthesis. This pathway may have evolved more than 2 billion years ago. At first, the oxygen simply dissolved in seas, lakes, wet mud, and other bacterial habitats. By about 1.5 billion years ago, significant amounts of dissolved oxygen were escaping into what had been an oxygen-free atmosphere. The accumulation of oxygen changed the atmosphere forever. And it made possible aerobic respiration—the most efficient pathway for releasing usable energy stored in organic compounds. Ultimately, the emergence of the noncyclic pathway allowed you and all other animals to be around today, breathing the oxygen that helps keep your cells alive.

In the light-dependent reactions, electrons of photosystem reaction centers are raised to higher energy levels, released, then transferred through nearby electron transport systems. Their energy drives ATP and NADPH formation. ATP can deliver energy and NADPH can deliver hydrogen and electrons to sites where organic compounds are built.

Oxygen, a by-product of photosynthesis, profoundly changed the early atmosphere and made aerobic respiration possible.

A CLOSER LOOK AT ATP FORMATION IN CHLOROPLASTS

As you probably have noticed, we saved the trickiest question for last. Exactly how does ATP form during the noncyclic pathway of photosynthesis? To arrive at the answer, let's walk through Figure 7.13.

Inside the thylakoid compartment of a chloroplast, many hydrogen ions (H^+) accumulate as a result of photolysis, the first step of the pathway. At this step, enzyme action splits water molecules into oxygen, hydrogen ions, and electrons. The two oxygen atoms combine at once to form O_2, which diffuses out of the chloroplast, then out of the cell. A primary acceptor molecule picks

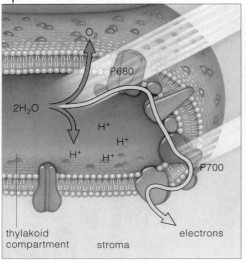

Therefore, through a combination of photolysis and electron transport, hydrogen ions are ever increasingly concentrated inside the compartment than they are in the stroma. The unequal distribution of these positively charged ions also creates a difference in electric charge across the membrane. An electric gradient, as well as a concentration gradient, has become established.

The combined force of the H^+ concentration gradient and electric gradient propels hydrogen ions through the interior of ATP synthases, a type of transport protein that spans the thylakoid membrane (Figure 7.13c). In other words, the ions flow out from the compartment, through ATP synthases, into the stroma. ATP synthases have built-in enzymatic machinery. The flow of ions

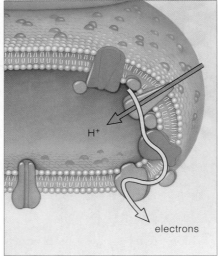
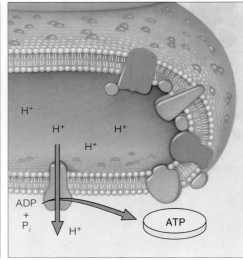

a Hydrogen ions derived from the splitting of water molecules (photolysis) accumulate inside the thylakoid compartment. (The two oxygen atoms combine to form O_2 and simply diffuse out of the chloroplast, then out of the photosynthetic cell.)

b More hydrogen ions accumulate in the compartment as certain components of electron transport systems accept excited electrons. They pick up hydrogen ions in the stroma and shunt them across the membrane at the same time.

c The ion accumulation sets up concentration and electric gradients across the thylakoid membrane. The ions follow the gradient and flow into the stroma, through the interior of ATP synthases embedded in the membrane. The flow drives ATP formation from ADP and P_i.

Figure 7.13 How ATP forms in chloroplasts during the noncyclic pathway of photosynthesis.

up the excited electrons and transfers them to a nearby electron transport system that is also positioned in the thylakoid membrane. However, as indicated in Figure 7.13a, the hydrogen ions derived from water remain in the thylakoid compartment.

Still more hydrogen ions enter the compartment when the electron transport systems are operating. This is true of both the cyclic and noncyclic pathways. As Figure 7.13b indicates, at the same time that certain components of the transport system accept electrons, they also pick up hydrogen ions from the stroma. They immediately shunt the ions across the membrane.

through them drives the machinery, which catalyzes the attachment of unbound phosphate to ADP. In this way, an ATP molecule forms.

The sequence of events just described is often called the *chemiosmotic theory* of ATP formation. As you will see in the next chapter, the same theory applies to ATP formation in mitochondria as well as in chloroplasts.

In chloroplasts, H^+ concentration and electric gradients form across the thylakoid membrane. The flow of ions from the thylakoid compartment to the stroma drives ATP formation.

7.6 LIGHT-INDEPENDENT REACTIONS

Light-independent reactions are the "synthesis" part of photosynthesis. ATP delivers the required energy for the reactions. NADPH delivers the required hydrogen and electrons. Carbon dioxide (CO_2) in the air around photosynthetic cells provides the carbon and oxygen.

We say the reactions are light-independent because they do not depend directly on sunlight. They proceed even in the dark, as long as ATP and NADPH hold out.

Capturing Carbon

Let's track a CO_2 molecule that diffuses into air spaces inside a leaf and ends up next to a photosynthetic cell. It diffuses into the cell, then into a chloroplast's stroma. An enzyme attaches the carbon atom of CO_2 to **RuBP** (ribulose bisphosphate), a compound with a backbone of five carbon atoms. The enzyme is called "rubisco" (for RuBP carboxylase). Its action produces an unstable six-carbon intermediate that splits into two molecules of **PGA** (phosphoglycerate). PGA is a stable molecule with a three-carbon backbone. The incorporation of a carbon atom from CO_2 into a stable organic compound is called **carbon fixation**.

Carbon fixation is the first step of a cyclic pathway, the **Calvin-Benson cycle**, that yields a sugar phosphate molecule and regenerates RuBP. For our purposes, we can simply focus on the carbon atoms of the substrates, intermediates, and end products, as Figure 7.14 shows.

Building the Glucose Subunits

Each PGA receives a phosphate group from ATP, then hydrogen and electrons from NADPH. The resulting intermediate is called **PGAL** (phosphoglyceraldehyde). To build *one* six-carbon sugar phosphate, the carbon from six CO_2 molecules must be fixed and twelve PGAL must form. Most of the PGAL is rearranged to form new RuBP, which can be used to fix more carbon. But two of the PGAL combine, thus forming glucose (six carbons) with an attached phosphate group.

When phosphorylated this way, glucose is primed to enter other reactions. Plants use it as a building block for their main carbohydrates—sucrose, cellulose, and starch. *The synthesis of these organic compounds by other pathways marks the end of the light-independent reactions.*

With six turns of the cycle, enough RuBP molecules form to replace the ones used in carbon fixation. The ADP, NADP+, and phosphate leftovers diffuse through the stroma, to the light-dependent reaction sites. There they are converted back to NADPH and ATP.

Photosynthetic cells convert newly formed sugar phosphates to sucrose or starch during daylight hours. Of all plant carbohydrates, sucrose is the most easily transportable. Starch is the most common storage form.

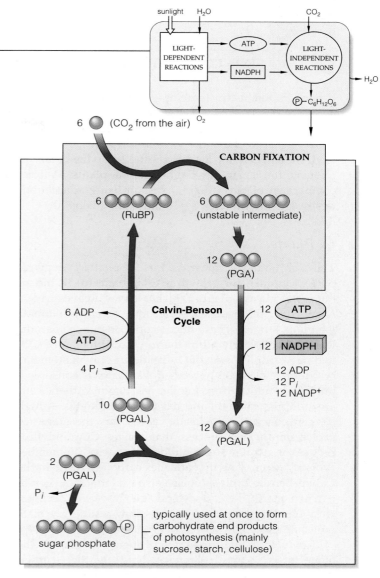

Figure 7.14 Summary of the light-independent reactions of photosynthesis. *Red* circles signify the carbon atoms of the key molecules. All of the intermediates have one or two phosphate groups attached. For simplicity, this diagram shows only the phosphate group on the resulting sugar phosphate. Remember also, many water molecules that formed in the light-dependent reactions enter this pathway, and six remain at its conclusion.

These cells also convert excess PGAL to starch. They briefly store starch as grains in the stroma. (You can see one of these grains in Figure 7.2c.) After the sun goes down, cells convert starch to sucrose, for export to other living cells in leaves, stems, and roots. Ultimately, the products and intermediates of photosynthesis end up as energy sources and building blocks for all of the lipids, amino acids, and other organic compounds that are required for growth, survival, and reproduction.

During the Calvin-Benson cycle, carbon is "captured" from carbon dioxide, a sugar phosphate forms during reactions that require ATP and NADPH, and RuBP (necessary to capture the carbon) is regenerated.

FIXING CARBON—SO NEAR, YET SO FAR

If sunlight intensity, air temperature, rainfall, and the composition of soil were uniform everywhere in the world, all year long, maybe photosynthetic reactions would proceed the same way in all plants. However, environments obviously differ—and so do the details of photosynthesis among their resident plants. A brief comparison of two different carbon-fixing adaptations to stressful environments will illustrate this point.

C4 Plants

Think of Kentucky bluegrass. Like all plants, it depends on CO_2 uptake for growth. CO_2 is plentiful in the air but is not always plentiful inside leaves. Leaves have a waxy cover that restricts water loss, except at **stomata** (singular, stoma). These microscopic openings span the leaf surface. Nearly all of the inward diffusion of CO_2 and outward diffusion of O_2 can occur only at stomata.

Stomata close on hot, dry days. Water is conserved, but CO_2 cannot diffuse into leaves. Photosynthetic cells are still busy, though, and oxygen accumulates. A high oxygen concentration inside leaves sets the stage for *photorespiration,* a process that wastes CO_2 and thus reduces a plant's sugar-building capacity. Remember rubisco, the enzyme that attaches carbon to RuBP in the Calvin-Benson cycle? Rubisco also can attach oxygen to it when the O_2 level rises and the CO_2 level declines. Only one (not two) PGA forms, along with a glycolate molecule that eventually is degraded back to CO_2.

Kentucky bluegrass is one of many **C3 plants**. The name refers to the first intermediate (the three-carbon PGA) of its carbon-fixing pathway. By contrast, the first intermediate formed when crabgrass, corn, and many other plants fix carbon is the four-carbon oxaloacetate. Hence *their* name, **C4 plants** (Figures 7.15 and 7.16*b*).

C4 plants maintain adequate amounts of CO_2 inside leaves even though they, too, close stomata on hot, dry days. *They fix carbon not once but twice,* in two types of photosynthetic cells. First mesophyll cells use the CO_2 to produce oxaloacetate, which is transferred to bundle-sheath cells around leaf veins. There CO_2 is released and fixed again—in the Calvin-Benson cycle. With this pathway, C4 plants get by with tinier stomata and lose less water, and they can produce more sugars than C3 plants can when conditions are hot, dry, and bright.

For example, 80 percent of the plants that evolved in Florida's heat are C4 species, compared to 0 percent in Manitoba, Canada. Kentucky bluegrass and other C3 species have greater advantage where temperatures drop below 25°C (they are not as sensitive to cold). Mix C3 and C4 species from different regions in a garden, and one or the other will do better during at least part of the year. That's why a bluegrass lawn that does well during cool spring weather in San Diego is overwhelmed during a hot summer by a C4 plant—crabgrass.

Especially in hot weather, photorespiration greatly lowers the photosynthetic efficiency of many important C3 crop plants, including tomatoes, rice, barley, wheat, soybeans, and potatoes. In some experiments, tomatoes were grown in hothouses at CO_2 concentrations high enough to eliminate photorespiration. The growth rate increased, in some cases by as much as five times. If photorespiration is so significantly wasteful, then why hasn't natural selection eliminated it? The answer may lie with RuBP carboxylase, the switch-hitting enzyme that sets the process in motion. The enzyme evolved in

Figure 7.15 (**a**) The internal structure of a leaf from corn (*Zea mays*), a typical C4 plant. Its photosynthetic mesophyll cells (*light green*) surround bundle-sheath cells (*light blue*) that surround tubes in leaf veins. The tubes transport water into leaves and photosynthetic products from them. (**b**) The carbon-fixing system that precedes the Calvin-Benson cycle in C4 plants.

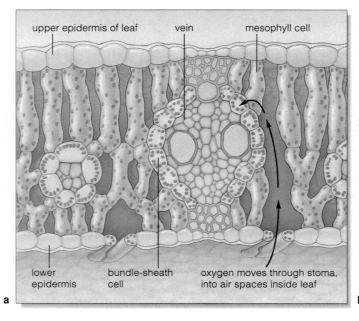

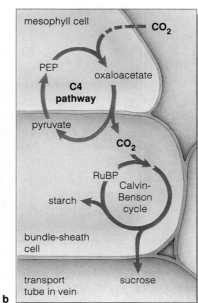

ancient times, when atmospheric levels of oxygen were still low and carbon dioxide levels high. Maybe its structure cannot undergo any mutation without adversely affecting its essential carbon-fixing activity. Or maybe the pathway that breaks down glycolate to carbon dioxide has proved so adaptive that it cannot be eliminated. Glycolate can be toxic at high concentrations.

The C4 pathway evolved separately in many lineages of flowering plants over the past 50 to 60 million years. By then, atmospheric CO_2 had declined to levels that gave C4 plants a selective advantage over C3 plants.

It will be interesting to see which pathway will be the most adaptive in years to come. The atmospheric levels of CO_2 have been rising for decades, and some ecologists are predicting that they will double during the next fifty years. If this happens, photorespiration will once again decline—possibly with beneficial effects on many of our vital crops.

CAM Plants

We see another splendid carbon-fixing adaptation in deserts and other dry environments. Think of a cactus plant, a succulent with juicy, water-storing tissues and thick surface layers that restrict water loss. It cannot open its stomata on blistering-hot days without losing precious water. Instead, it opens them and fixes CO_2 *at night*. Its cells store the resulting intermediate in their central vacuoles and then use it for photosynthesis the next day, when the stomata close.

Many plants are adapted this way. They are known as **CAM plants** (short for *Crassulacean Acid Metabolism*). Unlike C4 species, CAM plants do not fix carbon twice, in different types of cells. They fix it in the same cells, but at different times (Figure 7.16c).

During prolonged droughts, when many plants die, some CAM plants survive by keeping their stomata closed even at night. They repeatedly fix the CO_2 that forms during aerobic respiration. Not that much forms, but it is enough to allow these plants to maintain very low rates of metabolism. As you may have deduced,

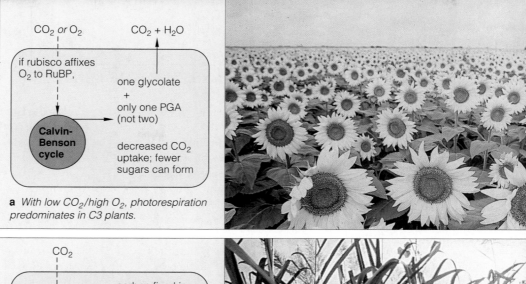

a With low CO_2/high O_2, photorespiration predominates in C3 plants.

b With low CO_2/high O_2, Calvin-Benson cycle predominates in C4 plants.

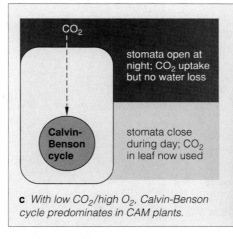

c With low CO_2/high O_2, Calvin-Benson cycle predominates in CAM plants.

Figure 7.16 Three ways of fixing carbon in hot, dry weather, when the CO_2 level is low and the O_2 level is high in leaves. (**a**) The C3 pathway is common among evergreen trees and shrubs as well as many nonwoody plants of temperate zones, such as sunflowers. (**b**) The C4 pathway is common among grasses and other plants that evolved in the tropics. Examples are corn, sorghum, crabgrass, and Bermuda grass, and the sugarcane shown here. (**c**) CAM plants open their stomata and fix carbon at night. Examples are pineapple, cacti and many other succulents (plants having a low surface-to-volume ratio), orchids, and Spanish moss.

CAM plants grow slowly. Try growing a cactus plant in Seattle or some other place with a mild climate and it will compete poorly with C3 and C4 plants.

C4 plants and CAM plants both have modified ways of fixing carbon for photosynthesis. The modifications counter the stress imposed by hot, dry conditions in their environments.

7.8 AUTOTROPHS, HUMANS, AND THE BIOSPHERE

As we conclude this chapter, remind yourself of where photosynthesis "fits" in the world of living things. Think about the mind-boggling numbers of single-celled and multicelled photoautotrophic organisms in which the light-trapping and sugar-building reactions are proceeding at this very moment, on land and in the sunlit waters of the Earth. The sheer volume of reactant molecules and product molecules that the photosynthesizers deal with might surprise you.

For example, drifting through the surface waters of the oceans and seas are uncountable numbers of single cells. You can't see them without a microscope—a row of 7 million cells of one aquatic species would be less than a quarter-inch long. Yet are they abundant! In some parts of the world, a cup of seawater may hold 24 million cells of one species, and that doesn't even include all the cells of *other* aquatic species.

Nearly all of the drifters are photoautotrophic bacteria, plants, and protistans. Together they are the pastures of the seas, the food base for diverse consumers. The pastures "bloom" in the spring, when seawater becomes warmer and enriched with nutrients that winter currents churn up from the deep. Then, populations burgeon, by rapid cell divisions. Until NASA gathered data from satellites in space, biologists had no idea that the number of cells and their distribution were so stupendous. For example, the satellite image in Figure 7.17*b* shows a springtime bloom in the surface waters of the North Atlantic Ocean. It stretches from North Carolina all the way past Spain!

Collectively, these single cells have enormous impact on the global climate. When they engage in photosynthesis, they sponge up nearly half the carbon dioxide we humans release each year, as when we burn fossil fuels or burn forests. Without the aquatic photoautotrophs, carbon dioxide in the air would accumulate more rapidly and possibly accelerate global warming, as described in Section 49.7. If our planet warms too much, all lowlands along the coasts of islands and continents may become submerged, and nations that are now the greatest food producers may be hit hard.

Amazingly, every day, humans dump tons of industrial wastes, fertilizers, and raw sewage into the ocean and thereby alter the living conditions for those drifting cells. How much of the noxious chemical brew will they continue to tolerate?

One final point: Photosynthesis is so central to our understanding of nature, it is sometimes easy to overlook other, less common energy-acquiring routes. Near hydrothermal vents on the floor of deep oceans, in hot springs, even in waste heaps of coal mines, we find diverse bacteria that are classified as **chemoautotrophs**. They obtain energy not from the sun's rays but rather from a variety of inorganic compounds in their environment. For example, certain species strip hydrogen and electrons from ammonium ions, and other species obtain them from iron and sulfur compounds. These organisms, too, exist in monumental numbers. They influence the global cycling of nitrogen and other vital elements through the biosphere. We will return later to their environmental effects. In this unit, we turn next to pathways by which cells release energy from glucose and other biological molecules, the chemical legacy of autotrophs everywhere.

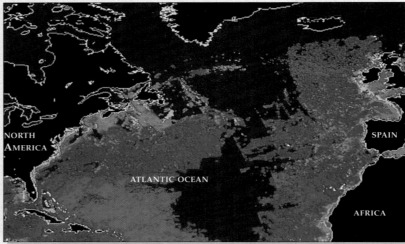

a Photosynthetic activity in winter. In this color-enhanced image, *red-orange* shows where chlorophyll concentrations are greatest.

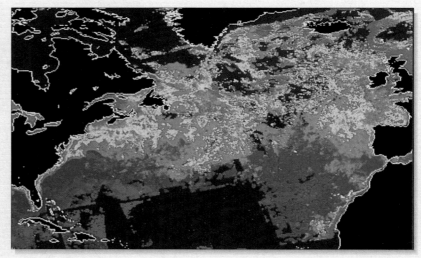

b Photosynthetic activity in spring.

Figure 7.17 Two satellite images that help convey the astounding magnitude of photosynthetic activity during the spring in the North Atlantic portion of the world ocean.

SUMMARY

Pigments and Radiation From the Sun:

1. Plants and other photoautotrophs use wavelengths of visible light as an energy source and carbon dioxide as the carbon source for building organic compounds. Animals and other heterotrophs get carbon and energy from organic compounds synthesized by plants and other autotrophs.

2. Photosynthesis is the main biosynthetic pathway by which carbon and energy enter the web of life. Figure 7.18 summarizes its two sets of reactions. The reactions start by trapping light energy ("photo") and end with assembly reactions ("synthesis").

   a. In plants, the light-dependent reactions take place at the thylakoid membrane system inside chloroplasts. These reactions produce ATP and NADPH.

   b. The light-independent reactions proceed outside the membrane system, in the chloroplast's stroma. They produce the sugar phosphates used in building sucrose, starch, and other end products of photosynthesis.

3. The sun continually radiates energy, which travels as waves through space. Of the entire electromagnetic spectrum, photoautotrophs harness the wavelengths of visible light, which consist of distinct packets of energy called photons.

   a. The shorter the wavelength (that is, the horizontal distance between crests of two successive waves), the more energetic its photons.

   b. We perceive wavelengths of visible light as colors that range from violet and blue (the most energetic) to red (the least energetic).

4. All but one group of photoautotrophs have chlorophyll *a* and various accessory pigments that collectively can absorb all the wavelengths of visible light.

   a. Chlorophylls absorb all wavelengths of visible light except green and yellow-green ones, which they transmit.

   b. Accessory pigments, including the carotenoids, anthocyanins, and phycobilins, absorb wavelengths that the chlorophylls cannot absorb.

5. Each pigment has a light-catching array of atoms that absorbs photons of specific energies. If photons of a given wavelength match the amount of energy required to boost electrons in that array to a higher energy level, they will be absorbed. If not, the wavelength will be transmitted and will impart color to the pigment.

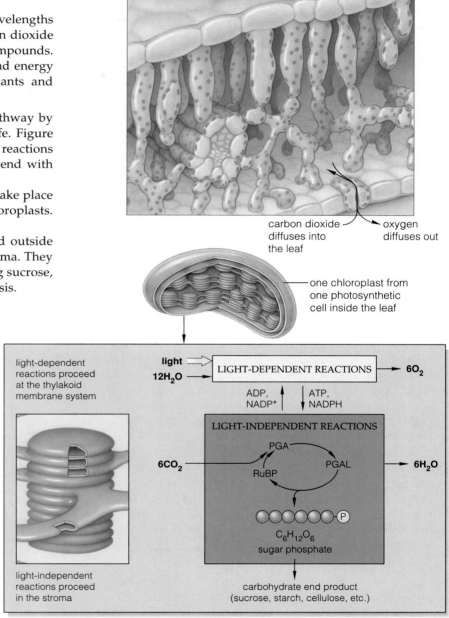

Figure 7.18 Summary of the main reactants, intermediates, and products of photosynthesis, corresponding to the equation:

$$12H_2O + 6CO_2 \xrightarrow{\text{LIGHT ENERGY}} 6O_2 + C_6H_{12}O_6 + 6H_2O$$

WATER   CARBON   OXYGEN   GLUCOSE   WATER
DIOXIDE

Starting with the light-dependent reactions, six water molecules are split, yielding twelve electrons that are necessary to form a sugar phosphate (glucose with a phosphate group attached). Other water molecules enter and leave reactions; a net $6H_2O$ remain at the end of the remaining pathway. For every four electrons, three ATP and two NADPH form. During the light-independent reactions, *each turn* of the Calvin-Benson cycle requires one CO_2, three ATP, and two NADPH. Each sugar phosphate that forms has a backbone of six carbon atoms, so its formation requires six turns of the cycle.

The Reactions of Photosynthesis:

1. Photosystems are clusters of pigments and proteins in thylakoid membranes (or in the plasma membrane of single-celled photoautotrophs). Thylakoid membranes are peppered with two types: photosystems I and II.

a. Photosystems and neighboring electron transport systems work together in pathways of ATP formation.

b. Photosystem I operates during a cyclic pathway of ATP formation.

c. Photosystems I and II operate together during the noncyclic pathway of ATP formation.

2. Here are key points concerning the light-dependent reactions at the thylakoid membrane of chloroplasts:

a. When photons are absorbed, a photosystem's reaction center (a special molecule of chlorophyll *a*) gives up excited electrons. A primary acceptor molecule transfers these to an electron transport system.

b. In the cyclic pathway, electrons travel from the reaction center of photosystem I (P700 chlorophyll), then give up their extra energy in a transport system, and then return to photosystem I.

c. In the noncyclic pathway, electrons travel from the reaction center of photosystem II (P680 chlorophyll), give up extra energy in a transport system, then enter photosystem I. Photon energy absorbed here excites the electrons further. They give up their extra energy in another transport system, and end up in NADPH.

d. At the start of the noncyclic pathway, photolysis occurs: water molecules are split into oxygen, hydrogen ions (H^+), and electrons. The electrons replace the ones that P680 initially gives up.

e. Photolysis and operation of the electron transport systems dump H^+ into the thylakoid membrane's inner compartment. Their action sets up concentration and electric gradients across the membrane that drive ATP formation at nearby transport proteins (ATP synthases) in the membrane.

3. Here are key points concerning the light-independent reactions in the stroma of chloroplasts:

a. ATP and NADPH formed during the first stage of photosynthesis are required for the reactions. The ATP delivers chemical energy and the NADPH delivers hydrogen and electrons to the stroma, where sugar phosphates form and then are combined into starch, cellulose, and other end products.

b. The sugar phosphates form in the Calvin-Benson cycle. This cyclic pathway begins when carbon from CO_2 in the air is affixed to RuBP, making an unstable intermediate that splits into two PGA. ATP transfers a phosphate group to each PGA. The resulting molecule receives H^+ and electrons from NADPH to form PGAL.

c. For every six carbon atoms that enter the cycle by way of carbon fixation, twelve PGAL form. Two PGAL are used to produce a six-carbon sugar phosphate. The remainder are used to regenerate the RuBP.

Carbon-Fixing Adaptations:

1. Rubisco, the carbon-fixing enzyme of the Calvin-Benson cycle, evolved when the atmosphere had far more carbon dioxide and far less oxygen. Today, when the O_2 level is higher than the CO_2 level in leaves, the enzyme attaches oxygen rather than carbon to RuBP. This results in formation of only one PGA (not two) and glycolate, a compound that can't be used to form sugars but instead is degraded to carbon dioxide and water. This wasteful process is called photorespiration.

2. Photorespiration predominates in C3 plants such as sunflowers under hot, dry conditions. Then, stomata close and O_2 from photosynthesis builds up in leaves to levels higher than CO_2 levels. Sugarcane and other C4 plants raise the CO_2 level by fixing carbon twice, in two cell types. CAM plants such as cacti raise it by fixing carbon at night, when stomata are open.

Review Questions

1. A cat eats a bird, which earlier speared and ate a caterpillar that had been chewing on a weed. Which of these organisms are the autotrophs? the heterotrophs? *CI*

2. Summarize the photosynthesis reactions as an equation. State the key events of both stages of reactions, then fill in the blanks on the diagram on the facing page. *7.1*

3. Which of the following pigments are most visible in a maple leaf in summer? Which become the most visible in autumn? *7.3*
   a. chlorophylls
   b. carotenoids
   c. anthocyanins
   d. phycobilins

4. Identify which of the following substances accumulates inside the thylakoid compartment of chloroplasts: glucose, chlorophyll, carotenoids, hydrogen ions, or fatty acids. *7.5*

5. Which is *not* used in the light-independent reactions: ATP, NADPH, RuBP, carotenoids, free oxygen, CO_2, or enzymes? *7.6*

6. How many carbon atoms from CO_2 must enter the Calvin-Benson cycle to produce one sugar phosphate? Why? *7.6*

7. A busily photosynthesizing plant takes up molecules of CO_2 that have incorporated radioactively labeled carbon atoms ($^{14}CO_2$). Identify the compound in which the labeled carbon will appear first: NADPH, PGAL, pyruvate, or PGA. *7.6*

Self-Quiz (*Answers in Appendix IV*)

1. Photosynthetic autotrophs use _____ from the air as a carbon source and _____ as their energy source.

2. In plants, light-*dependent* reactions proceed at the _____ .
   a. cytoplasm   c. stroma
   b. plasma membrane   d. thylakoid membrane

3. The light-*independent* reactions proceed in the _____ .
   a. cytoplasm   c. stroma
   b. plasma membrane   d. grana

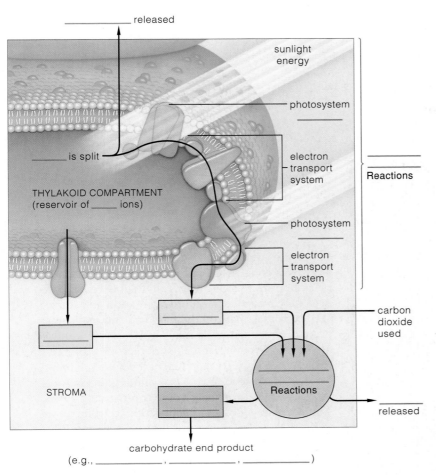

_____ released

sunlight energy

photosystem

_____ is split

electron transport system

THYLAKOID COMPARTMENT
(reservoir of _____ ions)

photosystem

electron transport system

} Reactions

carbon dioxide used

STROMA

Reactions

released

carbohydrate end product
(e.g., _____ , _____ , _____)

2. About 200 years ago, Jan Baptista van Helmont performed an experiment on the nature of photosynthesis. He wanted to know where growing plants acquired the raw materials necessary to increase in size. For his experiment, he planted a tree seedling weighing 5 pounds in a barrel filled with 200 pounds of soil. He watered the tree regularly. After five years passed, van Helmont again weighed the tree and the soil. At that time the tree weighed 169 pounds, 3 ounces. The soil weighed 199 pounds, 14 ounces. Because the tree's weight had increased so much and the soil's weight had decreased so very little, he concluded the tree gained weight as a result of the water he had added to the barrel.

Given your knowledge of the composition of biological molecules, why was van Helmont's conclusion misguided? On the basis of the current model of photosynthesis, provide a more plausible explanation of his results.

3. Like other accessory pigments, the carotenoids extend the effective range of light absorption beyond that of the main photosynthetic pigment, chlorophyll *a*. They also protect plants from *photo-oxidation*. This process begins when excitation energy in chlorophylls drives the conversion of oxygen into free radicals. As you know from Chapter 6, free radicals damage organic compounds and kill cells. When mutant plants that cannot produce carotenoids are grown in light, they bleach white and die. Given this observation, which molecules in the plant cells are among the first to go?

4. Captain Kirk and First Officer Spock land on a planet in a distant galaxy, where they find populations of a vibrantly purple, carbon-based life form. Spock suspects that the form secures the carbon and energy necessary for survival by a process similar to photosynthesis. How would he go about determining whether this is the case?

4. In the light-dependent reactions, _____ .
   a. carbon dioxide is fixed
   b. ATP and NADPH form
   c. CO₂ accepts electrons
   d. sugar phosphates form

5. When a photosystem absorbs light, _____ .
   a. sugar phosphates are produced
   b. electrons are transferred to ATP
   c. RuBP accepts electrons
   d. light-dependent reactions begin

6. The Calvin-Benson cycle starts when _____ .
   a. light is available
   b. light is not available
   c. carbon dioxide is attached to RuBP
   d. electrons leave a photosystem

7. In the light-independent reactions, ATP makes a phosphate-group transfer to _____ .
   a. RuBP   b. NADP⁺   c. PGA   d. PGAL

8. Match each event with its most suitable description.
   ____ RuBP used; PGA forms   a. photon absorption
   ____ ATP and NADPH used   b. noncyclic pathway
   ____ NADPH forms   c. CO₂ fixation
   ____ ATP and NADPH form   d. PGAL forms
   ____ only ATP forms   e. H⁺ and e⁻ to NADP⁺
   ____ energy matches amount   f. cyclic pathway
         needed to excite electrons

Critical Thinking Questions

1. Suppose a garden in your neighborhood is filled with red, white, and blue petunias. Explain the floral colors in terms of which wavelengths of light they are absorbing and reflecting.

Selected Key Terms

Readings

Davis, B. February 1992. "Going for the Green." *Discover* 13: 20. Describes work on artificial systems for photosynthesis.

Hendry, George. May 1990. "Making, Breaking, and Remaking Chlorophyll." *Natural History* 36–41.

Web Site See *http://www.wadsworth.com/biology* for practice quiz questions, hypercontents, BioUpdates, and critical thinking. The Wadsworth Biology Resource Center provides a wealth of information fully organized and integrated by chapter.

ENERGY-RELEASING PATHWAYS

The Killers Are Coming! The Killers Are Coming!

In 1990, descendants of "killer" bees that flew out of South America a few decades earlier buzzed across the border between Mexico and Texas. By 1995, they had invaded 13,287 square kilometers of Southern California and were busily setting up colonies.

When provoked, the bees behave in a terrifying way. For example, thousands flew into action simply because a construction worker started up a tractor a few hundred yards away from their hive. Agitated bees flew into a nearby subway station and started stinging passengers on the platform and inside the trains. They killed one person and injured a hundred others. Not too long ago, they put a couple of tree trimmers in Indio, California, in the hospital; they also killed two people in Texas and two in Arizona.

Where did these bees come from? In the 1950s, some queen bees had been shipped from Africa to Brazil for selective breeding experiments. Why? It happens that honeybees are big business. In addition to being a source of nutritious honey, bees are rented to commercial orchards— where their collective pollinating activities may make a significant contribution to the production of fruit. For example, if you position a screened cage around an orchard tree that has blossomed out, less than 1 percent of the tree's flowers will set fruit. Put a hive of honeybees in the same cage, and 40 percent of the flowers will set fruit.

Compared to their relatives in Africa, bees in Brazil are rather sluggish pollinators and honey producers. By cross-breeding the two varieties, researchers thought they might be able to come up with a strain of mild-mannered but zippier bees. They put local bees and imported bees together inside netted enclosures, complete with artificial hives. Then they let nature take its course.

Figure 8.1 One of the mild-mannered honeybees buzzing in for a landing on a flower, wings beating with energy provided by ATP. If this were one of its Africanized relatives protecting a hive, possibly you would not stay around to watch the landing. Both kinds of bees look alike. How can we tell them apart? From our own biased perspective, Africanized bees are the ones with an attitude problem.

Twenty-six African queen bees escaped. That was bad enough. Then beekeepers got wind of preliminary experimental results. After learning that the first few generations of offspring were more energetic but not overly aggressive, they imported hundreds of African queens and encouraged them to mate with the locals. And they set off a genetic time bomb.

Before long, African bees became established in commercial hives—and in wild bee populations. And their traits became dominant. The "Africanized" bees do everything other bees do, but they do more of it faster. Their eggs develop into adults more quickly. Adults fly more rapidly, outcompete other bees for nectar, and even die sooner.

When something disturbs their hives or swarms, Africanized bees become extremely agitated. They can remain that way for as long as eight hours. Whereas a mild-mannered honeybee might chase an intruding animal fifty yards or so, a squadron of Africanized bees will chase it a quarter of a mile. If they catch up to it, they collectively can sting it to death.

Doing things faster means having a continuous supply of energy and efficient ways of using it. An Africanized bee's stomach can hold thirty milligrams of sugar-rich nectar—which is enough fuel to fly sixty kilometers. That's more than thirty-five miles! Besides this, compared to other kinds of bees, the flight muscle cells of an Africanized bee have larger mitochondria. These organelles specialize in releasing a great deal of energy from sugars and other organic compounds, then converting it to the energy of ATP.

Whenever they tap into the stored energy of organic compounds, Africanized bees reveal their biochemical connection with other organisms. Study a primrose or puppy, a mold growing on stale bread, an amoeba in pondwater, or a bacterium living on your skin, and you will discover that their energy-releasing pathways differ in some details. But all of the pathways require characteristic starting materials. They yield predictable products and by-products. And they yield the universal energy currency of life—ATP.

In fact, throughout the biosphere, organisms put energy and raw materials to use in amazingly similar ways. *At the biochemical level, we find undeniable unity among all forms of life.* We will return to this idea in the concluding section of the chapter.

KEY CONCEPTS

1. All organisms can release energy stored in glucose and other organic compounds, then use it in ATP production. The energy-releasing pathways differ from one another. But the main types all start with the breakdown of glucose to pyruvate.

2. The initial breakdown reactions, known as glycolysis, can proceed in the presence of oxygen or in its absence. Said another way, these reactions can be the first stage of either aerobic or anaerobic pathways.

3. Two kinds of energy-releasing pathways are completely anaerobic, from start to finish. We call them fermentation and anaerobic electron transport. They proceed only in the cytoplasm, and none yields more than a small amount of ATP for each glucose molecule metabolized.

4. Another pathway, aerobic respiration, also starts in the cytoplasm. But it alone runs to completion in organelles called mitochondria. Compared with the other pathways, aerobic respiration releases far more energy from glucose.

5. Aerobic respiration has three stages. First, pyruvate forms from glucose (through glycolysis). Second, different reactions break down the pyruvate to carbon dioxide. These reactions liberate electrons and hydrogen, which coenzymes deliver to a transport system. Third, stepwise electron transfers through the system help set up the conditions that favor ATP formation. Free oxygen accepts the electrons at the end of the line and combines with hydrogen, thereby forming water.

6. Over evolutionary time, photosynthesis and aerobic respiration became linked on a global scale. The oxygen-rich atmosphere, a long-term outcome of photosynthetic activity, sustains aerobic respiration, which has become the dominant energy-releasing pathway. And most kinds of photosynthesizers use carbon dioxide and water from aerobic respiration as raw materials when they synthesize organic compounds:

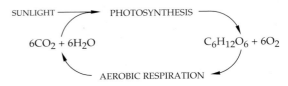

HOW CELLS MAKE ATP

Organisms stay alive by taking in energy. Plants and all other organisms that engage in photosynthesis get energy from the sun. Animals get energy secondhand, thirdhand, and so on, by eating plants and one another. Regardless of its source, energy must be in a form that can drive thousands of life-sustaining reactions. Energy that becomes converted into the chemical bond energy of adenosine triphosphate—ATP—serves that function.

Plants make ATP during photosynthesis. They and all other organisms also make ATP by breaking down carbohydrates (glucose especially), fats, and proteins. During the breakdown reactions, electrons get stripped from intermediates, then energy that is associated with the liberated electrons drives the formation of ATP. Electron transfers of the sort described in Section 6.6 are central to these energy-releasing pathways.

Comparison of the Main Types of Energy-Releasing Pathways

The first energy-releasing pathways evolved about 3.8 billion years ago, when conditions were very different on Earth. Because the atmosphere had little free oxygen, the pathways must have been *anaerobic*, which means they could run to completion without utilizing oxygen. Many bacteria and protistans still live in places where oxygen is absent or not always available. They make ATP by anaerobic routes, mainly fermentation pathways and anaerobic electron transport. Some cells in your own body can use an anaerobic route for short periods, but only when they are not receiving enough oxygen. Your cells, like most others, mainly use **aerobic respiration**, an oxygen-dependent pathway of ATP formation. With each breath you take, you are providing your actively respiring cells with a fresh supply of oxygen.

Make note of this point: *The main energy-releasing pathways all start with the same reactions in the cytoplasm.* During this initial stage of reactions, called **glycolysis**, enzymes cleave and rearrange each glucose molecule into two pyruvate molecules. Once this stage is over, the energy-releasing pathways differ. Most importantly, the aerobic pathway continues inside a mitochondrion (Figure 8.2). There, oxygen serves as the final acceptor of the electrons that were released at different reaction

a

steps. By contrast, anaerobic pathways start and end in the cytoplasm, where a substance other than oxygen is the final electron acceptor.

As you examine the energy-releasing pathways in sections to follow, keep in mind that the reaction steps do not proceed by themselves. Enzymes catalyze each step, and intermediate molecules formed at one step serve as substrates for the next enzyme in the pathway.

Overview of Aerobic Respiration

Of all the energy-releasing pathways, aerobic respiration gets the most ATP for each glucose molecule. Whereas anaerobic routes typically have a net yield of two ATP molecules, the aerobic route commonly yields thirty-six or more. If you were a bacterium, you wouldn't require much ATP. Being far larger, more complex, and highly active, you rely absolutely on the aerobic route's high yield. When a glucose molecule is the starting material, aerobic respiration can be summarized this way:

$$C_6H_{12}O_6 + 6O_2 \longrightarrow 6CO_2 + 6H_2O$$

GLUCOSE   OXYGEN   CARBON DIOXIDE   WATER

Figure 8.2 Where the aerobic and anaerobic pathways of ATP formation start and finish.

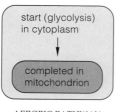

start (glycolysis) in cytoplasm

↓

completed in mitochondrion

AEROBIC PATHWAY

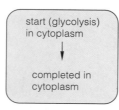

start (glycolysis) in cytoplasm

↓

completed in cytoplasm

ALL OTHER ENERGY-RELEASING PATHWAYS

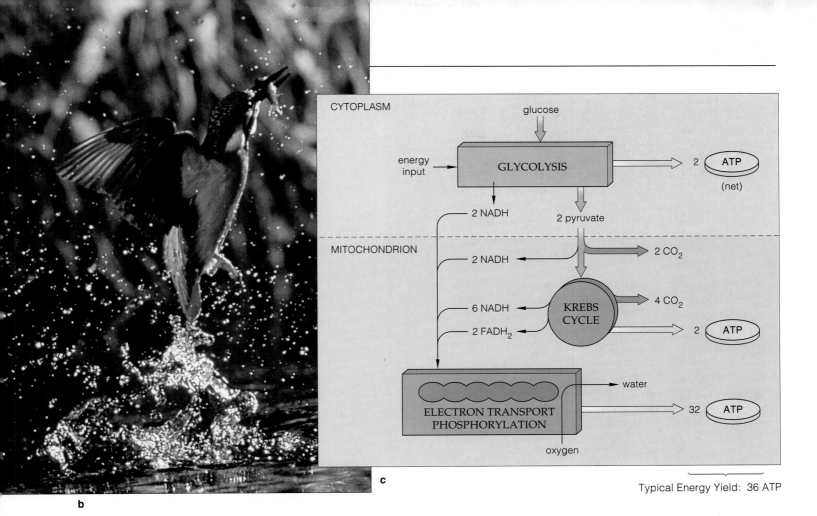

CYTOPLASM

glucose

energy input → GLYCOLYSIS → 2 ATP (net)

2 NADH

2 pyruvate

MITOCHONDRION

2 NADH ← 2 CO₂

6 NADH ← KREBS CYCLE → 4 CO₂

2 FADH₂ ← → 2 ATP

ELECTRON TRANSPORT PHOSPHORYLATION → water → 32 ATP

oxygen

c

Typical Energy Yield: 36 ATP

b

However, as you can see, the summary equation only tells us what the substances are at the start and finish of the pathway. In between are three reaction stages.

Let's use Figure 8.3 and the following descriptions as a brief overview of these reactions. The initial stage, again, is glycolysis. During the second stage, which is mainly a cyclic pathway of reactions called the **Krebs cycle**, pyruvate molecules are broken down to carbon dioxide and water by a series of enzyme-mediated steps. During both glycolysis and the Krebs cycle, coenzymes assist the enzymes by picking up electrons and hydrogen that are being stripped from the intermediates. Not much ATP forms in either stage. The large energy harvest comes *after* coenzymes deliver their cargo to an electron transport system.

In the third stage, the transport system functions as machinery for **electron transport phosphorylation**. It sets up hydrogen ion (H^+) concentration and electric gradients that drive ATP formation at nearby transport proteins. It is during this final stage that so many ATP molecules are produced. As it ends, oxygen inside the mitochondrion accepts the "spent" electrons from the

Figure 8.3 Overview of aerobic respiration, which proceeds through three stages. (**a, b**) Only this pathway delivers enough ATP to construct and maintain giant redwoods and other large, multicelled organisms. Only great amounts of ATP can sustain birds, bees, humans, and other highly active animals.

(**c**) Glucose partially breaks down to pyruvate in the first stage of reactions (glycolysis). In the second stage, which is mainly the Krebs cycle, pyruvate breaks down completely to carbon dioxide. Coenzymes (NAD^+ and FAD) pick up electrons and hydrogen stripped from intermediates at both stages.

In the final stage (electron transport phosphorylation), the loaded-down coenzymes (NADH and $FADH_2$) give up the electrons and hydrogen to a transport system. Energy released during the flow of electrons through the system drives ATP formation. Oxygen accepts the electrons at the end of the third stage.

From start (glycolysis) to finish, the typical net energy yield from each glucose molecule is thirty-six ATP.

last component of the transport system. Oxygen picks up hydrogen at the same time and thereby forms water.

Cells drive nearly all metabolic activities by releasing energy from glucose and other organic compounds and converting it to the chemical bond energy of ATP.

All of the main energy-releasing pathways start inside the cytoplasm with glycolysis, a stage of reactions that break down glucose to pyruvate.

The most common anaerobic pathways, which include the fermentation routes, end in the cytoplasm. Each has a net energy yield of two ATP.

Aerobic respiration, an oxygen-dependent pathway, runs to completion in the mitochondrion. From start (glycolysis) to finish, it commonly has a net energy yield of thirty-six ATP.

GLYCOLYSIS: FIRST STAGE OF THE ENERGY-RELEASING PATHWAYS

Let's track what happens to a glucose molecule in the first stage of aerobic respiration. Remember, the same things happen to glucose in the anaerobic routes.

As described earlier in Section 3.4, glucose is one of the simple sugars. Each molecule of it has six carbon, twelve hydrogen, and six oxygen atoms covalently bonded to one another. The carbon atoms make up the molecule's backbone, which we may represent this way:

During glycolysis, glucose or some other carbohydrate in the cytoplasm is partially broken down to **pyruvate**, a molecule with a backbone of three carbon atoms.

The first steps of glycolysis are *energy-requiring*. As Figure 8.4 shows, they proceed only when two ATP molecules each transfer a phosphate group to glucose and so donate energy to it. Such transfers, recall, are "phosphorylations." In this case, they raise the energy content of glucose to a level that is high enough to allow entry into the *energy-releasing* steps of glycolysis.

The first energy-releasing step cleaves the activated glucose into two molecules, which we can call PGAL (phosphoglyceraldehyde). Each PGAL is converted to an unstable intermediate, each of which allows ATP to form by giving up a phosphate group to ADP. The next intermediate in the sequence does the same thing.

Thus, a total of four ATP form by **substrate-level phosphorylation**. This metabolic event is defined as the direct transfer of a phosphate group from a substrate of a reaction to some other molecule, such as ADP. Remember, though, two ATP were invested to start the reactions. So the *net* energy yield is only two ATP.

Meanwhile, the coenzyme **NAD+** (nicotinamide adenine dinucleotide) picks up electrons and hydrogen liberated from each PGAL, thereby becoming NADH. When the NADH gives up its cargo at another reaction site, it becomes NAD+ once more. As is true of other enzyme helpers, then, NAD+ is reusable (Section 6.5).

In sum, glycolysis converts a bit of the energy stored in glucose to a transportable form of energy, in ATP. A coenzyme picks up electrons and hydrogen stripped from glucose. These have key roles in the next stage of reactions. So do the end products of glycolysis—the two molecules of pyruvate.

Glycolysis is an energy-releasing stage of reactions in which glucose or some other carbohydrate is partially broken down to two molecules of pyruvate.

A total of two NADH and four ATP molecules form at certain steps in the reaction sequence. However, subtracting the two ATP required to start the reactions puts the *net* energy yield of glycolysis at two ATP.

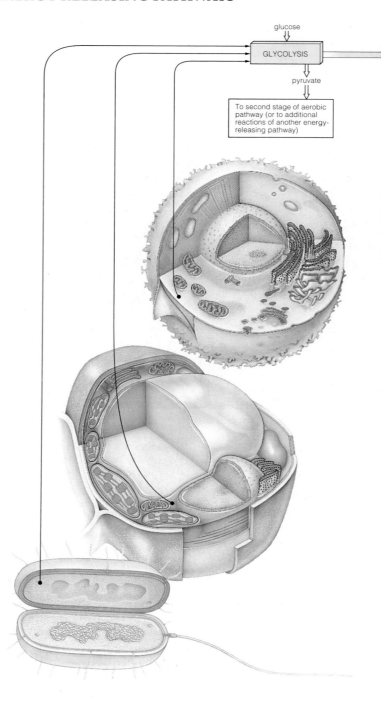

Figure 8.4 Glycolysis, first stage of the main energy-releasing pathways. The reaction steps proceed inside the cytoplasm of every living prokaryotic and eukaryotic cell.

In this example, glucose is the starting material. By the time the reactions end, two pyruvate, two NADH, and four ATP have been produced. Cells invest two ATP to start glycolysis, however, so the *net* energy yield of glycolysis is two ATP.

Depending on the type of cell and on environmental conditions, the pyruvate may be used in the second set of reactions of the aerobic pathway, which includes the Krebs cycle. Or it may be used in other reactions, such as those of fermentation pathways.

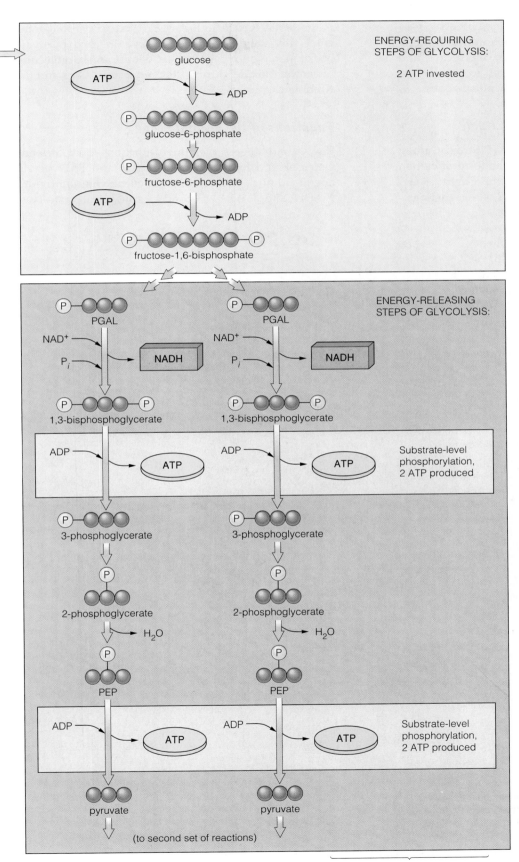

ENERGY-REQUIRING
STEPS OF GLYCOLYSIS:

2 ATP invested

a *Glycolysis starts with an energy investment of two ATP.* First, enzyme action promotes the transfer of a phosphate group from ATP to glucose, which has a backbone of six carbon atoms. With this transfer, the glucose molecule becomes slightly rearranged.

b Enzyme action promotes the transfer of a phosphate group from another ATP to the rearranged molecule.

c The resulting fructose-1,6-bisphosphate molecule splits at once into two molecules, each with a three-carbon backbone. We can call these two PGAL.

d During enzyme-mediated reactions, two NADH form after each PGAL gives up two electrons and a hydrogen atom to NAD^+. Each PGAL also combines with inorganic phosphate (P_i) present in the cytoplasm, then donates a phosphate group to ADP.

e *Thus two ATP have formed by the direct transfer of phosphate from two intermediate molecules that serve as substrates in the reactions.* With this formation of two ATP molecules, the original energy investment of two ATP is paid off.

f In the next two enzyme-mediated reactions, each of the two intermediate molecules releases a hydrogen atom and an —OH group, which combine to form water.

g The resulting intermediates (two molecules of 3-phosphoenolpyruvate, or PEP) are rather unstable. Each gives up a phosphate group to ADP. *Once again, two ATP have formed by substrate-level phosphorylation.*

h Thus the net energy yield from glycolysis is two ATP for each glucose molecule entering the reactions. The end products of glycolysis are two molecules of pyruvate, each with a three-carbon backbone.

NET ENERGY YIELD: 2 ATP

SECOND STAGE OF THE AEROBIC PATHWAY

Suppose two pyruvate molecules, formed by glycolysis, leave the cytoplasm and enter a **mitochondrion** (plural, mitochondria). In this organelle alone, the second and third stages of the aerobic pathway run to completion. Figure 8.5 shows its structure and functional zones.

Preparatory Steps and the Krebs Cycle

During the second stage, a bit more ATP forms. Carbon atoms depart from the pyruvate, in the form of carbon dioxide. And coenzymes latch onto the electrons and hydrogen stripped from intermediates of the reactions.

In a few preparatory steps, an enzyme removes a carbon atom from each pyruvate molecule. Coenzyme A, an enzyme helper, becomes **acetyl-CoA** when it combines with the remaining two-carbon fragment. It transfers this fragment to **oxaloacetate**, the entry point of the Krebs cycle. The name of this cyclic pathway honors Hans Krebs, who began working out its details

in the 1930s. Notice, in Figure 8.6, that *six* carbon atoms enter this stage of reactions (three in each pyruvate backbone). Notice also that *six* depart, in six molecules of carbon dioxide, during the preparatory steps and the Krebs cycle proper.

Functions of the Second Stage

The second stage serves three functions. First, it loads electrons and hydrogen onto NAD⁺ and **FAD** (flavin adenine dinucleotide, a different coenzyme) to produce NADH and FADH₂. Second, through substrate-level

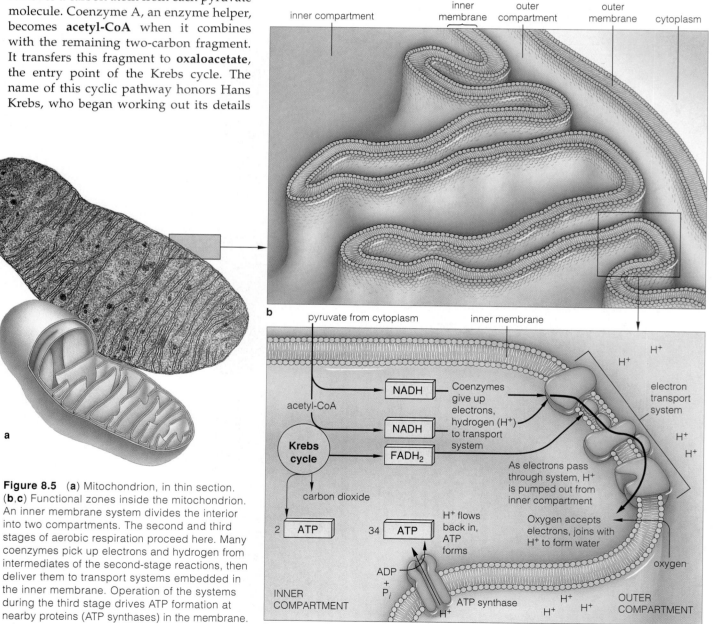

Figure 8.5 (**a**) Mitochondrion, in thin section. (**b,c**) Functional zones inside the mitochondrion. An inner membrane system divides the interior into two compartments. The second and third stages of aerobic respiration proceed here. Many coenzymes pick up electrons and hydrogen from intermediates of the second-stage reactions, then deliver them to transport systems embedded in the inner membrane. Operation of the systems during the third stage drives ATP formation at nearby proteins (ATP synthases) in the membrane.

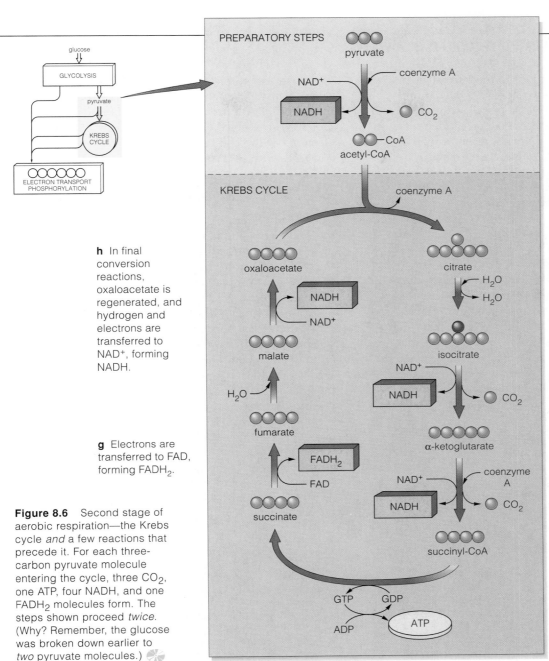

PREPARATORY STEPS

pyruvate

coenzyme A

NAD+

NADH

CO_2

CoA

acetyl-CoA

KREBS CYCLE

coenzyme A

oxaloacetate

citrate

H_2O
H_2O

NADH

NAD+

malate

isocitrate

H_2O

NAD+

NADH

CO_2

fumarate

α-ketoglutarate

FADH$_2$

FAD

NAD+

coenzyme A

NADH

CO_2

succinate

succinyl-CoA

GTP   GDP

ADP

ATP

glucose

GLYCOLYSIS

pyruvate

KREBS CYCLE

ELECTRON TRANSPORT PHOSPHORYLATION

a A pyruvate molecule enters a mitochondrion. It undergoes preparatory conversions before entering cyclic reactions (Krebs cycle).

b First, the pyruvate is stripped of a functional group (COO−), which departs as CO_2. Next, it gives up hydrogen and electrons to NAD+, forming NADH. A coenzyme joins with the remaining two-carbon fragment, forming acetyl-CoA.

c The acetyl-CoA is transferred to oxaloacetate, a four-carbon compound that is the point of entry into the Krebs cycle. The result is citrate, with a six-carbon backbone.

d Citrate enters conversion reactions in which a COO− group departs (as CO_2). Also, hydrogen and electrons are transferred to NAD+, forming NADH.

e Another COO− group departs (as CO_2) and another NADH forms. *At this point, three carbon atoms have been released, balancing out the three that entered the mitochondrion (in pyruvate).*

f After further reactions, ATP forms by substrate-level phosphorylation.

h In final conversion reactions, oxaloacetate is regenerated, and hydrogen and electrons are transferred to NAD+, forming NADH.

g Electrons are transferred to FAD, forming FADH$_2$.

Figure 8.6 Second stage of aerobic respiration—the Krebs cycle *and* a few reactions that precede it. For each three-carbon pyruvate molecule entering the cycle, three CO_2, one ATP, four NADH, and one FADH$_2$ molecules form. The steps shown proceed *twice*. (Why? Remember, the glucose was broken down earlier to *two* pyruvate molecules.)

phosphorylations, it produces two ATP. And third, it rearranges Krebs cycle intermediates into oxaloacetate. Cells have only so much oxaloacetate, and it must be regenerated to keep the cyclic reactions going.

The two ATP that form don't add much to the small yield from glycolysis. However, *many* coenzymes pick up electrons and hydrogen for transport to the sites of the third and final stage of the aerobic pathway:

Glycolysis:	2 NADH
Pyruvate conversion before Krebs cycle:	2 NADH
Krebs cycle:	2 FADH$_2$ + 6 NADH
Coenzymes sent to third stage:	2 FADH$_2$ + 10 NADH

Overall, these are the key points to remember about the second stage of aerobic respiration:

During the second stage of aerobic respiration, two pyruvate molecules from glycolysis enter a mitochondrion.

Each pyruvate molecule gives up a carbon atom, then its remnant enters the Krebs cycle. All of the carbon atoms of pyruvate eventually end up in carbon dioxide.

The preparatory steps and the cycle proper yield only two ATP. However, the reactions regenerate oxaloacetate, the entry point for the cycle. And many coenzymes pick up electrons and hydrogen that were stripped from substrates, for delivery to the final stage of the pathway.

THIRD STAGE OF THE AEROBIC PATHWAY

ATP production goes into high gear in the third stage of the aerobic pathway. Electron transport systems and neighboring proteins called ATP synthases serve as the production machinery. They are embedded in the inner membrane that divides the mitochondrion into two compartments (Figure 8.7). They interact with electrons and unbound hydrogen—that is, H^+ ions. Remember, coenzymes deliver this bounty from reaction sites of the first two stages of the aerobic pathway.

Electron Transport Phosphorylation

Briefly, during the final stage, electrons get transferred from one molecule of each transport system to the next in line. When certain molecules accept and then donate the electrons, they also pick up hydrogen ions in the inner compartment. Quickly afterward, they release them to the outer compartment. Their shuttling action sets up H^+ concentration and electric gradients across the inner mitochondrial membrane. Nearby in the membrane, the ions follow the gradients and flow back to the inner

compartment, through the interior of ATP synthases. The H^+ flow through these transport proteins drives formation of ATP from ADP and unbound phosphate. Free oxygen keeps ATP production going. It withdraws electrons at the end of the transport systems and then combines with H^+. Water is the result.

Summary of the Energy Harvest

In many types of cells, thirty-two ATP form during the third stage of aerobic respiration. Add these to the net yield from the preceding stages, and the total harvest is thirty-six ATP from one glucose molecule (Figure 8.8). That's a lot! An anaerobic pathway may use eighteen glucose molecules to produce the same amount of ATP.

Think of thirty-six ATP as a typical yield only. The actual amount depends on cellular conditions, as when cells require a given intermediate elsewhere and pull it out of the reaction sequence.

The yield also depends on how particular cells use the NADH that formed during glycolysis. Any NADH produced in the cytoplasm can't enter a mitochondrion. It can only deliver electrons and hydrogen *to* the outer mitochondrial membrane, where proteins shuttle them across to NAD^+ or to FAD molecules already in the mitochondrion. Both coenzymes deliver the electrons to transport systems of the inner membrane. However, FAD puts them at a *lower* entry point in the transport system, so *its* deliveries produce less ATP (Figure 8.8).

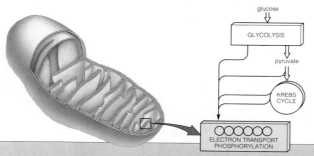

Figure 8.7 Electron transport phosphorylation, the third and final stage of aerobic respiration. The reactions proceed at electron transport systems and at ATP synthases, a type of transport protein, in the inner mitochondrial membrane. Each electron transport system consists of specific enzymes, cytochromes, and other proteins that act in sequence.

The inner membrane functionally divides the mitochondrion into two compartments. The third-stage reactions start in the inner compartment, when NADH and $FADH_2$ give up electrons and hydrogen to transport systems. Electrons are transferred *through* the system, but electron transport proteins drive unbound hydrogen (H^+) *to* the outer compartment:

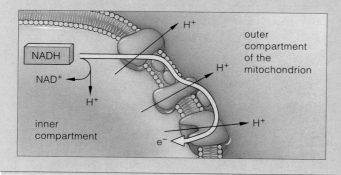

In very short order, there is a higher concentration of H^+ in the outer compartment compared to the inner one. Concentration and electric gradients now exist across the membrane. The ions follow the gradients and flow across the membrane, through the interior of the ATP synthases. Energy associated with the flow drives the formation of ATP from ADP and unbound phosphate (P_i). Hence the name, electron transport *phosphorylation*:

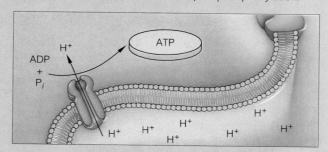

Do these metabolic events sound familiar? They should. As you may recall from Section 7.5, ATP forms in much the same way inside chloroplasts. According to the *chemiosmotic* theory, H^+ concentration and electric gradients across a cell membrane drive ATP formation. The theory applies also to mitochondria, although the ions flow in the opposite direction compared to chloroplasts.

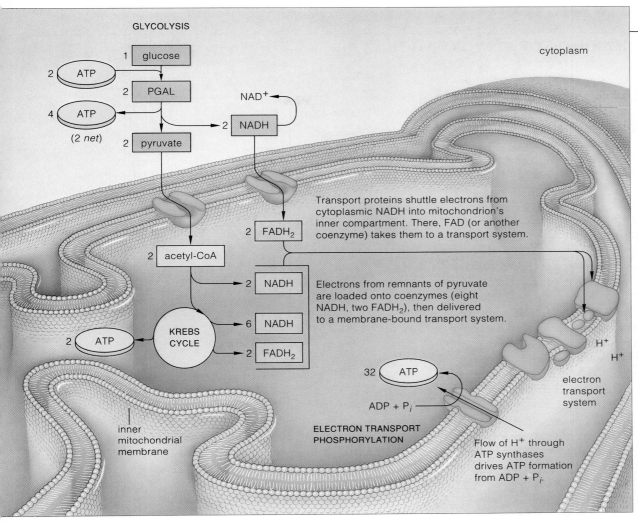

GLYCOLYSIS

a Two ATP formed in first stage in cytoplasm (during glycolysis, by *substrate-level* phosphorylations).

b NADH that formed in cytoplasm during first stage delivers electrons and hydrogen that help drive the formation of four ATP during third stage at the inner mitochondrial membrane (by *electron transport phosphorylations*).

c Two ATP form at second stage in mitochondrion (by *substrate-level* phosphorylations of Krebs cycle).

d Coenzymes from Krebs cycle and its preparatory steps deliver electrons and hydrogen that drive formation of twenty-eight ATP at third stage (by *electron transport phosphorylations* at the inner mitochondrial membrane).

36 ATP

TYPICAL NET
ENERGY YIELD

Figure 8.8 Summary of the harvest from the energy-releasing pathway of aerobic respiration. Commonly, thirty-six ATP form for each glucose molecule that enters the pathway. However, the net yield varies according to shifting concentrations of reactants, intermediates, and end products of the reactions. It also varies among different types of cells.

For example, cells differ in how they use the NADH from glycolysis. These NADH cannot enter mitochondria. They only deliver their cargo of electrons and hydrogen *to* certain transport proteins of the outer mitochondrial membrane. The proteins shuttle the electrons and hydrogen across the membrane, to NAD^+ or FAD already inside the mitochondrion, to form NADH or $FADH_2$.

Any NADH inside the mitochondrion delivers electrons to the highest possible point of entry into a transport system. When it does, enough H^+ can be pumped across the inner membrane to produce *three* ATP. By contrast, any $FADH_2$ delivers them to a lower entry point in the transport system. Fewer hydrogen ions can be pumped, so only *two* ATP can be produced.

In liver, heart, and kidney cells, for example, electrons and hydrogen enter the highest entry point of transport systems, so the overall energy harvest is thirty-eight ATP. More commonly, as in skeletal muscle and brain cells, they are transferred to FAD—so the overall harvest is thirty-six ATP.

One final point. Glucose, recall, has more energy (stored in more covalent bonds) than carbon dioxide or water does. When it breaks down to those more stable end products, about 686 kilocalories are released. Much of this energy escapes (as heat), but 7.5 kilocalories or so are conserved in each ATP molecule. Therefore, when 36 ATP form through the breakdown of a glucose molecule, the energy-conserving efficiency of aerobic respiration is $(36)(7.5)/(686) \times 100$, or 39 percent.

In the final stage of the aerobic pathway, coenzymes deliver electrons to transport systems of the inner mitochondrial membrane. As electrons move through the system, they set up H^+ gradients that drive ATP formation at nearby proteins in the membrane. Oxygen is the final electron acceptor.

Again, from start (glycolysis in the cytoplasm) to finish (in mitochondria), the pathway commonly has a net yield of thirty-six ATP for every glucose molecule metabolized.

8.5 ANAEROBIC ROUTES OF ATP FORMATION

So far, we have tracked the fate of a glucose molecule through the pathway of aerobic respiration. We turn now to its use as a substrate for fermentation pathways. Remember, these are anaerobic pathways; they do *not* use oxygen as the final acceptor of the electrons that ultimately drive the ATP-forming machinery.

Fermentation Pathways

Diverse kinds of organisms use fermentation pathways. Many are bacteria and protistans that make their homes in marshes, bogs, mud, deep-sea sediments, the animal gut, canned foods, sewage treatment ponds, and other oxygen-free settings. Some kinds of fermenters actually die if exposed to oxygen. The bacteria responsible for many diseases, including botulism and tetanus, are like this. Other kinds of fermenters, including the bacterial "employees" of yogurt manufacturers, are indifferent to the presence of oxygen. Still others can use oxygen, but they also can use a fermentation pathway when oxygen becomes scarce. Even your muscle cells do this.

As is true of aerobic respiration, glycolysis serves as the first stage of the fermentation pathways. Here also, enzymes split glucose and rearrange the fragments into two pyruvate molecules. Here again, two NADH form, and the net energy yield is two ATP. However, as you can see from Figure 8.9, the reactions do not completely break down glucose to carbon dioxide and water, and they produce no more ATP beyond the tiny yield from glycolysis. *The final steps serve only to regenerate NAD+, a coenzyme with central roles in the breakdown reactions.*

Fermentation yields enough energy to sustain many single-celled anaerobic organisms. It even helps carry some aerobic cells through times of stress. But it is not enough to sustain large, active, multicelled organisms,

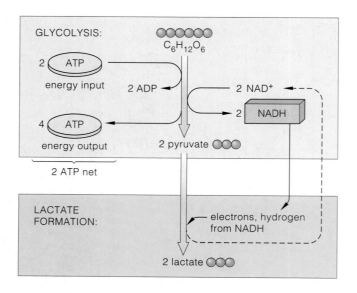

Figure 8.10 Lactate fermentation. In this anaerobic pathway, electrons end up in lactate, the reaction product.

this being one reason why you never will meet up with an anaerobic elephant.

LACTATE FERMENTATION With these points in mind, take a look at Figure 8.10, which tracks the main steps of **lactate fermentation**. During this anaerobic pathway, the *pyruvate* molecules from the first stage of reactions (glycolysis) accept the hydrogen and electrons from NADH. The transfer regenerates the NAD+ and, at the same time, converts each pyruvate to a three-carbon compound called lactate. You may hear people refer to this compound as "lactic acid." However, its ionized form (lactate) is far more common in cellular fluids.

Some bacteria, such as *Lactobacillus*, rely exclusively on this anaerobic pathway. Left to their own devices, their fermentation activities often spoil food. Yet certain fermenters have commercial uses, as when they break down glucose in huge vats where cheeses, yogurt, and sauerkraut are produced.

In humans, rabbits, and many other animals, some types of cells also can switch to lactate fermentation for a quick fix of ATP. When your own demands for energy are intense but brief—say, during a short race—muscle cells use this pathway. They cannot do so for long; they would throw away too much of glucose's stored energy for too little ATP. When the glucose stores are depleted, muscles fatigue and lose their ability to contract.

ALCOHOLIC FERMENTATION In the anaerobic route of **alcoholic fermentation**, each pyruvate molecule that forms during glycolysis is converted to an intermediate

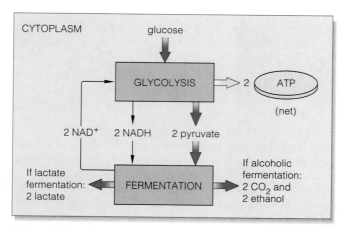

Figure 8.9 Overview of two fermentation routes. In this type of energy-releasing pathway, only the initial stage (glycolysis) has a net energy harvest, in the form of two ATP molecules. The remaining reactions serve to regenerate NAD+.

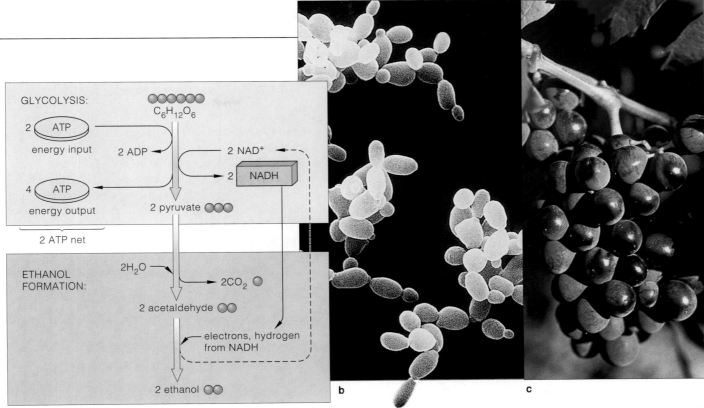

GLYCOLYSIS:

$C_6H_{12}O_6$

2 ATP
energy input

2 ADP

2 NAD^+

2 NADH

4 ATP
energy output

2 pyruvate

2 ATP net

ETHANOL
FORMATION:

$2H_2O$

$2CO_2$

2 acetaldehyde

electrons, hydrogen
from NADH

2 ethanol

a

b

c

Figure 8.11 (**a**) Alcoholic fermentation. In this anaerobic pathway, acetaldehyde, an intermediate of the reactions, is the final acceptor of electrons. Ethanol is the end product. Yeasts, single-celled organisms, use this pathway. (**b**) One species of *Saccharomyces* makes bread dough rise. Another (**c**) lives on sugar-rich tissues of ripened grapes.

form, acetaldehyde. The NADH transfers electrons and hydrogen to this form and so converts it to an alcoholic end product—ethanol (Figure 8.11).

Certain species of single-celled fungi called yeasts are renowned for their use of this pathway. One type, *Saccharomyces cerevisiae*, makes bread dough rise. Bakers mix the yeast with sugar, then blend both into dough. When yeast cells degrade the sugar, they release carbon dioxide. Bubbles of the gas expand the dough (make it rise). Oven heat forces the gas out of the dough, and a porous product remains.

Beer and wine producers use yeasts on a large scale. Vintners use wild yeasts living on grapes and cultivated strains of *S. ellipsoideus*, which remain active until the alcohol concentration in wine vats exceeds 14 percent. Wild yeasts die when the concentration passes 4 percent. Then, birds get drunk on naturally fermenting berries. That is why landscapers don't plant prodigious berry-producing shrubs near highways; drunk birds doodle into windshields. Wild turkeys get similarly tipsy when they gobble fermenting apples in untended orchards.

Anaerobic Electron Transport

Especially among the bacteria, we find less common energy-releasing pathways, some of which are topics of later chapters in the book. For example, many bacterial species have key roles in the global cycling of sulfur, nitrogen, and other crucial elements. Collectively, their metabolic activities influence nutrient availability for organisms everywhere.

For example, certain bacteria use **anaerobic electron transport**. Electrons stripped from some type of organic compound move on through transport systems of their plasma membrane. Commonly, an inorganic compound in the environment serves as the final electron acceptor. The net energy yield varies, but it is always small.

Even as you read this, some anaerobic bacteria that live in waterlogged soil are stripping electrons from a variety of compounds. They dump electrons on sulfate. Hydrogen sulfide, a putrid-smelling gas, is the result. The sulfate-reducing bacteria also live in many aquatic habitats that are enriched with decomposed organic material. They even live on the deep ocean floor, near hydrothermal vents. As described in Section 49.11, they form the food production base for unique communities.

In fermentation pathways, an organic substance that forms during the reactions serves as the final acceptor of electrons from glycolysis. The reactions regenerate NAD^+, which is required to keep the pathway operational.

In anaerobic electron transport, an inorganic substance (but not oxygen) usually serves as the final electron acceptor.

For each glucose molecule metabolized, anaerobic pathways typically have a net energy yield of two ATP, which only form during glycolysis.

ALTERNATIVE ENERGY SOURCES IN THE HUMAN BODY

So far, you have looked at what happens after a lone glucose molecule enters an energy-releasing pathway. Now you can start thinking about what cells do when they have too many or too few molecules of glucose.

Carbohydrate Breakdown in Perspective

THE FATE OF GLUCOSE AT MEALTIME   Consider what happens to you or any other mammal during a meal. Glucose and certain other small organic molecules are being absorbed across the gut lining, then the blood transports them through the body. A rise in the glucose level in blood prompts the pancreas to release insulin, a hormone that stimulates cells to take up glucose at a faster rate. The cells convert the windfall of glucose to glucose-6-phosphate and so "trap" it in the cytoplasm. (When phosphorylated, glucose cannot be transported back out, across the plasma membrane.) Look again at Figure 8.4, and you see that glucose-6-phosphate is the first activated intermediate of glycolysis.

If your glucose intake exceeds cellular demands for energy, ATP-producing machinery goes into high gear. Unless a cell is rapidly using ATP, its concentration of ATP can rise to a high level. Then, glucose-6-phosphate is diverted into a biosynthesis pathway that assembles glucose units into glycogen, a storage polysaccharide (Section 3.4). This is especially true of liver cells and muscle cells, which maintain the largest glycogen stores.

THE FATE OF GLUCOSE BETWEEN MEALS   When you are not eating, glucose is not entering your bloodstream and its level in the blood declines. If the decline were not countered, that would be bad news for the brain, your body's glucose hog. The brain constantly takes up more than two-thirds of the freely circulating glucose because its many hundreds of millions of cells simply use this sugar alone as their preferred energy source.

The pancreas responds to the decline by secreting glucagon, a hormone that prompts liver cells to convert glucose-6-phosphate back to glucose and send it back to the blood. Only liver cells do this; muscle cells won't give it up. The glucose level rises, and brain cells keep on trucking. Thus, *hormones control whether the body's cells use free glucose as an energy source or tuck it away.*

A word of caution: Don't let the preceding examples lead you to believe cells squirrel away large amounts of glycogen. In adult humans, glycogen makes up merely 1 percent or so of the body's total energy reserves, the energy equivalent of two cups of cooked pasta. Unless you eat on a regular basis, you will deplete the liver's small glycogen stores in less than twelve hours. Of the total energy reserves in, say, a typical adult American, 78 percent (about 10,000 kilocalories) is concentrated in body fat and 21 percent in proteins.

Energy From Fats

The question becomes this: How does the body access its huge reservoir of fats? A fat molecule, recall, has a glycerol head and one, two, or three fatty acid tails. Most fats that become stored in your body are in the form of triglycerides, with three tails each. Triglycerides accumulate in fat cells of adipose tissues, which form at the buttocks and other strategic places beneath the skin.

When blood glucose levels decline, triglycerides can be tapped as an energy alternative. Then, enzymes in fat cells cleave the bonds holding the glycerol and fatty acids together, and the breakdown products enter the bloodstream. Afterward, enzymes in the liver convert the glycerol to PGAL—an intermediate of glycolysis. Nearly all cells can take up the circulating fatty acids. Enzymes cleave the carbon backbone of the fatty acid tails and convert the fragments to acetyl-CoA—which can enter the Krebs cycle (Figures 8.6 and 8.12).

Each fatty acid tail has many more carbon-bound hydrogen atoms than glucose, so its breakdown yields much more ATP. In between meals or during sustained exercise, fatty acid conversions supply about half of the ATP that muscle, liver, and kidney cells require.

What happens if you eat too many carbohydrates? Exceed the glycogen-storing capacity of your liver and muscle cells, and the excess gets converted to fats. *Too much glucose ends up as excess fat.* Worse yet, 25 percent of the people in the United States are blessed with a combination of genes that allows them to eat as much as they like without gaining weight—but a diet far too rich in carbohydrates keeps the other 75 percent fat. For them, insulin levels remain elevated, which "tells" the body to store fat rather than use it for energy. We will return to this topic in Section 42.10.

Energy From Proteins

Eat more proteins than your body requires to grow and maintain itself, and its cells won't store them. Enzymes split these proteins into amino acid units. Then they remove the amino group ($-NH_3^+$) from each unit, and ammonia (NH_3) forms. What happens to the leftover carbon backbones? Depending on conditions in the cell, the outcome varies. These backbones can be converted to carbohydrates or fats. Or they may enter the Krebs cycle, as in Figure 8.12, where coenzymes can pick up hydrogen and electrons stripped away from the carbon atoms. The ammonia that forms undergoes conversions to become urea. This nitrogen-containing waste product would be toxic if it accumulated to high concentrations. Normally your body excretes ammonia, in urine.

As this brief discussion makes clear, maintaining and accessing the body's energy reserves is complicated

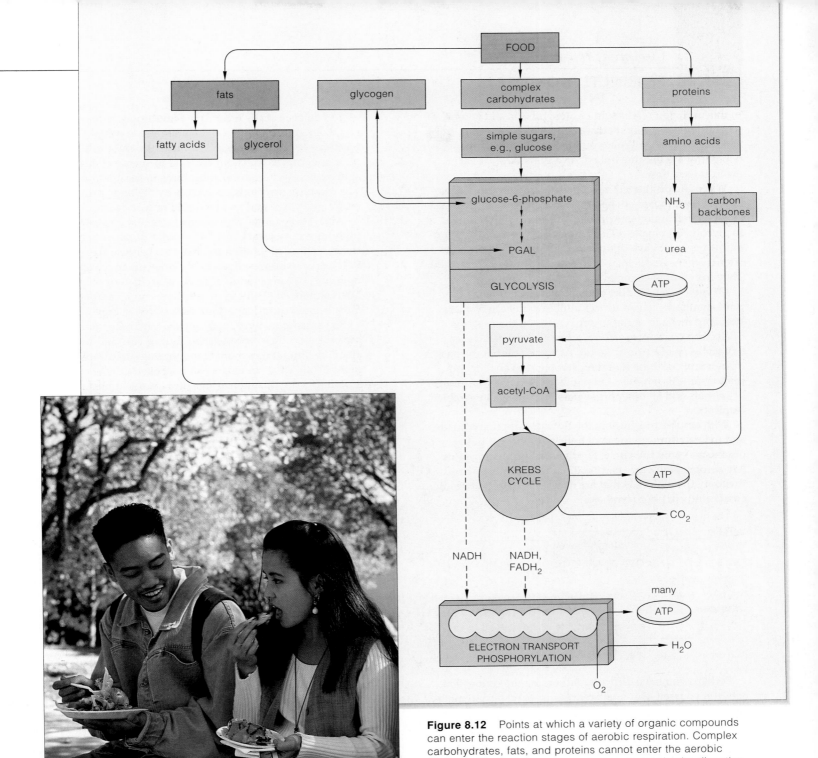

Figure 8.12 Points at which a variety of organic compounds can enter the reaction stages of aerobic respiration. Complex carbohydrates, fats, and proteins cannot enter the aerobic pathway directly. In humans and other mammals, the digestive system as well as individual cells must first break apart these large molecules into simpler, degradable subunits.

business. Hormonal controls over the disposition of glucose are special only because glucose is the fuel of choice for the all-important brain. However, as you will see in later chapters, providing all of your cells, organs, and organ systems with energy starts with the kinds and proportions of food you put in your mouth.

This concludes our look at aerobic respiration and other energy-releasing pathways. The section to follow may help you get a sense of how they fit into the larger picture of life's evolution and interconnectedness.

In humans and other mammals, the entrance of glucose or other organic compounds into an energy-releasing pathway depends on the kinds and proportions of carbohydrates, fats, and proteins in the diet as well as on the type of cell.

PERSPECTIVE ON LIFE

In this unit, you read about photosynthesis and aerobic respiration—the main pathways by which cells trap, store, and release energy. What you might not know is that the two pathways became linked, on a grand scale, over evolutionary time.

When life originated more than 3.8 billion years ago, the Earth's atmosphere had little free oxygen. The earliest single-celled organisms probably used reactions similar to glycolysis to make ATP. Without oxygen, fermentation pathways must have dominated. About 1.5 billion years later, oxygen-producing photosynthetic cells had emerged. They irrevocably changed the course of evolution.

Oxygen, a by-product of the noncyclic pathway of photosynthesis, began to accumulate in the atmosphere. Probably through mutations that affected the proteins of electron transport systems, some cells started using oxygen as an electron acceptor. At some point in the past, descendants of those fledgling aerobic cells abandoned photosynthesis entirely. Among them were the forerunners of animals and all other organisms that engage in aerobic respiration.

With aerobic respiration, the flow of carbon, hydrogen, and oxygen through the metabolic pathways of living organisms came full circle. For the final products of this key aerobic pathway—carbon dioxide and water—are precisely the materials that are necessary to build organic compounds in photosynthesis:

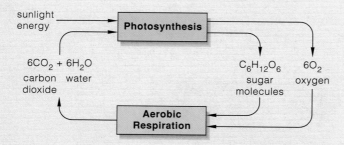

Perhaps you have difficulty fathoming the connection between yourself—an intelligent being—and such remote-sounding events as energy flow and the cycling of carbon, hydrogen, and oxygen. Is this really the stuff of humanity?

Think back, for a moment, on the structure of a water molecule. Two hydrogen atoms sharing electrons with an oxygen atom may not seem close to your daily life. And yet, through that sharing, water molecules show polarity—and they hydrogen-bond with one another. Their chemical behavior is a beginning for the organization of lifeless matter that leads to the organization of all living things.

For now you can imagine other molecules interspersed through water. The nonpolar kinds resist interaction with water; the polar kinds dissolve in it. On their own, the phospholipids among them assemble into a two-layered film. Such lipid bilayers, remember, serve as the very framework of all cell membranes, hence all cells. From the beginning, the cell has been the fundamental *living* unit.

The essence of life is not some mysterious force. It is metabolic control. With a cell membrane to contain them, reactions *can* be controlled. With mechanisms built into their membranes, cells can respond to energy changes and shifting concentrations of substances in the environment. The response mechanisms operate by "telling" proteins—enzymes—when and what to build or tear down.

And it is not some mysterious force that creates the proteins themselves. DNA, the slender double-stranded treasurehouse of inheritance, has the chemical structure—*the chemical message*—that allows molecule to reproduce molecule, one generation after the next. In your own body, DNA strands tell trillions of cells how countless molecules must be built or torn apart for their stored energy.

So yes, carbon, hydrogen, oxygen, and other atoms of organic molecules represent the stuff of you, and us, and all of life. But it takes more than molecules to complete the picture. Life exists as long as an unbroken flow of energy sustains its organization. Molecules are assembled into cells, cells into organisms, organisms into communities, and so on up through the biosphere. It takes energy inputs from the sun to maintain these levels of organization. And energy flows through time in one direction—from organized to less organized forms. Only as long as energy continues to flow into the web of life can life continue in all its rich diversity.

In short, life is no more *and no less* than a marvelously complex system of prolonging order. Sustained by energy transfusions from the sun, life continues onward, through its capacity for self-reproduction. For with the hereditary instructions contained in DNA, energy and materials can be organized, generation after generation. Even with the death of individuals, life elsewhere is prolonged. With each death, molecules are released and may be cycled once more, as raw materials for new generations.

In this flow of energy and cycling of material through time, each birth is affirmation of our ongoing capacity for organization, each death a renewal.

SUMMARY

1. Metabolic reactions run on the energy, inherent in phosphate groups, that ATP molecules deliver to them. Aerobic respiration, fermentation, and other pathways that release chemical energy from organic compounds, such as glucose, produce ATP.

2. After glucose enters these pathways, enzymes strip electrons and hydrogen from intermediates that form along the way. Coenzymes pick these up and deliver them to other reaction sites at which the pathway is completed. NAD^+ is the main coenzyme; the aerobic route also uses FAD. When loaded with electrons and hydrogen, they are designated NADH and $FADH_2$.

3. The main energy-releasing pathways all start with glycolysis, a stage of reactions that begin and end in the cytoplasm. The glycolytic reactions can be completed either in the presence of oxygen or in its absence.

   a. During glycolysis, enzymes break down a glucose molecule to two pyruvate molecules. Two NADH and four ATP form during the reactions.

   b. The *net* energy yield is two ATP (because two ATP had to be invested up front to get the reactions going).

4. Aerobic respiration continues on through two more stages: (1) the Krebs cycle and a few steps preceding it, and (2) electron transport phosphorylation. These stages proceed only inside the organelles called mitochondria, which occur only in eukaryotic cells.

5. The second stage of the aerobic pathway starts when an enzyme strips a carbon atom from each pyruvate. Coenzyme A binds the remaining two-carbon fragment (to form acetyl-CoA), then transfers it to oxaloacetate, the entry point of the Krebs cycle. The cyclic reactions, along with the steps immediately preceding them, load up ten coenzymes with electrons and hydrogen (eight NADH and two $FADH_2$). Two ATP form. Three carbon dioxide molecules are released for each pyruvate that entered this second stage.

6. The third stage of the aerobic pathway proceeds at a membrane that divides the interior of a mitochondrion into two compartments. Electron transport systems and ATP synthases are embedded in this inner membrane.

   a. Coenzymes deliver electrons from the first two stages to transport systems. In the outer compartment, hydrogen ions accumulate, so concentration and electric gradients form across the membrane.

   b. Hydrogen ions follow the gradients and flow from the outer to the inner compartment, through the interior of ATP synthases. Energy released during the ion flow drives the formation of ATP from ADP and unbound phosphate.

   c. Oxygen withdraws electrons from the transport system and at the same time combines with hydrogen ions to form water molecules. The oxygen is the final acceptor of electrons that initially resided in glucose.

7. Aerobic respiration has a typical net energy yield of thirty-six ATP for each glucose molecule metabolized. Yields vary, according to cell type and cell conditions.

8. Fermentation pathways as well as anaerobic electron transport also start with glycolysis, but they do not use oxygen; they are anaerobic, start to finish.

   a. Lactate fermentation has a net energy yield of two ATP, which form in glycolysis. The remaining reactions regenerate the NAD^+. The two NADH from glycolysis transfer electrons and hydrogen to two pyruvate from glycolysis. Two lactate molecules are the end products.

   b. Similarly, alcoholic fermentation has a net energy yield of two ATP from glycolysis, and its remaining reactions regenerate NAD^+. Enzymes convert pyruvate from glycolysis to acetaldehyde, and carbon dioxide is released. The NADH from glycolysis transfer electrons and hydrogen to the two acetaldehyde molecules, thus forming two ethanol molecules, the end products.

   c. Certain bacteria use anaerobic electron transport. Electrons are stripped from various organic compounds and travel through transport systems in the bacterial cell's plasma membrane. An inorganic compound in the environment often serves as the final electron acceptor.

9. In humans and other mammals, simple sugars such as glucose from carbohydrates, glycerol and fatty acids from fats, and carbon backbones of amino acids from proteins can enter ATP-producing pathways.

Review Questions

1. Is this true or false: Aerobic respiration occurs in animals but not plants, which make ATP only by photosynthesis. *8.1*

2. For this diagram of the aerobic pathway, fill in all the blanks and write in the number of molecules of pyruvate, coenzymes, and end products. Also write in the net ATP formed in each stage, and the net ATP formed from start (glycolysis) to finish. *8.1*

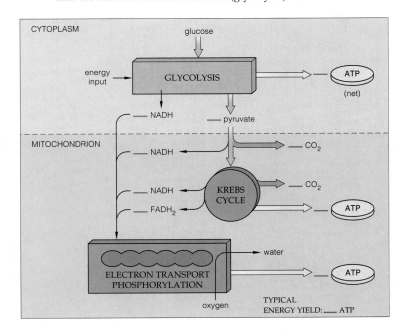

3. Is glycolysis energy-*requiring* or energy-*releasing*? Or do both kinds of reactions occur during glycolysis? 8.2

4. In what respect does *electron transport* phosphorylation differ from *substrate-level* phosphorylation? 8.2, *Figure 8.7*

5. Sketch the double-membrane system of the mitochondrion and show where transport systems and ATP synthases are located. 8.3

6. Name the compound that is the entry point for the Krebs cycle and state whether it directly accepts the pyruvate from glycolysis. For each glucose molecule, how many carbon atoms enter the Krebs cycle? How many depart from it, and in what form? 8.3

7. Is this statement true or false: Muscle cells cannot contract at all when deprived of oxygen. If true, explain why. If false, name the alternative(s) available to them. 8.5

Self-Quiz *(Answers in Appendix IV)*

1. Glycolysis starts and ends in the _____ .
   a. nucleus   c. plasma membrane
   b. mitochondrion   d. cytoplasm

2. Which of the following does *not* form during glycolysis?
   a. NADH   c. $FADH_2$
   b. pyruvate   d. ATP

3. The pathway of aerobic respiration is completed in the _____ .
   a. nucleus   c. plasma membrane
   b. mitochondrion   d. cytoplasm

4. In the last stage of aerobic respiration, _____ is the final acceptor of electrons that originally resided in glucose.
   a. water   c. oxygen
   b. hydrogen   d. NADH

5. _____ engage in lactate fermentation.
   a. *Lactobacillus* cells   c. Sulfate-reducing bacteria
   b. Muscle cells   d. a and b

6. In alcoholic fermentation, _____ is the final acceptor of the electrons stripped from glucose.
   a. oxygen   c. acetaldehyde
   b. pyruvate   d. sulfate

7. The fermentation pathways produce no more ATP beyond the small yield from glycolysis, but the remaining reactions _____ .
   a. regenerate ADP   c. dump electrons on an
   b. regenerate NAD^+ inorganic substance (not oxygen)

8. In certain organisms and under certain conditions, _____ can be used as an energy alternative to glucose.
   a. fatty acids   c. amino acids
   b. glycerol   d. all of the above

9. Match the event with its most suitable metabolic description.
   ____ glycolysis   a. ATP, NADH, $FADH_2$, CO_2,
   ____ fermentation   and water form
   ____ Krebs cycle   b. glucose to two pyruvate
   ____ electron transport c. NAD^+ regenerated, two ATP net
        phosphorylation   d. H^+ flows through ATP synthases

Critical Thinking

1. Diana suspects that a visit to her family doctor is in order. After eating carbohydrate-rich food, she always experiences sensations of being intoxicated and becomes nearly incapacitated—as if she had been drinking alcohol. She even wakes up with a hangover the next day. Having completed a course in freshman biology, Diana has an idea that something is affecting the way her body is metabolizing glucose. Explain why.

2. The cells of your body absolutely do not use nucleic acids as alternative energy sources. Suggest why.

3. The body's energy needs and its programs for growth depend on balancing the levels of amino acids in blood with proteins in cells. Cells of the liver, kidneys, and intestinal lining are especially important in this balancing act. When the levels of amino acids in blood decline, lysosomes in cells can rapidly digest some of their proteins (structural and contractile proteins are spared, except in cases of malnutrition). The amino acids released this way enter the blood and thereby help maintain the required levels.

   Suppose you embark on a body-building program. You already eat plenty of carbohydrates, but a nutritionist advises a protein-rich diet that includes protein supplements. Speculate on how the extra dietary proteins will be put to use, and in which tissues.

4. Each year, Canada geese lift off in precise formation from their northern breeding grounds. They head south to spend the winter months in warmer climates, and then they make the return trip in spring. As is true of other migratory birds, their flight muscle cells are efficient at using fatty acids as an energy source. (Remember, the carbon backbone of fatty acids can be cleaved into fragments that can be converted to acetyl-CoA for entry into the Krebs cycle.)

   Suppose a lesser Canada goose from Alaska's Point Barrow has been steadily flapping along for three thousand kilometers and is nearing Klamath Falls, Oregon. It looks down and notices a rabbit sprinting like the wind from a coyote with a taste for rabbit. With a stunning burst of speed, the rabbit reaches the safety of its burrow.

   Which energy-releasing pathway predominated in the rabbit's leg muscle cells? Why was the Canada goose relying on a different pathway for most of its journey? And why wouldn't the pathway of choice in goose flight muscle cells be much good for a rabbit making a mad dash from its enemy?

5. Reflect on this chapter's Introduction, then on Question 4. Now speculate on which energy-releasing pathway is predominating in an agitated Africanized bee chasing a farmer across a cornfield.

Selected Key Terms

acetyl-CoA 8.3   Krebs cycle 8.1
aerobic respiration 8.1   lactate fermentation 8.5
alcoholic fermentation 8.5   mitochondrion 8.3
anaerobic electron transport 8.5 NAD^+ 8.2
electron transport   oxaloacetate 8.3
   phosphorylation 8.1   pyruvate 8.2
FAD 8.3   substrate-level
glycolysis 8.1   phosphorylation 8.2

Readings

Levi, P. October 1984. "Travels with C." *The Sciences.* Journey of a carbon atom through the world of life.

Wolfe, S. 1995. *An Introduction to Molecular and Cellular Biology.* Belmont, California: Wadsworth. Exceptional reference text.

Web Site See *http://www.wadsworth.com/biology* for practice quiz questions, hypercontents, BioUpdates, and critical thinking. The Wadsworth Biology Resource Center provides a wealth of information fully organized and integrated by chapter.

FACING PAGE: *Human sperm, one of which will penetrate this mature egg and so set the stage for the development of a new individual in the image of its parents.*

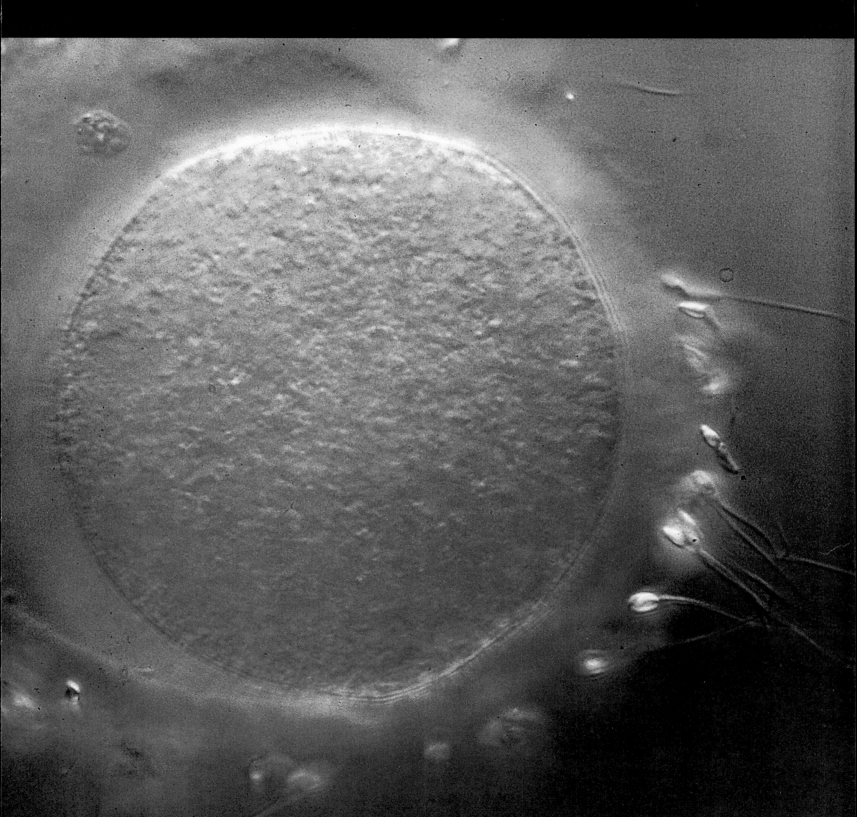

9

CELL DIVISION AND MITOSIS

Silver In the Stream of Time

Five o'clock, and the first rays from the sun dance over the wild Alagnak River of the Alaskan tundra. It is September, and life is ending and beginning in the clear, frigid waters. By the thousands, mature silver salmon have returned from the open ocean to spawn in their shallow native home. The females are tinged with red, the color of spawners, and they are dying.

This morning, a female salmon releases translucent pink eggs into a shallow "nest," hollowed out by her fins in the gravel riverbed (Figure 9.1). Moments later a male sheds a cloud of sperm, and fertilization follows. Trout and other predators eat most of the eggs, but some eggs survive and give rise to a new generation.

Within three years, the pea-size eggs have become streamlined salmon, fashioned from billions of cells. A few of their cells will develop into eggs or sperm. In time, on some September morning, they will take part in an ongoing story of birth, growth, death, and rebirth.

For you, as for salmon and all other multicelled species, growth as well as reproduction depends on *cell division*. Inside your mother, a fertilized egg divided in two, then the two into four, and so on until billions of cells were growing, developing in specialized ways, and dividing at different times to produce all of your genetically prescribed body parts. Your body now has roughly 65 trillion living cells—and many of them are still dividing. Every five days, for instance, divisions replace the tissue that lines your small intestine.

Understanding cell division—and, ultimately, how new individuals are put together in the image of their parents—begins with answers to three questions. *First*, what instructions are necessary for inheritance? *Second*, how are those instructions duplicated for distribution into daughter cells? *Third*, by what mechanisms are those instructions divided into daughter cells? We will require more than one chapter to consider the nature of

Figure 9.1 The last of one generation and the first of the next in Alaska's Alagnak River.

cell reproduction and other mechanisms of inheritance. Even so, the points made early in this chapter can help you keep the overall picture in focus.

Begin with the word **reproduction**. In biology, this means that parents produce a new generation of cells or multicelled individuals like themselves. The process starts in cells that are programmed to divide. And the ground rule for division is this: *Parent cells must provide their daughter cells with hereditary instructions, encoded in DNA, and enough metabolic machinery to start up their own operation.*

DNA, recall, contains instructions for synthesizing proteins. Some proteins are structural materials. Many are enzymes that speed the assembly of specific organic compounds, such as the lipids that cells use as building blocks and sources of energy. Unless a daughter cell receives the necessary instructions for making proteins, it simply will not be able to grow or function properly.

Also, the cytoplasm of a parent cell already contains enzymes, organelles, and other operating machinery. When a daughter cell inherits what looks like a blob of cytoplasm, it really is getting start-up machinery, which will keep it operating until it can use its inherited DNA for growing and developing on its own.

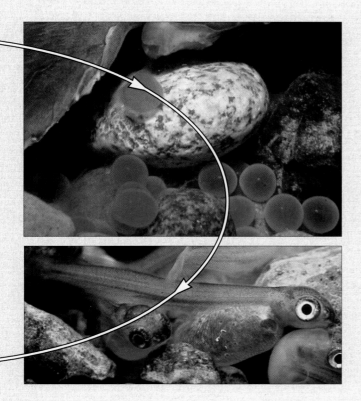

KEY CONCEPTS

1. The continuity of life depends on reproduction, by which parents produce a new generation of cells or multicelled individuals like themselves. Cell division is the bridge between generations.

2. When a cell divides, its two daughter cells must each receive a required number of DNA molecules as well as cytoplasm. For eukaryotic cells, a division mechanism called mitosis sorts out the DNA into two new nuclei. A separate mechanism divides the cytoplasm.

3. Mitosis is one part of the cell cycle. The other part is interphase, an interval when each new cell formed by mitosis and cytoplasmic division increases in mass, doubles the number of its components, then duplicates its DNA. The cycle ends when that cell divides.

4. In eukaryotic cells, many proteins with structural and functional roles are attached to the DNA. Each DNA molecule, with its attached proteins, is a chromosome.

5. Members of the same species have a characteristic number of chromosomes in their cells. The chromosomes differ from one another in length and shape, and they carry different portions of the hereditary instructions.

6. The body cells of humans and many other organisms have a diploid chromosome number; they contain two of each type of chromosome characteristic of the species.

7. Mitosis keeps the chromosome number constant, from one cell generation to the next. Therefore, if a parent cell is diploid, so will be the daughter cells.

8. Mitotic cell division is the basis of growth and tissue repair in multicelled eukaryotes. It also is the means by which single-celled eukaryotes and many multicelled eukaryotes reproduce asexually.

DIVIDING CELLS: THE BRIDGE BETWEEN GENERATIONS

Overview of Division Mechanisms

In plants, animals, and all other eukaryotic organisms, hereditary instructions are distributed among a number of DNA molecules. Before the cells of such organisms reproduce, they must undergo *nuclear* division. **Mitosis** and **meiosis** are two nuclear division mechanisms. Both sort out and package a parent cell's DNA into new nuclei, for the forthcoming daughter cells. Some other mechanism splits the cytoplasm into daughter cells.

Multicelled organisms grow by way of mitosis and the cytoplasmic division of body cells, which are called **somatic cells**. They also repair tissues that way. Nick yourself peeling a potato, and mitotic cell divisions will replace the cells that the knife sliced away. Besides this, many protistans, fungi, plants, and even some animals reproduce asexually by mitotic cell division.

By contrast, meiosis occurs only in **germ cells**, a cell lineage set aside for the formation of gametes (such as sperm and eggs) and sexual reproduction. As you will read in the next chapter, meiosis has much in common with mitosis, but the end result is different.

Prokaryotic cells, or bacteria, reproduce asexually by a different mechanism, called prokaryotic fission. We will consider the bacteria later, in Section 22.2.

Some Key Points About Chromosomes

In a nondividing cell, the DNA molecules are stretched out like thin threads, with many proteins attached to them. Each DNA molecule, with its attached proteins, is a **chromosome**. When a cell prepares for mitosis, each

threadlike chromosome is duplicated. It now consists of two DNA molecules, which will stay together until late in mitosis. As long as they remain attached, the two are called **sister chromatids** of the chromosome.

Figure 9.2 illustrates a eukaryotic chromosome in the unduplicated and duplicated states. Notice how the duplicated chromosome narrows down in one region along its length. This is the **centromere**, a small region with attachment sites for the microtubules that move the chromosome during nuclear division. Bear in mind, the sketch in Figure 9.2 is simplified. For example, the location of the centromere differs among chromosomes. And the two strands of a DNA molecule do not look like a ladder; they twist rather like a spiral staircase and are much longer than can be shown here.

Mitosis and the Chromosome Number

Each species has a characteristic **chromosome number**, which refers to the sum total of chromosomes in cells of a given type. Human somatic cells have 46, those of gorillas have 48, and those of pea plants have 14.

Actually, your 46 chromosomes are like volumes of two sets of books. Each set is numbered 1 to 23. For example, you have two "volumes" of chromosome 22— that is, *a pair of them*. Generally, both members of each pair have the same length and shape, and they carry the same portion of hereditary instructions for the same traits. Think of them as two sets of books on how to build a house. Your father gave you one set. Your mother had her own ideas about storage, plumbing, and so on, so she gave you a revised edition. Her set covers the same topics but says slightly different things about many of them.

We say the chromosome number is **diploid**, or $2n$, if a cell has two of each type of chromosome characteristic of the species. The body cells of humans, gorillas, pea plants, and a great many other organisms are like this. (By contrast, as explained in Chapter 10, their eggs and sperm have a *haploid* chromosome number, or n. This means they have only one of each type of chromosome characteristic of the species.)

With mitosis, a diploid parent cell can produce two diploid daughter cells. This doesn't mean each merely gets forty-six or forty-eight or fourteen chromosomes. If only the total mattered, one cell might get, say, two pairs of chromosome 22 and no pairs whatsoever of chromosome 9. However, neither cell would function properly *without two of each type of chromosome*.

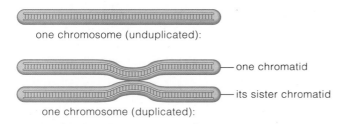

one chromosome (unduplicated):

one chromatid
its sister chromatid

one chromosome (duplicated):

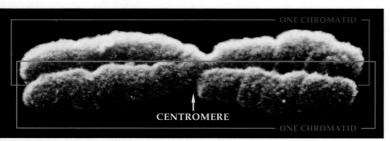

ONE CHROMATID

CENTROMERE

ONE CHROMATID

Figure 9.2   Sketch of a chromosome in the unduplicated and duplicated states. The scanning electron micrograph shows a human chromosome in the duplicated state; it consists of two sister chromatids, attached at the centromere.

Mitosis keeps the chromosome number constant, division after division, from one cell generation to the next. Thus, if a parent cell is diploid, its daughter cells will be diploid also.

THE CELL CYCLE

Mitosis is only one phase of the **cell cycle**. Such cycles start each time new cells are produced and end when those cells complete their own division. The cycle starts again for each new daughter cell (Figure 9.3). Usually, the longest phase of the cell cycle is **interphase**, which has three parts. During interphase, a cell increases its mass, roughly doubles the number of its cytoplasmic components, and then duplicates its DNA. The different parts of the cycle are often abbreviated this way:

G1 Of interphase, a "*Gap*" (interval) of cell growth before the onset of DNA replication

S   Of interphase, a time of "*Synthesis*" (DNA replication)

G2 Of interphase, a second "*Gap*" (interval) following DNA replication, when the cell prepares for division

M   *Mitosis*; nuclear division only, usually followed by cytoplasmic division

The cell cycle lasts about the same length of time for cells of a given type. Its duration differs among cells of different types. For example, all of the neurons (nerve cells) in your brain are arrested at interphase, and they usually will not divide again. By contrast, cells of a new sea urchin may double in number every two hours.

Adverse conditions may disrupt a cell cycle. When deprived of a vital nutrient, for instance, the free-living cells called amoebas do not leave interphase. Even so, if any cell proceeds past a certain point in interphase, the cycle normally will continue regardless of outside conditions, owing to built-in controls over its duration.

Figure 9.3 Eukaryotic cell cycle, generalized. The length of each part differs among different cell types.

[circular diagram of cell cycle]

G1
period of cell growth before the DNA is duplicated
(interphase begins in daughter cells)

S
period when the DNA is duplicated (that is, when chromosomes are duplicated)

Interphase

cytoplasm divided
telophase
anaphase
metaphase
prophase
(interphase ends in parent cell)

Mitosis

G2
period after DNA is duplicated; cell prepares for division

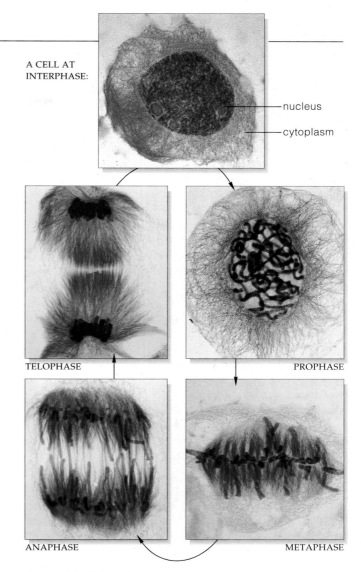

A CELL AT INTERPHASE:
nucleus
cytoplasm

TELOPHASE

PROPHASE

ANAPHASE

METAPHASE

Figure 9.4 Mitosis in a cell from the African blood lily, *Haemanthus*. The chromosomes are stained *blue*, and the many microtubules are stained *red*. Before reading further, take a moment to become familiar with the labels on the micrographs.

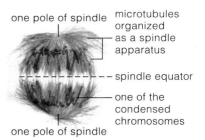

one pole of spindle
microtubules organized as a spindle apparatus
spindle equator
one of the condensed chromosomes
one pole of spindle

We turn now to mitosis and to how it maintains the chromosome number through turn after turn of the cell cycle. Figure 9.4 only hints at the divisional ballet that begins as a cell leaves interphase.

A cell cycle begins at interphase, when a new cell (formed by mitosis and cytoplasmic division) increases its mass, roughly doubles the number of its cytoplasmic components, then duplicates its chromosomes. The cycle ends when the cell divides.

THE STAGES OF MITOSIS—AN OVERVIEW

When a cell makes the transition from interphase to mitosis, it has stopped constructing new cell parts, and its DNA has been replicated. Within that cell, profound changes will now proceed smoothly, one after the other, through four stages. The sequential stages of mitosis are **prophase**, **metaphase**, **anaphase**, and **telophase**.

Figure 9.5 illustrates mitosis in an animal cell. By comparing the series of photographs against those of the plant cell in Figure 9.4, it becomes clear that chromosomes in both cells are moving about. They don't do so on their own. A **spindle apparatus** harnesses and moves them.

A spindle consists of microtubules arranged into two sets. Each set extends from one of the two poles (end points) of the spindle. The two sets overlap each other a bit at the spindle equator, or midway between the two poles. The formation of this bipolar, microtubular spindle establishes what will be the ultimate destinations of chromosomes during mitosis, as you will see shortly.

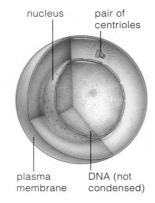

nucleus   pair of centrioles

plasma membrane   DNA (not condensed)

CELL AT INTERPHASE
The cell duplicates its DNA; then it prepares for nuclear division.

Figure 9.5 Mitosis. This nuclear division mechanism ensures that every daughter cell will have the same chromosome number as the parent cell. For clarity, the diagram shows only two pairs of chromosomes from a diploid (2n) animal cell. With only rare exceptions, the picture is more involved than this, as indicated by the micrographs of mitosis in a whitefish cell (*facing page*).

MITOSIS

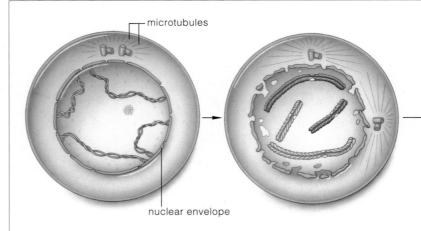

microtubules

nuclear envelope

EARLY PROPHASE
The DNA and its associated proteins have started to condense. The two chromosomes colored *purple* were inherited from the male parent. The other two (*blue*) are their counterparts, inherited from the female parent.

LATE PROPHASE
Chromosomes continue to condense. New microtubules are assembled, they move one of the two centriole pairs to the opposite end of the cell. The nuclear envelope starts to break up.

Prophase: Mitosis Begins

We know a cell is in prophase when its chromosomes become visible in the light microscope as threadlike forms. ("Mitosis" comes from the Greek *mitos*, meaning thread.) Each chromosome was duplicated earlier, in interphase. In other words, each now consists of two sister chromatids, joined at the centromere. Early on, the two sister chromatids twist and fold into a more compact form. By late prophase, all the chromosomes will be condensed into thicker, rod-shaped forms.

Meanwhile, in the cytoplasm, most microtubules of the cytoskeleton are breaking apart into their tubulin subunits (Section 4.8). The subunits reassemble near the nucleus, as *new* microtubules of the spindle. Many of these microtubules will extend from one spindle pole or the other to the centromere of a chromosome. The remainder will not interact at all with the chromosomes. They will extend from the poles and overlap each other.

While new microtubules are assembling, the nuclear envelope physically prevents them from interacting with the chromosomes inside the nucleus. However, the nuclear envelope starts breaking up as prophase ends.

Many cells have two barrel-shaped **centrioles**. Each centriole started duplicating itself during interphase, so there are two pairs of them when prophase is under way. Microtubules start moving one pair to the opposite pole of the newly forming spindle. Centrioles, recall, give rise to flagella or cilia. If you observe them in cells of an organism, you can bet that flagellated or ciliated cells (such as sperm cells) develop during its life cycle.

Transition to Metaphase

So much happens between prophase and metaphase that researchers give this transitional period its own name, "prometaphase." The nuclear envelope breaks up completely, into numerous tiny, flattened vesicles. Now the chromosomes are free to interact with microtubules that are extending toward them, from the poles of the forming spindle. Microtubules from both poles harness each chromosome and start pulling on it. The two-way pulling orients the chromosome's two sister chromatids toward opposite poles. Meanwhile, overlapping spindle

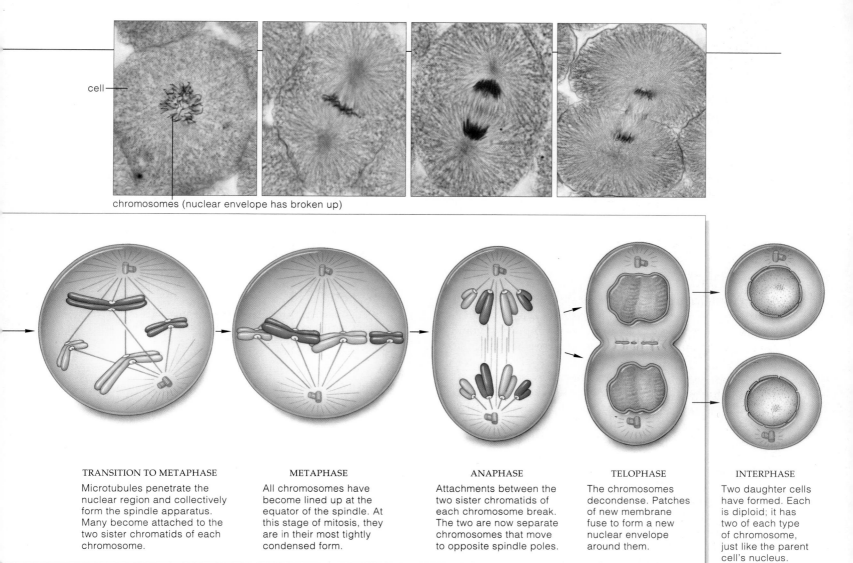

cell

chromosomes (nuclear envelope has broken up)

TRANSITION TO METAPHASE

Microtubules penetrate the nuclear region and collectively form the spindle apparatus. Many become attached to the two sister chromatids of each chromosome.

METAPHASE

All chromosomes have become lined up at the equator of the spindle. At this stage of mitosis, they are in their most tightly condensed form.

ANAPHASE

Attachments between the two sister chromatids of each chromosome break. The two are now separate chromosomes that move to opposite spindle poles.

TELOPHASE

The chromosomes decondense. Patches of new membrane fuse to form a new nuclear envelope around them.

INTERPHASE

Two daughter cells have formed. Each is diploid; it has two of each type of chromosome, just like the parent cell's nucleus.

microtubules ratchet past each other and push the poles of the spindle apart. The push–pull forces are balanced when the chromosomes reach the spindle's midpoint.

When all of the duplicated chromosomes are aligned midway between the poles of a completed spindle, we call this metaphase (*meta-* means "midway between"). The alignment is crucial for the next stage of mitosis.

From Anaphase Through Telophase

At anaphase, the sister chromatids of each chromosome separate from each other and move to opposite poles by two mechanisms. First, the microtubules attached to the centromere regions shorten and *pull* the chromosomes to the poles. Second, the spindle elongates as overlapping microtubules continue to ratchet past each other and *push* the two spindle poles even farther apart. Once each chromatid is separated from its sister, it has a new name. It is now a separate chromosome in its own right.

Telophase gets under way as soon as each of two clusters of chromosomes arrives at a spindle pole. The

chromosomes are no longer harnessed to microtubules, and they return to threadlike form. Vesicles of the old nuclear envelope fuse and form patches of membrane around the chromosomes. Patch joins with patch, and soon a new nuclear envelope separates each cluster of chromosomes from the cytoplasm. If the parent cell was diploid, each cluster contains two chromosomes of each type. With mitosis, remember, each new nucleus has the same chromosome number as the parent nucleus. Once two nuclei form, telophase is over—and so is mitosis.

Prior to mitosis, each chromosome in a cell's nucleus is duplicated, so that it consists of two sister chromatids.

Mitosis proceeds through four consecutive stages, called prophase, metaphase, anaphase, and telophase.

A microtubular spindle moves sister chromatids of every chromosome apart, to opposite spindle poles. Around each of two clusters of chromosomes, new nuclear envelope forms. Both of the daughter nuclei formed this way have the same chromosome number as the parent cell's nucleus.

A CLOSER LOOK AT THE CELL CYCLE

Close this book for a moment and reflect on what you have just learned from the two preceding sections. Be sure you have a clear picture of the flow of events in interphase and mitosis, up to the time of cytoplasmic division and the formation of new daughter cells. How easily you will get through many subsequent chapters depends on how well you understand the cell division story. If parts of the picture are still not clear, it may be worth your time to read Sections 9.2 and 9.3 once again, before continuing with the details presented here.

The Wonder of Interphase

If you could coax the DNA molecules from just one of your somatic cells to stretch in a single line, one after another, they would extend from your armpit past your fingertips. Salamander DNA is even more amazing. A single line of it would extend ten meters! The wonder is, enzymes and other proteins selectively scan all of a cell's DNA, switch protein-building instructions on and off, and even produce base-by-base copies of each DNA molecule—all during interphase.

G1, S, and G2 of interphase have distinct patterns of biosynthesis. During G1, most of the carbohydrates, lipids, and proteins for a cell's own use and for export are assembled. During S, the cell copies its DNA and synthesizes the proteins that will become organized into structural scaffolding for the condensed versions of chromosomes. Finally, during G2, the proteins that will drive mitosis to completion are produced.

Once S begins, events normally proceed at about the same rate in all cells of a species and continue all the way through mitosis. Given this observation, you might well assume the cycle has built-in, molecular brakes. Apply the brakes that operate in G1, and the cycle stalls in G1. Lift the brakes and the cycle runs to completion. Said another way, *control mechanisms govern the rate of cell division.*

Imagine a car losing its brakes just as it starts down a steep mountain road. As you will read later in the book, that is how cancer starts. Controls over division are lost, and the cell cycle simply cannot stop turning.

Chromosomes, Microtubules, and the Precision of Mitosis

Equally impressive is the precision with which mitosis parcels out DNA for forthcoming daughter cells. That precision depends upon chromosome organization and interactions among microtubules and motor proteins.

ORGANIZATION OF METAPHASE CHROMOSOMES Even during interphase, eukaryotic DNA has many proteins bound tightly to it. **Histones** are among them. Many

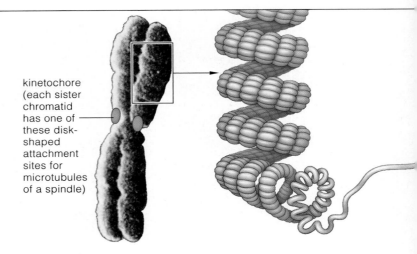

kinetochore (each sister chromatid has one of these disk-shaped attachment sites for microtubules of a spindle)

a A duplicated human chromosome during metaphase, when it is in its most condensed form. Interactions among certain chromosomal proteins keep its loops of DNA tightly packed in a "supercoiled" array.

Figure 9.6 One model of the levels of organization in a human metaphase chromosome.

histones are like spools for winding up small stretches of DNA. Each histone-DNA spool is a single structural unit called a **nucleosome**. Another histone stabilizes the spools. During mitosis (and meiosis also), interactions between histones and DNA make the chromosome coil back on itself repeatedly. The coiling greatly increases the chromosome's diameter. Other proteins besides the histones form a structural scaffold when the DNA folds further, into a series of loops (Figure 9.6).

A chromosome acquires its distinct shape and size late in prophase, when condensation is nearly complete. By then, each of its sister chromatids has at least one constricted region, the most prominent of which is the centromere. At the surface of the centromere is a disk-shaped structure, a **kinetochore**, which will serve as a docking site for spindle microtubules (Figure 9.6a).

When the threadlike DNA molecules are condensing into such compact chromosome structures, why don't they get tangled up? Actually, it appears that they do, but an enzyme called DNA topoisomerase chemically recognizes the tangling and puts things right. When researchers deliberately interfered with this enzyme's activity, the tangles persisted. Later, sister chromatids failed to separate at anaphase.

SPINDLES COME, SPINDLES GO Different species have between ten and tens of thousands of microtubules in the mitotic spindle. The plant cell shown in Figure 9.4 probably has 10,000 of them. In each case, the spindle is in exquisite balance with the cell's pool of tubulin subunits. Its microtubules are continually assembling and disassembling, so subunits are being withdrawn

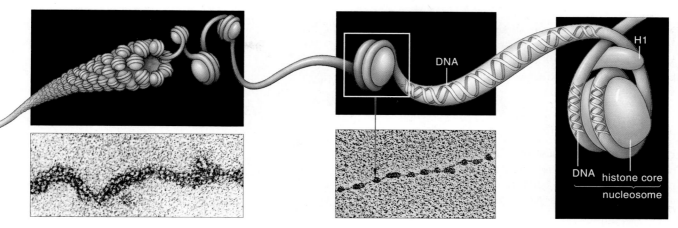

b At a deeper level of organization, the chromosomal proteins and the DNA are arranged as a cylindrical fiber (solenoid) thirty nanometers in diameter.

c Immerse a chromosome in saltwater and it loosens up to a beads-on-a-string organization. The "string" is one DNA molecule. Each "bead" is a nucleosome.

d A nucleosome consists of a double loop of DNA around a core of eight histones. Other histones stabilize the structural array.

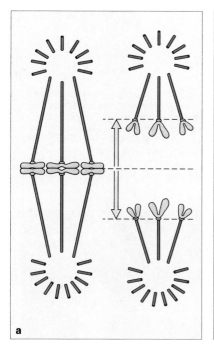

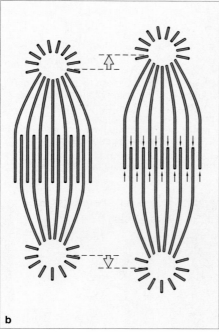

Figure 9.7 Models of two of the mechanisms that separate sister chromatids of a chromosome from each other at anaphase. In (**a**), the microtubules attached to the chromatids *shorten* and thereby decrease the distance between the kinetochores and spindle poles. In (**b**), overlapping microtubules ratchet past each other, thereby moving the spindle poles apart and increasing the distance between sister chromatids of each chromosome.

kinetochore is like a train chugging along a railroad track—except the track disassembles after it. Two motor proteins called dynein and kinesin have been isolated from kinetochores and may drive the sliding motion. As you read in Sections 4.8 and 4.9, **motor proteins** are a class of monomers that project from the surface of microtubules and some other cytoskeletal elements with roles in cell movements.

What about the microtubules that do not attach to kinetochores? Those extending from one pole actively slide past those extending from the other pole at the zone where they overlap (Figure 9.7*b*). Dynein and kinesin have been identified in this zone of overlap, also.

from and returned to the pool all the time. The balance can be tipped experimentally, as by exposing cells to the microtubule poison colchicine (Section 4.8). The spindle disassembles almost at once, but often it reassembles in seconds or minutes if the disrupting agent is removed.

At anaphase, some movements reduce the distance between kinetochores and the poles; others make the whole spindle lengthen (Figure 9.7). The kinetochores slide over the microtubules. Experimental observations show that the position of the microtubules stays put all through anaphase. The microtubules *do* shorten, in the manner shown in Figure 9.7*a*, so they must disassemble where a kinetochore passes over them. By analogy, the

Once the S stage of interphase begins, the cell cycle proceeds at about the same rate in all cells of a given type, all the way through mitosis. Molecular mechanisms control whether a cell enters S and thereby control the rate of cell division.

The condensed form of metaphase chromosomes arises by interactions among DNA and structural proteins, including histones, that associate with DNA throughout the cell cycle.

During mitosis, motor proteins act on microtubules of the spindle to produce chromosomal movements.

DIVISION OF THE CYTOPLASM

The cytoplasm usually divides at some time between late anaphase and the end of telophase. As you might gather from Figures 9.8 and 9.9, the mechanism of this **cytoplasmic division** (or cytokinesis, as it is commonly called) differs among organisms.

Cell Plate Formation in Plants

As described in Section 4.10, most plant cells are walled, which means their cytoplasm cannot be pinched in two. Cytoplasmic division of such cells involves **cell plate formation**, as shown in Figure 9.8. By this mechanism, vesicles chockful of wall-building materials fuse with

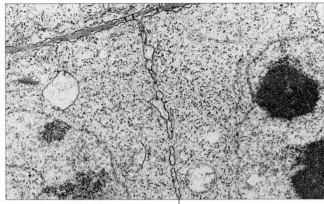

part of a newly forming cell plate ⌐

former spindle equator

a As mitosis ends, vesicles converge at the spindle equator. They contain cementing materials and structural materials for a new primary cell wall.

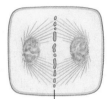

vesicles converging

b A cell plate starts forming as membranes of the vesicles fuse. Materials inside the vesicles get sandwiched between two new membranes that elongate along the plane of the cell plate.

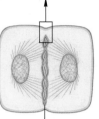

cell plate growing

c Cellulose is deposited on the inside of the "sandwich" and will form two cell walls. Other deposits will form a middle lamella and cement the walls together (Section 4.10).

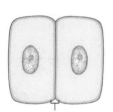

two new primary walls

d A cell plate grows at its margins until it fuses with the parent cell's plasma membrane. During plant growth, when cells expand and their walls are still thin, new material gets deposited on the old primary wall.

Figure 9.8 Cytoplasmic division of a plant cell, as brought about by cell plate formation.

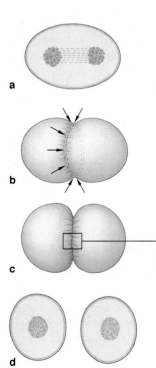

a

b

c

d

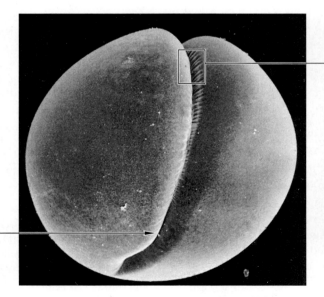

Figure 9.9 Cytoplasmic division of an animal egg. (**a**) Mitosis is completed, and the spindle is now disassembling. (**b**) Just beneath the plasma membrane of the parent cell, rings of microfilaments at the former spindle equator contract and close around the cell, rather like a purse string being tightened. (**c,d**) Contractions continue and in time will divide the cell in two. The micrographs show how the surface of the plasma membrane sinks inward. The depression defines the cleavage plane.

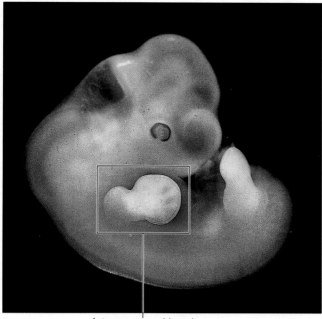

a   future arm and hand

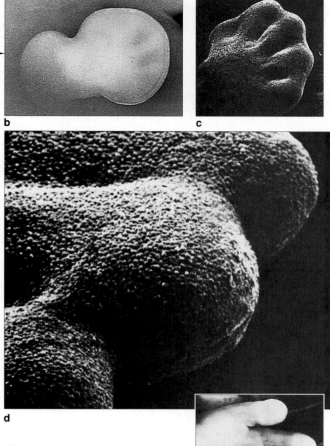

b   c

d

Figure 9.10 Development of the human hand by way of mitosis, cytoplasmic divisions, and other processes. Many individual cells resulting from mitotic cell divisions are visible in (**d**).

e

one another and with remnants from the microtubular spindle. Together, they form a disklike structure—a cell plate. At this location, deposits of cellulose accumulate and form a crosswall that divides the parent cell into two daughter cells.

Cleavage of Animal Cells

Unlike plant cells, an animal cell is not confined within a cell wall, and its cytoplasm typically "pinches in two." Consider a newly fertilized egg. Through a mechanism called **cleavage**, a shallow, ringlike depression forms at the cell surface, above the cell's midsection (Figure 9.9). The depression is known as the cleavage furrow. It is visible evidence of parallel arrays of microfilaments in the cytoplasm that connect with the plasma membrane. Microfilaments, remember, are threadlike cytoskeletal elements. These particular ones are organized so that they slide past one another (Section 4.9). When they do so, they pull the plasma membrane inward and thereby cut the cell in two.

This concludes our picture of mitotic cell division. Look now at your hands and try to envision all the cells in your palms, thumbs, and fingers. Imagine the divisions that produced all the generations of cells that preceded them when you were developing, early on, inside your mother (Figure 9.10). And be grateful for the astonishing precision of the mechanisms that led to their formation, for the alternatives can be terrible indeed. Why? Good health, and survival itself, depends absolutely on the proper timing and completion of cell

cycle events, including mitosis. Genetic disorders often arise from mistakes in the duplication or distribution of just one chromosome. As you will read in Section 15.6, cancer may rapidly destroy a mature tissue if controls are lost that otherwise prevent cells from dividing. The section that concludes this chapter describes a landmark case of such unchecked cell divisions.

Following mitotic cell division, a separate mechanism cuts the cytoplasm into two daughter cells, each with a daughter nucleus.

In plants, cytoplasmic division often involves the formation of a cell plate and a crosswall in between the adjoining, new plasma membranes of daughter cells.

Cytoplasmic division in animals may involve cleavage. Rings of microfilaments around a parent cell's midsection slide past one another in a way that pinches the cytoplasm in two.

Focus on Science

HENRIETTA'S IMMORTAL CELLS

Each human starts out as a single fertilized egg. By the time of birth, mitotic cell divisions and other processes have resulted in a human body of about a trillion cells. Even in an adult, billions of cells are still dividing. For example, cells of the stomach's lining divide every day. Liver cells usually do not divide, but if part of the liver becomes injured or diseased, repeated cell divisions will keep on producing more new cells until the damaged part is finally replaced.

In 1951, George and Margaret Gey of Johns Hopkins University were trying to develop a way to keep human cells dividing *outside* the body. With such isolated cells, these researchers and others could investigate basic life processes. They also could conduct studies of cancer and other diseases without having to experiment directly on patients and thereby gamble with human lives.

The Geys had samples of normal and diseased human cells, which local physicians had sent to them. But they just couldn't stop the descendants of those precious cells from dying out within a few weeks.

Mary Kubicek, a laboratory assistant, worked with the Geys in their efforts to start a self-perpetuating lineage of cultured human cells. She was about to give up after dozens of failed attempts. Even so, in 1951 she decided to prepare one more sample of cancer cells for culture. She gave the sample the code name **HeLa cells**, for the first two letters of the patient's first and last names.

The HeLa cells began to divide. And divide. And divide again! By the fourth day there were so many cells that Kubicek had to subdivide them into more culture tubes. As months passed, the culture continued to thrive.

Unfortunately, the tumor cells inside the patient's body were just as vigorous. Six months after the patient was first diagnosed as having cancer, tumor cells had spread to tissues throughout her body. Only eight months after the diagnosis, Henrietta Lacks, a young woman from Baltimore, was dead.

Although Henrietta passed away, some of her cells lived on in the Geys' laboratory as the first successful human cell culture. HeLa cells were soon shipped to other researchers, who passed cells on to others, and so on. HeLa cells came to live in laboratories all over the world. Some of the cultured cells even journeyed far into space, as components of experiments to be carried out aboard the *Discoverer XVII* satellite. Every year, hundreds of scientific papers describe research that is based on work with HeLa cells.

Henrietta was only thirty-one years old when runaway cell divisions quickly killed her. Yet now, many decades later, her legacy is still benefiting humans everywhere—in cellular descendants that are still alive and dividing, day after day after day.

SUMMARY

1. Through specific division mechanisms, summarized in Table 9.1, a parent cell provides each daughter cell with hereditary instructions (DNA) and cytoplasmic machinery necessary to start up its own operation.

  a. In eukaryotic cells, the nucleus divides by mitosis or meiosis. Cytoplasmic division typically follows.

  b. Prokaryotic cells divide by prokaryotic fission.

2. Each eukaryotic chromosome is one DNA molecule with numerous proteins attached. The chromosomes in a given cell differ in length, shape, and which part of the hereditary instructions they carry.

  a. "Chromosome number" refers to the sum total of chromosomes in the cells of a given type. Cells with a diploid chromosome number ($2n$) contain two of each kind of chromosome.

  b. Mitosis divides the nucleus into two equivalent nuclei, each having the same chromosome number as the parent cell. Therefore, it maintains the chromosome number from one cell generation to the next.

  c. Mitosis is the basis of growth and tissue repair among multicelled eukaryotes, and often of asexual reproduction among single-celled eukaryotes. (Meiosis occurs only in germ cells used in sexual reproduction.)

3. A cell cycle starts when a new cell forms. It proceeds through interphase and ends when the cell reproduces by mitosis and cytoplasmic division. In interphase, a cell increases in its mass and cytoplasmic components, then duplicates its chromosomes.

4. Each duplicated chromosome has two molecules of DNA, attached at the centromere. For as long as the two are attached, they are called sister chromatids.

5. Mitosis proceeds through four continuous stages:

  a. Prophase. Duplicated chromosomes, in threadlike form, start to condense. New microtubules start to assemble near the nucleus into a spindle apparatus. The nuclear envelope starts to break up.

  b. Metaphase. During the *transition* to metaphase, the nuclear envelope breaks up totally into vesicles. Microtubules of opposite poles of the forming spindle

Table 9.1 Summary of Cell Division Mechanisms	
Mechanisms	Functions
MITOSIS, CYTOPLASMIC DIVISION	In multicelled eukaryotes, the basis of bodily growth. In single-celled and many multicelled eukaryotes, the basis of asexual reproduction.
MEIOSIS, CYTOPLASMIC DIVISION	In single-celled and multicelled eukaryotes, the basis of gamete formation and of sexual reproduction.
PROKARYOTIC FISSION	In bacterial cells, the basis of asexual reproduction.

attach to only one of the two sister chromatids of each chromosome. *At* metaphase, all of the chromosomes are aligned at the spindle equator.

c. Anaphase. Microtubules pull sister chromatids of each chromosome away from each other, to opposite spindle poles. Each type of parental chromosome is represented by a daughter chromosome at both poles:

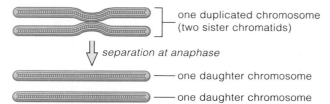

d. Telophase. The chromosomes decondense to the threadlike form. A new nuclear envelope forms around them. Each nucleus has the same chromosome number as the parent cell. Mitosis is completed.

6. Different mechanisms divide the cytoplasm near the end of nuclear division or afterward. In plant cells, a cell plate forms, then a crosswall and plasma membranes form. Animal cells pinch in two, as by cleavage.

Review Questions

1. Define the two types of division mechanisms that operate in eukaryotic cells. Does either one divide the cytoplasm? *9.1*

2. Define somatic cell and germ cell. *9.1*

3. What is a chromosome called when it is in the unduplicated state? In the duplicated state (with two sister chromatids)? *9.1*

4. Describe the microtubular spindle and its functions, then name and briefly describe the stages of mitosis. *9.3*

5. Briefly explain how cytoplasmic division differs in typical plant and animal cells. *9.5*

Self-Quiz (Answers in Appendix IV)

1. A somatic cell with two of each type of chromosome characteristic of the species has a(n) _____ chromosome number.
   a. diploid   c. abnormal
   b. haploid   d. a and c

2. A duplicated chromosome has _____ chromatid(s).
   a. one   c. three
   b. two   d. four

3. In a chromosome, a _____ is a constricted region with attachment sites for microtubules.
   a. chromatid   c. cell plate
   b. centromere   d. cleavage furrow

4. Interphase is the part of the cell cycle when _____ .
   a. a cell ceases to function
   b. a germ cell forms its spindle apparatus
   c. a cell grows and duplicates its DNA
   d. mitosis proceeds

5. After mitosis, the chromosome number of a daughter cell is _____ the parent cell's.
   a. the same as   c. rearranged compared to
   b. one-half   d. doubled compared to

6. Mitosis and cytoplasmic division function in _____ .
   a. asexual reproduction of single-celled eukaryotes
   b. growth, tissue repair, and sometimes asexual reproduction in many multicelled eukaryotes
   c. gamete formation in prokaryotes
   d. both a and b

7. Only _____ is not a stage of mitosis.
   a. prophase   b. interphase   c. metaphase   d. anaphase

8. Match each stage with the events listed.
   ____ metaphase   a. sister chromatids move apart
   ____ prophase   b. chromosomes start to condense
   ____ telophase   c. chromosomes decondense and daughter
   ____ anaphase   nuclei form
                      d. all duplicated chromosomes are aligned
                          at spindle equator

Critical Thinking

1. Suppose you have a means of measuring the amount of DNA in a single cell during the cell cycle. You first measure the amount during the G1 phase. At what points during the remainder of the cycle would you predict changes in the amount of DNA per cell?

2. A cell from a tissue culture has 38 chromosomes. After mitosis and cytoplasmic division, one daughter cell has 39 chromosomes, and the other has 37. Speculate on what might have occurred to cause the abnormal chromosome numbers. Generally speaking, how might the abnormality affect cell structure, function, or both?

3. Pacific yews (*Taxus brevifolius*) face extinction. People started stripping its bark and killing the trees when they heard that *taxol*, a chemical extract from the bark, may be useful for treating breast and ovarian cancer. (Synthesizing taxol in the laboratory may save the species.) Taxol prevents the disassembly of microtubules. What does this tell you about its potential as an anticancer drug?

4. X-rays and gamma rays emitted from some radioisotopes cause chemical damage to DNA, especially in cells engaged in DNA replication. High-level exposure can result in *radiation poisoning*. Hair loss and damage to the lining of the stomach and intestines are two early symptoms. Speculate why. Also speculate on why highly focused radiation therapy is used against some cancers.

Selected Key Terms

anaphase *9.3*	cytoplasmic	metaphase *9.3*
cell cycle *9.2*	division *9.5*	mitosis *9.1*
cell plate	diploid (chromosome	motor protein *9.4*
formation *9.5*	number) *9.1*	nucleosome *9.4*
centriole *9.3*	germ cell *9.1*	prophase *9.3*
centromere *9.1*	HeLa cell *9.6*	reproduction *CI*
chromosome *9.1*	histone *9.4*	sister chromatid *9.1*
chromosome	interphase *9.2*	somatic cell *9.1*
number *9.1*	kinetochore *9.4*	spindle apparatus *9.3*
cleavage *9.5*	meiosis *9.1*	telophase *9.3*

Reading

Murray, A., and M. Kirschner. March 1991. "What Controls the Cell Cycle?" *Scientific American* 264(3): 56–63.

Web Site See *http://www.wadsworth.com/biology* for practice quiz questions, hypercontents, BioUpdates, and critical thinking. The Wadsworth Biology Resource Center provides a wealth of information fully organized and integrated by chapter.

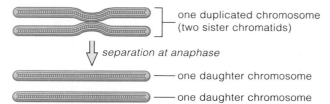

10 MEIOSIS

Octopus Sex and Other Stories

The couple clearly are interested in each other. First he caresses her with one tentacle, then another—and then another and another. She reciprocates with a hug here, a squeeze there. This goes on for hours. Finally the male reaches under his mantle, a fold of tissue that drapes around most of his body. He removes a packet of sperm from a reproductive organ and inserts it into an egg chamber underneath the female's mantle. For every sperm that fertilizes an egg, a new octopus may develop.

Unlike the one-to-one coupling between a male and female octopus, sex for the slipper limpet is a group activity. Slipper limpets are marine animals, relatives of those familiar snails on land. Before becoming transformed into a sexually mature adult, a slipper limpet must pass through a free-living stage of development called a larva. When a limpet larva is about to undergo transformation, it settles down on a pebble or shell or rock. If it settles down all by itself, it will become a female. Then, if another larva settles and develops on the

Figure 10.1 Variations in the reproductive modes among eukaryotic organisms.

(**a**) Some slipper limpets, busily perpetuating the species through group participation in sexual reproduction. The tiny crab in the foreground is merely a passerby, not a voyeur. (**b**) Live birth of an aphid, a type of insect that reproduces sexually in autumn but can switch to an asexual mode in summer.

With this chapter, we turn to the kinds of cells that serve as the bridge between generations of organisms. For many eukaryotic species, specialized phases of reproduction and development, including asexual episodes, loop out from the basic life cycle. Regardless of their specialized details, all of these life cycles turn on two basic events: *gamete formation* and *fertilization*.

first one, it will function right off as a male. However, if *another* male develops on top of it, that first male will gradually become a female. That third male *also* will become a female if a fourth male develops on top of it, and so on amongst ten or more limpets.

Slipper limpets typically live in such piles, with the bottom one always being the oldest female and the uppermost one being the youngest male (Figure 10.1*a*). Until they make the gender switch, the males release sperm; these fertilize a female's eggs, which then grow to become males and then, most likely, females—and so it goes, from one limpet generation to the next.

Even within a single species, we often come across variations in the mode of reproduction. For example, the life cycle of many sexually reproducing organisms also includes asexual episodes that are based on mitotic cell divisions. Consider the orchids, dandelions, and many other species of plants that are able to reproduce without engaging in sex. Consider the flatworms, a group of aquatic animals. They can split their small body into two roughly equivalent parts that each grow into a new flatworm.

Or consider the aphids. During the summer, nearly all aphids are females, which produce more females from *unfertilized* egg cells (Figure 10.1*b*). Only when autumn approaches do male aphids finally develop and do their part in the sexual phase of the life cycle. Even then, the females that manage to survive through the winter can do without males. Come summer, the females begin another round of producing offspring all by themselves.

These examples only hint at the immense variation in reproductive modes among eukaryotic organisms. And yet, despite the variation, *sexual* reproduction dominates their life cycles, and it inevitably involves certain events. Briefly, before the time of cell division, chromosomes are duplicated in reproductive cells. In animals, for instance, immature reproductive cells are called **germ cells**. In both males and females, the germ cells undergo meiosis and, later, cytoplasmic division. In time, their cellular descendants mature and may function as **gametes**, or sex cells. When gametes join at fertilization, they form the first cell of a new individual.

Meiosis, the formation of gametes, and fertilization are the hallmarks of sexual reproduction. As you will see in this chapter, these interconnected events contribute to the splendid diversity of life.

KEY CONCEPTS

1. Sexual reproduction proceeds through three events: meiosis, formation of gametes, and fertilization. Sperm and eggs are familiar gametes.

2. Meiosis, a nuclear division mechanism, occurs only in immature cells that are set aside for sexual reproduction. The germ cells of male and female animals are examples. Meiosis sorts out a germ cell's chromosomes into four new nuclei. After meiosis, gametes form by way of cytoplasmic division and other events.

3. Cells with a diploid chromosome number contain two of each type of chromosome characteristic of the species. The two function as a pair during meiosis. Typically, one chromosome of the pair is maternal, with hereditary instructions from a female parent. The other is paternal, with the same categories of hereditary instructions from a male parent.

4. Meiosis divides the chromosome number by half for each forthcoming gamete. Thus, if both parents are diploid ($2n$), the gametes that form are haploid (n). Later, the union of two gametes at fertilization restores the diploid number in the new individual ($n + n = 2n$).

5. During meiosis, each pair of chromosomes may swap segments. Each time they do, they exchange hereditary instructions about certain traits. Also, meiosis assigns one of every pair of chromosomes to a forthcoming gamete—but *which* gamete is its destination is a matter of chance. Hereditary instructions are further shuffled at fertilization. All three of these reproductive events lead to variations in traits among offspring.

6. In most species of plants, spore formation and other events intervene between meiosis and gamete formation.

10.1 COMPARISON OF ASEXUAL AND SEXUAL REPRODUCTION

When an orchid, flatworm, or aphid reproduces all by itself, what sort of offspring does it get? By the process of **asexual reproduction**, one parent alone produces offspring, and each offspring inherits the same number and kinds of genes as its parent. **Genes** are specific stretches of chromosomes—that is, of DNA molecules. Taken together, the genes for every species contain all of the heritable bits of information that are necessary to produce new individuals. Rare mutations aside, this means asexually produced offspring can only be clones, or genetically identical copies of the parent.

Inheritance gets much more interesting with **sexual reproduction**. This process involves meiosis, gamete formation, and fertilization (the union of two gametes). In humans and other species that use this process, the first cell of a new individual ends up with *pairs of genes*, on pairs of homologous chromosomes. Typically, one of each pair is maternal and the other paternal in origin.

If instructions encoded in every pair of genes were identical down to the last detail, sexual reproduction would produce clones, also. Just imagine—you, every single person you know, the entire human population might be a clone, with everybody looking alike. But the two genes of a pair may *not* be identical. Why not? The molecular structure of genes can change; this is what we mean by mutation. Depending on their structure, two genes that happen to be paired in a person's cells may "say" slightly different things about a trait. Each unique molecular form of the same gene is called an **allele**.

Such tiny differences affect thousands of traits. For example, whether your chin has a dimple depends on which pair of alleles you inherited at one chromosome location. One kind of allele at that location says "put a dimple in the chin," another kind says "no dimple." This leads us to a key reason why members of sexually reproducing species don't all look alike. *Through sexual reproduction, offspring inherit new combinations of alleles, which lead to variations in physical and behavioral traits.*

This chapter describes the cellular basis of sexual reproduction. More importantly, it starts us thinking about far-reaching effects of gene shufflings at different stages of the process. The process introduces variations in traits among offspring that may be acted upon by agents of natural selection. Thus, *variation in traits is a foundation for evolutionary change.*

Asexual reproduction produces genetically identical copies of the parent. Sexual reproduction introduces variations in the details of traits among offspring.

Sexual reproduction dominates the life cycle of eukaryotic species. Meiosis, formation of gametes, and fertilization are the basic events of this process.

10.2 HOW MEIOSIS HALVES THE CHROMOSOME NUMBER

Think "Homologues"

Think back on the preceding chapter and its focus on mitotic cell division. Unlike mitosis, **meiosis** divides chromosomes into separate parcels not once *but twice* prior to cell division. Unlike mitosis, it is the first step leading to the formation of gametes.

Gametes, recall, are sex cells such as sperm or eggs. In nearly all multicelled eukaryotic organisms, gametes arise from immature reproductive cells, such as germ cells, that form inside specialized structures and organs. Figure 10.2 shows examples of where gametes form.

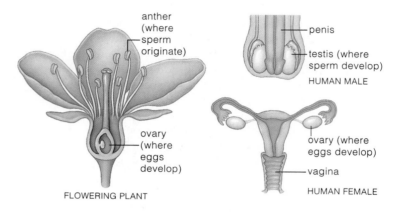

Figure 10.2 Examples of gamete-producing structures.

From Chapter 9, you know that the "chromosome number" is the sum total of chromosomes in cells of a given type. Germ cells start with the same chromosome number as somatic cells (the rest of the body's cells). If a cell has a **diploid number** (2n), it has *a pair* of each type of chromosome, often from two parents. In general, the chromosomes of each pair have the same length and shape. Their genes deal with the same traits. And they line up with each other during meiosis. We call them **homologous chromosomes** (*hom-* means alike).

As you can probably deduce from Figure 10.3, your own germ cells have 23 + 23 homologous chromosomes. After meiosis, 23 chromosomes—one of each type—end up in gametes. Thus meiosis halves the chromosome number, so that gametes have a **haploid number** (n).

Two Divisions, Not One

Meiosis resembles mitosis in some respects, even though the outcome is different. Before interphase gives way to meiosis, a germ cell duplicates its DNA. Each duplicated chromosome now consists of two DNA molecules. These

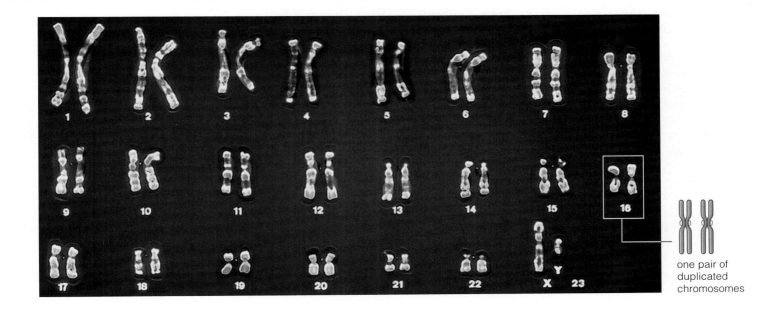

one pair of duplicated chromosomes

remain attached at a narrowed-down region, called the centromere. For as long as the two remain attached, they are known as **sister chromatids** of the chromosome:

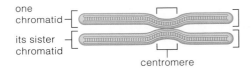

one chromatid —
its sister chromatid —
centromere

As in mitosis, the microtubules of a spindle apparatus move the chromosomes in prescribed directions.

With meiosis alone, however, *chromosomes proceed through two consecutive divisions, which end with the formation of four haploid nuclei.* The two divisions are called meiosis I and II:

DNA is replicated during interphase	**MEIOSIS I**	DNA is *not* replicated between divisions	**MEIOSIS II**
	PROPHASE I		PROPHASE II
	METAPHASE I		METAPHASE II
	ANAPHASE I		ANAPHASE II
	TELOPHASE I		TELOPHASE II

During meiosis I, each duplicated chromosome lines up with its partner, *homologue to homologue*; and then the partners are moved apart from each other:

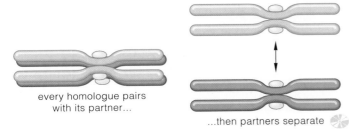

every homologue pairs with its partner...

...then partners separate

Figure 10.3 From a diploid cell of a human male, twenty-three pairs of homologous chromosomes. The sketch, corresponding to the two chromosomes inside the white box, indicates that the chromosomes are in the duplicated state. That is, each consists of two sister chromatids.

The cytoplasm typically starts to divide at some point after homologues have been separated from each other. The cytoplasmic division results in two daughter cells. Each daughter cell is haploid; it has only one of each type of chromosome. But remember, the chromosomes are still in the duplicated state.

Next, during meiosis II, *the two sister chromatids of each chromosome are separated from each other:*

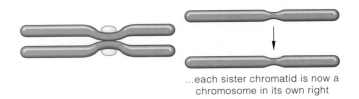

...each sister chromatid is now a chromosome in its own right

Four nuclei form, then the cytoplasm often divides once more. The final outcome is four haploid cells.

Figure 10.4 on the next two pages illustrates the key events of meiosis I and II.

Meiosis is a type of nuclear division mechanism that reduces the parental chromosome number by half, to the haploid number (*n*).

Meiosis proceeds only in immature cells, such as germ cells, that are set aside for sexual reproduction. It is the first step leading to the formation of gametes.

A VISUAL TOUR OF THE STAGES OF MEIOSIS

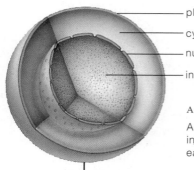

plasma membrane
cytoplasm
nuclear envelope
interior of nucleus

A GERM CELL AT INTERPHASE

Figure 10.4 Meiosis: the nuclear division mechanism by which a parental number of chromosomes in an immature reproductive cell is reduced by half (to the haploid number) for forthcoming gametes. In this case, only two of the pairs of homologous chromosomes of a germ cell are shown. The maternal chromosomes are shaded *purple.* The paternal chromosomes are shaded *blue.*

A germ cell that happens to have a diploid chromosome number (2*n*) is about to leave interphase and undergo the first division of meiosis. Its DNA is already duplicated, so each chromosome is in the duplicated state: it consists of two sister chromatids.

MEIOSIS I

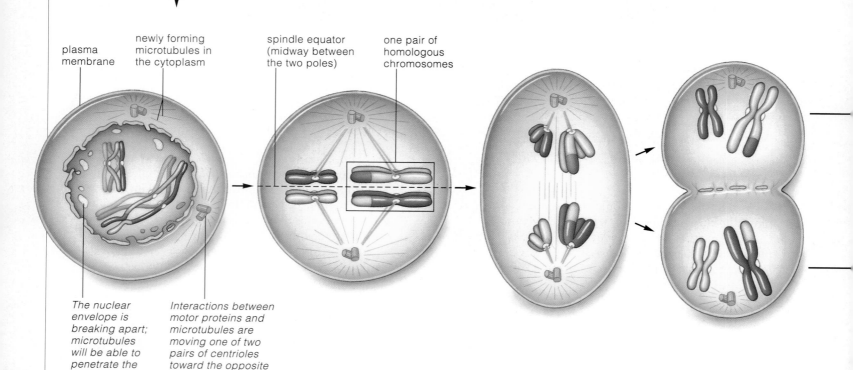

plasma membrane

newly forming microtubules in the cytoplasm

spindle equator (midway between the two poles)

one pair of homologous chromosomes

The nuclear envelope is breaking apart; microtubules will be able to penetrate the nuclear region.

Interactions between motor proteins and microtubules are moving one of two pairs of centrioles toward the opposite spindle pole.

PROPHASE I

Each duplicated chromosome is in threadlike form, but now it starts to twist and fold into more condensed form. It pairs with its homologue, and the two typically swap segments. The swapping, called crossing over, is indicated by the break in color on the pair of larger chromosomes. Each chromosome becomes attached to some microtubules of a newly forming spindle.

METAPHASE I

Motor proteins have been driving the movement of microtubules that became attached to the kinetochores of chromosomes. As a result, the chromosomes have been pushed and pulled into position, midway between the spindle poles. Now the spindle is fully formed, owing to dynamic interactions of motor proteins, microtubules, and the chromosomes themselves.

ANAPHASE I

Microtubules extending from the poles and overlapping at the spindle equator *lengthen* and push the poles apart. At the same time, the microtubules extending from the poles to the kinetochores of chromosomes *shorten,* and each chromosome is thereby pulled away from its homologous partner. These motions move the homologous partners to opposite poles.

TELOPHASE I

At some point, the cytoplasm of the germ cell divides. Two cells, each with a haploid chromosome number (*n*), result. That is, the cells have one of each type of chromosome that was present in the parent (2*n*) cell. All chromosomes are still in the duplicated state.

Of the four haploid cells that form by way of meiosis and cytoplasmic divisions, one or all may proceed to develop into gametes and function in sexual reproduction.

(In plants, the cells that form after meiosis has been completed may develop into spores, which take part in a stage of the life cycle that precedes the formation of gametes.)

MEIOSIS II

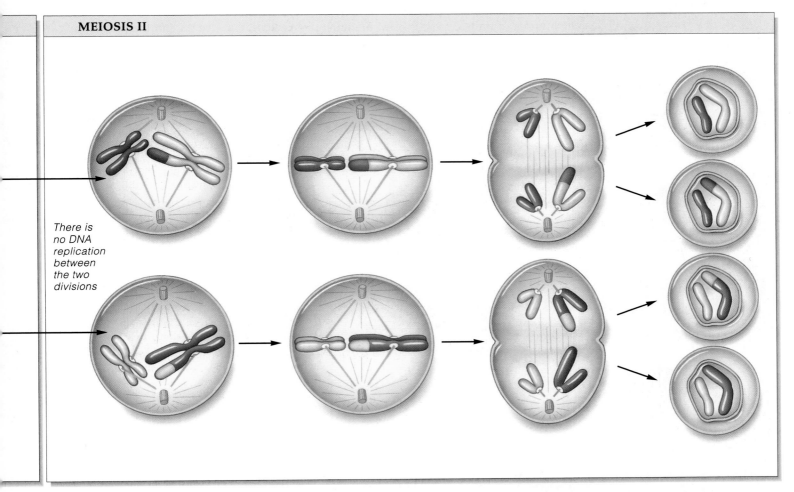

There is no DNA replication between the two divisions

PROPHASE II

In each of the two daughter cells, microtubules have already moved one member of the centriole pair to the opposite pole of the spindle during the transition to prophase II. Now, at prophase II, microtubules attach to the kinetochores of chromosomes, and motor proteins drive the movement of chromosomes toward the spindle's equator.

METAPHASE II

Now, in each daughter cell, interactions among motor proteins, spindle microtubules, and each duplicated chromosome have moved all of the chromosomes so that they are positioned at the spindle equator, midway between the two poles.

ANAPHASE II

The attachment between the two chromatids of each chromosome breaks. The former "sister chromatids" are now chromosomes in their own right. Motor proteins interact with kinetochore microtubules to move the separated chromosomes to opposite poles of the spindle.

TELOPHASE II

By the time telophase II is completed, there will be four daughter nuclei. At the time when the division of the cytoplasm is completed, each daughter cell will have a haploid chromosome number (*n*). All of those chromosomes will be in the unduplicated state.

KEY EVENTS OF MEIOSIS I

The preceding overview, in Sections 10.2 and 10.3, is enough to convey the overriding function of meiosis—that is, *the reduction of the chromosome number by half for forthcoming gametes.* However, as you will now read, two other events that take place during prophase and metaphase of meiosis I contribute greatly to the adaptive advantage of sexual reproduction.

That advantage, recall, is the production of offspring with new combinations of alleles. Those combinations translate into a new generation of individuals that display variation in the details of some number of traits.

a

b

one chiasma

e

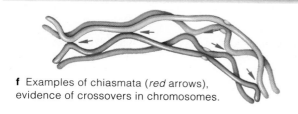

f Examples of chiasmata (*red* arrows), evidence of crossovers in chromosomes.

c Breaks occur in corresponding segments.

d Segments are exchanged and sealed in place.

g Homologues have new combinations of alleles.

Prophase I Activities

Prophase I of meiosis is a time of major gene shufflings. Consider Figure 10.5a, which shows two chromosomes condensed to threadlike form. All chromosomes in a germ cell condense this way. When they do, each is drawn close to its homologue by a process called synapsis. It is as if homologues become stitched together point by point along their length, with little space between. The intimate, parallel array favors **crossing over**, a molecular interaction between two of the *non*sister chromatids of a pair of homologous chromosomes. Nonsister chromatids break at the same places along their length and then exchange corresponding segments—that is, genes—at the break points. Figure 10.5c and d is a simplified picture of this interaction.

Figure 10.5 Key events of prophase I, the first stage of meiosis. For clarity, the diagram shows only one of the pairs of homologous chromosomes in the cell and one crossover event. *Blue* signifies the paternal chromosome; *purple* signifies its maternal homologue.

(**a**) Both chromosomes have been duplicated previously, at interphase. Early in prophase I, each duplicated chromosome is in threadlike form, attached at both ends to the nuclear envelope. Its two sister chromatids are so close together, they look like a single thread.

(**b**) Homologous chromosomes become zippered together, so all four chromatids are intimately aligned. A pair of sex chromosomes having different forms (such as X paired with Y) also become zippered together in a small region.

(**c,d**) One or more crossovers typically occur at intervals along the chromosomes. Each time, two nonsister chromatids break at identical sites. Then they swap segments at the breaks, and enzymes seal the broken ends. For clarity, these chromosomes are shown in condensed form and pulled apart, except at their ends. (Think of the blue chromosome as a spider and the purple one as its reflection in a mirror.) But remember, crossing over only proceeds while all four chromatids are extended like threads and tightly aligned together.

(**e,f**) As prophase I ends, the chromosomes continue to condense and become thicker, rodlike forms. They detach from the nuclear envelope and from each other except at contact points known as chiasmata. Each chiasma is indirect evidence that a crossover occurred at some place in the chromosomes. They don't indicate *which* point, because they slip down to the chromosome tips and vanish.

(**g**) Crossing over breaks up old combinations of alleles and puts new ones together in pairs of homologous chromosomes.

Gene swapping would be rather pointless if each type of gene never varied from one chromosome to the next. But remember, a gene can have slightly different forms—alleles. You can safely bet that some alleles on one chromosome will *not* be identical to those on the homologue. Therefore, every crossover represents a chance to swap a *slightly different version* of hereditary instructions for a particular trait.

We will look at the mechanism of crossing over in later chapters. For now, it is enough to remember this: *Crossing over leads to genetic recombination, which in turn leads to variation in the traits of offspring.*

Metaphase I Alignments

Major shufflings of whole chromosomes begin during the transition from prophase I to metaphase I, which is the second stage of meiosis. Suppose the shufflings are proceeding at this moment in one of your germ cells. By now, crossovers have made genetic mosaics out of the chromosomes, but put this aside in order to simplify tracking. Just call the twenty-three chromosomes you inherited from your mother the *maternal* chromosomes and their twenty-three homologues from your father the *paternal* chromosomes.

Kinetochore microtubules have already oriented one chromosome of each pair toward one spindle pole and its homologue toward the other pole (compare Section 9.4). Now they are moving all the chromosomes, which soon will become positioned at the spindle's equator.

Are all the maternal chromosomes attached to one pole and all the paternal ones attached to the other? Maybe, but probably not. Remember, the first contacts between kinetochore microtubules and chromosomes are random. Because of the random grabs, *the eventual positioning of a maternal or paternal chromosome at the spindle equator at metaphase I follows no particular pattern.* Carrying this one step further, either one of each pair of homologous chromosomes might end up at either pole of the spindle after they move apart at anaphase I.

Think about the possibilities when you are tracking merely three pairs of homologues. As you can see from Figure 10.6, by metaphase I, the homologues may be arranged in any one of four possible positions. In this case, 2^3 or eight combinations of maternal and paternal chromosomes are possible for forthcoming gametes.

Of course, a human germ cell has twenty-three pairs of homologous chromosomes, not just three. So a grand total of 2^{23} or *8,388,608 combinations* of maternal and paternal chromosomes is possible every time a germ cell gives rise to sperm or eggs!

Moreover, in each sperm or egg, many hundreds of alleles inherited from the mother might not "say" the exact same thing about hundreds of different traits as

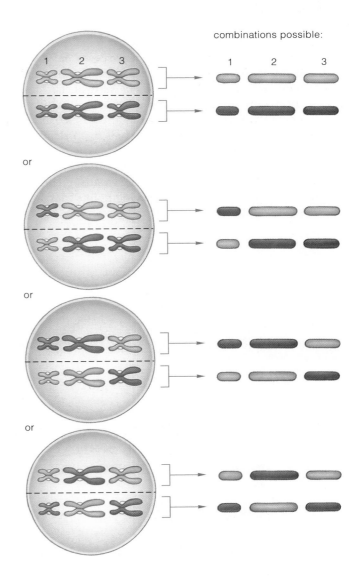

combinations possible:

Figure 10.6 The possible outcomes of the random alignment of three pairs of homologous chromosomes at metaphase I of meiosis. Three types of chromosomes are labeled 1, 2, and 3. Maternal chromosomes are *purple*; paternal ones are *blue*. With merely four possible alignments, eight combinations of maternal and paternal chromosomes are possible in forthcoming gametes.

the alleles inherited from the father. Are you beginning to get an idea of why such splendid mixes of traits show up even in the same family?

Crossing over is an interaction between a pair of homologous chromosomes. It breaks up old combinations of alleles and puts new ones together during prophase I of meiosis.

The random attachment and subsequent positioning of each pair of maternal and paternal chromosomes at metaphase I lead to different combinations of maternal and paternal traits in each generation of offspring.

FROM GAMETES TO OFFSPRING

The gametes that form after meiosis are not all the same in their details. For example, human sperm have one tail, opossum sperm have two, and roundworm sperm have none. Crayfish sperm look like pinwheels. Most eggs are microscopic, yet ostrich eggs tucked inside a shell are as large as a baseball. From appearance alone, you might not believe a plant gamete is even remotely like an animal's.

Later chapters contain details of how gametes form in the life cycles of representative organisms, including humans. Figure 10.7 and the following points may help you keep the details in perspective.

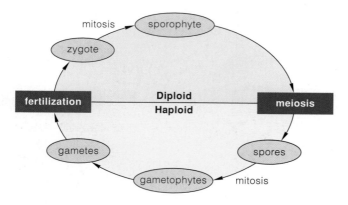

a Generalized life cycle for most kinds of plants

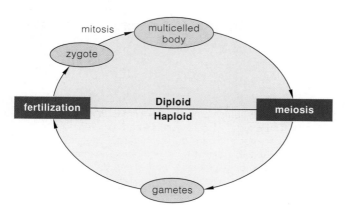

b Generalized life cycle for animals

Figure 10.7 Generalized life cycles for (**a**) most plants and (**b**) animals. The zygote is the first cell that forms when the nuclei of two gametes fuse together at fertilization.

For plants, a sporophyte (spore-producing body) develops, by way of mitotic cell divisions, from the zygote. Following meiosis, gametophytes (gamete-producing bodies) form. A pine tree is a sporophyte. Gametophytes develop in its cones.

Chapters 25, 26, and 45 include many illustrated examples of the life cycles of representative plants and animals.

Gamete Formation in Plants

For pine trees, apple trees, roses, dandelions, corn, and nearly all other familiar plants, certain events intervene between the times of meiosis and gamete formation. Among other things, spores form.

Spores are haploid cells, often with walls that help them resist episodes of drought, cold, or other adverse environmental conditions. When favorable conditions return, spores germinate and develop into a haploid body or structure that will produce gametes. Thus, *gamete*-producing bodies and *spore*-producing bodies develop during the life cycle of most kinds of plants. Figure 10.7*a* is a generalized diagram of these events.

Gamete Formation in Animals

In male animals, gametes form by a process that is known as spermatogenesis. Inside the male reproductive system, a diploid germ cell increases in size. It becomes a large, immature cell (a primary spermatocyte) that undergoes meiosis and cytoplasmic divisions. The outcome is four haploid daughter cells, which develop into immature cells called spermatids (Figure 10.8). The spermatids change in form, and each develops a tail. In this way, each becomes a **sperm**, a common type of mature male gamete.

In female animals, gametes form by a process that is known as oogenesis. In human females, for example, a diploid germ cell develops into an **oocyte**, or immature egg. Unlike a sperm cell, an oocyte accumulates many cytoplasmic components. As Figure 10.9 indicates, its daughter cells also differ in size and function. When the oocyte undergoes cytoplasmic division after meiosis I, one cell (the secondary oocyte) gets nearly all of the cytoplasm. The other cell, called the first polar body, is quite small. At some time afterward, both cells undergo meiosis II, then cytoplasmic division. One daughter cell of the secondary oocyte develops into a second polar body. The other receives most of the cytoplasm and develops into the gamete. A mature female gamete is called an ovum (plural, ova) or, more commonly, an **egg**.

Thus, after the first polar body divides, there are three polar bodies. These do not function as gametes. They serve as dumping grounds for chromosomes, so the egg will end up with a suitable (haploid) number of chromosomes at fertilization. Polar bodies do not have much in the way of nutrients or metabolic machinery. In time they degenerate.

More Shufflings at Fertilization

The chromosome number characteristic of the parents is restored at **fertilization**, the time when a male gamete penetrates a female gamete and their haploid nuclei

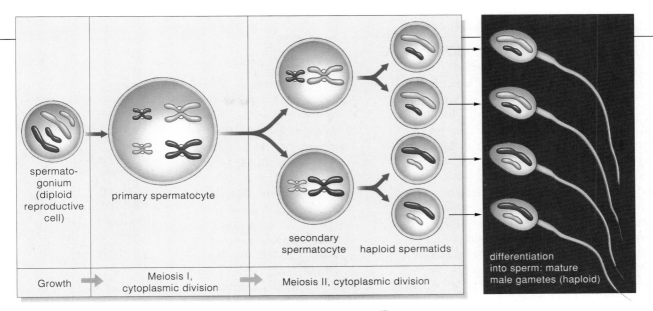

Figure 10.8 Generalized sketch of sperm formation in male animals.

spermato-
gonium
(diploid
reproductive
cell)

primary spermatocyte

secondary
spermatocyte

haploid spermatids

differentiation
into sperm: mature
male gametes (haploid)

Growth → Meiosis I, cytoplasmic division → Meiosis II, cytoplasmic division

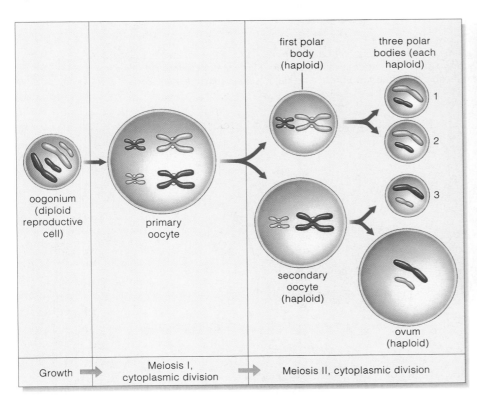

first polar
body
(haploid)

three polar
bodies (each
haploid)

1

2

3

oogonium
(diploid
reproductive
cell)

primary
oocyte

secondary
oocyte
(haploid)

ovum
(haploid)

Growth → Meiosis I, cytoplasmic division → Meiosis II, cytoplasmic division

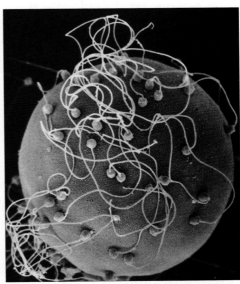

Figure 10.9 Egg formation in female animals. The generalized sketch is not at the same scale as Figure 10.8. Eggs are *far* larger than sperm, as the micrograph of sea urchin gametes shows. Also, the three polar bodies that form in meiosis are extremely small, compared to the egg.

fuse. If meiosis did not precede it, fertilization would double the chromosome number each new generation. Such changes in chromosome number would disrupt the hereditary instructions encoded in chromosomes, usually for the worse, for those instructions operate as a complex, fine-tuned package in each individual.

Fertilization contributes to variation in offspring. Reflect on the possibilities for humans alone. During prophase I, an average of two or three crossovers takes place in each human chromosome. Even without the crossovers, the random positioning of pairs of paternal and maternal chromosomes at metaphase I results in one of millions of possible chromosome combinations in each gamete. And of all the male and female gametes that are produced, which two actually get together is a matter of chance. The sheer number of combinations that can exist at fertilization is staggering!

Cumulatively, crossing over, the distribution of random mixes of homologous chromosomes into gametes, and fertilization contribute to variation in the traits of offspring.

MEIOSIS AND MITOSIS COMPARED

In this unit, our focus has been on two nuclear division mechanisms. Single-celled eukaryotic species reproduce asexually by way of mitosis, followed by cytoplasmic division. Many multicelled eukaryotic species rely on mitosis and cytoplasmic division during episodes of asexual reproduction in their life cycle, and all rely on it for growth and tissue repair. By contrast, meiosis occurs only in reproductive cells, such as germ cells that give rise to the gametes used in sexual reproduction. Figure 10.10 summarizes the basic similarities and differences between the two nuclear division mechanisms.

The end results of the two mechanisms differ in a crucial way. *Mitotic cell division only produces clones—genetically identical copies of a parent cell. But meiotic cell division, in conjunction with fertilization, promotes variation in traits among offspring.* First, crossing over at prophase I of meiosis puts new combinations of alleles in chromosomes. Second, the random assignment of either member of a pair of homologous chromosomes to either spindle pole at metaphase I puts different mixes of the maternal and paternal alleles into gametes. Third, different combinations of alleles are brought together by chance at fertilization. In later chapters, you will be reading about the ways in which both meiosis and fertilization contribute to the staggering diversity and evolution of sexually reproducing organisms.

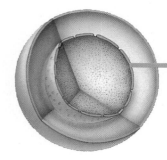

A *somatic cell* with a diploid chromosome number (2n) is at interphase. Before mitotic division begins, its DNA is replicated (all chromosomes are duplicated).

Figure 10.10 Summary of mitosis and meiosis. Both diagrams use a diploid (2n) animal cell as the example. They are arranged to help you compare similarities and differences between the division mechanisms. Maternal chromosomes are coded *purple*, and paternal chromosomes are coded *blue*.

MEIOSIS I

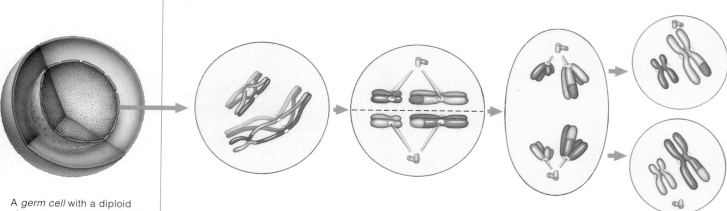

A *germ cell* with a diploid chromosome number (2n) is at interphase. Before mitotic division begins, its DNA is replicated (all chromosomes are duplicated).

PROPHASE I

Each duplicated chromosome (consisting of two sister chromatids) condenses to threadlike form, then rodlike form. *Crossing over* occurs. Each chromosome unzips from its homologue. Each gets attached to the spindle in transition to metaphase.

METAPHASE I

All chromosomes are now positioned at the spindle's equator.

ANAPHASE I

Each chromosome is separated from its homologue. They are moved to opposite poles of the spindle.

TELOPHASE I

When the cytoplasm divides, there are two cells. Each has a haploid (n) number of chromosomes, but these are still in the duplicated state.

MITOSIS

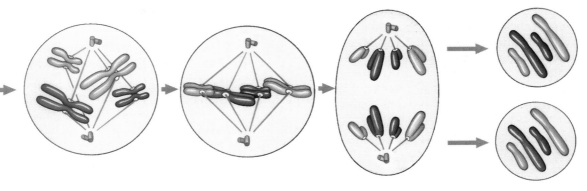

PROPHASE

Each duplicated chromosome (consisting of two sister chromatids) condenses from threadlike form to rodlike form. Each gets attached to the spindle during the transition to metaphase.

METAPHASE

All chromosomes are now positioned at the spindle's equator.

ANAPHASE

Sister chromatids of each chromosome are separated from each other. These new, daughter chromosomes are moved to opposite poles of the spindle.

TELOPHASE

When the cytoplasm divides, there are two cells. Each is diploid (2n)—*it has the same chromosome number as the parent cell.*

MEIOSIS II

There is no DNA replication between the two divisions.

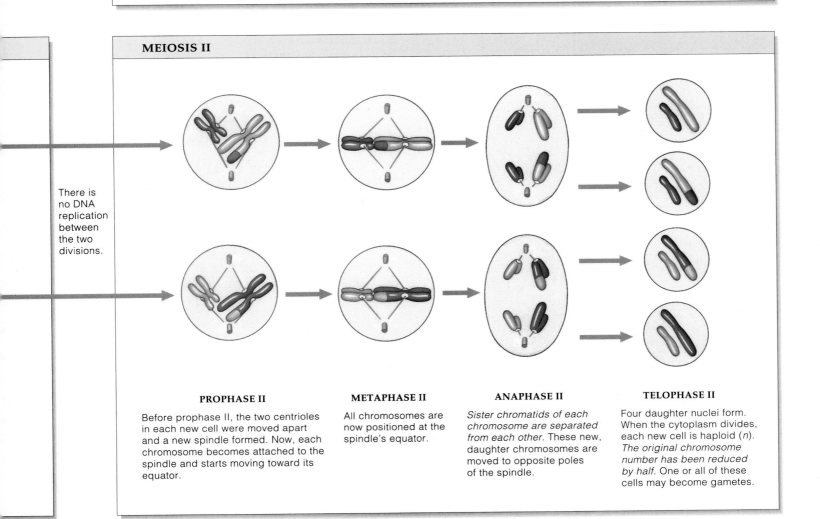

PROPHASE II

Before prophase II, the two centrioles in each new cell were moved apart and a new spindle formed. Now, each chromosome becomes attached to the spindle and starts moving toward its equator.

METAPHASE II

All chromosomes are now positioned at the spindle's equator.

ANAPHASE II

Sister chromatids of each chromosome are separated from each other. These new, daughter chromosomes are moved to opposite poles of the spindle.

TELOPHASE II

Four daughter nuclei form. When the cytoplasm divides, each new cell is haploid (n). *The original chromosome number has been reduced by half.* One or all of these cells may become gametes.

SUMMARY

1. The life cycle of each sexually reproducing species includes meiosis, gamete formation, and fertilization.

   a. Meiosis, a nuclear division mechanism, reduces the chromosome number of a parent germ cell by half. It *precedes* the formation of haploid gametes (typically, sperm in males, eggs in females).

   b. At fertilization, a sperm nucleus fuses with an egg nucleus. As Figure 10.11 shows, this event restores the chromosome number.

2. A germ cell with a diploid chromosome number (2*n*) has *two* of each type of chromosome characteristic of its species. Commonly, one of each pair of chromosomes is maternal (inherited from a female parent), and the other is paternal (from a male parent).

3. Each pair of maternal and paternal chromosomes shows homology, meaning the two are alike. In general, the two have the same length, same shape, and same sequence of genes. They interact during meiosis.

4. Chromosomes are duplicated during interphase. So before meiosis even begins, each consists of two DNA molecules that remain attached, as sister chromatids.

5. Meiosis consists of two consecutive divisions that require a microtubular spindle apparatus.

   a. In meiosis I, interactions among motor proteins, microtubules, and kinetochores of chromosomes move each type of chromosome away from its homologous partner.

   b. In meiosis II, similar interactions move the sister chromatids of each chromosome away from each other.

6. The following events characterize the first nuclear division (meiosis I):

   a. Crossing over occurs in prophase I. Two nonsister chromatids of each pair of homologous chromosomes commonly break at corresponding sites and exchange segments, which puts new combinations of alleles together. Alleles (slightly different molecular forms of the same gene) code for different forms of the same trait. Different combinations of alleles lead to variation in the details of a given trait among offspring.

   b. Also in prophase I, a microtubular spindle starts to form outside the nucleus, and the nuclear envelope starts to break up. If the cell has a duplicated pair of centrioles, one pair starts moving to the opposite pole of the newly forming spindle.

   c. At metaphase I, all of the pairs of homologous chromosomes have become positioned at the spindle equator. For each pair, either the maternal chromosome or its homologue can be oriented toward either pole.

   d. In anaphase I, interactions among motor proteins, microtubules, and the chromosomal kinetochores move each duplicated chromosome away from its homologue, toward opposite spindle poles.

7. The following events characterize the second nuclear division (meiosis II):

   a. At metaphase II, all the duplicated chromosomes are positioned at the spindle equator.

   b. Sister chromatids are moved apart in anaphase II. Each is now a separate, unduplicated chromosome.

   c. By the end of telophase II, four nuclei that have a haploid chromosome number (*n*) have been formed.

8. When the cytoplasm divides, there are four haploid cells. One or all of these may function as gametes (or as spores, in flowering plants).

9. Crossing over, the chance allocation of different mixes of pairs of maternal and paternal chromosomes to different gametes, and the chance of any two gametes meeting at fertilization all contribute to the immense variation in the details of traits among offspring.

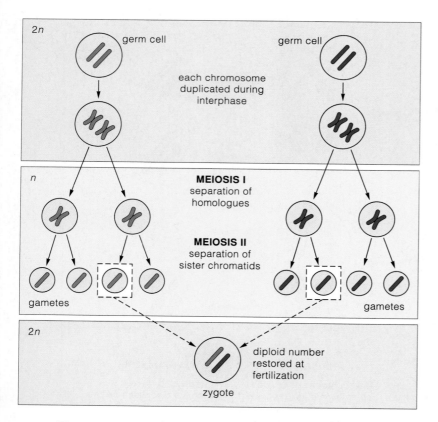

Figure 10.11 Summary of changes in the chromosome number at different stages of sexual reproduction, using diploid (2*n*) germ cells as the example. Meiosis reduces the chromosome number by half (*n*). Then the union of haploid nuclei of two gametes at fertilization restores the diploid number.

Review Questions

1. Genetically speaking, what is the key difference between the outcomes of sexual and asexual reproduction? *10.1, 10.6*

2. Suppose a diploid germ cell contains four pairs of homologous chromosomes, designated AA, BB, CC, and DD. How would the chromosomes of the gametes be designated? *10.2*

Figure 10.12 Example of the chin dimple trait (actually a fissure in the chin surface).

3. From each of his parents, actor Michael Douglas (Figure 10.12*a*) inherited a gene that influences the chin dimple trait. One form of the gene called for a dimple and the other didn't, but one is all it takes in this case. Figure 10.12*b* shows what his chin might have looked like if he had inherited two ordinary forms of the gene instead. What are alternative forms of the same gene called? *10.1*

4. To the right, the diploid chromosome numbers in somatic cells of a few kinds of organisms are listed. Figure out how many chromosomes would end up in gametes of each organism. *10.2*

Fruit fly, *Drosophila melanogaster*	8
Garden pea, *Pisum sativum*	14
Corn, *Zea mays*	20
Frog, *Rana pipiens*	26
Earthworm, *Lumbricus terrestris*	36
Human, *Homo sapiens*	46
Chimpanzee, *Pan troglodytes*	48
Amoeba, *Amoeba*	50
Horsetail, *Equisetum*	216

5. Define meiosis and then describe its main stages. In what key respects is meiosis *not* like mitosis? *10.3, 10.6*

6. Take a look at all of the chromosomes in the germ cell in the diagram at the right. Is this cell at anaphase I or anaphase II? *10.3, 10.6*

The cell is at anaphase ____ rather than anaphase ____. I know this because:

____
____
____
____ .

7. Outline the steps by which sperm and eggs form in animals. *10.5*

Self-Quiz (Answers in Appendix IV)

1. Sexual reproduction requires _____ .
   a. meiosis
   b. gamete formation
   c. fertilization
   d. all of the above

2. Meiosis is a division mechanism that produces _____ .
   a. two cells
   b. two nuclei
   c. four cells
   d. four nuclei

3. An animal cell having two rather than one of each type of chromosome has a _____ chromosome number.
   a. diploid
   b. haploid
   c. normal gamete
   d. both b and c

4. Meiosis _____ the parental chromosome number.
   a. doubles
   b. reduces
   c. maintains
   d. corrupts

5. Generally, a pair of homologous chromosomes _____ .
   a. carry the same genes
   b. are the same length, shape
   c. interact at meiosis
   d. all of the above

6. Before the onset of meiosis, all chromosomes are _____ .
   a. condensed
   b. released from protein
   c. duplicated
   d. both b and c

7. Each chromosome moves away from its homologue and ends up at the opposite spindle pole during _____ .
   a. prophase I
   b. prophase II
   c. anaphase I
   d. anaphase II

8. Sister chromatids of each chromosome move apart and end up at opposite spindle poles during _____ .
   a. prophase I
   b. prophase II
   c. anaphase I
   d. anaphase II

9. Match each term and its description.
   ____ chromosome number
   ____ alleles
   ____ metaphase I
   ____ interphase

   a. different molecular forms of the same gene
   b. none between meiosis I and II
   c. pairs of homologous chromosomes are aligned at spindle equator
   d. sum total of chromosomes in all cells of a given type

Critical Thinking

1. Assume you can measure the amount of DNA in a primary oocyte, then in a primary spermatocyte, which gives you a mass *m*. What mass of DNA would you expect to find in each mature gamete (egg and sperm) that forms after meiosis? What mass of DNA would you expect to find (1) in an egg fertilized by one of the sperm and (2) in that egg after the first DNA duplication?

2. Adam has a pair of alleles that influence whether a person is right- or left-handed. One allele says "left," and its partner says "right." Visualize one of his germ cells, in which chromosomes are being duplicated prior to meiosis. Visualize what happens to the chromosomes during anaphase I and II. (It may help to use index cards as models of the sister chromatids of each chromosome.) What fraction of Adam's sperm will carry the gene for right-handedness? For left-handedness?

3. Adam also has one allele for long eyelashes, and a partner allele (on the homologous chromosome) for short eyelashes. What fraction of his sperm will have these gene combinations:
   right-handed, long eyelashes   *left-handed, long eyelashes*
   right-handed, short eyelashes   *left-handed, short eyelashes*

Selected Key Terms

allele *10.1*	gamete *CI*	oocyte *10.5*
asexual reproduction *10.1*	gene *10.1*	sexual reproduction *10.1*
crossing over *10.4*	germ cell *CI*	sister chromatid *10.2*
diploid *10.2*	haploid *10.2*	sperm *10.5*
egg *10.5*	homologous chromosome *10.2*	spore *10.5*
fertilization *10.5*	meiosis *10.2*	

Readings

Klug, W., and M. Cummings. 1994. *Concepts of Genetics*. Fourth edition. New York: Macmillan.

Wolfe, S. 1995. *Introduction to Molecular and Cellular Biology*. Belmont, California: Wadsworth.

Web Site See *http://www.wadsworth.com/biology* for practice quiz questions, hypercontents, BioUpdates, and critical thinking. The Wadsworth Biology Resource Center provides a wealth of information fully organized and integrated by chapter.

11 OBSERVABLE PATTERNS OF INHERITANCE

A Smorgasbord of Ears and Other Traits

Basketball ace Charles Barkley has them. So does actor Tom Cruise. Actress Joan Chen doesn't, and neither did a monk named Gregor Mendel. To see how you fit in with these folks, use a mirror to check out your ears. Is the fleshy lobe at the base of each ear attached to the side of your head? If so, you and Barkley and Cruise have something in common. Or is the fleshy lobe not attached, so that you can flap it back and forth? If so, you are like Chen and Mendel (Figure 11.1).

Whether a person is born with attached or detached earlobes depends on a single kind of gene. That gene

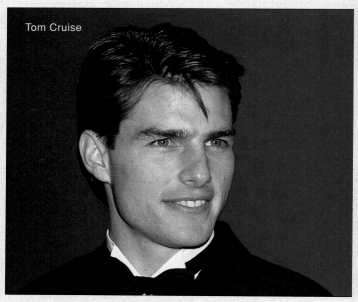

comes in slightly different molecular forms—alleles. Only one form has information about detached lobes. The information is put to use while a human body is developing inside the mother. It calls for a death signal, which is sent to all the cells positioned between the newly forming lobes and the head. Without the signal, the cells don't die, and earlobes don't detach.

We all have genes for thousands of traits, including earlobes, cheeks, lashes, and eyeballs. Most traits vary in their details from one person to the next. Remember, we inherit pairs of genes, on pairs of chromosomes. In some pairs, one allele has strong effects and overwhelms the other allele's contribution. The outgunned allele is said to be recessive to the dominant one. If you have detached earlobes, dimpled cheeks, long lashes, or large eyeballs, you have at least one and quite possibly two dominant alleles that affect the trait in a particular way.

Figure 11.1 Attached and detached earlobes of representative humans. This sampling provides observable evidence of a trait governed by a certain gene, which exists in different molecular forms in the human population. Do you have one or the other version of the trait? It depends on which molecular forms of the gene you inherited from your mother and father. As Gregor Mendel perceived, such easily observable traits can be used to identify patterns of inheritance that exist from one generation to the next.

When both alleles of a pair are recessive, nothing masks their effect on a trait. You get *attached* earlobes with one pair of recessive alleles (and *flat* feet with another, a *straight* nose with another, and so on).

How did we discover such remarkable things about our genes? It started with Gregor Mendel. By analyzing pea plants generation after generation, Mendel found indirect but *observable* evidence of how parents bestow units of hereditary information—genes—on offspring.

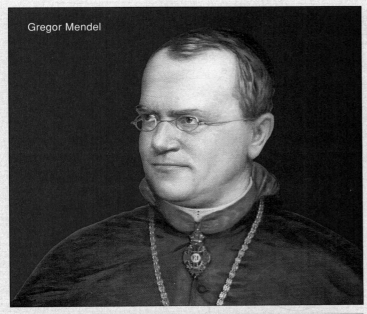

Gregor Mendel

Joan Chen

This chapter focuses on both the methods and the results of Mendel's experiments. They remain a classic example of how a scientific approach can pry open important secrets about the natural world. And to this day, they serve as the foundation for modern genetics.

Read

KEY CONCEPTS

1. Genes are units of information about heritable traits. Alleles, which are slightly different molecular forms of a gene, specify different versions of the same trait.

2. Each gene has a particular location on a particular chromosome of a species. Humans, pea plants, and other organisms with a diploid chromosome number inherit *pairs* of genes, at equivalent locations on pairs of homologous chromosomes.

3. When each pair of homologous chromosomes is moved apart at meiosis, their paired genes are moved apart also, and they end up in different gametes. Gregor Mendel found indirect evidence of this gene segregation when he crossbred plants showing different versions of the same trait, such as purple or white flowers.

4. Each pair of homologous chromosomes (and the genes they carry) is sorted out for distribution into one gamete or another independently of how the other pairs of homologous chromosomes are assorted. Mendel found indirect evidence of this when he tracked many plants having observable differences in two traits, such as flower color and height.

5. The contrasting traits that Mendel happened to study were specified by nonidentical alleles. One allele was dominant, in that its effect on a trait masked the effect of a recessive allele paired with it.

6. Not all traits have such clearly dominant or recessive forms. One allele of a pair may be fully or partially dominant over its partner or codominant with it. Also, two or more gene pairs often influence the same trait, and some single genes influence many traits. Besides this, environmental factors induce further variation in traits.

MENDEL'S INSIGHT INTO PATTERNS OF INHERITANCE

More than a century ago, people wondered about the basis of inheritance. It was common knowledge that sperm and eggs both transmit information about traits to offspring. But few suspected that the information is organized in units (genes). Instead, the idea was that a father's blob of information "blended" with a mother's blob at fertilization, like cream into coffee.

However, carried to its logical conclusion, blending would slowly dilute a population's shared pool of hereditary information until there was only a single version left of each trait. If that were so, why did, say, freckles keep showing up among the children of nonfreckled parents over the generations? Why weren't all the descendants of a herd of white stallions and black mares gray? The theory of blending scarcely explained the obvious variation in traits that people could observe with their own eyes. Nevertheless, few disputed the theory.

Charles Darwin was among the scholarly dissidents. According to a key premise of his theory of natural selection, individuals of a population show variation in heritable traits. Through the generations, variations that improve chances of surviving and reproducing occur with greater frequency than those that do not. The less advantageous variations may persist among fewer individuals or may even disappear. It is not that separate versions of a trait are "blended out" of the population. Rather, *each version of a trait may persist in a population, at frequencies that may change over time.*

Even before Darwin presented his theory, someone was gathering evidence that eventually would support his key premise. A monk, Gregor Mendel, had already guessed that sperm and eggs carry distinct "units" of information about heritable traits. By carefully analyzing traits of pea plants generation after generation, Mendel found indirect but *observable* evidence of how parents transmit genes to offspring.

Mendel's Experimental Approach

Mendel spent most of his adult life in a monastery in Brno, a city near Vienna that has since become part of the Czech Republic. The monastery of St. Thomas was somewhat removed from the European capitals, which were then the centers of scientific inquiry. Yet Mendel was not a man of narrow interests who accidentally stumbled onto principles of great import.

Having been raised on a farm, Mendel was aware of agricultural principles and their applications. He kept abreast of breeding experiments and developments described in the available literature. He was a member of the regional agricultural society. He also won several awards for developing improved varieties of vegetables and fruits. Shortly after entering the monastery, he spent two years studying mathematics at the University of Vienna. Few scholars of his time had such combined talents in plant breeding and mathematics.

a

carpel   stamen

→makes pollen

Figure 11.2 The garden pea plant (*Pisum sativum*), the organism Mendel chose for experimental tests of his ideas about inheritance. (**a**) This flower has been sectioned to show the location of its stamens and carpel. Sperm-producing pollen grains form in stamens. Eggs develop, fertilization takes place, and seeds mature inside the carpel.

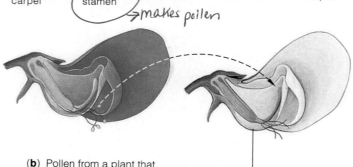

(**b**) Pollen from a plant that breeds true for purple flowers is brushed onto a floral bud of a plant that breeds true for white flowers and that had its own stamens snipped off.

(**c**) The cross-fertilized plant produces seeds, each of which is allowed to grow into a new plant.

(**d**) The flower color of the new plants can be used as visible evidence of patterns in how hereditary material might be transmitted from each parent plant.

e/e ex.:
E/e pairs of alleles
E/E

Shortly after his university training, Mendel began experimenting with the garden pea plant, *Pisum sativum* (Figure 11.2). This plant is self-fertilizing. Male as well as female gametes (call them sperm and eggs) develop in different parts of the same flower, where fertilization takes place. Nearly all the plants breed true for certain traits. In other words, successive generations are just like their parents in one or more traits, as when all of the offspring grown from seeds of self-fertilized, white-flowered parent plants have white flowers.

As Mendel knew, pea plants also will cross-fertilize when sperm and eggs from different plants are mixed together under controlled conditions. For his studies, he could open flower buds of a plant that bred true for a trait—say, white flowers—and snip out the stamens. (Stamens bear pollen grains in which sperm develop.) Then he could brush the "castrated" buds with pollen from a plant that bred true for a *different* version of the same trait—purple flowers. As Mendel hypothesized, he could use such clearly observable differences to track a given trait through many generations. If there were patterns to the trait's inheritance, *those patterns might tell him something about the hereditary material itself.*

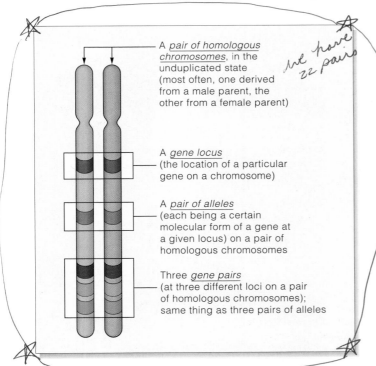

A *pair of homologous chromosomes*, in the unduplicated state (most often, one derived from a male parent, the other from a female parent)

we have 22 pairs

A *gene locus* (the location of a particular gene on a chromosome)

A *pair of alleles* (each being a certain molecular form of a gene at a given locus) on a pair of homologous chromosomes

Three *gene pairs* (at three different loci on a pair of homologous chromosomes); same thing as three pairs of alleles

Figure 11.3 A few genetic terms illustrated. Diploid organisms have pairs of genes, on pairs of homologous chromosomes. For example, you inherited one chromosome of each pair from your mother, and the other, homologous chromosome from your father.

Most genes can have slightly different molecular forms, called alleles. Different alleles specify different versions of the same trait. An allele at one location on a chromosome may or may not be identical to its partner on the homologous chromosome.

Some Terms Used in Genetics

Having read the chapter on meiosis, you already have insight into the mechanisms of sexual reproduction, which is more than Mendel had. Neither he nor anyone else of his era knew about chromosomes. So he could not have known that a chromosome number is reduced by half in gametes, then restored when gametes meet at fertilization. Even so, Mendel sensed what was going on. As we follow his thinking, let's simplify the story by substituting a few of the modern terms used in studies of inheritance (see also Figure 11.3):

1. **Genes** are units of information about specific traits, and they are passed from parents to offspring. Each gene has a specific location (locus) on a chromosome.

2. Cells with a diploid chromosome number ($2n$) have pairs of genes, on pairs of homologous chromosomes.

3. Mutation can change a gene's molecular structure and thus its information about a trait (as when the gene for flower color specifies purple and a mutated version specifies white). All the different molecular forms of the same gene are called **alleles**.

4. When offspring of genetic crosses inherit a pair of *identical* alleles for a trait, generation after generation, they are a **true-breeding lineage**. By contrast, when offspring of a genetic cross inherit a pair of *nonidentical* alleles for a trait, they are **hybrid offspring**.

5. When both alleles of a pair are identical, this is a *homozygous* condition. When the two are not identical, this is a *heterozygous* condition.

6. An allele is said to be *dominant* when its effect on a trait masks that of any *recessive* allele paired with it. We use capital letters for dominant alleles and lowercase letters for recessive ones; for instance, *A* and *a*.

7. Putting this all together, a **homozygous dominant** individual has a pair of dominant alleles (AA) for a trait under study. A **homozygous recessive** individual has a pair of recessive alleles (aa). And a **heterozygous** individual has a pair of nonidentical alleles (Aa).

8. Two terms help keep the distinction clear between genes and the traits they specify. **Genotype** refers to the particular genes an individual carries. **Phenotype** refers to an individual's observable traits.

9. When tracking the inheritance of traits through generations of offspring, these abbreviations apply:

P	parental generation
F_1	first-generation offspring
F_2	second-generation offspring

Mendel had an idea that in every generation, a plant inherits two "units" (genes) of information for a trait, one from each parent. To test this idea, he performed what we now call **monohybrid crosses**. The offspring of such a cross are heterozygous for the one trait being studied (which is what monohybrid means). The two parents breed true for different versions of the trait, so their offspring inherit a pair of nonidentical alleles.

Predicting the Outcome of Monohybrid Crosses

Mendel tracked many individual traits through two generations. For instance, in one series of experiments, he crossed true-breeding purple-flowered plants and true-breeding white-flowered ones. All plants grown from the seeds that resulted from this cross had purple flowers. Mendel allowed these plants to self-fertilize. Some plants grown from the seeds had white flowers!

If Mendel's hypothesis were correct—if each plant had inherited two units of information about flower color—then the unit for "purple" had to be dominant, because it had masked the unit for "white" in F_1 plants.

Let's rephrase his thinking. Germ cells of pea plants are diploid, with pairs of homologous chromosomes. Assume one parent is homozygous dominant (AA) and the other is homozygous recessive (aa) for flower color. After meiosis, a sperm or egg carries one allele for flower color (Figure 11.4). Thus, when a sperm fertilizes an egg, only one outcome is possible: $A + a = Aa$.

Before continuing, you should know Mendel crossed hundreds of plants and tracked thousands of offspring. Besides this, he counted and recorded the number of plants showing dominance or recessiveness. As you can see from Figure 11.5, an intriguing ratio emerged. On average, three of every four F_2 plants had the dominant phenotype, and one had the recessive phenotype.

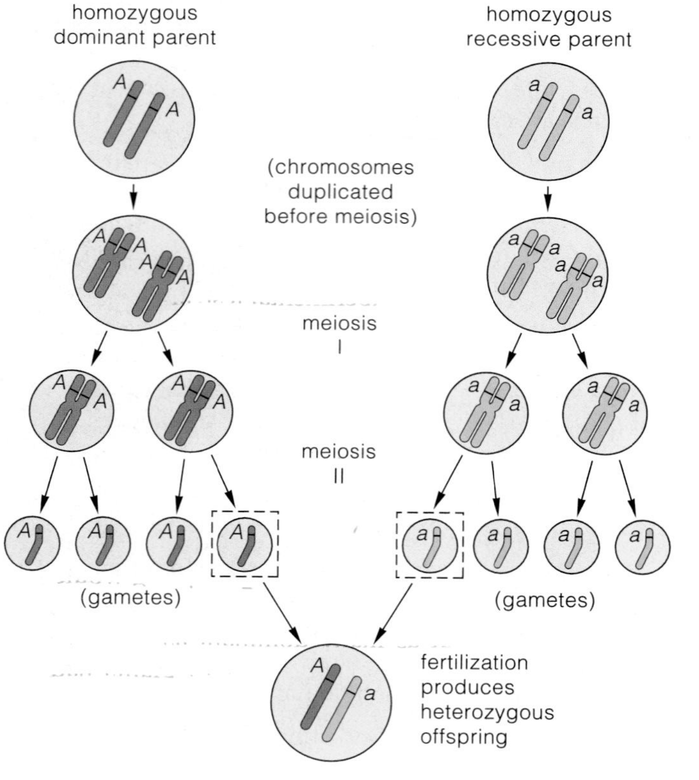

Figure 11.4 A monohybrid cross, showing how one gene of a pair segregates from the other. Two parents that breed true for two different versions of a trait can produce only heterozygous offspring.

Figure 11.5 (*Right*) Results from Mendel's monohybrid cross experiments with the garden pea plant (*P. sativum*). The numbers given are his counts of the F_2 plants that carried dominant or recessive hereditary "units" (alleles) for the trait. On average, the dominant-to-recessive ratio was 3:1.

Trait Studied	Dominant Form	Recessive Form	F_2 Dominant-to-Recessive Ratios:
seed shape	5,474 round	1,850 wrinkled	2.96:1
seed color	6,022 yellow	2,001 green	3.01:1
pod shape	882 inflated	299 wrinkled	2.95:1
pod color	428 green	152 yellow	2.82:1
flower color	705 purple	224 white	3.15:1
flower position	651 along stem	207 at tip	3.14:1
stem length	787 tall	277 dwarf	2.84:1

Average ratio for all traits studied: 3:1

female gametes

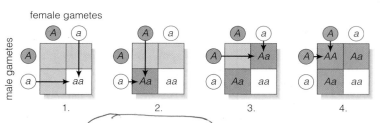

1.   2.   3.   4.

Figure 11.6 Punnett-square method of predicting the probable outcome of a genetic cross. Circles represent gametes. *Italic* letters on the gametes represent dominant or recessive alleles. The different squares show the different genotypes possible among offspring. In this case, gametes are from a self-fertilizing heterozygous (*Aa*) plant.

Figure 11.7 (*Right*) Results from one of Mendel's monohybrid crosses. On average, the dominant-to-recessive ratio among the second-generation (F_2) plants was 3 : 1.

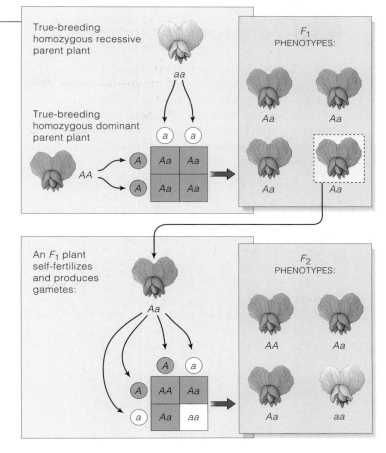

To Mendel, the ratio suggested that fertilization is a chance event, with a number of possible outcomes. And he had an understanding of probability, which applies to chance events *and therefore could help him predict the possible outcomes of crosses*. **Probability** simply means this: The chance that each outcome of a given event will occur is proportional to the number of ways it can be reached.

The **Punnett-square method**, explained in Figure 11.6 and applied in Figure 11.7, may help you visualize the possibilities. As you can see, if half of a plant's sperm or eggs were *a* and half were *A*, then four outcomes were possible each time a sperm fertilized an egg:

POSSIBLE EVENT:	PROBABLE OUTCOME:
sperm *A* meets egg *A*	1/4 *AA* offspring
sperm *A* meets egg *a*	1/4 *Aa*
sperm *a* meets egg *A*	1/4 *Aa*
sperm *a* meets egg *a*	1/4 *aa*

By this prediction, an F_2 plant had three chances in four of getting at least one dominant allele (purple flowers). It had one chance in four of getting two recessive alleles (white flowers). That is a probable phenotypic ratio of three purple to one white, or 3:1.

Mendel's observed ratios weren't *exactly* 3:1. You can see this for yourself by looking at the numerical results listed in Figure 11.5. Why did Mendel put aside the deviations? To understand why, flip a coin a couple of times. As we all know, a coin is just as likely to end up heads as tails. But often it ends up heads, or tails, several times in a row. So if you flip the coin only a few times, the observed ratio may differ greatly from the predicted ratio of 1:1. Flip the coin many, many times, and you are more likely to come close to the predicted ratio. Mendel understood the rules of probability—and he performed a large number of crosses. Almost certainly, this kept him from being confused by minor deviations from the predicted results of the experimental crosses.

Testcrosses

By running **testcrosses**, Mendel gained support for his prediction. In this type of experimental test, an organism shows dominance for a specified trait but its genotype is unknown, so it is crossed to a known homozygous recessive individual. Test results may reveal whether the organism is homozygous dominant or heterozygous.

Regarding the monohybrid crosses just described, Mendel tested his prediction that the purple-flowered F_1 offspring were heterozygous by crossing them with true-breeding, white-flowered plants. If they were all homozygous dominant, then all the F_2 offspring would show the dominant form of the trait. If heterozygous, there would be about as many dominant as recessive plants. Sure enough, about half of the F_2 plants had purple flowers (*Aa*) and half had white (*aa*). Can you construct two Punnett squares that show the possible outcomes of this testcross?

The results from Mendel's monohybrid crosses and testcrosses became the basis of a theory of **segregation**, which we state here in modern terms:

MENDEL'S THEORY OF SEGREGATION.   **Diploid cells have pairs of genes, on pairs of homologous chromosomes. During meiosis, the two genes of each pair are segregated from each other. As a result, they end up in different gametes.**

INDEPENDENT ASSORTMENT

By another series of experiments, Mendel attempted to explain how *two* pairs of genes might be assorted into gametes. He selected true-breeding plants that differed in two traits—for example, in flower color and height. With such **dihybrid crosses**, F_1 offspring inherit two gene pairs, each consisting of two nonidentical alleles.

Predicting the Outcome of Dihybrid Crosses

Let's diagram one of Mendel's dihybrid crosses. We can use A for flower color and B for height as the dominant alleles, and a and b as their recessive counterparts:

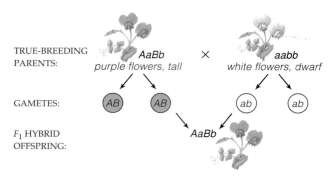

As Mendel would have predicted, the F_1 offspring from this cross are all purple-flowered and tall ($AaBb$). When

the F_1 plants mature and reproduce, how will the two gene pairs be assorted into gametes? The answer partly depends on the chromosomal locations of the gene pairs. Assume one pair of homologous chromosomes carries the Aa alleles and a *different* pair carries the Bb alleles. Next, think of how all chromosomes become positioned at the spindle equator during metaphase I of meiosis (Figures 8.4 and 11.8). The chromosome with the A allele might be positioned to move to either spindle pole (and then into one of four gametes). The same is true of its homologue. And the same is true of the chromosomes with the B and b alleles. Thus, after meiosis and gamete formation, four combinations of alleles are possible in the sperm or eggs: $1/4\ AB$, $1/4\ Ab$, $1/4\ aB$, and $1/4\ ab$.

Given the alternative alignments of chromosomes at metaphase I, several allelic combinations are possible at fertilization. Simple multiplication (four kinds of sperm times four kinds of eggs) tells us sixteen combinations of gametes are possible in the F_2 offspring of a dihybrid cross. Use the Punnett-square method to diagram the probabilities (Figure 11.9). Now add up all the possible phenotypes and you get 9/16 tall purple-flowered, 3/16 dwarf purple-flowered, 3/16 tall white-flowered, and 1/16 dwarf white-flowered plants. That is a probable phenotypic ratio of 9:3:3:1. Results from one dihybrid cross that Mendel described were close to this ratio.

Figure 11.8 Independent assortment. This example tracks two pairs of homologous chromosomes. An allele at one locus on a chromosome may or may not be identical with its partner allele on the homologous chromosome. At meiosis, either chromosome of a pair may become attached to either pole of the spindle. Thus, in this case, two different lineups are possible at metaphase I.

Nucleus of a diploid ($2n$) reproductive cell with only two pairs of homologous chromosomes

or

Possible alignments of the homologous chromosomes at metaphase I of meiosis, as shown by two diagrams:

The resulting alignments of chromosomes at metaphase II:

The combinations of alleles possible in the forthcoming gametes:

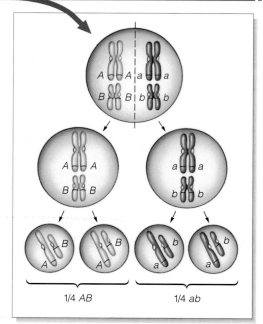

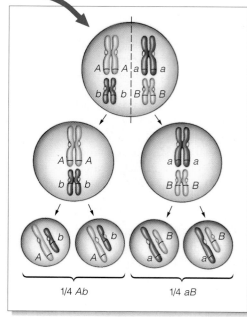

1/4 AB   1/4 ab   1/4 Ab   1/4 aB

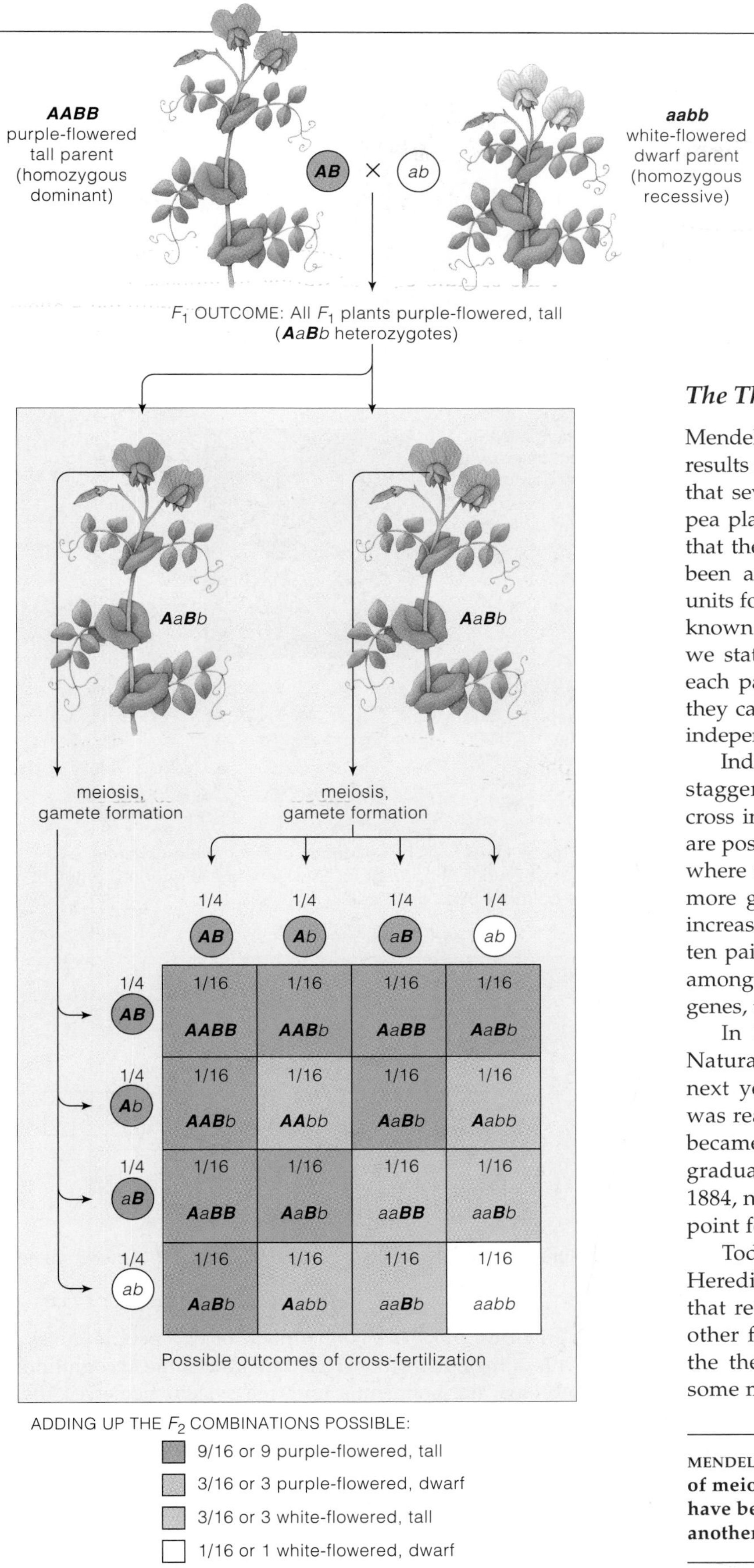

AABB
purple-flowered
tall parent
(homozygous
dominant)

AB × **ab**

aabb
white-flowered
dwarf parent
(homozygous
recessive)

F_1 OUTCOME: All F_1 plants purple-flowered, tall
(**AaBb** heterozygotes)

AaBb   **AaBb**

meiosis,
gamete formation

meiosis,
gamete formation

	1/4 **AB**	1/4 **Ab**	1/4 **aB**	1/4 **ab**
1/4 **AB**	1/16 **AABB**	1/16 **AABb**	1/16 **AaBB**	1/16 **AaBb**
1/4 **Ab**	1/16 **AABb**	1/16 **AAbb**	1/16 **AaBb**	1/16 **Aabb**
1/4 **aB**	1/16 **AaBB**	1/16 **AaBb**	1/16 **aaBB**	1/16 **aaBb**
1/4 **ab**	1/16 **AaBb**	1/16 **Aabb**	1/16 **aaBb**	1/16 **aabb**

Possible outcomes of cross-fertilization

ADDING UP THE F_2 COMBINATIONS POSSIBLE:

▣ 9/16 or 9 purple-flowered, tall

▣ 3/16 or 3 purple-flowered, dwarf

▣ 3/16 or 3 white-flowered, tall

▢ 1/16 or 1 white-flowered, dwarf

Figure 11.9 Results from Mendel's dihybrid cross between parent plants that bred true for different versions of two traits: flower color and plant height. *A* and *a* represent dominant and recessive alleles for flower color. *B* and *b* represent dominant and recessive alleles for plant height. As the Punnett square indicates, the probabilities of certain combinations of phenotypes among F_2 offspring occur in a 9:3:3:1 ratio, on the average.

The Theory in Modern Form

Mendel could do no more than analyze the numerical results from his dihybrid crosses, because he didn't know that seven pairs of homologous chromosomes carry the pea plant's "units" of inheritance. It just seemed to him that the two units for the first trait he was tracking had been assorted into gametes independently of the two units for the other trait. In time, his interpretation became known as the theory of **independent assortment**, which we state here in modern terms: By the end of meiosis, each pair of homologous chromosomes—and the genes they carry—have been sorted for shipment into gametes independently of how the other pairs were sorted out.

Independent assortment and hybrid crossing lead to staggering variety among offspring. In a monohybrid cross involving only a single gene pair, three genotypes are possible: AA, Aa, and aa. We can represent this as 3^n, where n is the number of gene pairs. When we consider more gene pairs, the number of possible combinations increases dramatically. Even if parents differ in merely ten pairs of genes, nearly 60,000 genotypes are possible among their offspring. If they differ in twenty pairs of genes, the number approaches 3.5 billion!

In 1865 Mendel presented his ideas to the Brünn Natural History Society. His ideas had little impact. The next year his paper was published, and apparently it was read by few and understood by no one. In 1871 he became an abbot of the monastery, and his experiments gradually gave way to administrative tasks. He died in 1884, never to know that his work would be the starting point for the development of modern genetics.

Today, Mendel's theory of segregation still stands. Hereditary material is indeed organized in units (genes) that retain their identity and are segregated from each other for distribution into different gametes. However, the theory of independent assortment has undergone some modification, as you will see in the next chapter.

MENDEL'S THEORY OF INDEPENDENT ASSORTMENT. **By the end of meiosis, the genes on pairs of homologous chromosomes have been sorted out for distribution into one gamete or another independently of gene pairs of other chromosomes.**

For the most part, Mendel studied traits having clearly dominant or recessive forms. As the remaining sections of this chapter will make clear, however, the expression of other traits is not as straightforward.

Incomplete Dominance

In **incomplete dominance**, one allele of a pair isn't fully dominant over its partner, so a heterozygous phenotype *somewhere in between* the two homozygous phenotypes emerges. Cross a true-breeding red snapdragon and a true-breeding white one. All F_1 offspring will have pink flowers. Cross two F_1 plants and expect red, pink, and white snapdragons in a predictable ratio (Figure 11.10). What causes this inheritance pattern? Red snapdragons have two alleles that allow them to make an abundance of red pigment molecules. White snapdragons have two mutated alleles that render them pigment-free. The pink ones are heterozygous. Their one red allele can specify enough pigment to make flowers pink but not red.

ABO Blood Types: A Case of Codominance

In **codominance**, a pair of nonidentical alleles specify two phenotypes, which are both expressed at the same time in heterozygotes. As an example, think about one of the recognition proteins at the surface of your red blood cells. This type of protein helps give red blood cells their unique identity, although it comes in slightly different molecular forms. A method of analysis known as *ABO blood typing* reveals which form a person has.

In humans, the gene that determines the protein's final structure has three alleles. Two alleles, I^A and I^B, are codominant when paired. The third, i, is recessive; a pairing with I^A or I^B masks its effects. Altogether, they represent a **multiple allele system**, which we define as the presence of three or more alleles of a gene among individuals of a population.

Before each protein molecule becomes positioned at the cell surface, it gets modified in the cytomembrane system (Section 4.5). There, an oligosaccharide chain is attached to it, then an enzyme specified by the gene of interest attaches a sugar monomer to the end of the chain. Alleles I^A and I^B code for two slightly different versions of the enzyme. The two attach different sugars, which give the protein its identity—either A or B.

Which alleles do you have? With either I^AI^A or I^Ai, you have type A blood. With I^BI^B or I^Bi, your blood is type B. With codominant alleles I^AI^B, it is AB—meaning you have both versions of the sugar-attaching enzyme. But if you are homozygous recessive (ii), the molecules never did get a sugar monomer attached to them. Your blood type is neither A nor B; that is what type "O" means. Figure 11.11 summarizes the possibilities.

homozygous parent   X   homozygous parent

All F_1 offspring are heterozygous for flower color:

Cross two of the F_1 plants, and the F_2 offspring will show three phenotypes in a 1:2:1 ratio:

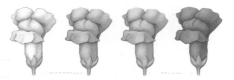

Figure 11.10 Visible evidence of incomplete dominance in heterozygous snapdragons, in which an allele for red pigment is paired with a "white" allele.

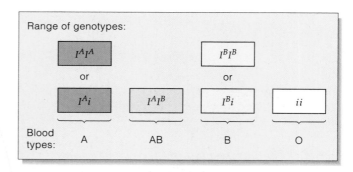

Range of genotypes:

I^AI^A		I^BI^B	
or		or	
I^Ai	I^AI^B	I^Bi	ii

Blood types: A   AB   B   O

Figure 11.11 Allelic combinations related to ABO blood typing.

During *transfusions*, the blood of two people mixes. Unless their red blood cells have the same recognition proteins, the recipient's immune system perceives the red blood cells from the donor as "nonself." It will act against those cells and may cause death (Chapter 40).

One allele may be fully dominant, incompletely dominant, or codominant with its partner on the homologous chromosome.

MULTIPLE EFFECTS OF SINGLE GENES

Expression of the alleles at just a single location on a chromosome may have positive or negative effects on two or more traits. This phenotypic outcome of a single gene's activity is known as **pleiotropy** (after the Greek *pleio-*, meaning more, and *-tropic*, meaning to change).

The genetic disorder *sickle-cell anemia* is a classic example of how the alleles at a single locus can have pleiotropic effects. The disorder arises from a mutated gene for beta-globin, one of two kinds of polypeptide chains in the hemoglobin molecule. Hemoglobin, recall, is the oxygen-transporting protein in red blood cells. We designate the mutated allele as Hb^S instead of Hb^A. Heterozygotes (Hb^A/Hb^S) usually show few symptoms of the disorder. Their red blood cells are able to produce enough normal hemoglobin molecules to compensate for the abnormal ones. In homozygotes (Hb^S/Hb^S), the red blood cells can only produce abnormal hemoglobin. This one abnormality may have drastic repercussions throughout the body, for it disrupts the concentration of oxygen in the bloodstream.

Humans, like most organisms, depend on the intake of oxygen for aerobic respiration. Oxygen in air flows into the lungs, then diffuses into the blood. There it binds to hemoglobin, which transports it through arteries, arterioles, and then small-diameter, thin-walled capillaries threading past all living cells in the body. A steep oxygen concentration gradient exists between blood and the fluid within the surrounding tissues, so most of the oxygen diffuses into the tissues, and on into cells. With this extensive cellular uptake, the concentration of oxygen in blood declines. The decline is most pronounced at high altitudes and during strenuous activity.

In the red blood cells of people who bear the mutant gene, abnormal hemoglobin molecules stick together into rodlike arrangements. The rods distort the cells into a sickle shape, as in Figure 11.12*b*. (A sickle is a farm tool having a crescent-shaped blade.) The distorted cells rupture easily, and their remnants clog and then rupture capillaries. When these oxygen transporters are swiftly destroyed, the cells in affected tissues become starved for oxygen. Also, their clumping effect leads to local failures in the capacity of the circulatory system to deliver oxygen and carry away carbon dioxide and other metabolic wastes.

Over time, ongoing expression of the mutant gene can cause widespread damage. Figure 11.12*c* tracks how the successive changes in phenotype may proceed.

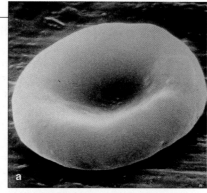

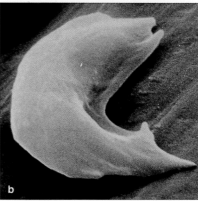

Figure 11.12 (**a**) A red blood cell from a person affected by sickle-cell anemia, which is a genetic disorder. This scanning electron micrograph shows the surface appearance of the affected individual's red blood cells when the blood is adequately oxygenated. (**b**) This scanning electron micrograph shows the sickle shape that red blood cells assume when the concentration of oxygen in blood is low. (**c**) This diagram summarizes the range of symptoms that are characteristic of an individual who is homozygous recessive for the disorder.

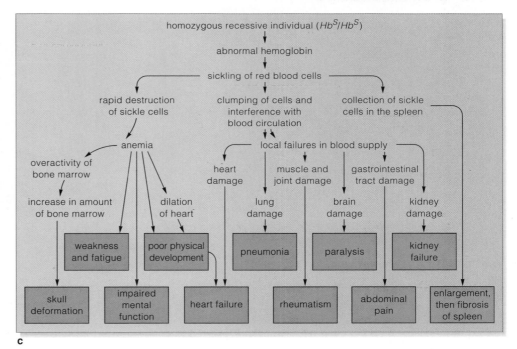

In chapters to come, you will read about other aspects of this genetic disorder.

The alleles at a single gene location may have positive or negative effects on two or more traits.

The effects may not be simultaneous. Rather, they may have repercussions over time. The gene may lead to an alteration in one trait, that change may alter another trait, and so on.

INTERACTIONS BETWEEN GENE PAIRS

Often a trait results from interactions among products of two or more gene pairs. For example, two alleles of one gene may mask expression of another gene's alleles, so some expected phenotypes may not appear at all. Such interactions between pairs of genes are called **epistasis** (meaning the act of stopping).

Hair Color in Mammals

Epistasis is common among the gene pairs responsible for the coloration of fur or skin in mammals. Consider the black, brown, or yellow fur of Labrador retrievers (Figure 11.13). The different colors arise from variations in the amount and distribution of melanin, a brownish black pigment. Enzymes and other products of many

gene pairs influence different steps in the production of melanin and its deposition in certain body regions.

The alleles of one gene specify an enzyme required to produce melanin. Expression of allele *B* (black) has a more pronounced effect and is dominant to *b* (brown). Alleles of a different gene control the extent to which molecules of melanin will be deposited in a retriever's hairs. Allele *E* permits full deposition. Two recessive alleles (*ee*) reduce deposition, and the fur will be yellow.

Interactions between these two gene pairs may not even be possible if a certain allelic combination exists at still another gene location. There, a gene (*C*) calls for tyrosinase, the first of several enzymes in the metabolic pathway that produces melanin. An individual having one or two dominant alleles (*CC* or *Cc*) can produce the

a BLACK LABRADOR

b YELLOW LABRADOR

c CHOCOLATE LABRADOR

Figure 11.13 The heritable basis of coat color among Labrador retrievers. The trait arises through interactions among the alleles of two pairs of genes.

One kind of gene is involved in melanin production. Allele *B* (black) of this gene is dominant to allele *b* (brown). A different kind of gene influences the deposition of melanin pigment in individual hairs. Allele *E* of this gene promotes deposition, but a pairing of recessive alleles (*ee*) of the gene blocks deposition, and a yellow coat results.

F_1 offspring of a dihybrid cross produce F_2 offspring in a 9:3:4 ratio, as the Punnett-square diagram to the right indicates.

The yellow Labrador in photograph (**b**) probably has genotype *BBee*, because it can produce melanin but cannot deposit pigment in hairs. After studying the photograph, can you say why? 🐾

HOMOZYGOUS PARENTS: $BBEE \times bbee$

F_1 PUPPIES: $BbEe$

ALLELIC COMBINATIONS POSSIBLE AMONG F_2 PUPPIES:

	BE	Be	bE	be
BE	BBEE	BBEe	BbEE	BbEe
Be	BBEe	BBee	BbEe	Bbee
bE	BbEE	BbEe	bbEE	bbEe
be	BbEe	Bbee	bbEe	bbee

RESULTING PHENOTYPES:

 9/16 or 9 black

3/16 or 3 brown

4/16 or 4 yellow

Figure 11.14 A rare albino rattlesnake. Like other animals that cannot produce melanin, it has pink eyes and its body is white, overall. (Eyes look pink because the absence of melanin from a tissue layer in the eyeball allows red light to be reflected from blood vessels in the eyes.) In birds and mammals, surface coloration results from pigments in feathers, fur, or skin. In fishes, amphibians, and reptiles, color-bearing cells give skin its surface coloration. Some of the cells contain melanin pigments or yellow-to-red pigments. Others contain crystals that reflect light and alter the effect of other pigments present.

The mutation affecting melanin production in the snake shown here had no effect on the production of yellow-to-red pigments and of light-reflecting crystals. Therefore, the snake's skin appears to be iridescent yellow as well as white.

a WALNUT COMB **b** ROSE COMB **c** PEA COMB **d** SINGLE COMB

Figure 11.15 Interaction between two genes that affect the same trait in domestic breeds of chickens. The initial cross is between a Wyandotte (with a rose comb, **b**, on the crest of its head) and a brahma (pea comb, **c**). With complete dominance at the locus for pea comb and at the locus for rose comb, products of the two gene pairs interact and give rise to a walnut comb (**a**). With complete recessiveness at both gene loci, products interact and give rise to a single comb (**d**).

functional enzyme. An individual bearing two recessive alleles (*cc*) cannot. When the biosynthetic pathway for melanin production is blocked, *albinism*—the absence of melanin—is the resulting phenotype (Figure 11.14).

Comb Shape in Poultry

In some cases, interaction between two gene pairs results in a phenotype that neither pair can produce alone. The geneticists W. Bateson and R. Punnett identified two interacting gene pairs (*R* and *P*) that affect comb shape in chickens. Allelic combinations of *rr* at one gene locus and *pp* at the other locus result in the least common phenotype, the single comb. The presence of dominant alleles (*R*, *P*, or both) results in varied phenotypes. The Figure 11.15 diagram shows the combinations of alleles that lead to rose, pea, or walnut combs.

Genes often interact, as when alleles of one gene mask the expression of another gene, and some expected phenotypes may not appear at all.

Regarding the Unexpected Phenotype

As Mendel demonstrated, the phenotypic effects of one or two pairs of certain genes show up in predictable ratios from one generation to the next. Besides this, certain interactions among two or more gene pairs also may produce phenotypes in predictable ratios, as the example of Labrador coat color in Section 11.6 clearly demonstrated. However, even when you are tracking a single gene through the generations, you may discover that the resulting phenotypes are not quite what you had expected.

Consider *camptodactyly*, a rare genetic abnormality that affects both the shape and the movement of fingers. Certain people who carry the gene for this heritable trait develop immobile, bent fingers on both hands. Other people develop immobile, bent fingers on the left or right hand only. Still others who carry the mutated gene develop fingers that are not affected either way.

What is the source of such confounding variation? Recall that most organic compounds are synthesized by a series of metabolic steps, *and different enzymes—each a gene product— regulate different steps.* Maybe one gene is mutated in one of a number of different ways. Maybe the product of another gene blocks the pathway or causes it to run nonstop, or not long enough. Or maybe poor nutrition or another factor that can vary in the individual's environment affects a crucial enzyme in the pathway. These are the sorts of variable factors that commonly introduce far less predictable variations in the phenotypes resulting from gene expression.

Continuous Variation in Populations

Generally, the individuals of a population display a range of small differences in most traits.

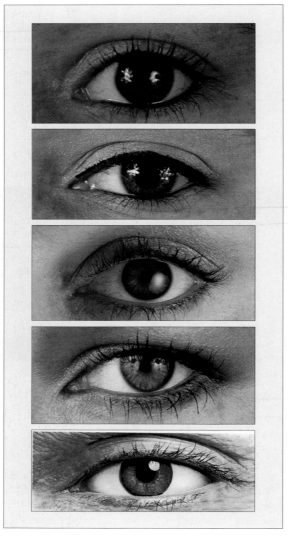

Figure 11.16 Samples from the range of continuous variation in human eye color. Different pairs of genes interact to produce and deposit melanin. Among other things, this pigment helps color the eye's iris. Different combinations of alleles result in small differences in eye color. Thus the frequency distribution for the eye-color trait appears to be continuous over a range from black to light blue.

This characteristic of populations, which is known as **continuous variation**, is largely an outcome of the number of genes affecting a trait and the number of environmental factors that influence their expression. In most cases, the greater the number of genes and environmental factors, the more continuous will be the expected distribution of all the versions of that trait.

Think about your own eye color. The colored part is the iris, a doughnut-shaped, pigmented structure beneath the cornea. As is true of all humans, the color of your iris is the cumulative outcome of a number of gene products. Those products take part in the stepwise production and distribution of melanin, the same light-absorbing pigment that influences coat color in mammals. Dark eyes that appear almost black have abundant deposits of melanin molecules in the iris—so much so that most of the light striking them is absorbed. Dark brown eyes have fewer deposits of melanin, and some light that is not absorbed is reflected from the iris. Light brown or hazel eyes have even fewer (Figure 11.16).

Green, gray, or blue eyes do not contain green, gray, or blue pigments. In these cases, the iris contains different quantities of melanin, but not very much of it. As a result, many or most of the blue wavelengths of light that do enter the eye are reflected out.

How might you describe the continuous variation of some trait within a group, such as the college students in Figure 11.17? They range from very short to very tall, with average heights much more common than either extreme. You can start out by dividing the full range of the different phenotypes into measurable categories. Next, count all the individual students in each category. This will give you the relative frequencies of phenotypes distributed across the range of measurable values.

The bar chart in Figure 11.17c plots the proportion of students in each category against the range of the measured phenotypes. The vertical bars that are the shortest

Number of individuals	1	4	8	10	16	16	15	15	14	13	13	11	9	8	8	5	1	2
Height (inches)	60	61	62	63	64	65	66	67	68	69	70	71	72	73	74	75	76	77

a Students at Brigham Young University, organized according to height, as a splendid example of continuous variation

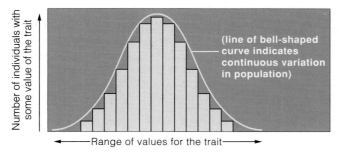

(line of bell-shaped curve indicates continuous variation in population)

Range of values for the trait

b Idealized bell-shaped curve for a population that displays continuous variation in some trait

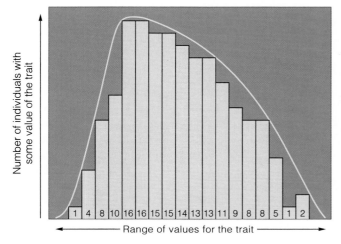

Range of values for the trait

c Specific bell-shaped curve corresponding to the distribution of the trait (height) illustrated by the photograph in (**a**)

Figure 11.17 Continuous variation in body height, a trait that is one of the characteristics of the human population.

(**a**) Suppose you want to find the frequency distribution for height in a group of 169 biology students at Brigham Young University. You decide on how finely the range of possible heights should be divided. Then you measure each student and assign her or him to the appropriate category. Finally, you divide the number in each category by the total number of all students in all categories.

(**b**) Often a bar graph is used to depict continuous variation in a population. In such graphs, the proportion of individuals in each category is plotted against the range of measured phenotypes. Notice the curved line above the bars. It is an idealized example of the kind of "bell-shaped" curve that emerges for populations showing continuous variation in a trait. The bell-shaped curve in (**c**) is a specific example of this type of diagram.

represent categories with the least number of students. The bar that is tallest represents the category with the greatest number of students. Finally, draw a graph line around all of the bars and you end up with a "bell-shaped" curve. Such curves are typical of populations that show continuous variation in a trait.

Enzymes and other products of genes regulate each step of most metabolic pathways. Mutations, gene interactions, and environmental conditions may affect one or more of the steps, and this leads to variations in phenotypes.

For most traits, the individuals of a population display continuous variation—that is, a range of small differences.

The greater the number of genes and environmental factors that can influence a trait, the more continuous will be the expected distribution of all the versions of that trait.

EXAMPLES OF ENVIRONMENTAL EFFECTS ON PHENOTYPE

We have mentioned, in passing, that environmental conditions often contribute to variable gene expression among individuals of a population. Before leaving this chapter, consider just two examples of environmentally induced variations in phenotype.

Possibly you have observed a Himalayan rabbit and probably a Siamese cat. Both of these furry mammals carry an allele that specifies a heat-sensitive version of one of the enzymes necessary for melanin production. At the surface of warm body regions, the enzyme is less active. Fur growing there is lighter in color than fur in cooler regions, which include the ears and other body parts that project away from the main body mass. The experiment shown in Figure 11.18 provided observable evidence of an environmental effect on gene expression.

The environment also influences genes that govern phenotypes of plants. You may have observed the color variation in the floral clusters of *Hydrangea macrophylla*, a species widely favored in home gardens (Figure 11.19). In this type of plant, the action of genes responsible for floral color can produce different phenotypes, depending on the acidity of the soil in which the plant is growing.

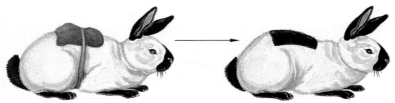

Figure 11.18 Example of the observable effect of different environmental conditions on gene expression in animals. A Himalayan rabbit normally has black hair only on its long ears, nose, tail, and lower leg limbs. For one experiment, a patch of a rabbit's white fur was plucked clean, and then an icepack was secured over the hairless patch. Where the colder temperature had been maintained, the hairs that grew back were black.

Himalayan rabbits are homozygous for the *ch* allele of a gene that codes for tyrosinase, an enzyme required to produce melanin. The allele specifies a heat-sensitive version of the enzyme, which is able to function only when the surrounding temperature is below about 33°C. When cells that give rise to hairs grow under warmer conditions, they cannot produce melanin and the hairs appear light. This happens in body regions that are massive enough to conserve a fair amount of metabolic heat. Ears and other slender extremities are cooler because they tend to lose metabolic heat more rapidly.

Figure 11.19 Effect of environmental conditions on gene expression in a favorite garden plant (*Hydrangea macrophylla*). Even plants that carry the same alleles have floral colors ranging from pink to blue, depending on the acidity of the soil in which they happen to be growing.

And so we conclude this chapter, which has dealt with heritable and environmental factors that give rise to variations in phenotype. What is the take-home lesson? Simply this: An individual's phenotype is an outcome of complex interactions among genes, enzymes and other gene products, and environmental factors.

Owing to gene mutations, cumulative gene interactions, and environmental effects on genes, individuals of a population show degrees of variation for many traits.

SUMMARY

Read

1. A gene is a unit of information about a heritable trait. Alleles of a gene are different molecular versions of that information. Through experimental crosses with pea plants, Mendel gathered evidence that diploid organisms have two genes for each trait and that genes retain their identity when transmitted to offspring.

2. Mendel performed monohybrid crosses between two true-breeding plants that displayed different versions of a single trait, such as flower color. The crosses provided indirect evidence that some forms of a gene may be dominant over other, recessive forms.

3. A homozygous dominant individual has inherited two dominant alleles (AA) for the trait being studied. A homozygous recessive has two recessive alleles (aa), and a heterozygote has two nonidentical alleles (Aa).

4. In Mendel's monohybrid crosses ($AA \times aa$), all F_1 offspring were Aa. Crosses between F_1 plants resulted in these combinations of alleles in F_2 offspring:

	A	a	
A	AA	Aa	AA (dominant)
			Aa (dominant)
a	Aa	aa	Aa (dominant)
			aa (recessive)

This produced the expected phenotypic ratio of 3:1.

5. Results from Mendel's monohybrid crosses led to the formulation of a theory of segregation. In modern terms, diploid organisms have pairs of genes, on pairs of homologous chromosomes. The two genes of each pair segregate from each other at meiosis, such that each gamete formed ends up with one or the other gene.

6. Mendel performed dihybrid crosses between two true-breeding plants that displayed different versions of two traits. Results were close to a 9:3:3:1 phenotypic ratio:

    9 dominant for both traits
    3 dominant for A, recessive for b
    3 dominant for B, recessive for a
    1 recessive for both traits

7. Mendel's dihybrid crosses led to the formulation of a theory of independent assortment. In modern terms, by the end of meiosis, the gene pairs of two homologous chromosomes have been sorted out for distribution into one gamete or another, independently of how the gene pairs of other chromosomes were sorted out.

8. Four factors commonly influence gene expression:
    a. Degrees of dominance may occur between some pairs of genes.
    b. The products of pairs of genes may interact to influence the same trait.
    c. One gene may have positive or negative effects on two or more traits, a condition called pleiotropy.
    d. Environmental conditions to which an individual is subjected may affect gene expression.

Review Questions

1. Define the difference between these terms: *11.1*
    a. gene and allele
    b. dominant allele and recessive allele
    c. homozygote and heterozygote
    d. genotype and phenotype

2. Define a true-breeding lineage. What is a hybrid? *11.1*

3. Distinguish between monohybrid and dihybrid crosses. What is a testcross, and why is it useful in genetic analysis? *11.2, 11.3*

4. Do segregation and independent assortment proceed during mitosis, meiosis, or both? *11.2, 11.3*

5. What do the vertical and horizontal arrows of this diagram represent? What do the bars and the curved line represent? *11.7*

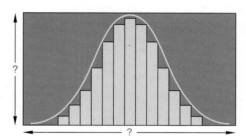

Self-Quiz *(Answers in Appendix IV)*

1. Alleles are _____ .
    a. different molecular forms of a gene
    b. different molecular forms of a chromosome
    c. self-fertilizing, true-breeding homozygotes

2. A heterozygote has a _____ for the trait being studied.
    a. pair of identical alleles
    b. pair of nonidentical alleles
    c. haploid condition, in genetic terms
    d. a and c

3. The observable traits of an organism are its _____ .
    a. phenotype   c. genotype
    b. sociobiology   d. pedigree

4. F_1 offspring of the monohybrid cross $AA \times aa$ are _____ .
    a. all AA c. all Aa
    b. all aa d. 1/2 AA and 1/2 aa

5. Second-generation offspring from a cross are the _____ .
    a. F_1 generation   c. hybrid generation
    b. F_2 generation   d. none of the above

6. Assuming complete dominance will occur, the offspring of the cross $Aa \times Aa$ will show a phenotypic ratio of _____ .
    a. 3:1   b. 9:1   c. 1:2:1   d. 9:3:3:1

7. Crosses between F_1 individuals resulting from the cross $AABB \times aabb$ lead to F_2 phenotypic ratios close to _____ .
    a. 1:2:1   b. 3:1   c. 1:1:1:1   d. 9:3:3:1

8. Match each example with the most suitable description.
    _____ dihybrid cross   a. *bb*
    _____ monohybrid cross   b. $AaBb \times AaBb$
    _____ homozygous condition   c. *Aa*
    _____ heterozygous condition   d. $Aa \times Aa$

Critical Thinking—Genetics Problems
(Answers in Appendix III)

1. One gene has alleles A and a. Another has alleles B and b. For each genotype listed, what type(s) of gametes will be produced? (Assume independent assortment occurs before gametes form.)

   a. $AABB$ c. $Aabb$
   b. $AaBB$ d. $AaBb$

2. Still referring to Problem 1, what will be the genotypes of the offspring from the following matings? Indicate the frequencies of each genotype among them.

   a. $AABB \times aaBB$ c. $AaBb \times aabb$
   b. $AaBB \times AABb$ d. $AaBb \times AaBb$

3. In one experiment, Mendel crossed a pea plant that bred true for green pods with one that bred true for yellow pods. All the F_1 plants had green pods. Which form of the trait (green or yellow pods) is recessive? Explain how you arrived at your conclusion.

4. Return to Problem 1, and assume you now study a third gene having alleles C and c. For each genotype listed, what type(s) of gametes will be produced?

   a. $AABBCC$ c. $AaBBCc$
   b. $AaBBcc$ d. $AaBbCc$

5. Mendel crossed a true-breeding tall, purple-flowered pea plant with a true-breeding dwarf, white-flowered plant. All F_1 plants were tall and had purple flowers. If an F_1 plant self-fertilizes, then what is the probability that a randomly selected F_2 offspring will be heterozygous for the genes specifying height and flower color?

6. At a certain gene location on a human chromosome, a dominant allele controls *tongue rolling*, an ability to curl up the sides of the tongue (Figure 11.20). People who are homozygous for a recessive allele at that locus cannot roll the tongue. At a different gene locus, a dominant allele controls whether the earlobes will be attached or detached (refer to Figure 11.1). These two pairs of genes assort independently. Suppose a tongue-rolling, detached-earlobed woman marries a man who has attached earlobes and cannot roll his tongue. Their first child has the father's phenotype. Given this outcome;

   a. What are the genotypes of the mother, father, and child?
   b. What is the probability that a second child of theirs will have detached earlobes and won't be a tongue roller?

7. Bill and his wife, Marie, hope to have children. Both have notably flat feet and long eyelashes, and they tend to sneeze a lot (hence the name of the *achoo syndrome*). A dominant allele gives rise to each of these traits: A (foot arch), E (eyelash length), and S (chronic sneezing). Bill is heterozygous and Marie is homozygous for all three traits.

   a. What is Bill's genotype? What is Marie's genotype?
   b. Marie becomes pregnant four times. What is the probability that each child will show all three of Bill's traits? Of Marie's traits?
   c. What is the probability that each child will have short lashes, high arches, and no chronic tendency to sneeze?

8. *DNA fingerprinting* is a method of identifying individuals based on locating unique base sequences in their DNA (Section 16.3). Before researchers refined the method, attorneys often relied on the ABO blood-typing system to settle disputes over paternity. Suppose, as a geneticist, you were called upon to testify during a paternity case in which the mother has type A blood, the child has type O blood, and the alleged father has type B blood. How would you respond to the following statements?

   a. Attorney of the alleged father: "The mother's blood is type A, so the child's type O blood must have come from the father. Because my client has type B blood, he simply could not be the father."
   b. Mother's attorney: "Because further tests prove this man is heterozygous, he must be the father."

9. Suppose you identify a new gene in mice. One of its alleles specifies white fur color. A second allele specifies brown fur color. You are asked to determine whether the relationship between the two alleles is one of simple dominance or incomplete dominance. What sorts of genetic crosses would give you the answer? On what types of observations would you base your conclusions?

10. Your sister moves away and gives you her purebred Labrador retriever, a female named Dandelion. Suppose you decide to breed Dandelion and sell puppies to help pay for your college tuition. Then you discover that two of her four brothers and sisters have a heritable hip disorder. If Dandelion mates with a male Labrador known to be free of the harmful allele, can you guarantee to a buyer that puppies will not carry the allele? Explain your answer.

11. A dominant allele W confers black fur on guinea pigs. If a guinea pig is homozygous recessive (ww), it has white fur. Fred would like to know whether his pet black-furred guinea pig is homozygous dominant (WW) or heterozygous (Ww). How might he determine his pet's genotype?

12. Red-flowering snapdragons are homozygous for the allele R^1. White-flowering snapdragons are homozygous for a different allele (R^2). Heterozygous plants (R^1R^2) bear pink flowers. What phenotypes should appear among F_1 offspring of the crosses listed, and what are the expected proportions for each phenotype?

   a. $R^1R^1 \times R^1R^2$ c. $R^1R^2 \times R^1R^2$
   b. $R^1R^1 \times R^2R^2$ d. $R^1R^2 \times R^2R^2$

Notice, in Problem 12, that in cases of incomplete dominance it is inappropriate to refer to either allele of a pair as dominant or recessive. When the phenotype of a heterozygous individual is halfway between those of the two homozygotes, then there is no dominance. Such alleles

Figure 11.20 A student at San Diego State University exhibiting the tongue-rolling trait for the benefit of a tongue-roll-challenged student.

are usually designated by superscript numerals, as shown here, rather than by uppercase letters for dominance and lowercase letters for recessiveness.

13. In chickens, two pairs of genes affect the comb type (Figure 11.15). When both genes are recessive, a chicken will have a single comb. A dominant allele of one gene, *P*, gives rise to a pea comb. A dominant allele of the other (*R*) gives rise to a rose comb. An epistatic interaction occurs when a chicken has at least one of both dominants, *P— R —* , which gives rise to a walnut comb. Predict the F_1 ratios resulting from a cross between two walnut-combed chickens that are heterozygous for both genes (*PpRr*).

14. As described in Section 11.5, a mutated allele gives rise to an abnormal form of hemoglobin (Hb^S instead of Hb^A). Homozygotes ($Hb^S Hb^S$) develop sickle-cell anemia. But heterozygotes ($Hb^A Hb^S$) show few outward symptoms.

Suppose a female's mother is unaffected by this genetic disorder, but her father is homozygous for the Hb^S allele. She marries a male who is heterozygous for the allele, and they plan to have children. For *each* pregnancy, state the probability that this couple will have a child who is:

  a. homozygous for the Hb^S allele

  b. homozygous for the Hb^A allele

  c. heterozygous $Hb^A Hb^S$

15. Certain dominant alleles are so vital for normal development that an individual who is homozygous recessive for a mutant recessive form of the allele cannot survive. Such recessive, *lethal alleles* can be perpetuated by heterozygotes. Consider the Manx allele (M^L) in cats. Homozygous cats ($M^L M^L$) die when they are still embryos inside the mother cat. In heterozygotes ($M^L M$), the spine develops abnormally, and the cats end up with no tail whatsoever (Figure 11.21).

Suppose two $M^L M$ cats mate. Among their *surviving* progeny, what is the probability that any one kitten will be heterozygous?

16. A recessive allele *a* is responsible for albinism, an inability to produce or deposit melanin in tissues. Humans and some other organisms can have this phenotype (Figure 11.22). In each of the following cases, what are the possible genotypes of the father, of the mother, and of their children?

  a. Both parents have normal phenotypes; some of their children are albino and others are unaffected.

Figure 11.21 Manx cat.

Figure 11.22 An albino male in India.

  b. Both parents are albino and have only albino children.

  c. The woman is unaffected, the man is albino, and they have one albino child and three unaffected children. (What gives rise to this 3:1 ratio?)

17. Kernel color in wheat plants is determined by two pairs of genes. Alleles of one pair show incomplete dominance over alleles of the other pair. For the gene pair at one locus on the chromosome, allele A^1 imparts one dose of red color to the kernel, whereas allele A^2 does not. At the second locus, allele B^1 gives one dose of red color to the kernel, whereas allele B^2 does not. A kernel with genotype $A^1 A^1 B^1 B^1$ is dark red. A kernel with genotype $A^2 A^2 B^2 B^2$ is white. All other genotypes have kernel colors in between these two extremes.

  a. If you cross a plant grown from a dark-red kernel with one grown from a white kernel, what genotypes and phenotypes would be expected among the offspring? In what proportions?

  b. If a plant with genotype $A^1 A^2 B^1 B^2$ self-fertilizes, what genotypes and what phenotypes would be expected among the offspring? In what proportions?

Selected Key Terms

allele *11.1*	hybrid offspring *11.1*
codominance *11.4*	incomplete dominance *11.4*
continuous variation *11.7*	independent assortment *11.3*
dihybrid cross *11.3*	monohybrid cross *11.2*
epistasis *11.6*	multiple allele system *11.4*
F_1 *11.1*	phenotype *11.1*
F_2 *11.1*	pleiotropy *11.5*
gene *11.1*	probability *11.2*
genotype *11.1*	Punnett-square method *11.2*
heterozygous *11.1*	segregation *11.2*
homozygous dominant *11.1*	testcross *11.2*
homozygous recessive *11.1*	true-breeding lineage *11.1*

Readings

Cummings, M. 1994. *Human Heredity*. Third edition. St. Paul: West.

Orel, V. 1984. *Mendel*. New York: Oxford University Press.

Web Site See *http://www.wadsworth.com/biology* for practice quiz questions, hypercontents, BioUpdates, and critical thinking. The Wadsworth Biology Resource Center provides a wealth of information fully organized and integrated by chapter.

CHROMOSOMES AND HUMAN GENETICS

Too Young to Be Old

Imagine being ten years old, trapped in a body that is rapidly getting a bit more shriveled, more frail—*old*—each day. You are only tall enough to peer over the top of the kitchen counter, and you weigh less than thirty-five pounds. Already you are bald and have a crinkled nose. Possibly you have a few more years to live. Yet, like Mickey Hayes and Fransie Geringer, you can still play and laugh with your friends (Figure 12.1).

Of every 8 million newborn humans, one is destined to grow old far too soon. That rare individual possesses a mutated gene on one of the forty-six chromosomes inherited from its mother or father. Through hundreds, thousands, then many billions of DNA replications and mitotic cell divisions, terrible information encoded in that gene was systematically distributed to every cell in the growing embryo, and later in the newborn. Its legacy will be accelerated aging and a greatly reduced life expectancy. These are the defining features of <u>Hutchinson-Gilford progeria syndrome</u>. There is no cure.

A **syndrome** is a set of symptoms that characterize a disorder. In this case, the mutation is the start of severe disruptions in interactions among the genes that bring about bodily growth and development. Outwardly apparent symptoms begin to emerge when the child is not even two years old. Skin that should be plump and resilient starts to thin. Skeletal muscles are weakened. Limb bones

that should lengthen and grow stronger start to soften. Hair loss becomes pronounced; extremely premature baldness is inevitable.

Most progeriacs die in their early teens as a result of strokes or heart attacks. These final insults are brought on by a hardening of the walls of arteries, a condition that is typical of advanced age.

There are no documented cases of progeria running in families, which suggests that the gene spontaneously mutates at random. It apparently is dominant over its normal partner on the homologous chromosome.

We began this unit of the book by looking at cell division, the starting point of inheritance. Then we

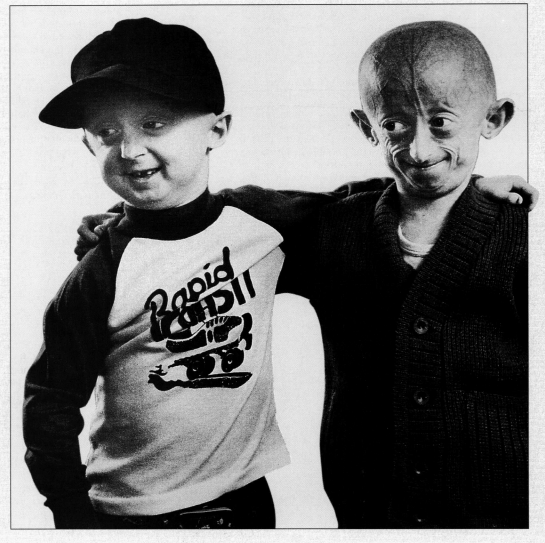

Figure 12.1 Two boys who met during a gathering of progeriacs at Disneyland, California, when they were not yet ten years old. Progeria, a heritable disorder, arises through a mutation in a dominant allele. Its defining characteristics are accelerated aging and a greatly reduced life expectancy.

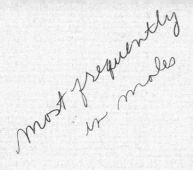
most frequently in moles

started thinking about how chromosomes—and the genes they carry—are shuffled during meiosis and at fertilization. In this chapter, we delve more deeply into patterns of chromosomal inheritance. If at times the described methods of analysis seem remote from the world of your interests, remember this: For better or worse, *an inherited collection of specific bits of information gives rise to traits that help define each organism*—and this includes yourself and other human individuals. When Mickey Hayes turned all of eighteen, he was the oldest living progeriac. Fransie was only seventeen years old when he died.

So in that sense, our story of Mickey and Fransie takes us back to the 1880s, and to the rediscovery of Gregor Mendel's studies of inheritance. In 1884 Mendel himself had just passed away, and his paper on pea plants had been gathering dust in a hundred libraries for nearly two decades. However, improvements in the resolving power of microscopes had rekindled efforts to locate the hereditary material within cells. By 1882, Walther Flemming had observed threadlike bodies—chromosomes—inside the nuclear region of dividing cells. By 1884, a question was taking shape: Could chromosomes be the hereditary material?

Then researchers realized each gamete has half the number of chromosomes of a fertilized egg. In 1887, August Weismann hypothesized that a special division process must reduce the chromosome number by half before gametes form. Sure enough, in that same year meiosis was discovered. Weismann began to promote this theory of heredity: The chromosome number is halved during meiosis, then restored at fertilization; therefore, a cell's hereditary material is half paternal and half maternal in origin. His theory was hotly debated among biologists, and it prompted a flurry of experimental crosses—just like the ones Mendel had carried out.

Finally, in 1900, researchers came across Mendel's paper while checking literature related to their own genetic crosses. To their surprise, their experimental results confirmed what Mendel's results had already suggested: Diploid cells have two units (genes) for each heritable trait, and the two units are segregated from each other before gametes form.

During the decades that followed, many researchers learned a great deal more about chromosomes. Let's turn now to a few high points of their work, which will serve as background for our understanding of human inheritance.

KEY CONCEPTS

1. One gene follows another in sequence along the length of a chromosome. Each gene has its own position in that sequence.

2. The combination of alleles in a chromosome does not necessarily remain intact through meiosis and gamete formation. During an event called crossing over, some alleles in the sequence swap places with their partners on the homologous chromosome. The alleles that are swapped may or may not be identical.

3. Allelic recombinations contribute to variations in the phenotypes of offspring.

4. The structure of a chromosome may change, as when a chromosome segment is deleted, duplicated, inverted, or moved to a new location. Also, the chromosome number may change as a result of an improper separation of duplicated chromosomes during meiosis or mitosis.

5. Changes in chromosome structure and in chromosome number are rare events. When they do occur, they often give rise to genetic abnormalities or disorders.

handwritten: 23 pairs = 22 autosomes, 1 sex chromo.

handwritten (left margin): study Vocab.

Genes and Their Chromosome Locations

Earlier chapters provided you with a general sense of the structure of chromosomes and their behavior in meiosis. To refresh your memory and get a glimpse of where you are going from here with your reading, take a moment to study the following list:

1. **Genes** are units of information about heritable traits. The genes of eukaryotic species are distributed among a number of chromosomes, and each has its own location (locus) in one type of chromosome.

2. A cell with a diploid chromosome number (2*n*) has inherited pairs of **homologous chromosomes**. All but one pair are identical in their length, shape, and gene sequence. The exception is a pairing of nonidentical sex chromosomes, such as X with Y. In all cases, the two members of a pair can interact during meiosis.

3. A pair of homologous chromosomes can carry identical or nonidentical alleles at a given locus. **Alleles**, which arise through mutation, are different molecular forms of the same gene. Although many different forms may have arisen in a population, a diploid cell can only have a pair of them.

4. A *wild-type* allele is the most common form of a gene, either in a natural population or in standard, laboratory-bred strains of a species. Any other form of the gene is called a mutated allele.

5. Think of genes on the same chromosome as being linked. The farther apart two linked genes are, the more vulnerable they are to **crossing over**, an event by which homologous chromosomes break and exchange corresponding segments. Crossing over results in **genetic recombination**, or nonparental combinations of linked alleles in gametes, then in offspring.

6. The random alignment of each pair of homologous chromosomes at metaphase I results in nonparental combinations of alleles in gametes, then in offspring.

7. Abnormal occurrences during meiosis or mitosis occasionally change the structure of chromosomes and the parental chromosome number.

As you will see, these are the points that will help you make sense of the basic patterns of human inheritance.

Autosomes and Sex Chromosomes

In all but one case, a pair of homologous chromosomes are exactly like each other in their length, shape, and gene sequence. Microscopists discovered the exception in the late 1800s. For many organisms, a distinctive chromosome is present in female *or* male individuals,

but not in both. For example, a diploid cell of a human male has one **X chromosome** and one **Y chromosome** (written as XY). That of a human female has two X chromosomes (XX). This inheritance pattern is common among many organisms, including all mammals and fruit flies. By contrast, in butterflies, moths, birds, and certain fishes, inheriting two identical sex chromosomes results in a male, and inheriting two nonidentical sex chromosomes results in a female.

Figure 12.2 shows examples of the human X and Y chromosomes. Notice that they are physically different; one is much shorter than the other. Besides this, they do not carry the same genes. Despite the differences, the two are able to synapse in a small region along their length, and this allows them to function as homologues during meiosis.

The human X and Y chromosomes fall in the more general category of **sex chromosomes**. The term refers to distinctive types of chromosomes which, in certain combinations, govern gender—that is, whether a new individual will develop into a male or a female. All other chromosomes in an individual's cells are the same in both sexes; they are called **autosomes**.

Karyotype Analysis

Today, microscopists can routinely analyze the physical appearance of a cell's sex chromosomes and autosomes. Chromosomes, recall, are in the most highly condensed state at metaphase of mitosis. At that time, their size, length, and centromere location are easiest to identify. In addition, after microscopists stain them in certain ways, the chromosomes of many species show distinct banding patterns. The most condensed chromosome regions often take up more stain and therefore appear as darker bands. Figure 10.3 gives an example.

A **karyotype** for an individual (or for a species) is a preparation of metaphase chromosomes, sorted out by their defining features. The next section explains how to construct a karyotype diagram: a cut-up, rearranged photograph in which all autosomes are lined up, largest to smallest, and sex chromosomes are positioned last.

Diploid cells have pairs of genes, on pairs of homologous chromosomes. At each gene locus, the alleles (alternative forms of a gene) may be identical or nonidentical.

As a result of crossing over and other events during meiosis, new combinations of alleles and parental chromosomes end up in offspring. Abnormal events during meiosis or mitosis also can change the structure and number of chromosomes.

Autosomes are the pairs of chromosomes that are the same in males and females of a species. One other pair, the sex chromosomes, govern a new individual's gender.

PREPARING A KARYOTYPE DIAGRAM

Karyotype diagrams can help answer questions about an individual's chromosomes. Chromosomes are in their most condensed form, and easiest to identify, in cells that are proceeding through metaphase of mitosis. Technicians don't count on finding a cell that happens to be dividing in the body when they go looking for it. Instead, they culture cells **in vitro** (literally, "in glass"). They put a small sample of blood, skin, bone marrow, or some other tissue in a glass container. In that container is a solution that stimulates cell growth and mitotic cell division for many generations.

Dividing cells can be arrested at metaphase by adding colchicine to a culture medium. As Section 4.8 describes, colchicine is an extract of the autumn crocus (*Colchicum autumnale*). Technicians and researchers use it to block spindle formation and so prevent duplicated chromosomes from separating during nuclear division. With suitable colchicine concentrations and exposure times, many metaphase cells can accumulate, and this increases the chances of finding candidates for karyotype diagrams.

Following colchicine treatment, the culture medium is transferred to the tubes of a centrifuge, a motor-driven spinning device (Section 5.2 and Figure 12.2*a*). Cells have greater mass and density than the surrounding solution, so the spinning force moves them farthest from the center of rotation, to the bottom of the attached tubes. Separation in response to a spinning force is called **centrifugation**.

Afterward, the cells are transferred to a saline solution. When immersed in this hypotonic fluid, they swell (by osmosis) and move apart. And so do the metaphase chromosomes. The cells are ready to be dropped onto a microscope slide, fixed (as by air-drying), and stained.

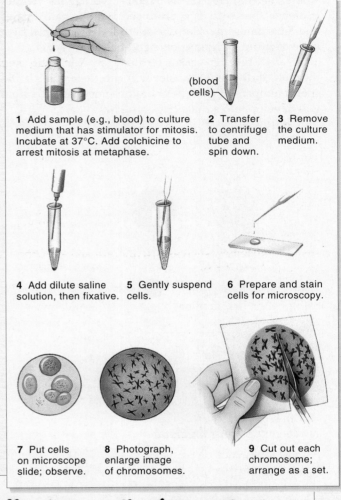

1 Add sample (e.g., blood) to culture medium that has stimulator for mitosis. Incubate at 37°C. Add colchicine to arrest mitosis at metaphase.

2 Transfer to centrifuge tube and spin down.

3 Remove the culture medium.

4 Add dilute saline solution, then fixative.

5 Gently suspend cells.

6 Prepare and stain cells for microscopy.

7 Put cells on microscope slide; observe.

8 Photograph, enlarge image of chromosomes.

9 Cut out each chromosome; arrange as a set.

a

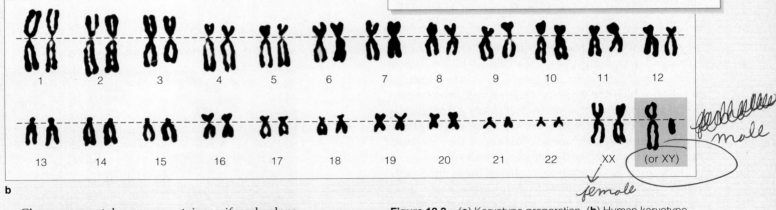

b

Chromosomes take up some stains uniformly along their length, and this allows identification of chromosome size and shape. With other staining procedures, horizontal bands show up along the length of the chromosomes of certain species. If researchers direct a ray of ultraviolet light at these chromosomes, the bands will fluoresce. You can see an example of this in Figure 10.3, which is another human karyotype diagram.

In the last steps of karyotype preparation, metaphase chromosomes are photographed, then the microscope

Figure 12.2 (**a**) Karyotype preparation. (**b**) Human karyotype. Human somatic cells have 22 pairs of autosomes and 1 pair of sex chromosomes (XX or XY). That is a diploid number of 46. These are metaphase chromosomes; each is in the duplicated state.

image is enlarged. The photographed chromosomes are cut apart, one at a time, then arranged according to their size, shape, and length of arms. Finally, all the pairs of homologous chromosomes are horizontally aligned by their centromeres, as shown in Figure 12.2*b*.

Analyzing human cells, as by karyotyping, has yielded evidence that each egg produced by a female carries one X chromosome. Half the sperm cells produced by a male carry an X chromosome and half carry a Y.

If an X-bearing sperm fertilizes an X-bearing egg, the new individual will develop into a female. If the sperm happens to carry a Y chromosome, the individual will develop into a male (Figure 12.3).

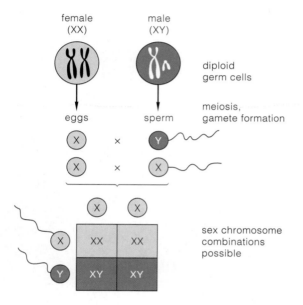

Figure 12.3 Pattern of sex determination in humans.

Among the very few genes on the Y chromosome is a "male-determining gene." As Figure 12.4 indicates, expression of this particular gene leads to the formation of testes, which are primary male reproductive organs. In the gene's absence, ovaries will form automatically. Ovaries are primary female reproductive organs. Testes and ovaries both produce sex hormones that influence the development of particular sexual traits.

A human X chromosome carries more than 2,300 genes. Like other chromosomes, it carries some genes associated with sexual traits, such as the distribution of body fat and hair. However, most of its genes deal with *nonsexual* traits, such as blood-clotting functions. These genes can be expressed in males as well as in females (males, remember, also carry one X chromosome).

A certain gene on the human Y chromosome dictates that a new individual will develop into a male. In the absence of the Y chromosome (and the gene), a female develops.

Figure 12.4 Boys, girls, and the Y chromosome.

For about the first four weeks of its existence, a human embryo has neither male nor female traits, even though it normally carries XY or XX chromosomes. However, ducts and other structures start forming that can go either way.

(**a–c**) In an XX embryo, the primary female reproductive organs (ovaries) start to form automatically—*in the absence of a Y chromosome.* By contrast, in an XY embryo, the primary male reproductive organs (testes) start to form during the next four to six weeks. Apparently, a gene region on the Y chromosome governs a fork in the developmental road that can lead to maleness.

The newly forming testes start to produce testosterone and other sex hormones. These hormones are crucial for the development of a male reproductive system. By contrast, in the XX embryo, the newly forming ovaries start to produce different kinds of sex hormones. These are crucial for the development of a female reproductive system.

The master gene for male sex determination is named SRY (short for *Sex-determining Region of the Y* chromosome). So far, the same gene has been identified in DNA from male humans, chimpanzees, rabbits, pigs, horses, cattle, and tigers. None of the females tested had the gene. Tests with mice indicate that the gene region becomes active about the time that testes are starting to develop.

The SRY gene resembles regions of DNA that are known to specify regulatory proteins. As described in Section 15.1, such proteins bind with certain parts of DNA and thereby turn genes on and off. It seems that the SRY gene product regulates a cascade of reactions that are necessary for male sex determination.

umbilical cord (lifeline between embryo and maternal tissues)

amnion (a protective, fluid-filled sac surrounding the embryo)

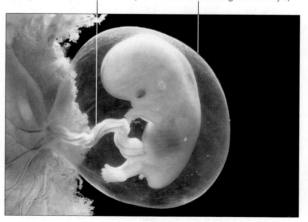

a A human embryo, eight weeks old. Male reproductive organs have already started to develop.

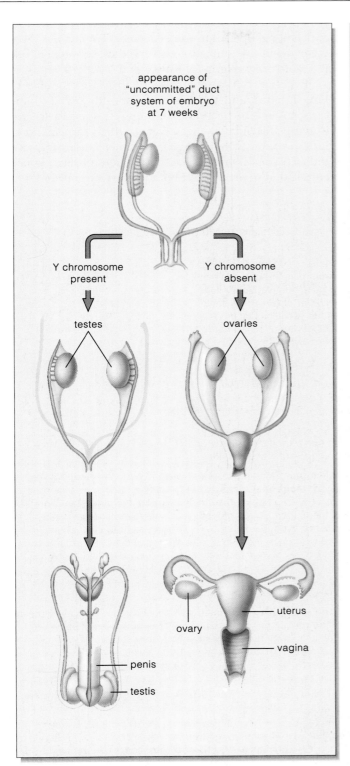

b The duct system in the early human embryo that may develop into the primary male reproductive organs *or* into the primary female reproductive organs.

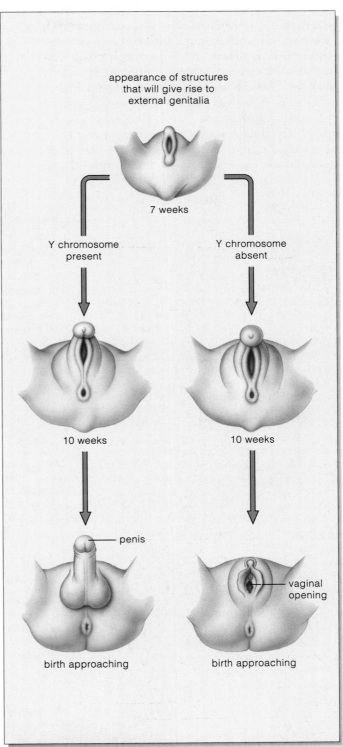

c External appearance of the newly forming reproductive organs in human embryos.

12.4 EARLY QUESTIONS ABOUT GENE LOCATIONS

Linked Genes—Clues to Inheritance Patterns

By the early 1900s, researchers were suspecting that each gene has a specific location on a chromosome. Through hybridization experiments involving mutant fruit flies (*Drosophila melanogaster*), Thomas Hunt Morgan and his coworkers helped confirm this. For example, they found evidence that a gene for eye color and a gene for wing size are located on the *Drosophila* X chromosome. Figure 12.5 describes one series of their experiments.

It seemed, during the early *Drosophila* experiments, that two mutant genes on the X chromosome (*w* for white eyes and *m* for miniature wings) were "linked." That is, they were traveling together during meiosis and ending up in the same gamete. For a time, they were called "sex-linked genes." Today, researchers use the more precise terms **X-linked** and **Y-linked genes**.

Eventually, researchers identified a large number of linked genes, and the ones on a specific chromosome came to be called a **linkage group**. *D. melanogaster*, for example, has four linkage groups, which correspond to its four pairs of homologous chromosomes. Similarly, Indian corn (*Zea mays*) has ten linkage groups, which correspond to ten pairs of homologous chromosomes.

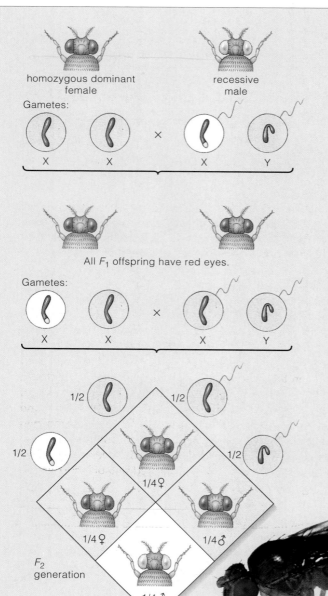

Figure 12.5 X-linked genes as clues to inheritance patterns.

In the early 1900s, the embryologist Thomas Morgan was studying inheritance patterns. He and his coworkers discovered an apparent genetic basis for the connection between gender and certain nonsexual traits. For example, human males and females both have blood-clotting mechanisms. Yet hemophilia, a blood-clotting disorder, shows up most often in males, not females, of a family lineage. This gender-specific outcome was not like anything Mendel saw in his hybrid crosses of pea plants. It made no difference which parent plant carried a recessive allele; the resulting phenotype was the same.

Morgan studied eye color and other nonsexual traits of fruit flies (*Drosophila melanogaster*). These flies can be raised in bottles on cornmeal, molasses, and agar. A female lays hundreds of eggs in a few days, and offspring can themselves reproduce in less than two weeks. Morgan could track hereditary traits through nearly thirty generations of thousands of flies in a year's time.

At first, all flies were wild type for eye color; they had brick-red eyes. Then, as a result of an apparent mutation in a gene controlling eye color, a white-eyed male appeared in one of the bottles.

Morgan established true-breeding strains of white-eyed males and females. Then he did paired, **reciprocal crosses**. (In the first of such paired crosses, one parent displays the trait of interest. In the second, the other parent displays it.) For the first cross, Morgan allowed white-eyed males to mate with homozygous red-eyed females. All F_1 offspring had red eyes. Of the F_2 offspring, however, only some of the males had white eyes. In the second cross, white-eyed females were mated with true-breeding red-eyed males. Half of the F_1 offspring were red-eyed females and half were white-eyed males. Of the F_2 offspring, 1/4 were red-eyed females, 1/4 white-eyed females, 1/4 red-eyed males, and 1/4 white-eyed males!

The seemingly odd results implied a relationship between an eye-color gene and gender. Probably the gene was located on a sex chromosome. But which one? Because females (XX) could be white-eyed, the recessive allele would have to be on one of their X chromosomes. Suppose white-eyed males (XY) also carry the recessive allele on their X chromosome—and suppose there is no corresponding eye-color allele on the Y chromosome. If that were so, then the males would have white eyes, for they have no dominant allele to mask the effect of the recessive one.

The diagram at left illustrates the expected results when Morgan's idea of an X-linked gene is combined with Mendel's concept of segregation. By proposing that one particular gene is located on an X chromosome but not on the Y, Morgan was able to explain the outcome of his reciprocal crosses. His experimental results matched the predicted outcomes.

Image captions within Figure 12.5:
homozygous dominant female × recessive male
Gametes: X X × X Y
All F_1 offspring have red eyes.
Gametes: X X × X Y
1/2 × 1/2
1/2 ... 1/2
1/4 ♀   1/4 ♂
1/4 ♀   1/4 ♂
F_2 generation

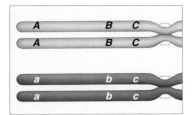 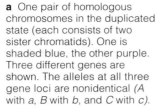

a One pair of homologous chromosomes in the duplicated state (each consists of two sister chromatids). One is shaded blue, the other purple. Three different genes are shown. The alleles at all three gene loci are nonidentical (*A* with *a*, *B* with *b*, and *C* with *c*).

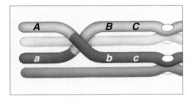

b In prophase I of meiosis, two nonsister chromatids break while tightly aligned. (All four chromatids are shown pulled apart for clarity, as in Figure 8.5.) They swap segments, then enzymes seal the broken ends. The breakage and exchange represent one crossover event.

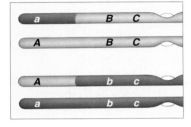

c The outcome of the crossover is genetic recombination between two of four chromatids. (They are shown after meiosis, as unduplicated, separate chromosomes.)

Figure 12.6 Simplified diagram of crossing over. This event occurs in prophase I of meiosis (Figure 8.4).

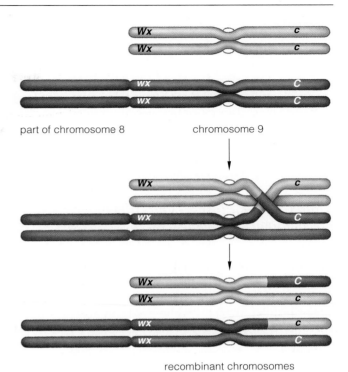

part of chromosome 8   chromosome 9

recombinant chromosomes

Figure 12.7 Harriet Creighton and Barbara McClintock's correlation of a cytological difference with a genetic difference in chromosome 9 from a strain of Indian corn (*Zea mays*).

Crossing Over and Genetic Recombination

If linked genes always stayed together through meiosis, then a dihybrid cross between true-breeding parents should always have a predictable outcome. Specifically, the most frequent phenotypes among the F_2 offspring should be those of the original parents. (Here you may wish to review Section 11.3 and Figure 12.6.) However, a number of puzzling results from the early *Drosophila* experiments did not match this expectation.

For example, Morgan crossed a *Drosophila* female that was recessive for white eyes and miniature wings with a wild-type male (red eyes and long wings). As expected, all F_1 males had white eyes and miniature wings, and all females were wild type. After crossing F_1 flies, Morgan analyzed 2,441 of the F_2 offspring. A significant number had white eyes and long wings or red eyes and miniature wings! As you will see in the next section, 900 (or 36.9 percent) were recombinants. As Morgan hypothesized, physical exchanges must have occurred between X chromosomes during meiosis.

It was not until 1931 that two genetic researchers, Harriet Creighton and Barbara McClintock, discovered evidence of such exchange. They were experimenting with two *Z. mays* chromosomes that differed physically in a way that could be distinguished with a microscope. Such distinguishable features are **cytological markers** for the genes being studied.

The researchers used corn that was heterozygous for two genes on chromosome 9. Two alleles of one gene specify colored (*C*) or colorless (*c*) seeds. One allele of the other gene specifies the synthesis of two forms of starch (*Wx*), and the other allele specifies only one (*wx*). One chromosome 9 that was used in an experiment had genotype *cWx* and a normal appearance (Figure 12.7). Another chromosome 9 had genotype *Cwx* and was longer. An abnormal event caused a piece of a different chromosome to become attached to one of its ends.

During meiosis, crossing over sometimes occurred between the two gene loci on a pair of the cytologically different chromosomes. The outcome was two kinds of genetic recombinants: *CWx* and *cwx*. As Creighton and McClintock realized, when genetic recombination had occurred, cytological features of the chromosomes also had changed. When *wx* ended up with *c* instead of *C*, so did the extra length. They observed no such correlation in F_1 generations that retained the parental genotype.

Genes on the same chromosome belong to the same linkage group and do not assort independently during meiosis. Crossing over between homologous chromosomes disrupts gene linkages and results in the production of nonparental combinations of genes in chromosomes.

Correlations between specific genes and cytological markers provide evidence of genetic recombination.

RECOMBINATION PATTERNS AND CHROMOSOME MAPPING

As we now know, crossing over is not a rare event. In fact, for humans and most other eukaryotic species, meiosis cannot even be completed properly unless each pair of homologous chromosomes takes part in at least one crossover. The preceding section briefly described experimental evidence of this remarkable event. Let us now consider a few specific examples of the ways in which crossing over disrupts gene linkages and what researchers do with this information.

How Close Is Close? A Question of Recombination Frequencies

Figure 12.8 shows Morgan's cross between a wild-type *Drosophila* male and a female that was recessive for white eyes and for miniature wings. Again, the two parental genotypes showed up equally among the F_1 offspring (50 percent white-eyed, miniature-winged males and 50 percent wild-type females). Of the 2,441 F_2 offspring analyzed, 36.9 percent were recombinants.

Morgan's group extended their investigations to other X-linked genes. In one series of experiments, a true-breeding white-eyed, yellow-bodied mutant female was crossed with a wild-type male (with red eyes and a gray body). As expected, 50 percent of the F_1 offspring showed one or the other parental phenotype. In this case, however, only 129 of 2,205 of the F_2 offspring that were analyzed—or 1.3 percent—were recombinants.

From the results of many such experimental crosses, it appeared that the genes controlling eye color and wing size were not as "tightly linked" as those for eye color and body color. Also, the two parental genotypes were represented in approximately equal numbers of the F_2 offspring, and the same was true of the two kinds of recombinant genotypes. What was going on here? As Morgan hypothesized, certain alleles tend to remain together during meiosis more often than others *because they are positioned closer together on the same chromosome.*

Imagine any two genes at two different locations on the same chromosome. *The probability that a crossover*

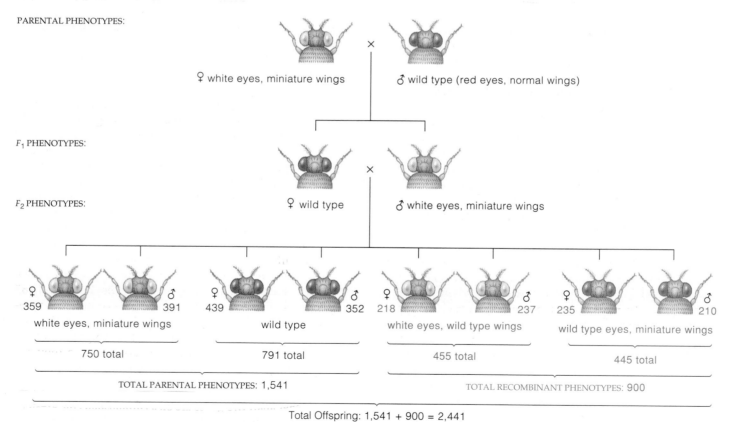

PARENTAL PHENOTYPES:

♀ white eyes, miniature wings   ×   ♂ wild type (red eyes, normal wings)

F_1 PHENOTYPES:

F_2 PHENOTYPES:

♀ wild type   ×   ♂ white eyes, miniature wings

| ♀ 359 ... 391 ♂ | ♀ 439 ... 352 ♂ | ♀ 218 ... 237 ♂ | ♀ 235 ... 210 ♂ |

white eyes, miniature wings   wild type   white eyes, wild type wings   wild type eyes, miniature wings

750 total   791 total   455 total   445 total

TOTAL PARENTAL PHENOTYPES: 1,541   TOTAL RECOMBINANT PHENOTYPES: 900

Total Offspring: 1,541 + 900 = 2,441

PERCENT RECOMBINANTS: 900/2,441 × 100 = 36.9%

Figure 12.8 Experimental crosses with *Drosophila melanogaster* that provided indirect evidence of how crossing over disrupts linkage groups. In this example, Morgan's group tracked a mutant gene for white eyes and a different mutant gene for miniature wings. As you probably know, the symbol ♀ signifies female and the symbol ♂ signifies male.

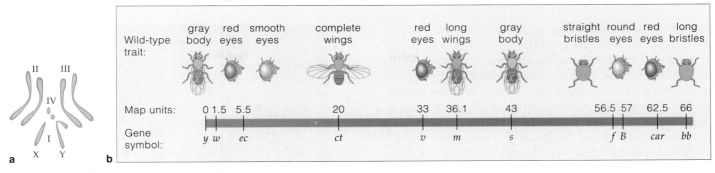

Figure 12.9 (**a**) *Drosophila melanogaster* chromosomes. (**b**) A linkage map for a number of genes on chromosome I (here, the X chromosome). As you can see, more than one gene can influence the same trait, such as eye color.

will disrupt their linkage is proportional to the distance that separates them. Suppose genes *A* and *B* are twice as far apart as two other genes, *C* and *D*:

We would expect crossing over to disrupt the linkage between *A* and *B* much more often.

Two genes are very closely linked when the distance between them is small; their allelic combinations nearly always end up in the same gamete. Gene linkage is more vulnerable to crossover if the distance between genes is greater. When two genes are very far apart, crossing over disrupts linkage so often that those genes assort independently of each other into gametes.

Linkage Mapping

After using testcrosses to analyze crossover patterns, Alfred Sturtevant, one of Morgan's students, proposed that the percentage of recombinants in gametes might be a quantitative measure of the relative positions of genes along a chromosome. This investigative approach is now called **linkage mapping**.

The relative, linear distance between any two linked genes on a genetic map is expressed in map units. One map unit corresponds to a crossover frequency of 1 percent, twenty map units correspond to a crossover frequency of 20 percent, and so on. For instance, take a look at Figure 12.9, which is a linkage map for one of the *D. melanogaster* chromosomes. The amount of gene recombination to be expected between "smooth eyes" and "long wings" would be 30.6 percent (5.5 map units subtracted from 36.1 map units).

Linkage maps do not show *actual* physical distances between genes. The most accurate approximations have been calculated only for closely linked genes. Why? As the map distance increases, multiple crossovers occur and skew recombination frequencies. Even so, the *map distance* between two genes as estimated by genetic crosses generally correlates with the *physical distance*, or length of DNA between them, as calculated by other methods. Exceptions to this generalization are known to occur in the centromere region.

Of the several thousand known genes in the four types of *Drosophila* chromosomes, the positions of about a thousand have been mapped. What about the 50,000 to 100,000 genes on the twenty-three types of human chromosomes? Unlike fruit flies, humans do not lend themselves to experimental crosses. Nevertheless, some tight gene linkages have been identified by tracking the resulting phenotypes, one generation after another, in certain families.

For example, *color blindness* and a blood-clotting disorder, *hemophilia*, are disorders that are caused by recessive alleles at two gene loci on the X chromosome. One female carried both alleles, although she herself was symptom-free. So was her father, so she must have inherited a normal X chromosome from him (remember, males have only one X and one Y). The X chromosome she inherited from her mother must have carried both mutant alleles. The female gave birth to six sons. Three sons developed color blindness and hemophilia, and two were unaffected. Here was phenotypic evidence that recombination had not occurred. But her sixth son started life as a fertilized, recombinant egg; he was color-blind only. Thus, for the two mutant alleles in this family, the recombination frequency is 1/6 (or 0.167 percent). Many such affected families would have to be examined to get a good estimate of the map distance between the two genes. Generally, these genes do not have high recombination frequencies because they are very closely linked at one end of the X chromosome.

The farther apart two genes are on a chromosome, the greater will be the frequency of crossing over and therefore of genetic recombination between them.

Linkage mapping is a method of measuring the relative, linear distances between genes on the same chromosome. Such maps roughly correspond to actual physical distances, or lengths of DNA, as calculated by other methods.

HUMAN GENETIC ANALYSIS

Some organisms, including pea plants and fruit flies, are ideal for genetic analysis. They grow and reproduce rapidly in small spaces, under controlled conditions. It does not take very long to track a trait through many generations. Humans are another story. We live under variable conditions in diverse environments. We select our own mates and reproduce if and when we want to. Humans live as long as the geneticists who study them, so tracking traits through generations can be tedious. Most human families are so small, there are not enough offspring for easy inferences about inheritance.

Constructing Pedigrees

To get around the problems associated with analyzing human inheritance, geneticists put together pedigrees. A **pedigree** is a chart that shows genetic connections among individuals. Genetic researchers construct them by using standardized methods and definitions, as well as standardized symbols to represent individuals, as shown in Figure 12.10.

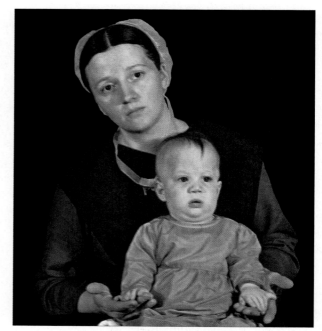

An affected child with six digits on each hand

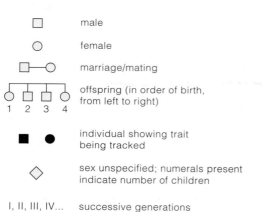

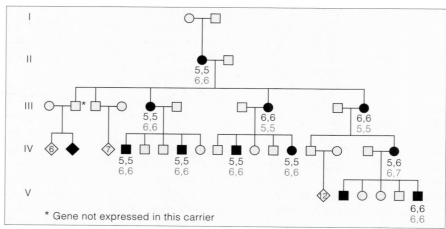

a Standardized symbols in pedigrees

b Pedigree for a family in which polydactyly recurs

Figure 12.10 (**a**) Some of the symbols used in constructing pedigrees. (**b**) One pedigree for *polydactyly*, a condition in which a person has extra fingers, extra toes, or both. Expression of the gene for this trait can vary from one individual to the next. *Black* numerals signify the number of fingers on each hand where data were available; *blue* numerals signify the number of toes on each foot.

When analyzing pedigrees, geneticists rely on their knowledge of probability and of Mendelian inheritance patterns, which might yield clues to the genetic basis for a trait. For instance, they might determine that the responsible allele is dominant or recessive, or that it is located on an autosome or a sex chromosome.

Gathering a great many family pedigrees increases the numerical base for analysis. When a trait shows a simple Mendelian inheritance pattern, a geneticist may have confidence in predicting the probability of its occurrence among the children of prospective parents. We will return to this topic later in the chapter.

Regarding Human Genetic Disorders

Table 12.1 lists some of the heritable traits that have been studied in detail. A few of these are abnormalities, or deviations from the average condition. Said another way, a **genetic abnormality** is nothing more than a rare, uncommon version of a trait, as when a person is born

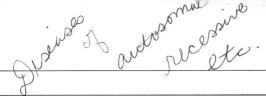
Diseases of autosomal recessive etc.

Table 12.1 Examples of Human Genetic Disorders and Genetic Abnormalities

Disorder or Abnormality*	Main Consequences	Disorder or Abnormality*	Main Consequences
AUTOSOMAL RECESSIVE INHERITANCE		**X-LINKED DOMINANT INHERITANCE**	
Albinism 11.6; CT 11.9	Absence of pigmentation	Faulty enamel trait 12.8	Problems with teeth
Blue offspring CT 18.9	Bright blue skin coloration	**X-LINKED RECESSIVE INHERITANCE**	
Cystic fibrosis CT 4.12	Excessive glandular secretions leading to tissue, organ damage	Color blindness 12.5, 36.6	Inability to distinguish among all or some colors of the spectrum of visible light
Ellis-van Creveld Syndrome 18.8	Extra fingers, toes, short limbs	Duchenne muscular dystrophy 12.8 CT 15.7	Muscles waste away
Galactosemia 12.7	Brain, liver, eye damage	Hemophilia, 11.5, 12.8	Impaired blood-clotting ability
Phenylketonuria (PKU) 6.2; 12.11	Mental retardation	Testicular feminization syndrome 37.2	XY individual but having some female traits, sterility
Sickle-cell anemia 11.5; 14.1; 18.6	Adverse plleiotropic effects on organs throughout body	X-linked anhidrotic ectodermal dysplasia 15.4	In human females, mosaic patches of skin with or without sweat glands
AUTOSOMAL DOMINANT INHERITANCE		**CHANGES IN CHROMOSOME NUMBER**	
Achondroplasia 12.7	One form of dwarfism	Down syndrome 12.9	Mental retardation, heart defects
Achoo syndrome CT 11.9	Chronic sneezing	Klinefelter syndrome 12.9	Sterility, retardation
Amyotrophic lateral sclerosis (ALS) 12.8	Loss of all muscle function	Turner syndrome 12.9	Sterility; abnormal ovaries, abnormal sexual traits
Camptodactyly 11.7	Rigid, bent little fingers	XYY condition 12.9	Mild retardation or free of symptoms
Familial hypercholesterolemia CI 16	High cholesterol levels in blood; eventually clogged arteries	**CHANGES IN CHROMOSOME STRUCTURE**	
Huntington disorder 12.7	Nervous system degenerates progressively, irreversibly	Cri-du-chat syndrome 12.10	Mental retardation, abnormally formed larynx
Polydactyly 12.6	Extra fingers, toes, or both	Fragile X syndrome 12.10	Mental retardation
Progeria CI12; 12.7	Drastic premature aging		
Tay-Sachs disorder 12.7	Progressive deterioration of the nervous system		

*Italic numbers indicate sections in which a disorder is described. CI signifies Chapter Introduction. CT signifies an end-of-chapter Critical Thinking question.

with six toes on each foot instead of five. Whether an individual or society at large views an abnormal trait as disfiguring or merely interesting is subjective. As the classic novel *The Hunchback of Notre Dame* so clearly emphasized, there is nothing inherently life-threatening or even ugly about it.

By comparison, a **genetic disorder** is an inherited condition that sooner or later causes mild to severe medical problems. The alleles underlying severe genetic disorders do not abound in populations, for they put individuals at great risk. Why do they not disappear entirely? There are two reasons. First, rare mutations introduce new copies of the alleles into the population. Second, in heterozygotes, the harmful allele is paired with a normal one that might cover its functions, so it still can be passed on to offspring.

You may hear someone refer to a genetic disorder as a disease, but the terms are not always interchangeable. A disease results from infection by bacteria, viruses, or some other environmental agent that invades the body and multiplies inside its tissues. Illness follows only if the infection leads to tissue damage that interferes with normal body functions. When a person's genes increase susceptibility to infection or weaken the response to it, the resulting illness might be called a **genetic disease**.

With these qualifications in mind, we will turn next to examples of inheritance in the human population. As you will see, genetic analyses of family pedigrees have often revealed simple Mendelian inheritance patterns for certain traits. Researchers have traced many of the traits to dominant or recessive alleles on an autosome or X chromosome. They have traced others to changes in the structure or number of chromosomes.

For many genes, pedigree analysis might reveal simple Mendelian inheritance patterns that will allow inferences about the probability of their transmission to children.

A genetic abnormality simply is a rare or less common version of an inherited trait. A genetic disorder is an inherited condition that results in mild to severe medical problems.

Autosomal Recessive Inheritance

For some traits, inheritance patterns reveal two clues that point to a recessive allele on an autosome. *First,* if both parents are heterozygous, any child of theirs will have a 50 percent chance of being heterozygous and a 25 percent chance of being homozygous recessive, as Figure 12.11 indicates. *Second,* if the parents are both homozygous recessive, any child of theirs will be, also.

About 1 in 100,000 newborns are homozygous for a recessive allele that causes *galactosemia.* They cannot produce functional molecules of an enzyme that stops a product of lactose breakdown from accumulating to toxic levels. Lactose normally is converted to glucose and galactose, then to glucose-1-phosphate (which is broken down by glycolysis or converted to glycogen). In affected persons, the full conversion is blocked:

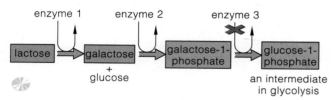

A high blood level of galactose can damage the eyes, liver, and brain. Malnutrition, diarrhea, and vomiting are early symptoms. A high galactose level, the telling symptom, can be detected in urine samples. Untreated galactosemics often die in childhood. But if affected individuals are placed on a restricted diet that excludes dairy products, they can grow up symptom-free.

People who are homozygous for another autosomal recessive allele develop *Tay-Sachs disorder.* They appear normal at birth, but their brain and spinal cord start to deteriorate before they reach their first birthday. Mental retardation, blindness, and loss of neural and muscle function follow. Affected children usually die between ages three and four. The mutant allele responsible for the disorder causes a deficiency in an enzyme that is necessary for the metabolism of sphingolipids, a type of lipid that is an especially abundant component of the plasma membrane of cells in nerves and the brain.

Autosomal Dominant Inheritance

Two clues of a different sort indicate that an autosomal dominant allele is responsible for a trait. *First,* the trait typically appears in each generation, for the allele is usually expressed even in heterozygotes. *Second,* if one parent is heterozygous and the other is homozygous recessive, there is a 50 percent chance that any child of theirs will be heterozygous (Figure 12.12).

A few dominant alleles cause severe disorders, yet they persist in populations. Some are perpetuated by

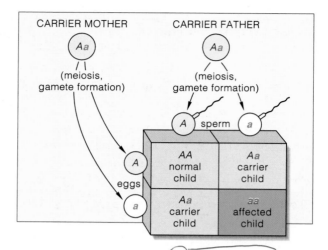

Figure 12.11   One pattern for autosomal recessive inheritance. In this example, both parents are heterozygous carriers of the recessive allele (shown in *red*).

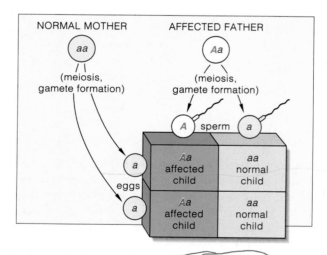

Figure 12.12   One pattern for autosomal dominant inheritance. In this example, the dominant allele (*red*) is fully expressed in carriers.

spontaneous mutations. This is true of progeria, the rare aging disorder described in the chapter introduction. In other cases, expression of the dominant allele may not interfere with reproduction, or the affected individuals have children before symptoms become severe.

For example, Huntington disorder is a condition characterized by a gradual increase in involuntary movements, a progressive deterioration of the nervous system, and eventual death. The symptoms may not

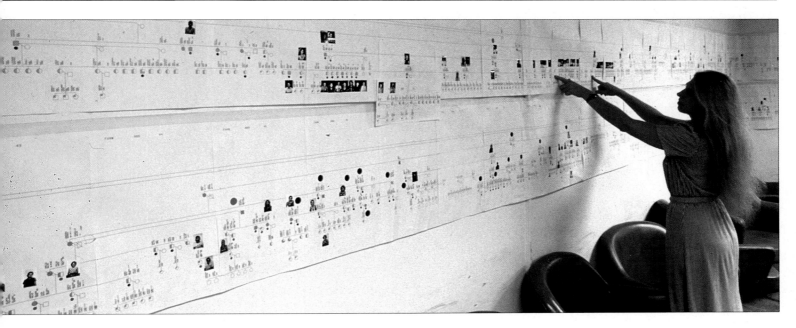

Figure 12.13 From the human genetic researcher Nancy Wexler, a pedigree for Huntington disorder, by which the nervous system progressively degenerates. Wexler's team constructed an extended family tree for nearly 10,000 people in Venezuela. Wexler has a special interest in Huntington disorder; she herself has a 50 percent chance of developing it.

Figure 12.14 A painting by Velázquez of Infanta Margarita Teresa of the Spanish court and her maids, including the achondroplasic woman at the far right.

even start to show up until the affected individual is past age thirty. For most people, reproduction already has occurred by then. Affected individuals usually die in their forties or fifties, before they may realize that they have passed on the mutant allele to their children.

Analysis of affected and unaffected individuals in one notably extended family in Venezuela revealed that a dominant allele on human chromosome 4 is the genetic culprit responsible for Huntington disorder. Part of the pedigree for this family is shown in Figure 12.13.

As another example, *achondroplasia* affects about 1 in 10,000 people. The homozygous dominant condition commonly leads to stillbirth. Heterozygotes are able to reproduce. However, while they are young and their limb bones are forming, the cartilage components of those bones cannot form properly. As an outcome, at maturity, affected individuals have abnormally short arms and legs, relative to their other body parts. Figure 12.14 shows a famous example. Adult achondroplasics are less than 4 feet, 4 inches tall. The dominant allele often has no other phenotypic effects than this.

Most individuals affected by an autosomal recessive disorder have heterozygous, symptom-free parents. If both parents are homozygous for the allele, all of their children will be, also.

Most often, an autosomal dominant disorder does not skip generations, because the responsible allele is expressed even in heterozygotes. If one parent is heterozygous and the other is homozygous recessive, each child they may produce has a 50 percent chance of being heterozygous.

Genetic disorders arising from a pair of autosomal dominant alleles are rare in populations. They persist because of rare, spontaneous mutations or because they do not interfere with reproduction.

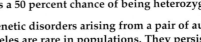

An X- or Y-linked trait, recall, arises from expression of one or more genes on the X or Y chromosome. In most cases, evidence for Y-linked inheritance is questionable or nonexistent. Better documentation exists for X-linked inheritance patterns, a few of which will be described in the paragraphs to follow.

X-Linked Recessive Inheritance

Distinctive clues often show up when a recessive allele on an X chromosome causes a trait. First, males show the recessive phenotype far more often than females do. A recessive allele can be masked in females, who may inherit a dominant allele on their other X chromosome. The allele cannot be masked in males, who have only one X chromosome (Figure 12.15). Second, a son cannot inherit the recessive allele from his father. A daughter can. If a daughter is heterozygous, there is a 50 percent chance each son of hers will inherit the allele.

Hemophilia A, a blood-clotting disorder, is a case of X-linked recessive inheritance. In most people, a blood-clotting mechanism quickly stops bleeding from minor injuries. Reactions that result in clot formation require the products of several genes. If any of the genes is mutated, its defective product will allow bleeding to proceed for an abnormally long time. About 1 in 7,000 males inherits the gene for hemophilia A. Clotting time is close to normal in heterozygous females.

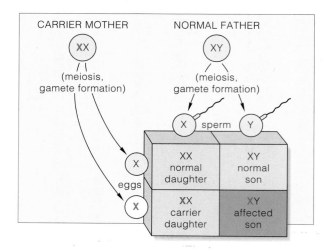

Figure 12.15 One pattern for X-linked inheritance. Assume the mother carries a recessive allele on one X chromosome (*red*).

The frequency of hemophilia A was unusually high in royal families of nineteenth-century Europe. Queen Victoria of England was a carrier (Figure 12.16). At one time, the recessive allele was present in eighteen of her sixty-nine descendants. A hemophilic great-grandchild, Crown Prince Alexis, was a focus of political intrigue that helped trigger the Russian Revolution of 1917.

Another, rare case of X-linked recessive inheritance is *Duchenne muscular dystrophy*. Muscles slowly enlarge

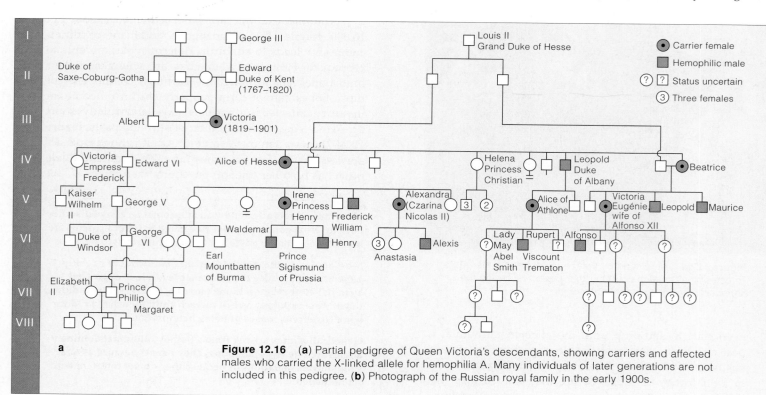

Figure 12.16 (**a**) Partial pedigree of Queen Victoria's descendants, showing carriers and affected males who carried the X-linked allele for hemophilia A. Many individuals of later generations are not included in this pedigree. (**b**) Photograph of the Russian royal family in the early 1900s.

with fat and connective tissue deposits, but the muscle cells atrophy (waste away). At first, children with this form of dystrophy appear normal. Between the second and tenth birthdays, muscle weakening makes them progressively more imbalanced and clumsier. After they are twelve years old, affected persons no longer can walk. In time, the chest and head muscles deteriorate. Typically they die of respiratory failure during their early twenties. At present, there is no cure.

X-Linked Dominant Inheritance

The *faulty enamel trait* is one of the few known cases of X-linked dominant inheritance. With this disorder, the hard, thick enamel coating that normally protects teeth fails to develop properly (Figure 12.17). The inheritance pattern is like that for X-linked recessive alleles, except the allele is also expressed in heterozygous females. The phenotype tends to be less pronounced in females than in males. A son cannot inherit the dominant allele for the trait from an affected father, but all daughters will. If a woman is heterozygous, she will transmit the allele to half of her offspring, regardless of their sex.

A Few Qualifications

Don't take the preceding examples of autosomal and X-linked traits too seriously. We include them not to turn

 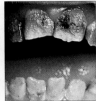

Figure 12.17 The discolored, abnormal tooth enamel of two individuals affected by the faulty enamel trait.

you into an armchair geneticist, but rather to give you a general idea of the kinds of clues that hold meaning for trained geneticists. Before diagnosing a case, geneticists often find it necessary to pool together many pedigrees. Typically they make detailed analyses of clinical data and keep abreast of research. Why? More than one type of gene might be responsible for a given phenotype. Geneticists already know of dozens of conditions that arise from a mutated gene on an autosome *or* a mutated gene on the X chromosome. They know of autosomal genes that show dominance in males and recessiveness in females, so the traits appear to be due to X-linked recessive inheritance even though they are not.

Also, don't assume that errant genes automatically relegate a person to life's sidelines. More than twenty years ago, Stephen Hawking noticed his muscles were weakening. In time he had trouble swallowing and speaking. Motor neurons in his brain and spinal cord were deteriorating; scar tissue was forming along his spinal cord. His muscles could not get proper signals from the nervous system and they began to waste away. Such are the symptoms of *amyotrophic lateral sclerosis*, or ALS. The disorder, which killed baseball's Lou Gehrig, affects about 1 in 1,000 people. Possibly a virus issues the death warrant, but a mutated allele of an autosomal dominant gene, EAAT2, may be the executioner. The gene specifies a membrane transport protein that is supposed to sponge up excess glutamate in the brain and spinal cord. Glutamate serves as a communication signal between nerve cells (neurons) that control the body's muscles. In excess amounts, it kills the cells.

Hawking's affliction has not immobilized his mind, which continues to dance freely around some of the most challenging questions we humans have dared to ask. Through his acclaimed research in astrophysics and his eloquent yet accessible writings, Hawking has changed the way we view time and the universe.

b The Russian royal family members. All are believed to have been executed near the end of the Russian Revolution. They were recently exhumed from their hidden graves, but DNA fingerprinting indicates the remains of Alexis and one daughter, Anastasia, are not among them.

Czarina Alexandra (a carrier; descendant of Queen Victoria)

Czar Nikolas II (free of allele for hemophilia A)

Alexis (hemophilic son)

A recessive X-linked allele may be masked in heterozygous females but not in males. A father can pass on the allele to daughters but not sons. A heterozygous daughter has a 50 percent chance of passing on the allele to any son of hers.

An X-linked dominant allele is expressed each generation, even in heterozygotes. If one parent is heterozygous and the other is homozygous recessive, there is a 50 percent chance that any child of theirs will be heterozygous.

CHANGES IN CHROMOSOME NUMBER

Occasionally, abnormal events occur before or during cell division, then gametes and new individuals end up with the wrong chromosome number. The consequences range from minor to lethal physical changes.

Categories and Mechanisms of Change

With **aneuploidy**, individuals have one extra or one less chromosome. This condition is a major cause of human reproductive failure. Quite possibly it affects half of all fertilized eggs. Autopsies show that most of the human embryos that were miscarried (spontaneously aborted before pregnancy reached full term) were aneuploids.

With **polyploidy**, individuals have three or more of each type of chromosome. About one-half of all species of flowering plants are polyploid (Section 19.4). So are some insects, fishes, and other animals. Polyploidy is lethal for humans. It may disrupt interactions between the genes of autosomes and sex chromosomes at key steps in the complex pathways of development and reproduction. All but 1 percent of human polyploids die before birth. The rare newborns die within a month.

Chromosome numbers can change during mitotic or meiotic cell divisions. Suppose a cell cycle goes through DNA duplication and mitosis but is arrested before the cytoplasm divides. The cell is *tetra*ploid, with four of each type of chromosome. Suppose one or more pairs of chromosomes fail to separate in mitosis or meiosis, an event called **nondisjunction**. Some or all forthcoming cells will have too many or too few chromosomes, as in the example in Figure 12.18. For certain experiments, developmental biologists commonly expose living cells to colchicine to induce nondisjunction (Section 12.2).

The chromosome number also may become changed at fertilization. Visualize a normal gamete uniting by chance with an $n + 1$ gamete (one extra chromosome). The new individual will be "trisomic" ($2n + 1$); it will have three of one type of chromosome and two of every other type. What if the other gamete is $n - 1$? In that case, the new individual will be "monosomic" ($2n - 1$).

Change in the Number of Autosomes

Most changes in the number of autosomes arise through nondisjunction during gamete formation. Let's consider one of the most common of the resulting disorders. A trisomic 21 newborn, with three chromosomes 21, will show the effects of *Down syndrome*. The symptoms vary greatly, but most affected individuals show moderate to severe mental impairment. About 40 percent develop heart defects. Abnormal development of the skeleton means older children have shortened body parts, loose joints, and poorly aligned bones of the hips, fingers, and toes. Muscles and muscle reflexes are weaker than normal, and speech and other motor skills develop slowly. With special training, trisomic 21 individuals often take part in normal activities (Figure 12.19). As a group, they are cheerful, affectionate people who derive great pleasure from socializing.

Down syndrome is one of many disorders that can be detected by prenatal diagnosis (Section 12.11). Before detection procedures were widespread, about 1 in 700 newborns of all ethnic groups were trisomic 21. Now the number is closer to 1 in 1,100. The risk is greater if pregnant women are more than thirty-five years old, as indicated by the graph in Figure 12.19b.

Figure 12.18 Nondisjunction. In this example, a pair of chromosomes fails to separate at anaphase I of meiosis. The chromosome number changes in gametes. (Make a sketch of nondisjunction at anaphase II. What will the chromosome numbers be in gametes?)

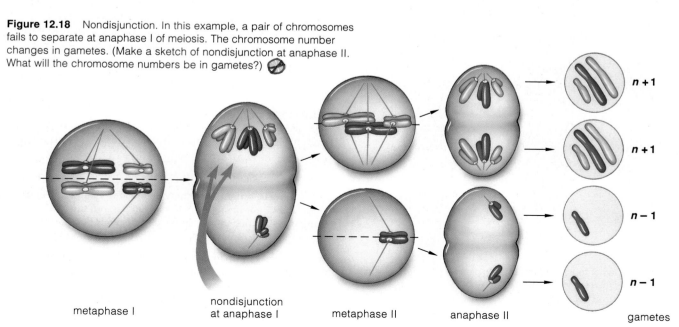

metaphase I nondisjunction at anaphase I metaphase II anaphase II gametes

$n + 1$
$n + 1$
$n - 1$
$n - 1$

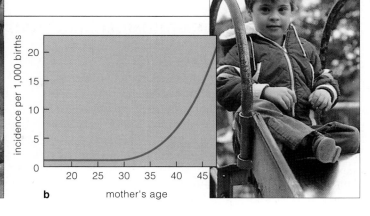

Figure 12.19 Down syndrome. (**a**) Karyotype revealing a trisomic 21 condition (*red* arrows). (**b**) Relation between the frequency of Down syndrome and mother's age at the time of childbirth. Results are from a study of 1,119 affected children who were born in Victoria, Australia, between 1942 and 1957. The young lady above was a lively participant in the Special Olympics, held annually in San Mateo, California.

Change in the Number of Sex Chromosomes

Most sex chromosome abnormalities arise as a result of nondisjunction during meiosis and gamete formation. Let's look at a few phenotypic outcomes.

TURNER SYNDROME Inheriting one X chromosome without a partner X or Y chromosome gives rise to *Turner syndrome*, which affects 1 in 2,500 to 10,000 or so newborn girls. Nondisjunction affecting sperm accounts for 75 percent of the cases. We see fewer people with Turner syndrome compared to people having other sex chromosome abnormalities. The likely reason is that at least 98 percent of all X0 zygotes spontaneously abort early in pregnancy. Approximately 20 percent of all spontaneously aborted embryos in which chromosome abnormalities were detected have been X0.

Despite the near-lethality, X0 survivors are not as disadvantaged as other aneuploids. They grow up well proportioned, albeit short—4 feet, 8 inches tall, on the average. Generally, their behavior is normal during childhood. But most Turner females are infertile. They do not have functional ovaries and so cannot produce eggs or sex hormones. Without sex hormones, breast enlargement and the development of other secondary sexual traits are reduced. Possibly as a result of their arrested sexual development and small size, females often become passive and are easily intimidated by peers during their teens. Some patients have benefited from hormone therapy and corrective surgery.

KLINEFELTER SYNDROME One out of every 500 to 2,000 liveborn males has two X chromosomes and one Y chromosome. This XXY condition results mainly from nondisjunction in the mother (about 67 percent of the time, compared to 33 percent in the father). Symptoms of the resulting *Klinefelter syndrome* develop after the onset of puberty.

XXY males are taller than average and are sterile or nearly so. Their testes usually are much smaller than average; the penis and scrotum are not. Facial hair is often sparse, and breasts may be somewhat enlarged. Injections of the hormone testosterone can reverse the feminized traits but not the low fertility. Some XXY males show mild mental impairment, although many fall within the normal range of intelligence. Except for low fertility, many show no outward symptoms at all.

XYY CONDITION About 1 in 1,000 males has one X and two Y chromosomes. *XYY males* tend to be taller than average. Some may be mildly retarded, but most are phenotypically normal. At one time, XYY males were thought to be genetically predisposed to become criminals. The erroneous conclusion was based on small numbers of cases in highly selected groups, such as prison inmates. Investigators often knew who the XYY males were, and this may have biased their evaluations. There were no **double-blind studies**, by which different investigators gather data independently of one another and then match them up only after both sets of data are completed. In this case, the same investigators gathered the karyotypes and the personal histories. Fanning the stereotype was a sensationalized report in 1968 that a mass-murderer of young nurses was XYY. He wasn't.

In 1976, a Danish geneticist issued a report on a large-scale study based on records of 4,139 tall males, twenty-six years old, who had reported to their draft board. Besides giving results of physical examinations and intelligence tests, the records provided clues to socioeconomic status, educational history, and any criminal convictions. Only twelve of the males were XYY, which left more than 4,000 for the control group. The only significant finding was that tall, mentally impaired males who engage in criminal activity are more likely to get caught—irrespective of karyotype.

Most changes in chromosome number arise as a result of nondisjunction during meiosis and gamete formation.

CHANGES IN CHROMOSOME STRUCTURE

Rare genetic disorders or abnormalities also arise when the physical structure of a chromosome changes. Such aberrations often occur spontaneously in nature. They also are induced in the laboratory through exposure to chemical substances or irradiation. Either way, they often can be detected through microscopic examination and karyotype analysis of cells undergoing mitosis or meiosis.

For our purposes, we can limit our review to four major categories of chromosome aberrations that may have serious or even lethal consequences for the human individual:

1. With a **deletion**, a segment of a chromosome is lost.

2. With **duplication**, the same linear stretch of DNA within a chromosome is repeated, often several to many times in the same chromosome or a different one.

3. With an **inversion**, a linear stretch of DNA within a chromosome becomes oriented in the reverse direction, with no molecular loss.

4. With **translocation**, a stretch of one chromosome's DNA physically moves to another location in the same chromosome or a different one, with no molecular loss.

Deletion

Viral attack, irradiation (especially ionizing radiation), chemical assaults, or some other environmental factor can trigger a deletion, or the loss of some segment of a chromosome (Figure 12.20).

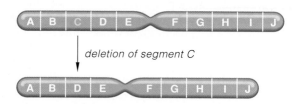

Figure 12.20 Simple diagram of a chromosome deletion.

Any chromosome region is vulnerable to deletion. The loss may even occur at one end, which makes the chromosome unstable. Wherever it happens, a deletion permanently excises one or more of the chromosome's genes. The loss of genetic information usually causes problems for the individual.

For example, one deletion from human chromosome 5 results in mental retardation and the development of an abnormally shaped larynx. When affected infants cry, the sounds they produce are like a cat meowing. Hence the name of the disorder, *cri-du-chat* (cat-cry). Figure 12.21 shows an affected child.

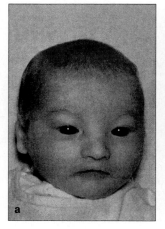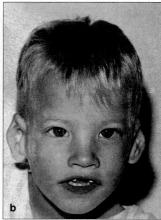

Figure 12.21 Cri-du-chat syndrome. (**a**) A male infant who will develop the cri-du-chat syndrome as a result of a deletion on the short arm of chromosome 5. The photograph was taken just after birth. Notice the ears positioned low on the side of the head relative to the eyes. (**b**) The same boy, photographed four years later. By this age, affected individuals no longer make the mewing sounds typical of the syndrome.

You might be thinking that species with diploid cells have an advantage when deletions do occur. After all, the presence of the same segment on the homologous chromosome may mediate the effect of the loss. But the sword of chance cuts both ways. If a missing gene's partner on the homologous chromosome is normal, all well and good. If it is a harmful recessive allele, nothing will mask or compensate for *its* effects.

Duplications

Even normal chromosomes contain duplications, which are gene sequences that are repeated several to many times. Figure 12.22 shows different forms that such a duplication can take. Often the same gene sequence has been repeated thousands of times.

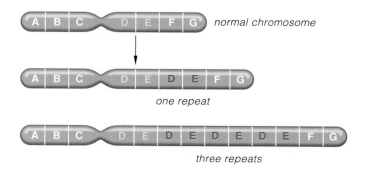

Figure 12.22 Simple diagrams of chromosome duplications.

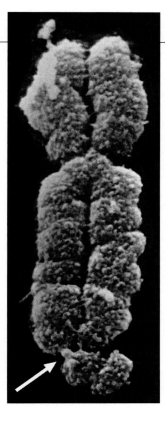

Figure 12.23 Indirect, cytological evidence of the fragile X syndrome. An abnormal constriction in a certain segment of the long arm of the X chromosome (*arrow*) is associated with this genetic disorder.

Consider the *fragile X syndrome*, which affects 1 of every 1,500 males and 1 of every 2,500 females. The disorder correlates with multiple copies of a nucleotide sequence in one gene (FMR-1) on the X chromosome. Nearly all people have about 29 of the repeats; however, affected people have 700 or even more. When the number of repeats exceeds a certain threshold, gene expression apparently gets reduced. In cultured human cells, most X chromosomes with the unstable array of repeats have an odd constricted region (Figure 12.23).

Duplications that have harmful or lethal effects tend not to be conserved over the course of evolution. An individual's growth, development, and maintenance activities all depend on intricate interactions among a number of gene products, so mutations that drastically alter genes are usually selected against. Even so, many other duplications apparently have done their bearers no harm. Although they are relatively rare events, they have accumulated over millions, even billions of years and are now built into the DNA of all species.

Biologists speculate that duplicates of some gene sequences with neutral effects could have an adaptive advantage. In effect, a copy could free up a gene for chance mutations that might turn out to be useful, for the normal gene would still issue the required product. One or more duplicated gene sequences could become slightly modified, and then the products of those genes could function in slightly different or new ways.

Several kinds of duplicated, modified genes seem to have had pivotal roles in evolution. Consider the gene regions for the polypeptide chains of the hemoglobin molecule, as shown in Section 3.7. In humans and other primates, these regions contain copies of remarkably similar gene sequences. They produce whole families of polypeptide chains with slight structural differences. Each of the resulting hemoglobin molecules performs transport functions with slightly different efficiencies, depending on cellular conditions.

Inversion

During an inversion, a segment that loops out from a chromosome becomes reinserted at the same place, but in the reverse order. The reversal alters the position and order of the chromosome's genes (Figure 12.24).

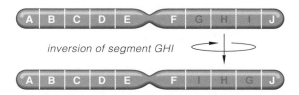

inversion of segment GHI

Figure 12.24 Simple diagram of a chromosome inversion.

Together with duplications, certain inversions and translocations probably helped put primate ancestors of humans on a unique evolutionary road. Of the twenty-three pairs of human chromosomes, eighteen are nearly identical to their counterparts in the chimpanzee and in the gorilla. The other five pairs differ at inverted and translocated regions.

Translocation

In most translocations, part of a chromosome exchanges places with a corresponding part of a *non*homologous chromosome, as in Figure 12.25.

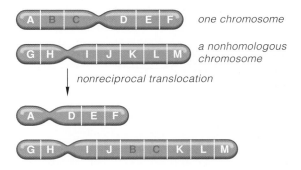

one chromosome

a nonhomologous chromosome

nonreciprocal translocation

Figure 12.25 Simple diagram of a chromosome translocation.

Chromosome 14, for instance, may end up with part of chromosome 8 (and chromosome 8 with a part of 14). In this case, normal controls over the segment's genes are lost at the new location, and a form of cancer results.

On rare occasions, a segment of a chromosome may get lost, inverted, moved to a new location, or duplicated.

Most chromosome aberrations are harmful or lethal. But over evolutionary time, many have been conserved; they confer adaptive advantages or have done their bearers no harm.

12.11 PROSPECTS IN HUMAN GENETICS

With the arrival of their newborn, parents typically ask, "Is our baby normal?" Quite naturally, they want their baby to be free of genetic disorders, and most of the time it is. But what are the options when it is not?

We do not approach diseases and heritable disorders the same way. Diseases follow infection by bacteria and other agents from the outside, and we attack them with antibiotics, surgery, and other weapons. But how do we attack an "enemy" within our genes? Do we institute regional, national, or global programs to identify people who might be carrying harmful alleles? Do we tell them they are "defective" and run a risk of bestowing their disorder on their children? Who decides which alleles are "harmful"? Should society bear the cost of treating genetic disorders before and after birth? If so, should society also have a say in whether affected embryos will be born at all, or whether they should be aborted?

Questions such as these are only the tip of an ethical iceberg. And we do not have answers that are universally acceptable throughout our society.

PHENOTYPIC TREATMENTS   Often, the symptoms of genetic disorders can be either minimized or suppressed by exerting dietary controls, making adjustments to specific environmental conditions, and even intervening surgically or by way of hormone replacement therapy.

For example, dietary control works in PKU, or phenylketonuria. A certain gene specifies an enzyme that converts one amino acid to another —phenylalanine to tyrosine. If an individual is homozygous recessive for a mutated form of the gene, the first of these amino acids accumulates inside the body. If excess amounts are diverted into other pathways, then phenylpyruvate and other compounds may form. High levels of phenylpyruvate in the blood can impair the functioning of the brain. When they restrict their intake of phenylalanine, affected persons are not required to dispose of excess amounts, and they can therefore lead normal lives. Among other things, they can avoid soft drinks and other food products that are sweetened with aspartame, a compound that contains phenylalanine.

Environmental adjustments help counter or minimize the symptoms of some disorders, as when albinos avoid exposure to direct sunlight. Surgical reconstructions also can minimize many problems. For example, surgeons can close up a form of *cleft lip* in which a vertical fissure cuts through the lip and extends into the roof of the mouth.

GENETIC SCREENING   Through large-scale screening programs in the general population, affected persons or carriers of a harmful allele often can be detected early enough to start preventive measures before symptoms develop. For example, most hospitals in the United States routinely screen newborns for PKU, so today it is less common to see people with symptoms of this disorder.

GENETIC COUNSELING   If a first child or close relative has a severe heritable problem, prospective parents may worry about their next child. They may request help in evaluating their options from a qualified professional counseler. *Genetic counseling* often includes diagnosis of parental genotypes, detailed pedigrees, and biochemical testing for hundreds of known metabolic disorders. Geneticists, too, may be contacted to help predict risks for disorders that follow simple Mendelian inheritance, although not all disorders do this. During the genetic counseling, prospective parents also must be reminded that the same risk applies to each pregnancy.

PRENATAL DIAGNOSIS   Methods of *prenatal diagnosis* can be used to determine the sex as well as more than a hundred genetic conditions of fetuses ("prenatal" means "before birth").

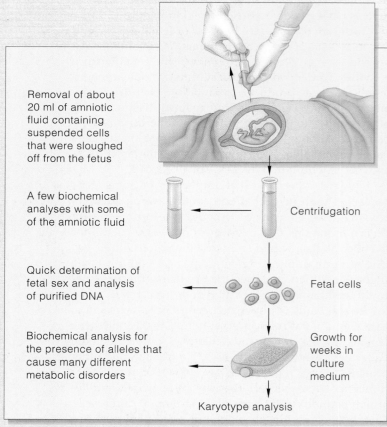

Removal of about 20 ml of amniotic fluid containing suspended cells that were sloughed off from the fetus

A few biochemical analyses with some of the amniotic fluid

Centrifugation

Quick determination of fetal sex and analysis of purified DNA

Fetal cells

Biochemical analysis for the presence of alleles that cause many different metabolic disorders

Growth for weeks in culture medium

Karyotype analysis

Figure 12.26   Steps in amniocentesis.

For example, suppose a woman who is forty-five years old and pregnant worries that her forthcoming child may develop Down syndrome. With **amniocentesis**, a clinician withdraws a tiny sample of the fluid inside the amnion, a membranous sac around the fetus. Some cells that the fetus has sloughed off are suspended in the sample. These are cultured and then analyzed, as by karyotyping. Figure 12.26 is a simplified diagram of the procedure.

By contrast, with **chorionic villi sampling** (CVS), a clinician withdraws some cells from the chorion, which is a fluid-filled, membranous sac that surrounds the amnion. It is possible to perform CVS weeks before amniocentesis. It can yield results as early as the ninth week of pregnancy.

Both procedures carry risks. Either one may accidentally cause infection or puncture the fetus. Also, a 1994 report from the Centers for Disease Control pointed out that mothers who request CVS might run a 0.3 percent risk that their future child will have missing or underdeveloped fingers and toes. Mothers-to-be probably should be counseled to balance that finding against (1) the overall risk of 3 percent that any child will have some kind of birth defect, (2) the severity of genetic disorder that her forthcoming child might be at risk of developing, and (3) how old she is at the time of her pregnancy.

REGARDING ABORTION   What happens if prenatal diagnosis does reveal a serious problem? Do prospective parents opt for induced **abortion**—that is, the expulsion of the embryo from the uterus? We can only say here that they must weigh their awareness of the crushing severity of the genetic disorder against ethical and religious beliefs. Worse still, they must play out their personal tragedy on a larger stage, dominated now by a nationwide battle between fiercely vocal "pro-life" and "pro-choice" factions. We return to this volatile issue in Sections 43.13 and 45.15.

PREIMPLANTATION DIAGNOSIS   Another procedure, *preimplantation diagnosis*, relies on **in-vitro fertilization**. "In vitro," recall, literally means "in glass." Sperm and eggs that have been taken from prospective parents are quickly transferred to an enriched medium in a glass

petri dish. One or more eggs may become fertilized. In two days, mitotic cell divisions may convert a fertilized egg into a ball of eight cells, such as the one shown in Figure 12.27a.

According to one view, the tiny, free-floating ball is a *pre*-pregnancy stage. Like the unfertilized eggs discarded monthly from a woman, the ball is not attached to the

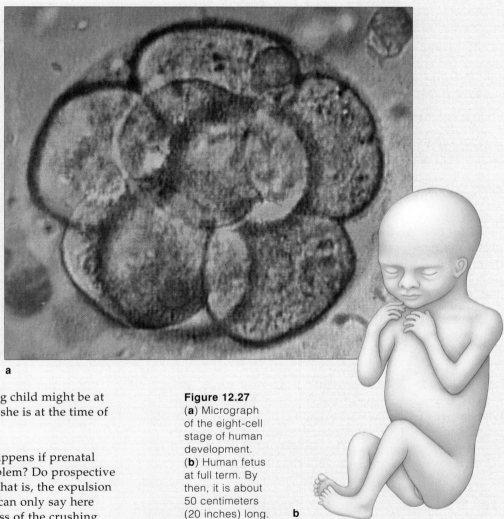

a

Figure 12.27
(**a**) Micrograph of the eight-cell stage of human development.
(**b**) Human fetus at full term. By then, it is about 50 centimeters (20 inches) long.   **b**

uterus. All cells in the ball have the same genes and are not yet committed to giving rise to specialized cells of a heart, lungs, and other organs. Doctors take one of the undifferentiated cells and analyze its genes for suspected disorders. If the cell has no detectable genetic defects, the ball is inserted into the uterus.

Several couples who are at risk of passing on muscular dystrophy, cystic fibrosis, and other disorders have opted for the procedure. Several *"test-tube" babies* have been born in good health and are free of the harmful genes.

SUMMARY

1. Genes, the units of instruction for heritable traits, are arranged one after the other along chromosomes.

2. Human cells are diploid ($2n$), with twenty-three pairs of homologous chromosomes that interact during meiosis. Each pair has the same length, shape, and gene sequence (except for an XY pairing).

3. Human females have two X chromosomes. Males have one X paired with one Y. All other chromosomes are autosomes (the same in both females and males). A gene on the Y chromosome determines gender.

4. Pedigrees, or charts of genetic connections through lines of descent, often provide clues to inheritance of a trait. Certain patterns are characteristic of dominant or recessive alleles on autosomes or on the X chromosome.

5. Genes on the same chromosome represent a linkage group. However, crossing over (breakage and exchange of segments between homologues) disrupts linkages, as summarized in Figure 12.28. The farther apart two gene loci are along the length of a chromosome, the greater will be the frequency of crossovers between them.

6. A chromosome's structure may become altered on rare occasions. A segment might be deleted, inverted, moved to a new location (translocated), or duplicated.

7. On rare occasions, the parental chromosome number may change. Gametes and new individuals may end up with one more or one less chromosome than the parents (aneuploidy) or with three or more of each type of chromosome (polyploidy). Nondisjunction at meiosis (prior to gamete formation) accounts for most of these chromosome abnormalities.

8. Crossing over adds to potentially adaptive variation in traits among members of a population. By contrast, most changes in chromosome number or structure are harmful or lethal. Over evolutionary time, such changes occasionally have proved to be neutral or have offered adaptive advantage, and so have accumulated in DNA.

Review Questions

1. What is a gene? What are alleles? *12.1*

2. Distinguish between: *12.1, 12.2*
   a. homologous and nonhomologous chromosomes
   b. sex chromosomes and autosomes
   c. karyotype and karyotype diagram

3. Define genetic recombination, and describe how crossing over can bring it about. Also give an example of cytological evidence that a crossover has occurred in a cell. *12.1, 12.4*

4. Define pedigree. Also explain why a genetic abnormality is not the same as a genetic disorder or genetic disease. *12.6*

5. Contrast a typical pattern of autosomal recessive inheritance with that of autosomal dominant inheritance. *12.7*

6. Contrast a typical pattern of X-linked recessive inheritance with that of X-linked dominant inheritance. *12.8*

7. Define aneuploidy and polyploidy. Make a sketch of an example of nondisjunction. *12.9*

8. Distinguish among a chromosomal deletion, duplication, inversion, and translocation. *12.10*

Self-Quiz *(Answers in Appendix IV)*

1. _____ segregate during _____ .
   a. Homologues; mitosis
   b. Genes on one chromosome; meiosis
   c. Homologues; meiosis
   d. Genes on one chromosome; mitosis

2. The probability of a crossover occurring between two genes on the same chromosome is _____ .
   a. unrelated to the distance between them
   b. increased if they are closer together on the chromosome
   c. increased if they are farther apart on the chromosome
   d. impossible

3. Chromosome structure can be altered by _____ .
   a. deletions   d. translocations
   b. duplications   e. all of the above
   c. inversions

4. Nondisjunction can be caused by _____ .
   a. crossing over in mitosis
   b. segregation in meiosis
   c. failure of chromosomes to separate during meiosis
   d. multiple independent assortments

Figure 12.28
Summary of the results of hybrid crosses when gene linkage is an example of complete and when crossing over affects the outcome.

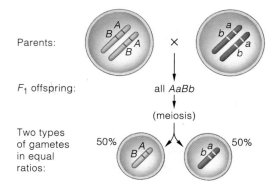

Complete gene linkage (no crossovers); half the gametes have on parental genotype and half have the other.

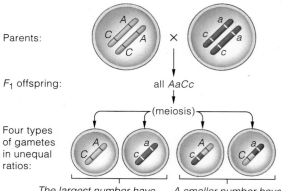

The largest number have the parental genotype

A smaller number have recombinant genotypes

5. A gamete affected by nondisjunction would have _____ .
   a. a change from the normal chromosome number
   b. one extra or one missing chromosome
   c. the potential for a genetic disorder
   d. all of the above

6. Genetic disorders can be caused by _____ .
   a. altered chromosome number   c. mutation
   b. altered chromosome structure   d. all of the above

7. Which of the following events contributes to phenotypic variation in a population?
   a. independent assortment
   b. crossing over
   c. changes in chromosome structure and number
   d. all of the above

8. Match the chromosome terms appropriately.
   ____ crossing over
   ____ deletion
   ____ nondisjunction
   ____ translocation
   ____ karyotype

   a. number and defining features of individual's metaphase chromosomes
   b. chromosome segment moves to a nonhomologous chromosome
   c. disrupts gene linkages at meiosis
   d. causes gametes to have abnormal chromosome numbers
   e. loss of a chromosome segment

Critical Thinking—Genetics Problems
(Answers in Appendix III)

1. Human females are XX and males are XY.
   a. Does a male inherit the X from his mother or father?
   b. With respect to X-linked genes, how many different types of gametes can a male produce?
   c. If a female is homozygous for an X-linked gene, how many types of gametes can she produce with respect to that gene?
   d. If a female is heterozygous for an X-linked gene, how many types of gametes can she produce with respect to that gene?

2. Suppose one allele of a Y-linked gene results in nonhairy ears in males. Another allele results in rather long hairs, a condition called *hairy pinnae* (Figure 12.29).
   a. Why would you *not* expect females to have hairy pinnae?
   b. Any son of a hairy-eared male will also be hairy-eared, but no daughter will be. Explain why.

3. Suppose you carry two linked genes with alleles *Aa* and *Bb*, respectively, as in Figure 12.30. If the crossover frequency between these two genes is zero, what genotypes would be expected among the gametes you produce, and with what frequencies?

4. In *D. melanogaster*, a gene governs wing length. A dominant allele at this gene locus results in the formation of long wings. A recessive allele results in the formation of vestigial (short) wings, as shown in Figure 12.31. You cross a homozygous dominant, long-winged fly with a homozygous recessive,

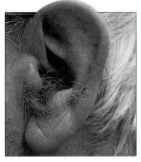

Figure 12.29 Hairy pinnae.

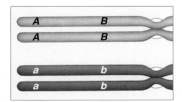

Figure 12.30 Two of your linked genes, on a pair of homologous chromosomes.

Figure 12.31 Vestigial-winged *D. melanogaster.*

vestigial-winged fly. You ask a technician to expose the fertilized eggs to a level of x-rays known to induce mutation and deletions. Later, when irradiated fertilized eggs develop into adults, most of the flies are heterozygous and have long wings. A few have vestigial wings. What might explain these results?

5. Say you have alleles for lefthandedness and straight hair on one chromosome and alleles for righthandedness and curly hair on the homologous chromosome. If the two loci for the alleles are very close together along the chromosome, how likely is it that crossing over will occur between them? If they are distant from each other, is a crossover more or less likely to occur?

6. Individuals affected by *Down syndrome* typically have an extra chromosome 21. In other words, their body cells contain a total of 47 chromosomes.
   a. At which stages of meiosis I and II could a mistake occur that could result in the altered chromosome number?
   b. In a few cases, 46 chromosomes are present. Included in this total are two normal-appearing chromosomes 21 and a longer-than-normal chromosome 14. Explain how these few individuals can have a normal chromosome number.

7. For the mugwump, a tree-dwelling mammal, the male is XX and the female is XY. However, sex-linked genes are found to have the same effect as in humans. For instance, a recessive X-linked allele *c* produces *red-green color blindness*. If a normal female mugwump mates with a phenotypically normal male mugwump whose mother was color blind, what is the probability that a son from that mating will be color blind? A daughter?

8. One type of childhood *muscular dystrophy* is a recessive, X-linked trait. A slowly progressing loss of muscle function leads to death, usually by age twenty or so. Unlike color blindness, this disorder is restricted to males. Suggest why.

Selected Key Terms

abortion *12.11*	homologous chromosome *12.1*
allele *12.1*	inversion *12.10*
amniocentesis *12.11*	in-vitro *12.2*
aneuploidy *12.9*	in-vitro fertilization *12.11*
autosome *12.1*	karyotype *12.1*
centrifugation *12.2*	linkage group *12.4*
chorionic villi sampling *12.11*	linkage mapping *12.5*
crossing over *12.1*	nondisjunction *12.9*
cytological marker *12.4*	pedigree *12.6*
deletion *12.10*	polyploidy *12.9*
double-blind study *12.9*	reciprocal cross *12.4*
duplication *12.10*	sex chromosome *12.1*
gene *12.1*	syndrome *CI*
genetic abnormality *12.6*	translocation *12.10*
genetic disease *12.6*	X chromosome *12.1*
genetic disorder *12.6*	X- or Y-linked gene *12.4*
genetic recombination *12.1*	Y chromosome *12.1*

Readings

Russell, P. 1992. *Genetics*. Third edition. New York: Harper Collins.

Travis, J. 30 November 1996. "Clue to Lou Gehrig's Disease Emerges." *Science News* 150: 340.

Web Site See *http://www.wadsworth.com/biology* for practice quiz questions, hypercontents, BioUpdates, and critical thinking. The Wadsworth Biology Resource Center provides a wealth of information fully organized and integrated by chapter.

13 DNA STRUCTURE AND FUNCTION

Cardboard Atoms and Bent-Wire Bonds

One might have wondered, in the spring of 1868, why Johann Friedrich Miescher was collecting cells from the pus of open wounds and, later, from the sperm of a fish. Miescher, a physician, wanted to identify the chemical composition of the nucleus. These particular cells have very little cytoplasm, which makes it easier to isolate the nuclear material for analysis.

Miescher finally succeeded in isolating an organic compound with the properties of an acid. Unlike other substances in cells, it incorporated a notable amount of phosphorus. Miescher called the substance nuclein. He had discovered what came to be known many years later as **deoxyribonucleic acid**, or **DNA**.

The discovery did not cause even a ripple through the scientific community. At the time, no one really knew much about the physical basis of inheritance—that is, *which chemical substance encodes the instructions for reproducing parental traits in offspring*. Few even suspected that the cell nucleus might hold the answer. For a time, researchers generally believed that hereditary instructions had to be encoded in the structure of some unknown class of proteins. After all, heritable traits are spectacularly diverse. Surely the molecules encoding information about those traits were structurally diverse also. Proteins are put together from potentially limitless combinations of twenty different amino acid subunits, so they almost certainly could function as the sentences (genes) in each cell's book of inheritance.

By the early 1950s, however, the results of many ingenious experiments clearly indicated that DNA was the substance of inheritance. Moreover, in 1951, Linus Pauling did something that no one had done before. Through his training in biochemistry, a talent for

model building, and a few great educated guesses, he deduced the three-dimensional structure of the protein collagen. Pauling's discovery was truly electrifying. If someone could pry open the secrets of proteins, then why not assume the same might be done for DNA? And once the structural details of the DNA molecule were understood, wouldn't they provide clues to its biological functions? *Who would go down in history as having discovered the very secrets of inheritance?*

Scientists around the world started scrambling after that ultimate prize. Among them were James Watson, a young postdoctoral student from Indiana University, and Francis Crick, an exuberant researcher working at Cambridge University. Exactly how could DNA, a molecule consisting of only four kinds of subunits, hold genetic information? Watson and Crick spent long hours arguing over everything they had read about the size, shape, and bonding requirements of the subunits of DNA. They fiddled with cardboard cutouts of the subunits. They even badgered chemists to help them identify any potential bonds they might have overlooked. Then they assembled models from bits of metal, held together with wire "bonds" bent at seemingly suitable angles.

Figure 13.1 James Watson and Francis Crick posing in 1953 by their newly unveiled structural model of DNA.

In 1953, they put together a model that fit all of the pertinent biochemical rules and all of the facts about DNA they had gleaned from other sources (Figures 13.1 and 13.2). They had discovered the structure of DNA. And the breathtaking simplicity of that structure enabled them to solve another long-standing riddle— *how life can show unity at the molecular level and yet give rise to such spectacular diversity at the level of whole organisms.*

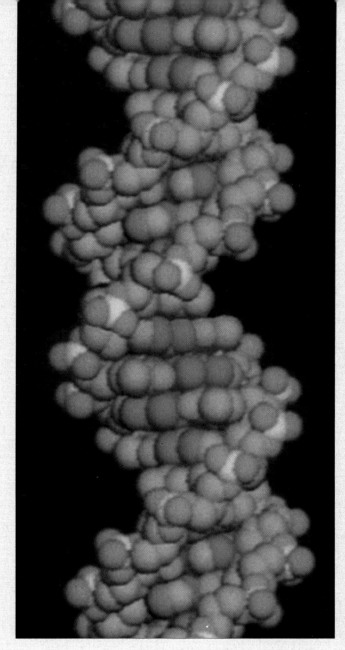

Figure 13.2 A more recent, computer-generated model of DNA. Although more sophisticated in appearance, it is much the same as the prototype Watson and Crick put together decades ago.

With this chapter, we turn to investigations that led to our current understanding of DNA. The story is more than a march through details of its structure and function. *It also is revealing of how ideas are generated in science*. On the one hand, having a shot at fame and fortune quickens the pulse of men and women in any profession, and scientists are no exception. On the other hand, science proceeds as a community effort, with individuals sharing not only what they can explain but also what they do not understand. Even when an experiment fails to produce the anticipated results, it might turn up information that others can use or lead to questions that others can answer. Unexpected results, too, might be clues to something important about the natural world.

KEY CONCEPTS

1. In all living cells, DNA molecules are the storehouses of information about heritable traits.

2. In a DNA molecule, two strands of nucleotides twist together, like a spiral stairway. Each strand consists of four kinds of nucleotides, which are the same except for one component—a nitrogen-containing base. The four bases are adenine, guanine, thymine, and cytosine.

3. Great numbers of nucleotides are arranged one after another in each strand of the DNA molecule. In at least some regions, the order in which one kind of nucleotide follows another is unique for each species. Hereditary information is encoded in that particular sequence.

4. Hydrogen bonds connect the bases of one strand of the DNA molecule to bases of the other strand. As a rule, adenine pairs (hydrogen-bonds) with thymine, and guanine with cytosine.

5. Before a cell divides, its DNA is replicated with the assistance of enzymes and other proteins. Each double-stranded DNA molecule starts unwinding. A new, complementary strand is assembled bit by bit on the exposed bases of each parent strand, according to the base-pairing rule.

13.1 DISCOVERY OF DNA FUNCTION

Early and Puzzling Clues

The year was 1928. Frederick Griffith, an army medical officer, was attempting to develop a vaccine against *Streptococcus pneumoniae*, a bacterium that causes a form of pneumonia. (When introduced into a person's body, vaccines can mobilize internal defenses against a real attack. Many are preparations of either killed or weakened bacterial cells.) Griffith never did develop a vaccine. But his work unexpectedly opened a door to the molecular world of heredity.

Griffith isolated and cultured two different strains of the bacterium. He noticed that colonies of one strain had a rough surface appearance, but those of the other strain appeared smooth. He designated the two strains R and S and used them in a series of four experiments:

1. Laboratory mice were injected with live R cells. As Figure 13.3 indicates, they did not develop pneumonia. *The R strain was harmless.*

2. Mice were injected with live S cells. The mice died. Blood samples taken from them teemed with live S cells. *The S strain was pathogenic* (disease-causing).

3. S cells were killed by exposure to high temperature. Mice injected with these cells did not die.

4. Live R cells were mixed with heat-killed S cells and injected into mice. The mice died—and blood samples from them teemed with *live* S cells!

What was going on in the fourth experiment? Maybe heat-killed S cells in the mixture weren't really dead. But if that were true, then mice injected with heat-killed S cells alone (experiment 3) would have died. Maybe harmless R cells in the mixture had mutated into a killer form. But if that were true, then mice injected with the R cells alone (experiment 1) would have died.

The simplest explanation was as follows: *Heat killed the S cells but did not destroy their hereditary material— including the part that specified "how to cause infection."* Somehow, that material had been transferred from dead S cells to living R cells, which put it to use.

Further experiments showed the harmless cells had indeed picked up information on causing infections and were permanently transformed into pathogens. After hundreds of generations, descendants of transformed bacterial cells were still infectious!

The unexpected results of Griffith's experiments intrigued Oswald Avery and his fellow biochemists. In time, they were even able to transform harmless bacterial cells with *extracts* of killed pathogens. Finally in 1944, after rigorous chemical analyses, they were able to report that the hereditary substance in their extracts probably was DNA—not proteins, as was then widely believed. To give experimental evidence of this conclusion, they reported that they had added certain protein-digesting enzymes to some extracts, but cells exposed to those extracts were transformed anyway. To other extracts they had added an enzyme that digests DNA but not proteins. Doing so blocked hereditary transformation.

Despite these impressive experimental results, most biochemists refused to give up on the proteins. Avery's findings, they said, probably applied only to bacteria.

Confirmation of DNA Function

By the early 1950s molecular detectives, including Max Delbrück, Alfred Hershey, Martha Chase, and Salvador Luria, were using viruses as experimental subjects. The viruses they had selected, called **bacteriophages**, infect *Escherichia coli* and other bacteria.

Viruses are as biochemically simple as you can get. Although they are not alive, they do contain hereditary information about building more new virus particles. At some point after they infect a host cell, viral enzymes take over a portion of the cell's metabolic machinery— which starts churning out substances that are necessary to construct new virus particles.

genetic material
viral coat
sheath
baseplate
tail fiber

1 Mice injected with live cells of harmless strain R.

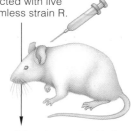

Mice do not die. No live R cells in their blood.

2 Mice injected with live cells of killer strain S.

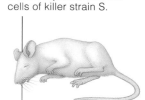

Mice die. Live S cells in their blood.

3 Mice injected with heat-killed S cells.

Mice do not die. No live S cells in their blood.

4 Mice injected with live R cells *plus* heat-killed S cells.

Mice die. Live S cells and live R cells in their blood.

Figure 13.3 Results of Griffith's experiments with a harmless and a pathogenic strain of *Streptococcus pneumoniae*, as described in the text above.

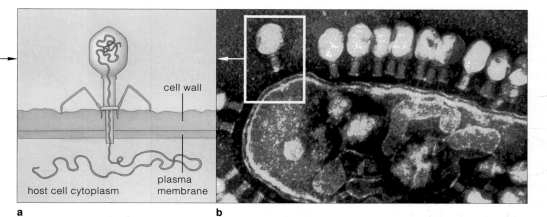

a

b

Figure 13.4 (**a**) *Far left*: Structural organization of a T4 bacteriophage. The diagram shows the genetic material of this type of virus being injected into the cytoplasm of a host cell. In this case, it is DNA (the *blue*, threadlike strand). (**b**) Electron micrograph of T4 virus particles infecting a bacterium (*Escherichia coli*) that has just become an unwilling host. 🌀

Figure 13.5 Two examples of the landmark experiments that pointed to DNA as the substance of heredity. In the 1940s, Alfred Hershey and his colleague, Martha Chase, were studying the biochemical basis of inheritance. They were aware that certain bacteriophages consisted of proteins and DNA. *Did the proteins, DNA, or both contain the viral genetic information?*

To find a possible answer, Hershey and Chase designed two experiments, based on two known biochemical facts. First, the proteins of bacteriophages incorporate sulfur (S) but not phosphorus (P). Second, their DNA incorporates phosphorus but not sulfur.

(**a**) In one experiment, some bacterial cells were grown on a culture medium that incorporated the radioisotope ^{35}S. Therefore, when the bacterial cells synthesized proteins, they would take up that radioisotope of sulfur—which would serve as a tracer. (Here you may wish to review Section 2.2.) After the cells became labeled with the tracer, bacteriophages were allowed to infect them. As the infection ran its course, viral proteins were synthesized inside the host cells. These proteins also became labeled with ^{35}S. And so did the new generation of virus particles.

Next, the labeled bacteriophages were allowed to infect a new batch of unlabeled bacteria that were suspended in a fluid culture medium. Afterward, Hershey and Chase whirred the fluid in a kitchen blender. Whirring dislodged the viral protein coats from the cells. The particles became suspended in the fluid medium. Analysis revealed the presence of labeled protein in the fluid—but no protein *inside* the bacterial cells.

(**b**) For the second experiment, Hershey and Chase cultured more bacterial cells. The phosphorus that was available to them for synthesizing DNA included the radioisotope ^{32}P. Later, bacteriophages were allowed to infect the cells.

As predicted, viral DNA synthesized inside the infected cells became labeled, as did the new generation of virus particles. Next, the labeled particles were allowed to infect bacteria that were suspended in a fluid medium. Then they were dislodged from the host cells. Analysis revealed that the labeled viral DNA was not in the fluid. DNA remained *inside* the host cells, where its hereditary information had to be put to use to make more virus particles. Here was strong evidence that DNA is the genetic material of this type of virus. 🌀

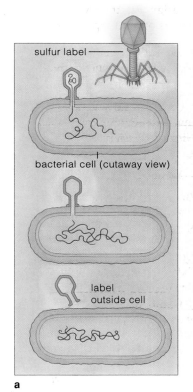

sulfur label

bacterial cell (cutaway view)

label outside cell

a

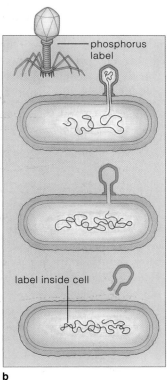

phosphorus label

label inside cell

b

By 1952, researchers knew that some bacteriophages consist only of DNA and a protein coat. Also, electron micrographs revealed that the main part of the viruses remains *outside* the cells they are infecting (Figure 13.4). Possibly, such viruses were injecting genetic material alone *into* host cells. If that were true, was the material DNA, protein, or both? Through many experiments, researchers accumulated strong evidence that DNA, not proteins, serves as the molecule of inheritance. Figure 13.5 describes two of these landmark experiments.

Information for producing the heritable traits of single-celled and multicelled organisms is encoded in DNA.

Components of DNA

Long before the bacteriophage studies were under way, biochemists knew that DNA contains only four types of nucleotides that are the building blocks of nucleic acids. A **nucleotide** consists of a five-carbon sugar (which is deoxyribose in DNA), a phosphate group, and one of the following nitrogen-containing bases:

adenine	guanine	thymine	cytosine
(A)	(G)	(T)	(C)

As you can see from Figure 13.6, all four types of nucleotides in DNA have their component parts joined together in much the same way. However, T and C are pyrimidines, which are single-ring structures. A and G are purines, which are larger, bulkier molecules; they have double-ring structures.

By 1949, Erwin Chargaff, a biochemist, had shared two crucial insights into the composition of DNA with the scientific community. First, the amount of adenine relative to guanine differs from one species to the next. Second, the amount of adenine in DNA always equals that of thymine, and the amount of guanine always equals that of cytosine. We may show this as:

$$A = T \quad \text{and} \quad G = C$$

The relative proportions of the four kinds of nucleotides were tantalizing clues. In some way, those proportions almost certainly were related to the arrangement of the two chains of nucleotides in a DNA molecule.

The first convincing evidence of that arrangement came from Maurice Wilkins's laboratory in England. One of Wilkins's colleagues, Rosalind Franklin, had obtained especially good **x-ray diffraction images** of DNA fibers. By this process, a beam of x-rays is directed at a molecule, which scatters the beam in patterns that can be captured on film. The pattern itself consists only of dots and streaks; in itself, it doesn't reveal molecular structure. However, the images of those patterns can be used to calculate the positions of the molecule's atoms.

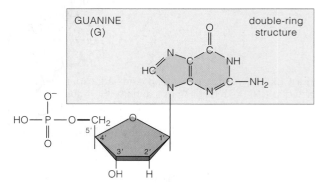

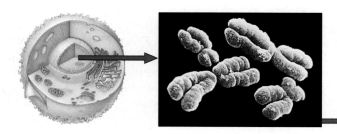

Figure 13.6 The four kinds of nucleotide subunits of DNA. Small numerals on the structural formulas identify the carbon atoms to which other parts of the molecule are attached.

All chromosomes contain DNA. What does DNA contain? Four kinds of nucleotides. A nucleotide has a five-carbon sugar (shaded *red*), which has a phosphate group attached to the fifth carbon of its carbon ring structure. It also has one of four kinds of nitrogen-containing bases (*blue*) attached to its first carbon atom. The four nucleotides differ only in *which* base is attached to that carbon atom.

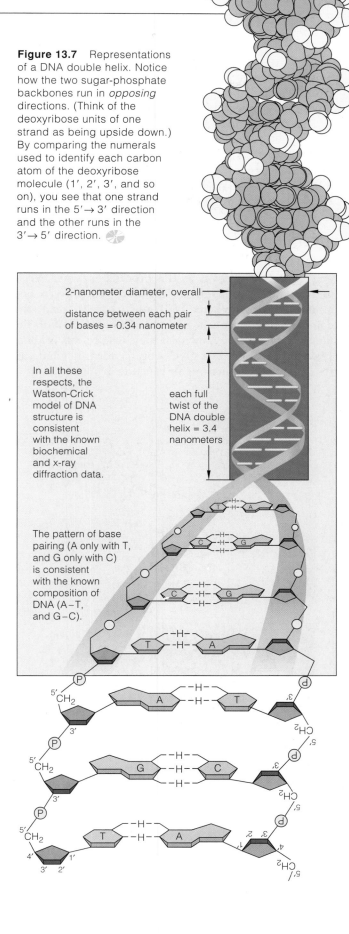

Figure 13.7 Representations of a DNA double helix. Notice how the two sugar-phosphate backbones run in *opposing* directions. (Think of the deoxyribose units of one strand as being upside down.) By comparing the numerals used to identify each carbon atom of the deoxyribose molecule (1', 2', 3', and so on), you see that one strand runs in the 5'→3' direction and the other runs in the 3'→5' direction.

2-nanometer diameter, overall

distance between each pair of bases = 0.34 nanometer

In all these respects, the Watson-Crick model of DNA structure is consistent with the known biochemical and x-ray diffraction data.

each full twist of the DNA double helix = 3.4 nanometers

The pattern of base pairing (A only with T, and G only with C) is consistent with the known composition of DNA (A–T, and G–C).

DNA does not readily lend itself to x-ray diffraction. However, researchers could rapidly spin a suspension of DNA molecules, spool them onto a rod, and gently pull them into gossamer fibers, like cotton candy. If the atoms in DNA were arranged in a regular order, x-rays directed at a fiber should scatter in a regular pattern that could be captured on film.

Calculations based on Franklin's images strongly indicated that the DNA molecule had to be long and thin, with a 2-nanometer diameter. Some molecular configuration was being repeated every 0.34 nanometer along its length, and another one, every 3.4 nanometers.

Could the sequence of nucleotide bases be twisting, like a circular stairway? Certainly Pauling thought so. After all, he discovered the helical shape of collagen. He and everybody else—including Wilkins, Watson, and Crick—were thinking "helix." Watson later wrote, "We thought, why not try it on DNA? We were worried that *Pauling* would say, why not try it on DNA? Certainly he was a very clever man. He was a hero of mine. But we beat him at his own game. I still can't figure out why."

Pauling, it turned out, made a big chemical mistake. His model had hydrogen bonds at phosphate groups holding DNA's structure together. That does happen in highly acidic solutions. It doesn't happen in cells.

Patterns of Base Pairing

As Watson and Crick perceived, DNA consists of *two* strands of nucleotides, held together at their bases by hydrogen bonds. The bonds form when the two strands run in opposing directions and twist together into a double helix (Figure 13.7). Two kinds of base pairings form along the length of the molecule: A—T and G—C. This bonding pattern permits variation in the order of bases in any given strand. For example, in even a tiny stretch of DNA from a rose, gorilla, human, or any other organism, the sequence might be:

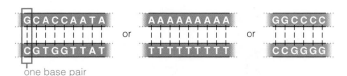

In fact, even though all DNA molecules show the same bonding pattern, each species has unique base sequences in its DNA. *This molecular constancy and variation among species is the foundation for the unity and diversity of life.*

The pattern of base pairing between the two strands in DNA is constant for all species—A with T, and G with C. However, the DNA molecules of each species show unique differences in the sequence of base pairs along their length.

DNA REPLICATION AND REPAIR

How a DNA Molecule Gets Duplicated

The discovery of DNA structure was a turning point in studies of inheritance. Until then, no one could explain **DNA replication**, or how the molecule of inheritance is duplicated before the cell divides. The Watson-Crick model suggested at once how this might be done.

Enzymes can easily break hydrogen bonds between the two nucleotide strands of a DNA molecule. When they act on the molecule, one strand unwinds from the other, thereby exposing some nucleotide bases. Cells have stockpiles of free nucleotides, and these pair with exposed bases. Each parent strand remains intact, and a companion strand is assembled on each one according to this base-pairing rule: A to T, and G to C. As soon as a stretch of a new, partner strand forms on a stretch of the parent strand, the two twist into a double helix.

Because the parent strand is conserved, each "new" DNA molecule is really half old, half new (Figure 13.8). That is why biologists sometimes refer to this process as *semiconservative* replication.

DNA replication requires a large team of molecular workers. For instance, one kind of enzyme unwinds the two nucleotide strands, as in Figure 13.9. At the same time, many other proteins bind to unwound portions and hold them apart while replication proceeds. **DNA polymerases** are key players. These enzymes attach free nucleotides to a growing strand. Other enzymes, the **DNA ligases**, seal new short stretches of nucleotides into a continuous strand. Some of these enzymes have uses in recombinant DNA technology (Section 16.1).

Monitoring and Fixing the DNA

DNA polymerases, DNA ligases, and other enzymes also engage in **DNA repair**. By this process, enzymes excise and repair altered parts of the base sequence in one strand of a double helix. DNA polymerases "read" the complementary sequence on the other strand. With the aid of other repair enzymes, they restore the original sequence. Section 13.4 gives you an idea of what can happen if the excision-repair function is impaired.

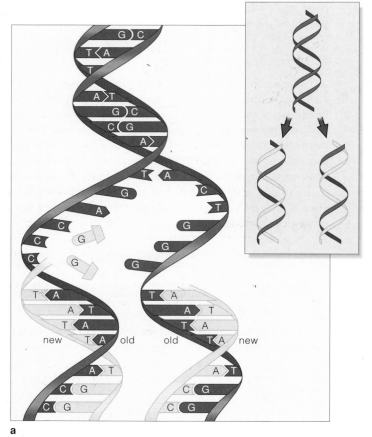

a

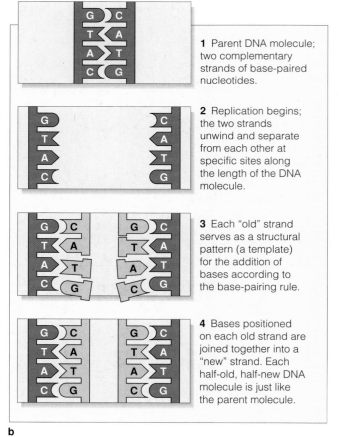

1 Parent DNA molecule; two complementary strands of base-paired nucleotides.

2 Replication begins; the two strands unwind and separate from each other at specific sites along the length of the DNA molecule.

3 Each "old" strand serves as a structural pattern (a template) for the addition of bases according to the base-pairing rule.

4 Bases positioned on each old strand are joined together into a "new" strand. Each half-old, half-new DNA molecule is just like the parent molecule.

b

Figure 13.8 (**a**) Overview of the semiconservative nature of DNA replication. The original two-stranded DNA molecule is shown in *blue*. Each parent strand remains intact, and a new strand (*yellow*) is assembled on each one. (**b**) A closer look at the base additions.

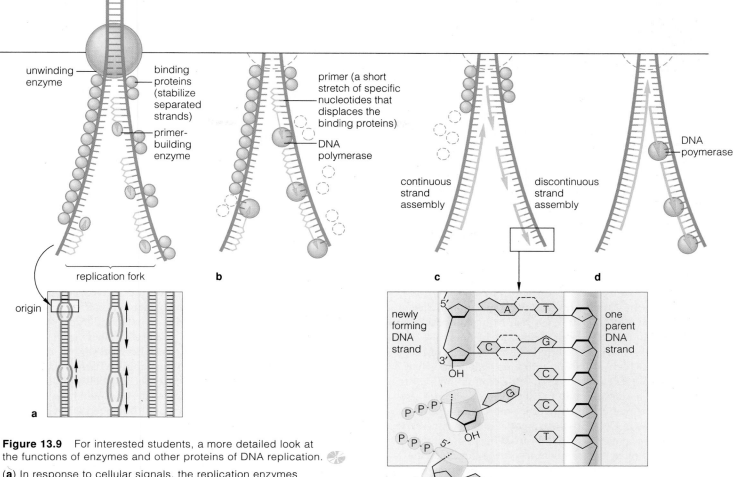

Figure 13.9 For interested students, a more detailed look at the functions of enzymes and other proteins of DNA replication.

(**a**) In response to cellular signals, the replication enzymes become active at many short, specific base sequences called *origins*. Organized complexes of these enzymes as well as other proteins unwind the two strands of the parent DNA molecule, prevent them from rewinding, and assemble a new strand on each one. They do this only in *replication forks*. These are limited, V-shaped regions that advance in both directions away from an origin. Enzymes rewind the half-old, half-new molecules while they are being completed, behind each advancing replication fork.

(**b**) At each fork, some binding proteins are displaced when enzymes synthesize primers at intervals along both of the parent templates. DNA polymerases recognize the primers as "start" tags. They join nucleotide monomers together behind the primers. Exposed bases of a parent template dictate which kind of nucleotide will be added in sequence according to the base-pairing rule: A with T, and G with C.

(**c**) As discovered by Reiji Okazaki, strand assembly is a *continuous* process on one parent template. It is *discontinuous* on the other strand, where nucleotides must be assembled in short stretches only. Why? They can only be joined together in the 5'→3' direction, because this leaves one of the —OH groups of the growing sugar-phosphate backbone exposed. This exposed group is the only kind of site where nucleotide units can be joined together.

(**d**) Primers are removed. DNA polymerases fill in the gaps but cannot make the final connection between the stretches of nucleotides. DNA ligases must seal the tiny nicks that remain. The result is continuous strands on both parent templates.

Free nucleotides brought up for strand assembly also provide the energy that drives the replication process. Each has three phosphate groups attached. The action of DNA polymerase splits away two of these groups, and some of the released energy is used to attach the nucleotide to a growing strand.

Regarding the Chromosomal Proteins

Earlier, in Section 9.4, you read that eukaryotic DNA has a tremendous number of histones and other protein molecules bound tightly to it. Many of these proteins interact with the DNA and form the scaffolding for the condensed organization of the metaphase chromosome. You may well wonder: Wouldn't the proteins interfere with DNA replication and other functions? Apparently, some parts of the protein scaffold intervene *between* genes, not in regions with protein-building information. Do proteins organize the DNA into "domains" having distinct functions? Does the organization make it easier for enzymes to replicate the DNA, even to "read" the genes as the first step in protein synthesis? These are just two of the possibilities being probed by the new generation of molecular detectives.

DNA is replicated prior to cell division. Enzymes unwind its two strands. Each strand remains intact throughout the process—it is conserved—and enzymes assemble a new, complementary strand on each one.

Enzymes involved in replication also repair the DNA where base-pairing errors have crept into the nucleotide sequence.

13.4 WHEN DNA CAN'T BE FIXED

1992 was an unforgettable year for Laurie Campbell. She finally turned eighteen. And in that same year she just happened to notice a peculiar mole on her skin. This one was suspiciously black. It had an odd lumpiness about it, a ragged border, and an encrusted surface. Laurie quickly made an appointment with her family doctor, who just as quickly ordered a biopsy. The mole turned out to be a *malignant melanoma*—the deadliest form of skin cancer.

Laurie was lucky. She detected the cancer in its earliest stage, before it could spread through her body. Ever since, she consistently checks out the appearance of other moles that pepper her skin. She is painfully aware of having become a statistic—one of 500,000 people in the United States alone who develop skin cancer in any given year, and one of the 23,000 with malignant melanoma. She knows now that 7,500 die each year from skin cancer, and that 5,600 of them die from malignant melanoma.

Laurie is smart. She plotted out the position of every mole on her body. Once a month, this body map is her guide for a quick but thorough self-examination. Figure 13.10 shows examples of what she looks for. Laurie also schedules a medical examination every six months.

Changes in DNA are the triggers for skin cancer. The ultraviolet wavelengths in light from the sun, tanning lamps, and other environmental sources can cause the abnormal molecular changes. Among other things, the wavelengths can promote covalent bonding between two adjacent thymine bases in a nucleotide strand. In this way, the two nucleotides to which the bases belong are combined into an abnormal, bulky structure called a thymine dimer within the DNA.

Normally, a DNA repair mechanism gets rid of such bulky lesions. The mechanism requires at least seven gene products. Mutation in one or more of the required genes can skew the repair machinery. Thymine dimers can accumulate in skin cells of individuals with this type of mutation. The accumulation may trigger development of lesions, including skin cancer.

The risk of skin cancer is greater for some than others. At one extreme, a newborn destined to develop *xeroderma pigmentosum* literally faces a dim future. The individuals affected by this genetic disorder cannot be exposed to sunlight, even briefly, without risking disfiguring skin tumors and possible early death from cancer.

You are at risk if you have moles that are chronically irritated, as by shaving or abrasive clothing. You are at risk if your skin, including your lip surface, is chronically chapped, cracked, or sore. You are at risk if your family has a history of cancer or if you have undergone radiation therapy. And, like Laurie, you are at risk if you have pale skin and burn easily in the sun. Even on slightly overcast days, fair-skinned people must apply a high SPF (Sun Protection Factor) sunblock, especially between 10 A.M. and 2 P.M. They should wear wide-brimmed hats, long sleeves, and long pants or skirts. Damaged DNA and skin cancer are the reality, in spite of the ill-advised, socially promoted allure of a golden tan.

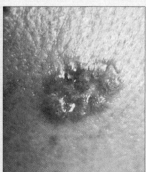 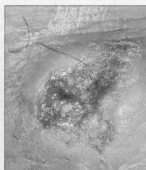 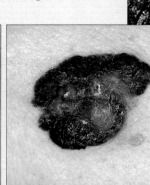

a Basal cell carcinoma   **b** Squamous cell carcinoma   **c** Malignant melanoma

Figure 13.10 Some examples of what can happen when repair enzymes cannot fix changes in the nucleotide sequence of DNA. (**a**) *Basal cell carcinoma*, the most common form of skin cancer. This slow-growing, raised lump may be uncolored, reddish-brown, or black. (**b**) *Squamous cell carcinoma*, the second most common skin cancer. These pink growths, firm to the touch, grow rapidly under the surface of skin exposed to the sun. (**c**) *Malignant melanoma*, which spreads most rapidly. The malignant cells form very dark, encrusted lumps. They may itch like an insect bite or bleed easily. (**d**) Laurie Campbell, avoiding the sun—and melanoma.

d

SUMMARY

1. The hereditary information of cells and multicelled organisms is encoded in DNA (deoxyribonucleic acid).

2. DNA consists of nucleotide subunits. Each of these has a five-carbon sugar (deoxyribose), one phosphate group, and one of four kinds of nitrogen-containing bases (adenine, thymine, guanine, or cytosine).

3. A DNA molecule consists of two nucleotide strands twisted together into a double helix. The bases of one strand pair (hydrogen-bond) with bases of the other.

4. The bases of the two strands in a DNA double helix pair in constant fashion. Adenine pairs with thymine (A to T), and guanine with cytosine (G to C). *Which* base pair follows the next (A—T, T—A, G—C, or C—G) varies along the length of the strands.

5. Overall, the DNA of one species includes a number of unique stretches of base pairs that set it apart from the DNA of all other species.

6. During DNA replication, enzymes unwind the two strands of a double helix and assemble a new strand of complementary sequence on each parent strand. Two double-stranded molecules result. One strand of each molecule is "old" (it is conserved); the other is "new."

7. Some of the enzymes involved in DNA replication also repair DNA where base-pairing errors have been introduced into the nucleotide sequence.

Review Questions

1. Name the three molecular parts of a nucleotide in DNA. Name the four different bases that occur in these nucleotides. *13.2*

2. What kind of bond joins two DNA strands in a double helix? Which nucleotide base-pairs with adenine? With guanine? *13.2*

3. Explain how DNA molecules can show constancy and variation from one species to the next. *13.2*

Self-Quiz *(Answers in Appendix IV)*

1. Which is *not* a nucleotide base in DNA?
   a. adenine   c. uracil   e. guanine
   b. thymine   d. cytosine

2. What are the base-pairing rules for DNA?
   a. A–G, T–C   b. A–C, T–G   c. A–U, C–G   d. A–T, G–C

3. A DNA strand having the sequence C–G–A–T–T–G would be complementary to the sequence _____ .
   a. C–G–A–T–T–G   c. T–A–G–C–C–T
   b. G–C–T–A–A–G   d. G–C–T–A–A–C

4. One species' DNA differs from others in its _____ .
   a. sugars   c. base sequence
   b. phosphate groups   d. all of the above

5. When DNA replication begins, _____ .
   a. the two DNA strands unwind from each other
   b. the two DNA strands condense for base transfers
   c. two DNA molecules bond
   d. old strands move to find new strands

6. DNA replication requires _____ .
   a. free nucleotides   c. many enzymes
   b. new hydrogen bonds   d. all of the above

7. Match the DNA terms appropriately.
   ____ DNA polymerase   a. two nucleotide strands
        and DNA ligase   twisted together
   ____ constancy in   b. A with T, G with C
        base pairing   c. hereditary material
   ____ replication   duplicated
   ____ DNA double helix   d. replication enzymes

Critical Thinking

1. Chargaff's data suggested that adenine pairs with thymine, and guanine pairs with cytosine. What other data available to Watson and Crick suggested that adenine-guanine and cytosine-thymine pairs normally do not form?

2. One of Matthew Meselson and Frank Stahl's experiments supported the semiconservative model of DNA replication. The researchers made "heavy" DNA by growing *Escherichia coli* in a medium enriched with ^{15}N, a heavy isotope of nitrogen. They prepared "light" DNA by growing *E. coli* in the presence of ^{14}N, the more common isotope. An available technique helped them identify which replicated molecules were heavy, light, or hybrid (one heavy strand, one light). Use two pencils of two different colors, one for heavy strands and one for light strands. Starting with a DNA molecule having two heavy strands, sketch the daughter molecules that would form after one replication in an ^{14}N-containing medium. Now sketch the four DNA molecules that would result if these daughter molecules were replicated a second time in the ^{14}N medium.

3. Mutations (permanent changes in base sequences of genes) are the original source of genetic variation. This variation is the raw material of evolution. Yet how can both statements be true, given that cells have efficient mechanisms to repair DNA before mutations can become established?

4. As indicated in Section 4.11, a pathogenic strain of *E. coli* has acquired an ability to produce a dangerous toxin that has caused medical problems and fatalities. This is especially true of young children who have ingested undercooked, contaminated beef. Develop a hypothesis to explain how a normally harmless bacterium such as *E. coli* can become a pathogen.

Selected Key Terms

adenine (A) *13.2*   DNA repair *13.3*
bacteriophage *13.1*   DNA replication *13.3*
cytosine (C) *13.2*   guanine (G) *13.2*
deoxyribonucleic acid (DNA) *CI*   nucleotide *13.2*
DNA ligase *13.3*   thymine (T) *13.2*
DNA polymerase *13.3*   x-ray diffraction image *13.2*

Readings

Watson, J. 1978. *The Double Helix.* New York: Atheneum. Highly personal view of scientists and their methods, interwoven into an account of how DNA structure was discovered.

Wolfe, S. 1995. *Introduction to Molecular and Cellular Biology.* Belmont, California: Wadsworth. Comprehensive, current, and accessible.

Web Site See *http://www.wadsworth.com/biology* for practice quiz questions, hypercontents, BioUpdates, and critical thinking. The Wadsworth Biology Resource Center provides a wealth of information fully organized and integrated by chapter.

14

FROM DNA TO PROTEINS

Beyond Byssus

Picture a mussel, of the sort shown in Figure 14.1. Hard-shelled but soft of body, it is using its muscular foot to probe a wave-scoured rock. At any moment, pounding waves can whack the mussel into the water, hurl it repeatedly against the rock with shell-shattering force, and so offer up a gooey lunch for gulls.

By chance, the mussel's foot comes across a crevice in the rock. The foot moves, broomlike, and sweeps the crevice clean. It presses down, forcing air out from underneath it, then arches up. The result is a vacuum-sealed chamber, rather like the one that forms when a plumber's rubber plunger is being squished down and up to unclog a drain. Into this vacuum chamber the mussel spews a fluid, consisting of keratin and other proteins, which bubbles into a sticky foam. Now, by curling its foot into a small tubular shape and pumping the foam through it, the mussel forms sticky threads about as wide as a human whisker. As a final touch, it varnishes the threads with another type of protein and thereby ends up with an adhesive called byssus, which anchors the mussel to the rock.

Byssus is the world's premier underwater adhesive. Nothing that humans have manufactured even comes close. (Sooner or later, water chemically degrades or deforms synthetic adhesives.) Byssus truly fascinates biochemists, dentists, and surgeons looking for better ways to

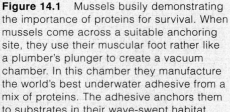

Figure 14.1 Mussels busily demonstrating the importance of proteins for survival. When mussels come across a suitable anchoring site, they use their muscular foot rather like a plumber's plunger to create a vacuum chamber. In this chamber they manufacture the world's best underwater adhesive from a mix of proteins. The adhesive anchors them to substrates in their wave-swept habitat.

do tissue grafts and to rejoin severed nerves. Genetic engineers insert mussel DNA into yeast cells, which reproduce in large numbers and serve as "factories" for translating mussel genes into useful quantities of proteins. This exciting work, like the mussel's own byssus building, starts with one of life's universal concepts: *Every protein is synthesized in accordance with instructions contained in DNA.*

You are about to trace the steps leading from DNA to protein. Many enzymes are players in this pathway, and so is another kind of nucleic acid besides DNA. The same steps produce *all* proteins, from mussel-inspired adhesives to the keratin in your hair and fingernails to the insect-digesting enzymes of a Venus flytrap.

Start out by thinking of each cell's DNA as a book of protein-building instructions. The alphabet used to create the book is simple enough: A, T, G, and C (for the nucleotide bases adenine, thymine, guanine, and cytosine). How do you get from this alphabet to a protein? The answer starts with DNA's structure.

DNA, recall, is a double-stranded molecule. Which kind of nucleotide base follows the next along the length of a strand—that is, the **base sequence**—differs from one kind of organism to the next. As you read in the preceding chapter, before a cell divides, its DNA is replicated and the two strands unwind entirely from each other. However, at other times in a cell's life, the two strands unwind only in certain regions to expose particular base sequences—genes. Most of those genes contain instructions for building proteins.

It takes two steps, **transcription** and **translation**, to carry out a gene's protein-building instructions. In eukaryotic cells, transcription proceeds in the nucleus. In this step, a selected base sequence in DNA serves as a structural pattern—as a template—for assembling a strand of **ribonucleic acid** (**RNA**) from the cell's pool of free nucleotides. Afterward, the RNA moves into the cytoplasm, where translation proceeds. In this second step, RNA directs the assembly of amino acids into polypeptide chains. The newly formed chains become folded into the three-dimensional shapes of proteins.

In short, DNA guides the synthesis of RNA, then RNA guides the synthesis of proteins:

$$\text{DNA} \xrightarrow{\textit{transcription}} \text{RNA} \xrightarrow{\textit{translation}} \text{PROTEIN}$$

The newly synthesized proteins will have structural and functional roles in cells. Some even will have roles in building more DNA, RNA, and proteins.

KEY CONCEPTS

1. Life cannot exist without enzymes and other proteins. Proteins consist of polypeptide chains, which consist of amino acids. The sequence of amino acids corresponds to a gene, which is a sequence of nucleotide bases in a DNA molecule.

2. The path leading from genes to proteins has two steps, called transcription and translation.

3. In transcription, the double-stranded DNA molecule is unwound at a gene region, then an RNA molecule is assembled on the exposed bases of one of the strands.

4. In translation, a certain type of RNA directs the linkage of one amino acid after another, in the sequence required to produce a specific kind of polypeptide chain.

5. With few exceptions, the genetic "code words" by which DNA instructions are translated into proteins are the same in all species of organisms.

6. A mutation is a permanent change in a gene's base sequence. Such changes are the original source of genetic variation in populations.

7. Mutations give rise to alterations in protein structure, protein function, or both. The alterations may lead to small or large differences in traits among the individuals of a population.

DISCOVERING THE CONNECTION BETWEEN GENES AND PROTEINS

GARROD'S HYPOTHESIS During the early 1900s a physician, Archibald Garrod, was puzzling over certain illnesses. They appeared to be heritable, for they kept recurring in the same families. They also appeared to be metabolic disorders, for blood or urine samples from the affected patients contained abnormally high levels of a substance that was known to be produced at a certain step in a metabolic pathway.

Most likely, the enzyme that operated at the *next* step in the metabolic pathway was defective. Something was preventing it from chemically recognizing or interacting properly with the substance that is supposed to be its substrate. If Garrod's reasoning were correct, then the metabolic pathway would be blocked from that step onward, as Figure 14.2 indicates.

Garrod's hypothesis could explain why the molecules of a particular substance were accumulating in excess amounts in the body fluids of affected individuals. He suspected that the metabolic activities of his patients differed from unaffected individuals in one important

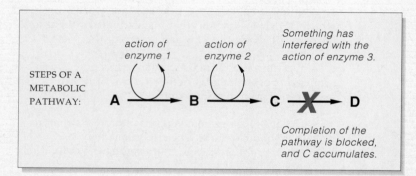

STEPS OF A METABOLIC PATHWAY:

action of enzyme 1   action of enzyme 2   *Something has interfered with the action of enzyme 3.*

A ⟶ B ⟶ C ✗ D

Completion of the pathway is blocked, and C accumulates.

Figure 14.2 How a defective enzyme can block completion of a metabolic pathway.

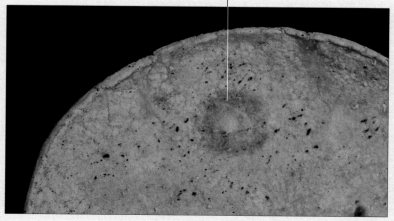

fungal colony on a tortilla, and isn't *that* appetizing

Figure 14.3 Colony of the red bread mold (*Neurospora crassa*) and other fungal species on a stale tortilla.

respect: They had inherited a single metabolic defect. Thus, Garrod concluded, specific "units" of inheritance (genes) must function through the synthesis of specific enzymes.

BEADLE, TATUM, AND A BREAD MOLD Thirty-three years later two researchers, George Beadle and Edward Tatum, were experimenting with the red bread mold (*Neurospora crassa*). This fungus is a common spoiler of baked goods (Figure 14.3). But it also lends itself to genetic experiments and has become an organism of choice in the laboratory. *N. crassa* can be grown easily on an inexpensive culture medium that contains only sucrose, mineral salts, and biotin, which is one of the B vitamins. The fungal cells can synthesize all of the other nutrients they require, including other vitamins.

Suppose a fungal enzyme that takes part in a synthesis pathway is defective as a result of a gene mutation. The researchers suspected that this had happened in some of the strains of *N. crassa* that they were studying. One strain grew only when it was supplied with vitamin B_6, another with vitamin B_1, and so on.

Chemical analysis of cell extracts revealed a different defective enzyme in each mutant strain. In other words, *each inherited mutation corresponded to a defective enzyme*. Here was evidence favoring the "one-gene, one-enzyme" hypothesis.

CLUES FROM GEL ELECTROPHORESIS The one-gene, one-enzyme hypothesis was refined as a result of investigations into the genetic basis of *sickle-cell anemia*. This heritable disorder arises from the presence of an abnormal version of a protein, hemoglobin, in the red blood cells of affected individuals. The abnormal hemoglobin is designated HbS instead of HbA (Section 11.5).

In 1949, the biochemists Linus Pauling and Harvey Itano subjected molecules of HbS and HbA to **gel electrophoresis**. This laboratory procedure uses an electric field to move molecules through a viscous gel and separate them according to their size, shape, and net surface charge. Often the gel is sandwiched between glass or plastic plates to form a viscous slab. The two ends of the slab are suspended in two salt solutions that are connected by electrodes to a power source (Figure 14.4). When voltage is applied to the apparatus, the molecules present in the gel migrate through the electric field according to their individual charge, and they move away from one another in the gel. Later on, the molecules can be pinpointed by staining the gel after a predetermined period of electrophoresis.

Pauling and Itano carefully layered a mixture of HbS and HbA molecules on the gel at the top of the slab. As the molecules gradually migrated down through the gel, they separated into distinct bands. The band that moved fastest carried the greatest

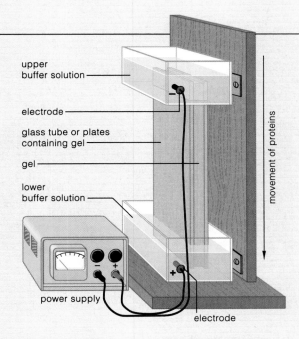

upper
buffer solution

electrode

glass tube or plates
containing gel

gel

lower
buffer solution

power supply

electrode

movement of proteins

Figure 14.4 One type of apparatus that is used for gel electrophoresis studies.

beta chain

beta chain

alpha chain

alpha chain

a Arrangement of the four polypeptide chains of a hemoglobin molecule, and a closer view of one of the beta chains.

Figure 14.5 A single amino acid substitution that starts with a gene mutation and ends with the symptoms of sickle-cell anemia, which are described in Section 11.5.

surface charge, and this turned out to be composed of HbA molecules. HbS molecules moved more slowly.

ONE GENE, ONE POLYPEPTIDE Later, Vernon Ingram pinpointed the difference between HbS and HbA. Recall that hemoglobin contains four polypeptide chains (Figure 14.5). Two are designated alpha and the other two beta. An abnormal HbS chain arises from a gene mutation that affects protein synthesis. The mutation causes valine instead of glutamate to be added as the sixth amino acid of the beta chain (Figure 14.5c). Whereas glutamate carries an overall negative charge, valine has no net charge—and so HbS behaved differently in the electrophoresis studies.

As a result of the one mutation, HbS hemoglobin has a "sticky" (hydrophobic) patch. In blood capillaries, where oxygen concentrations are at their lowest, hemoglobin molecules interact at the sticky patches. They aggregate into rods and distort red blood cells, and the consequences adversely affect organs throughout the body (Section 11.5).

The discovery of the genetic difference between alpha and beta chains of hemoglobin suggested that *two* genes code for hemoglobin—one for each kind of polypeptide chain. More importantly, it further suggested that genes must code for proteins in general, not just for enzymes.

And so a more precise hypothesis emerged: *The amino acid sequences of polypeptide chains—the structural units of proteins—are encoded in genes.*

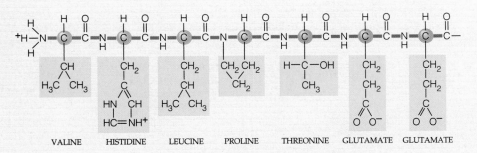

VALINE   HISTIDINE   LEUCINE   PROLINE   THREONINE   GLUTAMATE   GLUTAMATE

b The normal sequence of amino acids at the start of the beta chain of the HbA molecule

VALINE   HISTIDINE   LEUCINE   PROLINE   THREONINE   VALINE   GLUTAMATE

c The single amino acid substitution (coded *yellow*) that gives rise to the abnormal beta chain of the HbS molecule

The Three Classes of RNA

Before turning to the details of protein synthesis, let's clarify one point. The introduction to this chapter might have left you with the impression that protein synthesis requires only one class of RNA molecules. Actually, it requires three. Transcription of most genes produces **messenger RNA**, or **mRNA**—the only class of RNA that carries *protein-building* instructions. Transcription of some other genes produces **ribosomal RNA**, or **rRNA**, a major component of ribosomes. Ribosomes, recall, are structural units upon which polypeptide chains can be assembled. Transcription of still other genes produces **transfer RNA**, or **tRNA**. This delivers amino acids one by one to a ribosome in the order specified by mRNA.

How RNA Is Assembled

An RNA molecule is almost but not quite like a single strand of DNA. RNA, too, consists of only four types of nucleotides. Each nucleotide has a five-carbon sugar ribose (not DNA's deoxyribose), a phosphate group, and a base. Three of RNA's bases—adenine, cytosine, and guanine—are the same as in DNA. However, the fourth type of base is **uracil**, not thymine (Figure 14.6). Like thymine, uracil can pair with adenine. This means a new RNA strand can be put together on a DNA region according to base-pairing rules (Figure 14.7).

Transcription resembles DNA replication in another respect. Enzymes add the nucleotide units to a growing

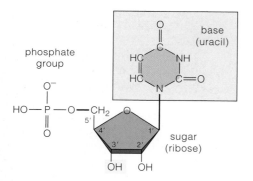

Figure 14.6 Structural formula for one of the four types of RNA nucleotides. The three others have a different base (adenine, guanine, or cytosine instead of uracil, shown here). Compare Figure 13.6, which shows DNA's four nucleotides. Notice that the sugars of DNA and RNA differ at one group only (*yellow*).

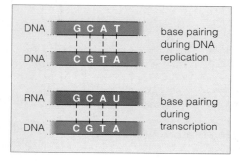

Figure 14.7 Comparison of how nucleotide bases pair with one another in DNA and in RNA.

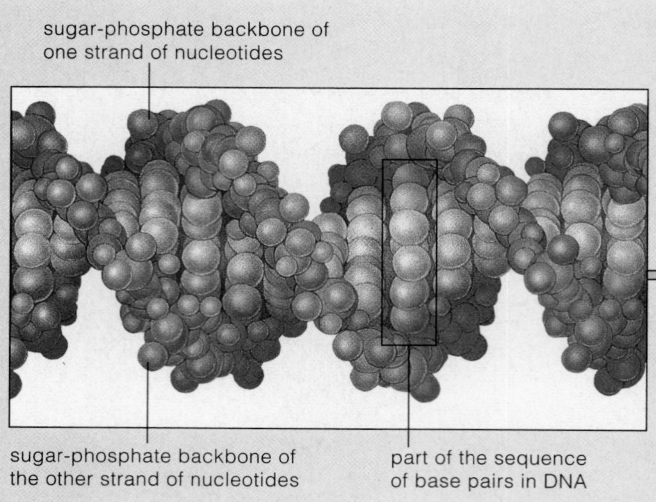

sugar-phosphate backbone of one strand of nucleotides

sugar-phosphate backbone of the other strand of nucleotides

part of the sequence of base pairs in DNA

a This sketch shows a gene region in part of a DNA double helix. In this region, the base sequence of one of the two nucleotide strands (not both) is about to be transcribed into an RNA molecule.

Figure 14.8   The process of gene transcription, by which an RNA molecule is assembled on a DNA template.

RNA strand one at a time, in the 5' → 3' direction. (Here you may wish to refer to Figure 13.9c.)

Transcription *differs* from DNA replication in three key respects. First, only a selected stretch of one DNA strand, not the whole molecule, serves as the template. Second, different enzymes, called **RNA polymerases**, catalyze the addition of nucleotides to the 3' end of a growing strand. Third, transcription results in a single, free strand of RNA nucleotides.

Transcription starts at a **promoter**, a base sequence in DNA that signals the start of a gene. Proteins help position an RNA polymerase on the DNA so it binds to the promoter. The enzyme then moves along the DNA strand, joining nucleotides one after another (Figure 14.8). When it reaches a base sequence that serves as a stop signal, the RNA is released as a free transcript.

Finishing Touches on the mRNA Transcripts

In eukaryotic cells alone, a newly formed molecule of mRNA is unfinished; it must undergo modifications before its protein-building instructions can be put to use. Just as a dressmaker might snip off some threads or add some bows on a dress before it leaves the shop, so do eukaryotic cells tailor their pre-mRNA.

Enzymes immediately attach a cap to the 5' end of a pre-mRNA molecule. The cap is a nucleotide that has a methyl group and phosphate groups bonded with it.

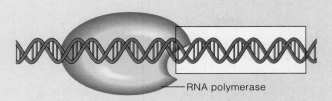

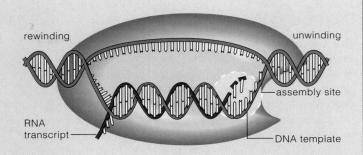

b A molecule of RNA polymerase binds with a promoter region in the DNA. It will recognize the base sequence positioned "downstream" from that site as a template for linking together the nucleotides adenine, cytosine, guanine, and uracil into a strand of RNA.

d All through transcription, the DNA double helix becomes unwound just in front of the RNA polymerase. Short lengths of the newly forming RNA strand temporarily wind up with the DNA template strand. Then they unwind from it, and the two strands of DNA wind up together again.

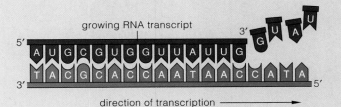

c During transcription, RNA nucleotides are base-paired, one after another, with exposed bases on the DNA template.

e At the end of the gene region, the last stretch of the RNA is unwound and released from the DNA template.

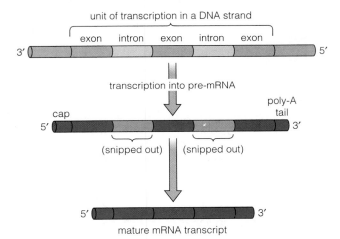

Figure 14.9 Transcription and modification of newly formed mRNA in the nucleus of eukaryotic cells. The cap simply is a nucleotide with functional groups attached. The poly-A tail is a string of adenine nucleotides.

Also, enzymes attach a tail of about 100 to 200 adenine-containing nucleotides to the 3' end of most pre-mRNA transcripts. Hence the name, "poly-A tail" (for multiple adenine units). The tail gets wound up with proteins. Later, in the cytoplasm, the cap will assist the binding of mRNA to a ribosome. Also, enzymes will gradually destroy the wound-up tail from the tip on back, then the mRNA. Such tails "pace" enzyme access to mRNA.

Apparently they help keep protein-building messages intact for as long as the cell requires them.

Besides these alterations, the mRNA message itself is modified. Most eukaryotic genes have one or more **introns**, base sequences that do not get translated into an amino acid sequence. The introns intervene between **exons**, the only parts of mRNA that get translated into protein. As Figure 14.9 shows, introns are transcribed right along with exons, but enzymes snip them out before the mRNA leaves the nucleus in mature form.

It could be that some introns are evolutionary junk, the leftovers of past mutations that led nowhere. Yet other introns are sites where instructions for building a particular protein can be snipped apart and spliced back together in various ways. The alternative splicing allows different cells in your body to use the same gene to make different versions of a pre-mRNA transcript, and therefore different versions of the resulting protein. We will return to this topic in the next chapter.

During gene transcription, a sequence of exposed bases in one of the two strands of a DNA molecule serves as the template for assembling a single strand of RNA. The assembly follows base-pairing rules (adenine only with uracil, cytosine only with guanine).

Before leaving the nucleus, each new mRNA transcript, or pre-mRNA, undergoes modification into final form.

DECIPHERING THE mRNA TRANSCRIPTS

The Genetic Code

Like a strand of DNA, an mRNA molecule is a linear sequence of nucleotides. What are the protein-building "words" encoded in that sequence? Gobind Khorana, Marshall Nirenberg, and other investigators came up with the answer. They deduced that RNA polymerases "read" nucleotide bases *three at a time*, as triplets. In an mRNA strand, such base triplets are called **codons**.

Figure 14.10 will give you an idea of how the order of different codons in an mRNA strand dictates the order in which particular amino acids will be added to a growing polypeptide chain.

Count the codons listed in Figure 14.11, and you see there are sixty-four kinds. Notice how most of the twenty kinds of amino acids correspond to more than one codon. Glutamate corresponds to the code words GAA *or* GAG, for example. Also notice how AUG has dual functions. It codes for the amino acid methionine, and it also is an initiation codon, the START signal for translating the mRNA transcript at a ribosome. That is,

a Base sequence of a gene region in DNA:

b Part of an mRNA strand, transcribed from the DNA:

c What the amino acid sequence will be when the mRNA is translated into a polypeptide chain:

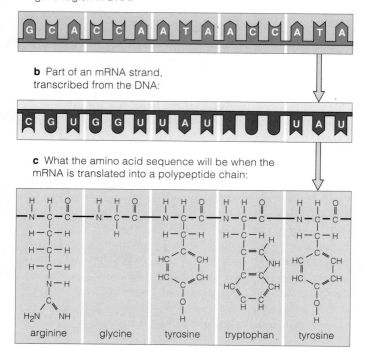

arginine   glycine   tyrosine   tryptophan   tyrosine

Figure 14.10 The steps from genes to proteins. (**a**) This diagram represents a region of a DNA double helix that was unwound during transcription. (**b**) The exposed bases on one DNA strand served as a template for assembling an mRNA strand. In the new mRNA transcript, every three nucleotide bases equaled one codon. Each codon calls for one amino acid in a polypeptide chain. (**c**) Referring to Figure 14.7, can you fill in the blank codon for tryptophan in the chain?

First Letter	Second Letter				Third Letter
	U	C	A	G	
U	phenylalanine	serine	tyrosine	cysteine	U
	phenylalanine	serine	tyrosine	cysteine	C
	leucine	serine	STOP	STOP	A
	leucine	serine	STOP	tryptophan	G
C	leucine	proline	histidine	arginine	U
	leucine	proline	histidine	arginine	C
	leucine	proline	glutamine	arginine	A
	leucine	proline	glutamine	arginine	G
A	isoleucine	threonine	asparagine	serine	U
	isoleucine	threonine	asparagine	serine	C
	isoleucine	threonine	lysine	arginine	A
	methionine (or START)	threonine	lysine	arginine	G
G	valine	alanine	aspartate	glycine	U
	valine	alanine	aspartate	glycine	C
	valine	alanine	glutamate	glycine	A
	valine	alanine	glutamate	glycine	G

Figure 14.11 The genetic code. The codons in mRNA are nucleotide bases, "read" in blocks of three. Sixty-one of the base triplets correspond to specific amino acids. Three others serve as signals that stop translation. The left column of the diagram shows the first of the three nucleotides in each codon in mRNA. The middle columns show the second nucleotide. The right column shows the third. Reading from left to right, for instance, the triplet UGG corresponds to tryptophan. Both UUU and UUC correspond to phenylalanine.

the "three-bases-at-a-time" selections start at the first AUG in the nucleotide sequence of the transcript. The codons UAA, UAG, UGA do not correspond to amino acids. They are STOP signals that prevent the further addition of amino acids to a new polypeptide chain.

The set of sixty-four different codons is the **genetic code**. It is the basis of protein synthesis in all organisms.

Roles of tRNA and rRNA

In a cell's cytoplasm are pools of free amino acids and free tRNA molecules. The tRNAs each have a molecular "hook," an attachment site for amino acids. They also have an **anticodon**, a nucleotide triplet that is able to

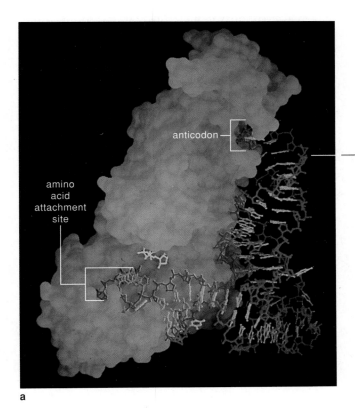

a

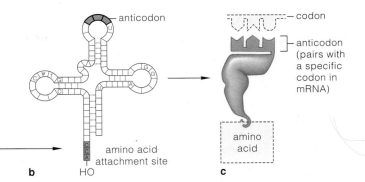

anticodon

codon

anticodon (pairs with a specific codon in mRNA)

amino acid attachment site

HO

b

amino acid

c

Figure 14.12 (**a**) Computer-generated, three-dimensional model of one type of tRNA molecule. The tRNA (*reddish brown*) is shown attached to a bacterial enzyme (*green*), along with an ATP molecule (*gold*). This particular enzyme catalyzes the attachment of amino acids to tRNAs.

(**b**) Structural features common to all tRNAs. (**c**) Simplified model of tRNA that you will come across in illustrations to follow. The "hook" that is sketched at one end is the site to which a specific amino acid can become attached.

One small ribosomal subunit, with a platform beneath the surface of the site shown by the *red* arrow.

+

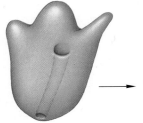

One large ribosomal subunit, with a tunnel through part of its interior.

In an intact ribosome, the platform and tunnel are aligned, as indicated by this side view.

Figure 14.13 Model of eukaryotic ribosomes. Polypeptide chains are assembled on the platform of the small ribosomal subunit. Newly forming chains may move through the tunnel of the large ribosomal subunit.

base-pair with codons (Figure 14.12). When tRNAs bind to the codons, they position their attached amino acids automatically, in the order specified by mRNA.

A cell has sixty-four kinds of codons, but it is able to utilize fewer kinds of tRNAs. How do they match up? According to base-pairing rules, adenine must pair with uracil, and cytosine with guanine. However, for codon-anticodon interactions, the rules loosen up at the third base. As an example, CCU, CCC, CCA, and CCG all specify proline. All three can pair with the tRNA that carries proline. Such freedom in the codon-anticodon pairing at a third base is known as the "wobble effect."

Even before anticodons interact with the codons of an mRNA strand, that strand must bind to specific sites on the surface of ribosomes. As shown in Figure 14.13, each ribosome has two subunits. These are assembled inside the nucleus from rRNA and protein components, some of which show enzyme activity. At some point,

the subunits are shipped separately to the cytoplasm. There they will combine as functional units only during translation of the message encoded in mRNA.

The nucleotide sequence of both DNA and mRNA encodes protein-building instructions. The genetic code is a set of sixty-four base triplets (nucleotide bases, read in blocks of three). A codon is a base triplet in mRNA.

Different combinations of codons specify the amino acid sequence of different polypeptide chains, start to finish.

mRNAs are the only molecules that carry protein-building instructions from DNA to the cytoplasm.

tRNAs deliver amino acids to ribosomes, where they base-pair with codons in the order specified by mRNA. Their action translates mRNA into a sequence of amino acids.

rRNAs are components of ribosomes, the structures upon which amino acids are assembled into polypeptide chains.

STAGES OF TRANSLATION

Translation of the mRNA proceeds in the cytoplasm. It has three stages: initiation, elongation, and termination.

In *initiation*, a tRNA that can start transcription and an mRNA transcript are both loaded onto a ribosome. First, the initiator tRNA binds with the small ribosomal subunit. Its anticodon can base-pair with AUG, the start codon in the transcript. The AUG also binds with the small subunit. Second, a large ribosomal subunit binds with the small subunit and thereby forms the initiation complex (Figure 14.14a). The next stage can begin.

In *elongation*, a new polypeptide chain forms as the mRNA passes between the ribosomal subunits, like a thread being moved through the eye of a needle. Again,

ribosomal components that function as enzymes join amino acids in the sequence dictated by the codons of mRNA. As you can see from Figure 14.14b, they catalyze the formation of a peptide bond between the growing polypeptide chain and each new amino acid delivered to the intact ribosome. (Here you may wish to refer to the diagram of peptide bond formation in Section 3.6.)

In *termination*, a stop codon is reached and there is no tRNA with a corresponding anticodon. Now release factors bind to the A site and trigger enzyme action that detaches the mRNA and the chain from the ribosome (Figure 14.14c). As Section 4.5 indicates, the detached chain may simply join the pool of free proteins in the cytoplasm. Or even before it is finished, the chain may enter the cytomembrane system, starting with the inner

Figure 14.14 Translation, the second step of protein synthesis.

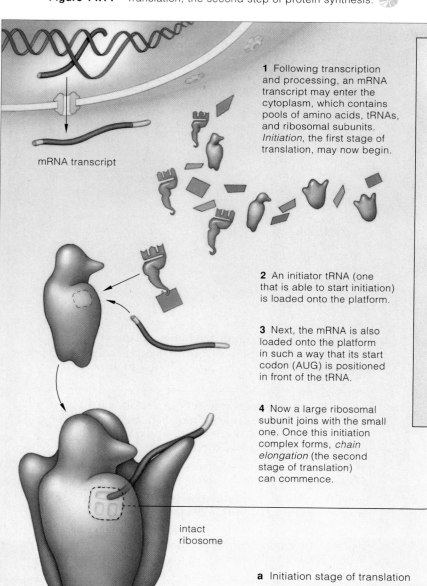

mRNA transcript

1 Following transcription and processing, an mRNA transcript may enter the cytoplasm, which contains pools of amino acids, tRNAs, and ribosomal subunits. *Initiation*, the first stage of translation, may now begin.

2 An initiator tRNA (one that is able to start initiation) is loaded onto the platform.

3 Next, the mRNA is also loaded onto the platform in such a way that its start codon (AUG) is positioned in front of the tRNA.

4 Now a large ribosomal subunit joins with the small one. Once this initiation complex forms, *chain elongation* (the second stage of translation) can commence.

intact ribosome

a Initiation stage of translation

binding site for mRNA

P (first binding site for tRNA)

A (second binding site for tRNA)

1 Close-up of binding sites on the platform of the small ribosomal subunit. It shows the relative positions of the binding sites for an mRNA transcript and for tRNAs that deliver amino acids to the intact ribosome.

A U G C U G
U A C

amino acid 1

G A C

amino acid 2

2 The initiator tRNA is already positioned in the first tRNA binding site (*P*). Its anticodon matches up with the start codon (AUG) of the mRNA strand, which already is in position also. Another tRNA is about to move into the second binding site (*A*). This tRNA is one that can bind with the codon that follows the start signal.

b Elongation stage of translation

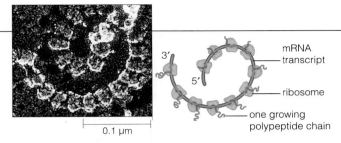

mRNA transcript

ribosome

one growing polypeptide chain

0.1 μm

Figure 14.15 From a eukaryotic cell, a polysome, or numerous ribosomes simultaneously translating the same mRNA molecule.

spaces of rough ER. Many of the newly formed chains take on final form in the system before they are shipped to their ultimate destinations inside or outside the cell.

Often, numerous ribosomes and tRNAs translate the same mRNA transcript simultaneously. The transcript threads through all the ribosomes, which are arranged one after another in assembly-line fashion, as in Figure 14.15. Such close arrays of ribosomes on the transcript are often called **polysomes**. Their presence indicates that a cell is rapidly producing a number of copies of a polypeptide chain from the same mRNA transcript.

Translation is initiated through the convergence of a small ribosomal subunit, an initiator tRNA, an mRNA transcript, and then a large ribosomal subunit.

More tRNAs deliver amino acids to the ribosome in the order dictated by the sequence of mRNA codons, to which the tRNA anticodons base-pair. A polypeptide chain grows as peptide bonds form between every two amino acids.

Translation is over when a stop codon triggers events that cause the chain and the mRNA to detach from the ribosome.

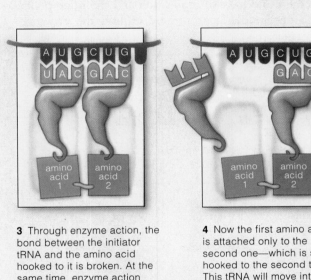

3 Through enzyme action, the bond between the initiator tRNA and the amino acid hooked to it is broken. At the same time, enzyme action catalyzes the formation of a peptide bond between the two amino acids. After these bonding events are completed, the initiator tRNA will be released from the ribosome.

4 Now the first amino acid is attached only to the second one—which is still hooked to the second tRNA. This tRNA will move into the P site on the ribosomal platform, sliding the mRNA with it by one codon. When it does so, the third codon will become aligned above the A site.

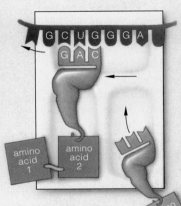

5 A third tRNA is about to move into the A site. Its anticodon is capable of base-pairing with the third codon of the mRNA transcript. Next, through enzyme action, a peptide bond will form between amino acids 2 and 3.

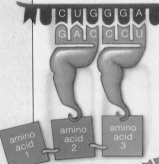

6 Steps 3 through 5 are repeated again and again. The polypeptide chain continues to grow until a stop codon is reached in the mRNA transcript. That is when termination, the last stage of protein synthesis, can begin, as shown in (**c**).

1 Once a stop codon is reached, the mRNA transcript is released from the ribosome.

2 The newly formed polypeptide chain also is released.

c Chain termination stage of translation

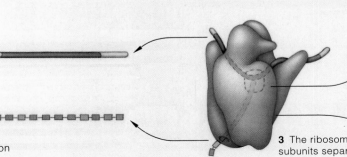

3 The ribosomal subunits separate.

HOW MUTATIONS AFFECT PROTEIN SYNTHESIS

Whenever a cell puts its genetic code into action, it is making precisely those proteins that it requires for its structure and functions. If something changes a gene's code words, the resulting protein may also change. If the protein is central to cell architecture or metabolism, we can expect the outcome to be an abnormal cell, such as those sickled red blood cells that make HbS instead of HbA hemoglobin.

Gene sequences do change. Sometimes one base gets substituted for another in the nucleotide sequence. At other times, an extra base is inserted into it or a base is lost. Such small-scale changes in the nucleotide sequence of a DNA molecule are **gene mutations**. There is some leeway here, for more than one codon often specifies the same amino acid. For instance, if a mutation changes UCU to UCC, it probably would not have dire effects, because both codons specify serine (Figure 14.11). More often, though, gene mutations give rise to proteins with skewed or blocked functions.

Common Types of Gene Mutations

Figures 14.16 and 14.17 show examples of two common types of gene mutations. In the first, adenine is wrongly paired with a cytosine unit in the DNA strand. Repair enzymes probably will detect the error, then remove and replace one of the bases. However, they may "fix" the mismatch by substituting the wrong base for that pair. The outcome of such a **base-pair substitution** may be the replacement of one amino acid with a different one during protein synthesis. This is what happened, recall, in individuals who carry the mutated gene that gives rise to sickle-cell anemia (Section 14.1).

In the second example, an extra base is inserted into a gene region of DNA. Remember, polymerases read a nucleotide sequence in blocks of three. This insertion shifts the three-at-a-time reading frame; hence the name "frameshift mutation." The gene now has a different message, and an abnormal protein will be synthesized.

Frameshift mutations fall within broader categories of gene mutation, called **insertions** and **deletions**. In such cases, one to several base pairs are inserted into a DNA molecule or deleted from it.

As a final example, Barbara McClintock discovered that mutations can result when **transposable elements** are on the move. These are DNA regions that move spontaneously from one location to another in the same DNA molecule or to a different one. Often they will inactivate the genes into which they become inserted. As Figure 14.18 indicates, the unpredictability of such jumps can cause interesting variations in phenotype.

Causes of Gene Mutations

Many gene mutations arise spontaneously while DNA is being replicated. This should not come as a surprise, given the rapid pace of replication and the huge pools of free nucleotides concentrated around the growing

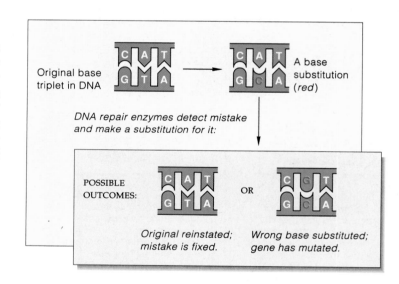

Figure 14.16 Example of a base-pair substitution.

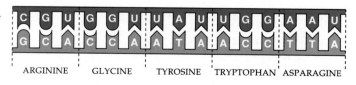

Resulting mRNA transcript of DNA:

Part of parental DNA template:

Resulting amino acid sequence: ARGININE   GLYCINE   TYROSINE   TRYPTOPHAN   ASPARAGINE

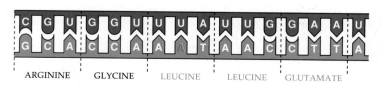

Now mRNA has altered message:

Parental template has one base insertion (red):

Resulting amino acid sequence: ARGININE   GLYCINE   LEUCINE   LEUCINE   GLUTAMATE

Figure 14.17 Example of an insertion, a type of mutation in which an extra base gets inserted into a gene region of DNA. This insertion has caused a *frameshift*; it has changed the reading frame for base triplets in the DNA and in the mRNA transcript of that region. As a result, the wrong amino acids will be called up when the mRNA transcript is translated into protein.

Figure 14.18 Barbara McClintock, who won a Nobel Prize for her insight that some genes can move from one site to another in DNA molecules, as transposable elements. In her hands is an ear of Indian corn (*Zea mays*). Its kernels sent her on the road to discovery.

In the kernels, all cells have the same pigment-coding genes. But some kernels are colorless or spottily colored. In the ancestor of the plant from which this ear of corn was plucked, a gene in a germ cell left its position in a DNA molecule, invaded another DNA molecule, and shut down a pigment gene. The plant inherited the mutation. As cell divisions proceeded in the growing plant, none of the mutated cell's descendants was able to synthesize pigment molecules. Each gave rise to colorless kernel tissue. Later, in some cells, the movable gene slipped out of the gene specifying pigment. And all the descendants of *those* cells produced pigment—and colored kernel tissue.

strands. Proofreading and repair enzymes detect most of them but, like most people, they are good but not perfect. A low number of mistakes do slip past these enzymes with predictable frequency.

Each gene has a characteristic **mutation rate**, which is the probability it will mutate spontaneously during a specified interval, such as each DNA replication cycle. (This is not the same as mutation *frequency*, the number of times a gene mutation has occurred in a population, as in 1 million gametes that produced 500,000 people.)

Mutation rates vary. For bacteriophages and bacteria, it is about 10^{-5} to 10^{-7} per generation, or once every 100,000 to 100 million replication cycles. For eukaryotes in general, it ranges between 10^{-4} and 10^{-6} per gene per generation. Consider the gene for hemoglobin. The rate at which it mutates to H^a, the recessive allele that causes hemophilia, is 3×10^{-5}. The rate at which the red-eye gene in *Drosophila melanogaster* mutates to the white-eye allele is 4×10^{-5}. It might be that differences in proofreading functions give rise to such variations in rates. Or maybe some mutations are silent, meaning we just can't detect them by ordinary genetic tests.

Not all mutations are spontaneous. Many result after exposure to mutagens, or mutation-causing agents in the environment. Ultraviolet radiation, especially the 260-nanometer wavelength in sunlight, is mutagenic. This is the wavelength DNA absorbs most strongly, and it often induces crosslinks to form between pyrimidine neighbors on the same DNA strand. Skin cancers are one outcome (Section 13.4). Other mutagens are gamma rays and x-rays. They ionize water and other molecules around the DNA, so free radicals form. These molecular fragments, which have an unpaired electron, can attack the structure of DNA. Such **ionizing radiation** causes base substitutions or breaks in one or both strands.

Natural and synthetic chemicals in the environment can accelerate the rate of spontaneous mutations. For example, **alkylating agents** might transfer methyl or ethyl groups to reactive sites on the bases or phosphate groups of DNA. At an alkylated site, the DNA becomes more susceptible to breaks and base-pair disruptions that invite mutation. Many cancer-causing agents, or **carcinogens**, operate by alkylating DNA.

The Proof Is in the Protein

When you think about the examples of mutation rates, you can deduce that spontaneous gene mutations are rare in terms of a human lifetime. If one arises in a somatic cell, any good or bad consequences will not endure, for it cannot be passed on to offspring. If the gene mutation arises in a germ cell or gamete, however, it may enter the evolutionary arena. The same is true of a mutation in an asexually reproducing organism or cell. In all such cases, the test is this: *A protein specified by a heritable mutation may have harmful, neutral, or beneficial effects on the ability of an individual to function in the prevailing environment.* As you will read in the next unit of the book, the outcomes of gene mutations can have powerful evolutionary consequences.

A gene mutation is an alteration in one to several bases in the nucleotide sequence of DNA.

Each gene has a spontaneous and characteristic mutation rate, which may be accelerated by exposure to harmful radiation and certain chemicals in the environment.

A protein specified by a mutated gene may have harmful, neutral, or beneficial effects on the ability of an individual to function in the prevailing environment.

SUMMARY

1. Cells, and multicelled organisms, cannot stay alive without enzymes and other proteins. A protein consists of one or more polypeptide chains, each of which is composed of a linear sequence of amino acids.

a. The amino acid sequence of a polypeptide chain corresponds to a gene region in a double-stranded DNA molecule. Each gene is a sequence of nucleotide bases in one of the two strands. The bases are adenine, thymine, guanine, and cytosine (A, T, G, and C).

b. For most genes, that sequence corresponds to a linear sequence of specific amino acids for a particular polypeptide chain. (Some genes specify tRNA or rRNA, not the mRNA that is translated into proteins.)

2. The path from genes to proteins has two steps, called transcription and translation:

$$DNA \xrightarrow{\text{transcription}} RNA \xrightarrow{\text{translation}} PROTEIN$$

a. During transcription, the double-stranded DNA is unwound at a gene region. Enzymes use its exposed bases as a template, or a structural pattern, to assemble a strand of ribonucleic acid (RNA) from the cell's pool of free nucleotides.

b. During translation, three different classes of RNAs interact in the synthesis of polypeptide chains, which later twist, fold, and often become modified into the final, three-dimensional shape of the protein.

c. Figure 14.19 is a visual summary of this flow of genetic information from DNA to proteins, as it occurs in eukaryotic cells. The DNA is transcribed into RNA in the nucleus, but RNA is translated in the cytoplasm. Prokaryotic cells (bacteria) lack a nucleus; transcription *and* translation proceed in their cytoplasm.

d. Understanding of the connection between genes and proteins started with studies of mutations that affected particular enzymes known to catalyze steps in metabolic pathways. Comparisons between normal and abnormal proteins, hemoglobin especially, led to the hypothesis that the amino acid sequence of polypeptide chains is encoded in genes.

3. Here are the key points concerning transcription:

a. When RNA is transcribed from exposed bases of DNA, base-pairing rules govern its assembly. Guanine pairs with cytosine, as in DNA replication, but uracil (not thymine) pairs with adenine in RNA:

DNA:	thymine	adenine	guanine	cytosine
RNA:	adenine	**uracil**	cytosine	guanine

b. Different gene regions in DNA serve as templates for assembling different RNA molecules.

c. Messenger RNA (mRNA) is the only class of RNA that has protein-building instructions.

d. Ribosomal RNA (rRNA) becomes a component of ribosomes, the physical units upon which polypeptide chains will be assembled.

e. Transfer RNA (tRNA) is the vehicle of translation; it will latch onto free amino acids in the cytoplasm and delivers them to ribosomes. It will do so in a sequence that corresponds to the sequential message of mRNA.

f. RNA transcripts of eukaryotic cells are processed into final form before being shipped from the nucleus. For example, mRNA's noncoding portions (introns) are excised, and its coding portions (exons) are spliced together. Only mature mRNA transcripts get translated.

4. Here are the key points concerning translation:

a. mRNA interacts with tRNAs and ribosomes so that amino acids get linked in the sequence required to produce a specific kind of polypeptide chain.

b. Translation is based on the genetic code, a set of sixty-four base triplets. A triplet is a series of nucleotide bases that ribosomal proteins "read" in blocks of three.

c. In mRNA, a base triplet is a codon. An anticodon is a complementary triplet in tRNA. Some combination of codons specifies what the amino acid sequence of a polypeptide chain will be, start to finish.

5. Translation proceeds through three stages:

a. Initiation. A small ribosomal subunit, an initiator tRNA, and an mRNA transcript converge. The small subunit then binds with a large ribosomal subunit.

b. Chain elongation. tRNAs deliver amino acids to the ribosome. Their anticodons base-pair with codons in mRNA. The amino acids become linked by peptide bonds to form a new polypeptide chain.

c. Chain termination. A stop codon in mRNA causes the chain and the mRNA to detach from the ribosome.

6. Gene mutations are potentially heritable, small-scale alterations in the nucleotide sequence of DNA.

a. Many gene mutations arise spontaneously during DNA replication. Others arise after the DNA is exposed to mutagens, such as ultraviolet radiation and other mutation-causing agents in the environment.

b. A base-pair substitution (replacement of one base pair by a different one) affects a single codon, but one amino acid substitution may alter protein function. Insertions of one or more bases into a gene or deletions from it can shift the reading frame to specify different amino acids. Transposable (movable) elements typically inactivate genes into which they become inserted.

c. Each gene has a characteristic mutation rate: the probability that it will spontaneously mutate in some specified time interval, such as a DNA replication cycle.

7. A protein specified by a mutated gene might have harmful, neutral, or beneficial effects on the individual. The outcome depends on prevailing conditions in the internal and external environments. Somatic mutations affect individuals only. Mutations in reproductive cells are heritable and can enter the evolutionary arena.

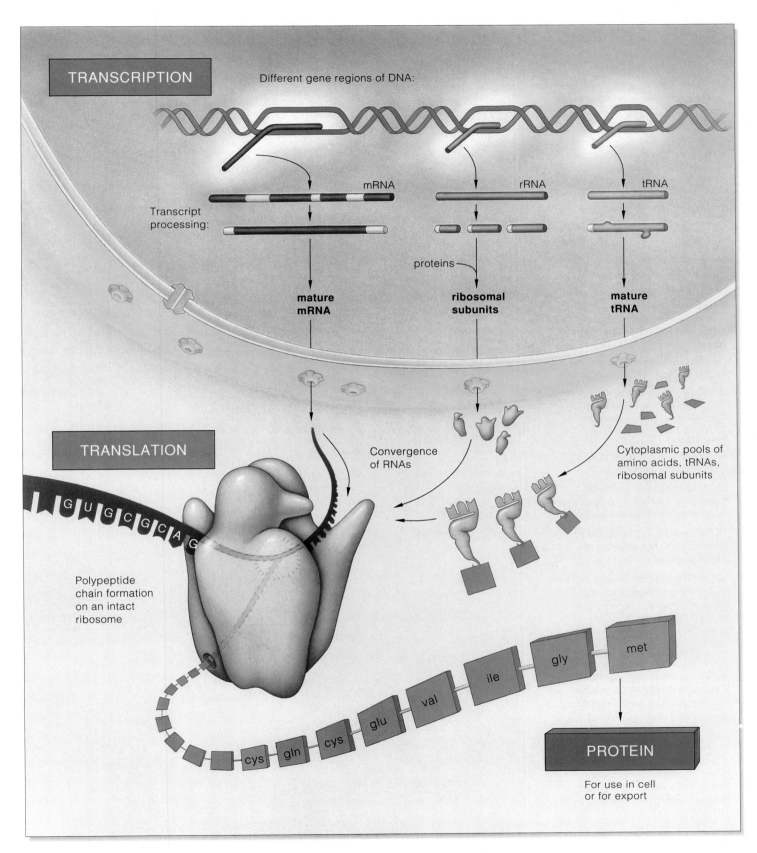

TRANSCRIPTION

Different gene regions of DNA:

mRNA

rRNA

tRNA

Transcript processing:

proteins

mature mRNA

ribosomal subunits

mature tRNA

TRANSLATION

Convergence of RNAs

Cytoplasmic pools of amino acids, tRNAs, ribosomal subunits

G U G C G C A G

Polypeptide chain formation on an intact ribosome

cys   gln   cys   glu   val   ile   gly   met

PROTEIN

For use in cell or for export

Figure 14.19 Visual summary of the steps of protein synthesis—transcription and translation—as it proceeds in all eukaryotic cells.

Review Questions

1. Are the polypeptide chains of proteins assembled on DNA? If so, state how. If not, state how they are assembled, and on which molecules. *CI, 14.2*

2. Define gene transcription and translation, the two stages of events by which proteins are synthesized. Both stages proceed in the cytoplasm of prokaryotic cells. Where does each stage proceed in eukaryotic cells? *CI*

3. Briefly state how gel electrophoresis, a common laboratory procedure, works. How did it yield a clue that small differences in normal and abnormal versions of the same protein may lead to big differences in how the proteins function? *14.1*

4. Name the three classes of RNA and briefly describe their functions. *14.2, 14.3*

5. In what key respect does the sequence of nucleotide bases in RNA differ from those in DNA? *14.2*

6. How does the process of gene transcription resemble DNA replication? How does it differ? *14.2*

7. The pre-mRNA transcripts of eukaryotic cells contain introns and exons. Are the introns or exons snipped out before the transcript leaves the nucleus? *14.2*

8. Distinguish between codon and anticodon. *14.3*

9. Cells use the set of sixty-four codons in the genetic code to build polypeptide chains from twenty kinds of amino acids. Do different codons specify the same amino acid? If so, in what respect do they differ? *14.3*

10. Name the three stages of translation and briefly describe the key events of each one. *14.4*

11. Review Figure 14.19. Then, on your own, fill in the blanks of the diagram at the right.

12. Define gene mutation. Do all mutations arise spontaneously? Do environmental agents serve as the trigger for change in each case? *14.5*

13. Define and state the possible outcomes of the following types of mutation: a base-pair substitution, a base insertion, and an insertion of a transposable element at a new location in the DNA. *14.5*

14. Define and explain the difference between mutation rate and mutation frequency. What determines whether an altered product of a mutation will have helpful, neutral, or harmful effects? *14.5*

Self-Quiz (*Answers in Appendix IV*)

1. Garrod, then Beadle and Tatum, deduced that the individual, heritable mutations they investigated corresponded to _____ .
   a metabolic disorders
   c. defective tRNAs
   b. defective enzymes
   d. both a and b

2. DNA contains many different genes that are transcribed into different _____ .
   a proteins
   c. mRNAs, tRNAs, and rRNAs
   b. mRNAs
   d. all are correct

3. An RNA molecule is _____ .
   a. a double helix
   c. always double-stranded
   b. usually single-stranded
   d. usually double-stranded

4. An mRNA molecule is produced by _____ .
   a. replication
   c. transcription
   b. duplication
   d. translation

5. Each codon calls for a specific _____ .
   a. protein
   c. amino acid
   b. polypeptide
   d. carbohydrate

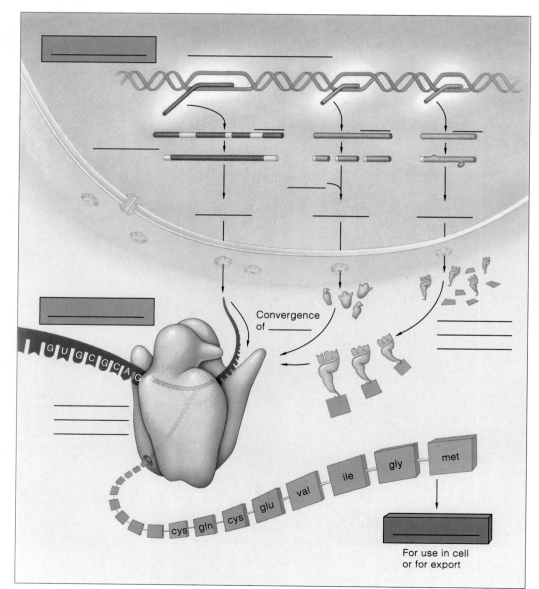

Convergence of _____

cys gln cys glu val ile gly met

For use in cell or for export

6. Referring to Figure 14.11, use the genetic code to translate the mRNA sequence UAUCGCACCUCAGGAGACUAG. Notice that the first codon in the frame is UAU. Which amino acid sequence is being specified?

a. TYR—ARG—THR—SER—GLY—ASP—STOP

b. TYR—ARG—THR—SER—GLY

c. TYR—ARG—TYR—SER—GLY—ASP—STOP

d. none of the above

7. Anticodons pair with _____ .
   a. mRNA codons
   b. DNA codons
   c. tRNA anticodons
   d. amino acids

8. In translation, an initiation complex consists of _____ .
   a. an initiator tRNA
   b. the start codon of mRNA
   c. a small ribosomal subunit
   d. a large ribosomal subunit
   e. all of the above

9. A polysome is _____ .
   a. the initiation site for translation
   b. a mutated ribosome
   c. a number of ribosomes on the same transcript
   d. a chromosomal duplication during transcription

10. Match the terms with the suitable description.
    ____ alkylating agent
    ____ chain elongation
    ____ exon
    ____ genetic code
    ____ anticodon
    ____ intron
    ____ codon

    a. coding part of mRNA transcript
    b. base triplet coding for amino acid
    c. second stage of translation
    d. base triplet that pairs with codon
    e. one kind of environmental agent that can induce mutation in DNA
    f. set of sixty-four codons for mRNA
    g. noncoding part of mRNA transcript

Critical Thinking

1. Noah has isolated a tRNA with a mutation in the anticodon 3′-AAU instead of 3′-AUU. What effect will the mutation have on protein synthesis in cells having the mutated tRNA?

2. Carefully examine the diagram of the genetic code in Figure 14.11. Identify the similarities and differences among all of the codons that specify the same amino acid. Can you discern any patterns in the similarities and differences? Researchers did, and on the basis of their insight, they formulated a hypothesis that the original genetic code in the first living cells on Earth may have been a simpler, *two-letter* code that specified fewer amino acids. In what respects do the patterns in Figure 14.11 seem to support their hypothesis?

3. A DNA polymerase made an error during the replication of an important gene region of DNA. None of the DNA repair enzymes detected or repaired the damage. A portion of the DNA strand with the error is shown here:

After the DNA molecule is replicated and two daughter cells have formed, one cell is carrying a mutation and the other cell is normal. Develop a hypothesis to explain this observation.

4. This diagram shows the nucleotide sequence for a gene:

When the gene is transcribed, this mRNA is produced:

5′ AUGCCCGCCUUUGCUACUUGGUAG 3′

An insertion, indicated in *red*, occurs in the gene:

What effect will this particular mutation have on the structure of the protein product?

5. In the bacterium *E. coli*, the end product (E) of the following metabolic pathway is absolutely essential for life:

enzyme 1   enzyme 3   enzyme 4

A
   →   C   →   D   →   E
B

enzyme 2

Ginetta, a geneticist, is attempting to isolate mutations in the genes for the four enzymes of this pathway. She has been able to isolate mutations in the genes for enzymes 1 and 2. In each case, the mutant *E. coli* cells synthesize a reduced amount of the pathway's end product E. However, Ginetta has not been able to isolate *E. coli* cells that have mutations in the genes for enzymes 3 and 4. Develop a hypothesis to explain why.

Selected Key Terms

alkylating agent *14.5*
anticodon *14.3*
base-pair substitution *14.5*
base sequence *CI*
carcinogen *14.5*
codon *14.3*
deletion (of base) *14.5*
exon *14.2*
gel electrophoresis *14.1*
gene mutation *14.5*
genetic code *14.3*
insertion (of base) *14.5*
intron *14.2*

ionizing radiation *14.5*
mRNA (messenger RNA) *14.2*
mutation rate *14.5*
polysome *14.4*
promoter (RNA) *14.2*
ribonucleic acid (RNA) *CI*
RNA polymerase *14.2*
rRNA (ribosomal RNA) *14.2*
transcription *CI*
translation *CI*
transposable element *14.5*
tRNA (transfer RNA) *14.2*
uracil *14.2*

Readings

Crick, F. October 1966. "The Genetic Code: III." *Scientific American.*

Nowak, R. 4 February 1994. "Mining Treasures From Junk DNA." *Science* 263: 608–610.

Russell, P. 1992. *Genetics.* Third edition. New York: Harper Collins. Chapter 18 has a good introduction to gene mutation.

Web Site See *http://www.wadsworth.com/biology* for practice quiz questions, hypercontents, BioUpdates, and critical thinking. The Wadsworth Biology Resource Center provides a wealth of information fully organized and integrated by chapter.

15 CONTROLS OVER GENES

Here's to Suicidal Cells!

Every day, a portion of you disappears as millions of body cells of your skin, intestines, thymus gland, and elsewhere commit suicide. Fortunately, millions of freshly dividing cells are just as rapidly replacing them and are thereby contributing to the survival of your various parts and, ultimately, you.

For all multicelled organisms, the very first cell of a new individual contains marching orders that will take its descendants through a program of growth, development, and reproduction, then on to eventual death. As part of that program, many cells heed calls to self-destruct when they finish a prescribed function. They also can execute themselves when they become altered in ways that might pose a threat to the body as a whole, as by infection or cancerous transformation. **Apoptosis** (pronounced APP-oh-TOE-sis) is the name for this form of cell death. As Figure 15.1a shows, it starts with molecular signals that activate and thus unleash lethal weapons of self-destruction that were stockpiled earlier within the cell itself. Apoptosis is not the same as necrosis, which is the passive death of many cells that results from severe tissue damage.

Figure 15.1b shows a cell in the act of suicide. First it shrank away from its neighbors. Now its cytoplasm

seems to be roiling, and its surface repeatedly bubbles outward and inward. No longer are the chromosomes extended through the nucleoplasm; they have bunched together near the nuclear envelope. The nucleus, then the whole cell, breaks apart. Phagocytic white blood cells patrol the body's tissues, and when they encounter suicidal cells or remnants of them, the patrolling cells swiftly engulf them.

The timing of cell death is predictable in some cells, such as the pigment-packed keratinocytes that form the densely packed sheets of dead cells that are continually sloughed off and replaced at the surface of your skin. They have a three-week life span, more or less. And yet keratinocytes and other body cells, even the kinds that are supposed to last a lifetime, can be induced to die ahead of schedule.

All it takes is sensitivity to specific signals that can activate weapons of death. Those weapons are protein-cleaving enzymes, the **ICE-like proteases**. They are named after the first to be discovered (*I*nterleukin-1 *C*onverting *E*nzyme). Think of them as folded pocket-knives or lethal Ninja weapons. When popped open, they can chop apart structural proteins, including the building blocks of cytoskeletal elements and

signal to die

a

Figure 15.1 (**a**) Computer model of the suicidal response of a cell to an external triggering signal. In this case, a membrane receptor directs the message to ICE-like proteases. When activated, these enzymes chop up the cytoskeleton and activate other enzymes that slice up the DNA. (**b**) A cell caught in the act of suicide, as evidenced by its surface blebs.

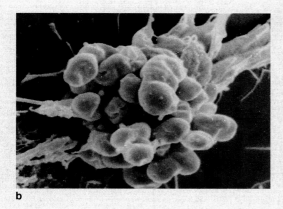

b

nucleosomes that organize the DNA (Figure 15.1*a*). They will do so, for example, if a cell is deprived of the signalling molecules called growth factors, if it loses contact with its neighbors, or if it receives altered signals about when to grow, divide, or cease dividing. The weapons also will be unleashed in the presence of certain regulatory proteins that can induce apoptosis.

The knives remain sheathed in cancer cells, which are supposed to—but don't—kill themselves on cue. As described at the end of this chapter, cancer arises as a result of mutations in several genes that govern cell growth and division. Researchers already know that the gene coding for p53 has been tampered with or inactivated in many masses of transformed cells. The normal form of the protein induces apoptosis when a cell's DNA is damaged. Intriguingly, p53 is absent or malfunctioning in more than half of the cancer patients that the researchers have investigated to date.

If cancer cells live long enough, they may end up with a number of mutations that do more than allow their descendants to divide uncontrollably. The cells may also end up ignoring the death signal if they manage to break away from their home tissue. Then they will be free to establish new colonies of cancer cells in distant tissues.

This fleeting glimpse of Ninja knives and cancerous transformations invites us to reflect on how lucky we are when proper gene controls are in place and our cells are operating as they should. As you will see in this chapter, controls govern whether any particular gene will be transcribed and translated in a particular cell. They also govern which products of genes will be activated or inhibited at any given time. The extent to which you and all other organisms on Earth depend on controls over genes and gene products is just astounding.

KEY CONCEPTS

1. In cells, a variety of controls govern when, how, and to what extent genes are expressed. The control elements operate in response to changing chemical conditions and to signals from the outside environment.

2. Control is exerted by way of regulatory proteins and other molecules that operate before, during, or after transcription. Different kinds interact with DNA, with RNA that has been transcribed from the DNA, or with gene products—that is, with the resulting polypeptide chains or final proteins.

3. Prokaryotic cells depend on rapid control over short-term shifts in nutrient availability and other aspects of their surrounding environment. Commonly, they rely on a small number of regulatory proteins that exert rapid, on-off control of transcription.

4. All eukaryotic cells depend on controls over short-term shifts in diet and levels of activity. In multicelled species, they also depend on controls over an intricate, long-term program of growth and development.

5. Controls over eukaryotic cells come into play when new cells contact one another in developing tissues, and as cells start interacting with their neighbors by way of hormones and other signaling molecules.

6. Although all cells of a multicelled organism inherit the same genes, different cell types activate or suppress many of those genes in different ways. The controlled, selective use of genes leads to synthesis of the proteins that give each type of cell its distinctive structure, function, and products.

At this very moment, bacteria are feeding on nutrients in your gut. Red blood cells are binding, transporting, or giving up oxygen, and great numbers of epithelial cells in your skin are busily synthesizing the protein keratin. Like cells everywhere, they are functioning by virtue of the protein products of genes.

Cells don't express all of their genes all of the time. Instead, cells use some genes and their products only once. They use other genes at certain times, all of the time, or not at all. *Which genes are being expressed depends on the type of cell, its moment-by-moment adjustments to changing chemical conditions, which signals from the outside it happens to be receiving—and its built-in control systems.*

For example, availability of nutrients shifts rapidly and often for the enteric bacteria, which inhabit animal intestines. Like other prokaryotic cells, they can rapidly transcribe certain genes and synthesize many nutrient-digesting enzyme molecules when nutrients happen to be moving past, and they can restrict synthesis when nutrients are scarce. By contrast, the composition and solute concentrations of the fluid bathing your cells do not shift drastically. And few of your cells exhibit rapid shifts in transcription.

The control systems consist of certain molecules, such as **regulatory proteins**, that come into play during transcription or translation, or after translation. Various components of the systems have a capacity to interact with DNA, RNA, or gene products (either newly minted polypeptide chains or final proteins, such as enzymes). Some operate in response to extracellular signals such as hormones. Others operate in response to changing concentrations of substances within the cell.

Negative control systems block the activity of their target molecule, and **positive control systems** promote it. Thus, one regulatory protein inhibits transcription of a particular gene when it binds with DNA, but the action of a different regulatory protein enhances that gene's transcription. Bear in mind, regulatory proteins do not always act alone. For instance, as you will read shortly, some types release their grip on a target when a different type comes along and interacts with them.

Summing up, these are the points to keep in mind as you read through the rest of this chapter:

Cells exert control over when, how, and to what extent each of their genes is expressed.

The expression of a given gene depends on the type of cell and its functions, on chemical conditions, and on signals from the outside environment.

Many regulatory proteins and other molecules exert control over gene expression through their interactions with DNA or RNA. Others exert control through their interaction with gene products, either polypeptide chains or final proteins.

Let's first consider some examples of gene control in prokaryotic cells—that is, bacteria. When nutrients are plentiful and other environmental conditions also favor growth, bacteria tend to grow and divide indefinitely. Gene controls promote the rapid synthesis of enzymes that have roles in nutrient digestion and other growth-related activities. Transcription is rapid, and translation is initiated even before mRNA transcripts are finished. Remember, bacteria have no nucleus; nothing separates their DNA from ribosomes in the cytoplasm.

When a nutrient-degrading pathway utilizes several enzymes, all genes for those enzymes are transcribed, often into one continuous mRNA molecule. The genes are not transcribed when conditions turn unfavorable. The rest of this section provides two cases of this all-or-nothing control of transcription.

Negative Control of Transcription

All mammals house the enteric bacterium *Escherichia coli*. It lives on glucose, lactose (a sugar in milk), and other ingested nutrients. Like other adult mammals, you probably don't drink milk around the clock. When you do, *E. coli* cells in your gut rapidly transcribe three genes for enzymes with roles in breakdown reactions that begin with lactose.

A promoter precedes the three genes, which are next to one another. A **promoter**, recall, is a base sequence that signals the start of a gene. Another sequence, an **operator**, intervenes between a promoter and bacterial genes. It is a binding site for a **repressor**, a regulatory protein that can block transcription (Figure 15.2). An arrangement in which a promoter and operator service more than one gene is an **operon**. Elsewhere in *E. coli* DNA, a different gene codes for this repressor, which can bind with the operator *or* with a lactose molecule.

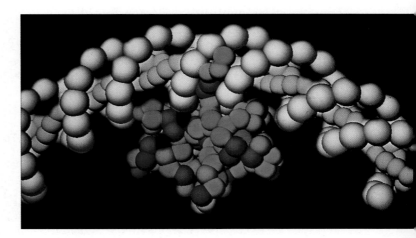

Figure 15.2   Model of a repressor protein (*green*) binding to an operator at a site in a bacterial DNA molecule (*blue*).

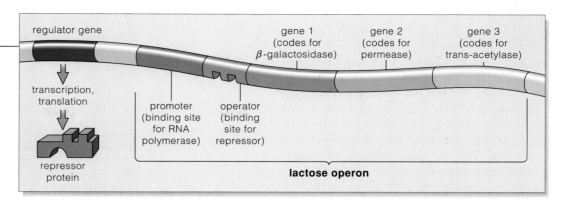

a A repressor protein exerts negative control over three genes of the lactose operon by binding to the operator and inhibiting transcription.

regulator gene

gene 1 (codes for β-galactosidase)

gene 2 (codes for permease)

gene 3 (codes for trans-acetylase)

transcription, translation

promoter (binding site for RNA polymerase)

operator (binding site for repressor)

repressor protein

lactose operon

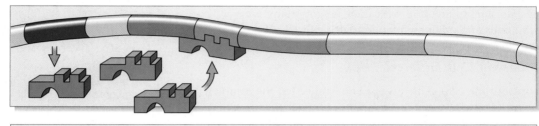

b When the concentration of lactose is low, the repressor is free to block transcription. Being bulky, it overlaps the promoter and prevents binding by RNA polymerase. The enzymes (not needed) are not produced.

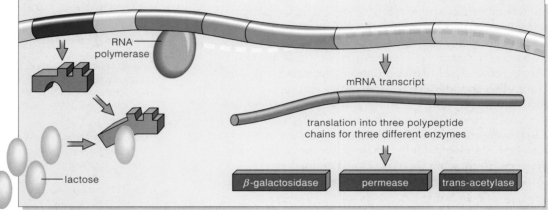

c At high concentration, lactose is an inducer of transcription. It binds to and distorts the shape of the repressor—which now cannot bind to the operator. The promoter is exposed and the genes can be transcribed.

RNA polymerase

mRNA transcript

translation into three polypeptide chains for three different enzymes

β-galactosidase

permease

trans-acetylase

lactose

Figure 15.3 Negative control of the lactose operon. The first gene of this operon codes for an enzyme that splits lactose, a disaccharide, into two subunits (glucose and galactose). The second codes for an enzyme that transports lactose into cells. The third enzyme functions in metabolizing certain sugars.

When the lactose concentration is low, a repressor binds with the operator, as in Figure 15.3. Being a large molecule, it overlaps the promoter, so transcription is blocked. Thus, *lactose-degrading enzymes are not built if they are not required.* When the concentration of lactose is high, odds are greater that a lactose molecule will bind with the repressor. Binding alters the repressor's shape, so it cannot bind with the operator—and RNA polymerase can now transcribe the genes. Thus, *lactose-degrading enzymes are synthesized only when required.*

Positive Control of Transcription

E. coli cells pay far more attention to glucose than to lactose. They transcribe genes for glucose breakdown continually, at faster rates. Even if lactose is present, the lactose operon isn't used much—*unless glucose is absent.*

At such times, a regulatory protein called CAP acts on the operon. CAP is an **activator protein**, a key player in a positive control system. To understand how it works, you have to know that the lactose operon's promoter is not good at binding RNA polymerase. It does a better job when CAP adheres to it first. But CAP won't do this unless it is activated by a small molecule called cAMP.

Among other things, cAMP is produced from ATP, which *E. coli* can produce by glucose breakdown (that is, by way of glycolysis). Not much cAMP is available in the cell when glucose is plentiful and glycolysis is proceeding full bore. When these conditions prevail, the activator protein does not become primed to adhere to the promoter, and transcription of the lactose operon genes slows almost to a standstill. Now suppose that glucose is scarce and lactose becomes available. cAMP can accumulate, CAP-cAMP complexes can form—and the lactose operon genes can be transcribed.

Prokaryotic cells, which must respond rapidly to changing conditions, commonly rely on a small number of regulatory proteins that exert rapid, on-off control of transcription.

CONTROLS IN EUKARYOTIC CELLS

Like bacteria, all eukaryotic cells depend on controls over short-term shifts in diet and level of activity. If the cells are merely one of hundreds, millions, or even trillions of cells in the body of a multicelled organism, they also require more intricate controls. For them, gene activity changes as a program of development unfolds, as new cells contact one another in the developing tissues, and as cells start interacting with their neighbors by way of hormones and other signaling molecules.

A Case of Cell Differentiation

Consider this: All cells in your body inherited the same genes, for they descended from the same fertilized egg. Many of the genes specify proteins that are basic to any cell's structure and functioning. That is why the protein subunits of ribosomes are the same from one cell to the next, as are many enzymes. *Yet nearly all of your cells became specialized in composition, structure, and function.* This process of **cell differentiation** proceeds while all multicelled species develop. It arises as embryonic cells and their descendants activate and suppress some of the total number of genes in unique, selective ways. Thus, only immature red blood cells use the genes for making hemoglobin. Only certain white blood cells use the genes for making weapons called antibodies.

Selective Gene Expression at Many Levels

Eukaryotic cells exert highly selective control of gene expression at many levels. Figure 15.4 and the next few paragraphs will give you a sense of what goes on at these levels. You will come across specific examples in the sections that follow.

CONTROLS RELATED TO TRANSCRIPTION   Many genes concerned with housekeeping tasks in eukaryotic cells are under positive controls that promote continuous, low levels of transcription. For many other genes, we see less of the rapid, all-or-nothing control over gene transcription that is common among bacteria. Why? In multicelled organisms, the internal environment (that is, tissue fluids and blood) affords much more stable operating conditions for individual cells. Most often, transcription rates rise or fall by degrees in response to slight shifts in concentrations of signaling molecules, substrates, and products in that environment.

As the examples in Figure 15.4*a* indicate, some gene sequences are repeatedly duplicated or rearranged in genetically programmed ways prior to transcription. Besides this, programmed chemical modifications often shut down many genes. So does the orderly packaging of DNA by histones and other chromosomal proteins,

a   CONTROLS RELATED TO TRANSCRIPTION.   At any given time, most genes of a multicelled organism are shut down, either permanently or temporarily. The genes necessary for a cell's everyday tasks are under positive controls that promote ongoing, low levels of transcription. With the help of these controls, a cell is assured of having enough enzymes and other proteins to carry out its most basic functions. Transcription of many other genes often shifts only slightly. In this case, controls work to assure chemical responsiveness even when concentrations of specific substances rise or fall only slightly.

Also, even before some genes are transcribed, a portion of their base sequences may be amplified, rearranged, or chemically modified in temporary or reversible ways. These are not mutations; they are heritable, programmed events that affect the manner in which particular genes will be expressed (if at all).

1. *Gene amplification.* Some cells that require enormous numbers of certain molecules temporarily increase the number of the required genes. In response to a molecular signal, multiple rounds of DNA replication sometimes produce hundreds or thousands of gene copies prior to transcription! This happens in immature amphibian eggs and glandular cells of some insect larvae (Section 15.4).

2. *DNA rearrangements.* In a few cell types, many base sequences are alternative "choices" for parts of a gene. Prior to transcription, they are snipped out of the DNA. Combinations of the snippets are spliced together as the gene's final base sequence. For instance, this happens when B lymphocytes, a special class of white blood cells, are forming (Section 40.9). The cells transcribe and then translate their uniquely rearranged DNA into staggeringly diverse versions of protein weapons called antibodies, which act specifically against one of staggeringly diverse kinds of pathogens and other foreign agents in the body.

3. *Chemical modification.* Numerous histones and other proteins interact with eukaryotic DNA in highly organized fashion. The DNA-protein packaging, as well as chemical modifications to particular sequences, influences gene activity. Usually, only a small fraction of a cell's genes are available for transcription. A more dramatic shutdown is called X chromosome inactivation in female mammals, as described in Section 15.4.

Figure 15.4   Examples of the levels of control over gene expression in eukaryotes.

as you may realize after reflecting on the organization of eukaryotic chromosomes (Section 9.4).

POST-TRANSCRIPTIONAL CONTROLS   Control also occurs after gene transcription (Figure 15.4*b–d*). Many controls govern transcript processing, transport of mature RNAs from the nucleus, and translation rates. Others deal with modification of the new polypeptide chains. Still others deal with activating, inhibiting, and degrading

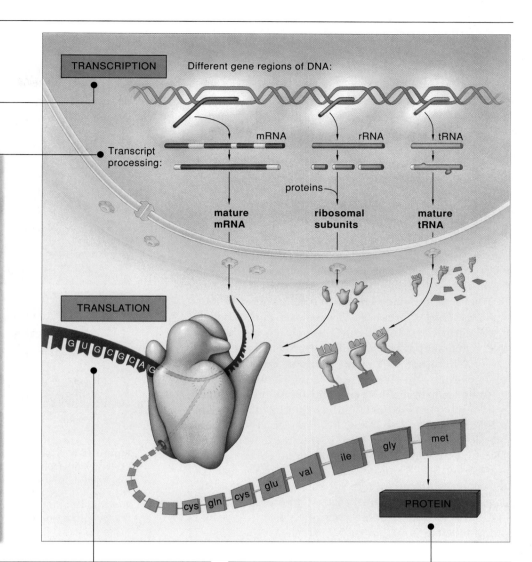

TRANSCRIPTION

Different gene regions of DNA:

mRNA   rRNA   tRNA

Transcript processing:

proteins

mature mRNA   **ribosomal subunits**   **mature tRNA**

TRANSLATION

G U G C G C A G

cys gln cys glu val ile gly met

PROTEIN

b TRANSCRIPT PROCESSING CONTROLS. As you read earlier in Section 14.2, pre-mRNA transcripts undergo modifications before they leave the nucleus.

For example, introns are removed and exons are spliced together in more than one way. In other words, a transcript from a single gene may undergo *alternative splicing*.

Pre-mRNA transcribed from a gene that specifies a contractile protein, troponin-1, is like this. Enzymes excise different portions of the pre-mRNA transcript in different cells. When the exons are spliced together, the final protein-building message is slightly different from one cell to the next. The resulting proteins are all very similar, but each is unique in a certain region of its amino acid sequence. The proteins function in slightly different ways, which may account for the subtle variations we observe in the functioning of different types of muscles in the body.

c CONTROLS OVER TRANSLATION. Diverse controls govern when, how rapidly, and how often a given mRNA transcript will be translated. Sections 37.2, 44.3, and other parts of the book provide elegant examples.

The stability of a transcript influences the number of protein molecules that can be produced from it. Enzymes destroy transcripts from the poly-A tail on up (Section 14.2). The tail's length and its attached proteins affect the pace of degradation. Also, after leaving the nucleus, some transcripts are inactivated, temporarily or permanently. For example, in unfertilized eggs, many transcripts are inactivated and stored in the cytoplasm. These "masked messengers" will not be available for translation until after fertilization, when great numbers of protein molecules will be required for the early cell divisions of the new individual.

d CONTROLS FOLLOWING TRANSLATION. Before they can become fully functional, many polypeptide chains must pass through the cytomembrane system (Section 4.5). There they undergo modification, as by having specific oligosaccharides or phosphate groups attached to them.

Also, a variety of control mechanisms govern the activation, inhibition, and stability of the enzymes and other molecules that are involved in protein synthesis. Allosteric control of tryptophan synthesis, as described earlier in Section 6.5, is an example.

the final proteins. Consider enzymes alone, the proteins that catalyze nearly all metabolic reactions. In addition to selectively transcribing and translating genes for those enzymes, control systems activate and inhibit enzyme molecules that already exist. Just imagine the coordination that governs which of thousands of types of enzymes become stockpiled, deployed, or degraded in a given interval. *That coordination governs all short-term and long-term aspects of cell structure and function.*

In each multicelled organism, gene controls underlie basic, short-term housekeeping tasks in cells and more intricate, long-term patterns of bodily growth and development.

All of those cells inherit the same genes, yet most become specialized in composition, structure, and function. This process of cell differentiation arises as different populations of cells activate and suppress some of their genes in highly selective, unique ways.

EVIDENCE OF GENE CONTROL

By some estimates, the cells of a multicelled organism rarely use more than 5 to 10 percent of their genes at a given time. One way or another, controls are keeping most of the genes repressed. In addition, which genes are being expressed varies, depending on the stage of growth and development that the organism is passing through. The following examples hint at the kinds of control mechanisms that operate at different stages.

Transcription in Lampbrush Chromosomes

While the unfertilized eggs of amphibians mature, they grow extremely fast. Growth requires numerous copies of enzymes and other proteins. The germ cells that give rise to the eggs stockpile large numbers of RNAs and ribosomes, all of which will have roles in rapid protein synthesis. During prophase I of meiosis, chromosomes in the germ cells decondense in such a way that the DNA loops out profusely from their protein scaffold.

The chromosomes look so bristly at this time, they are said to have a "lampbrush" configuration (Figure 15.5). At such times, histones and other proteins that structurally organize the DNA have loosened their grip, so that the genes specifying RNA molecules are now accessible. Histones, recall, are part of nucleosomes, the basic unit of organization in eukaryotic chromosomes (Section 9.4). Each nucleosome consists of a stretch of DNA looped around a core of histone molecules. The diagram in Figure 15.6b is based on photomicrographs that show the nucleosome packaging being loosened during transcription.

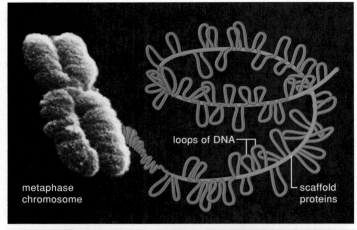

a

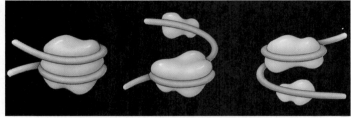

b

one nucleosome (a bit of DNA looped around a histone core)   one nucleosome loosened up

Figure 15.6 Changes in DNA organization that may promote transcription. (**a**) At certain times, DNA decondenses into loops that extend from attachment sites on scaffold proteins. (**b**) The tight DNA-histone packing in nucleosomes also may loosen up. Transcription proceeds mainly in regions where chromosome packaging has become most relaxed.

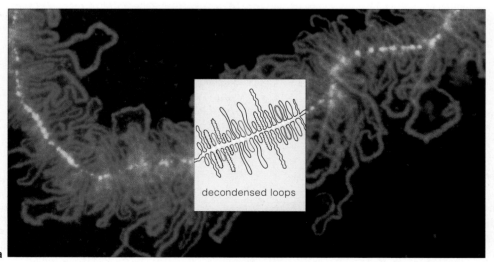

a

decondensed loops

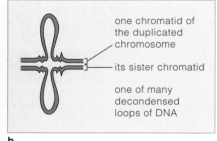

b

one chromatid of the duplicated chromosome

its sister chromatid

one of many decondensed loops of DNA

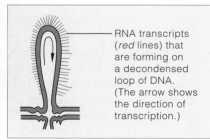

c

RNA transcripts (*red* lines) that are forming on a decondensed loop of DNA. (The arrow shows the direction of transcription.)

Figure 15.5 Lampbrush chromosome from a germ cell of a newt (*Notophthalmus viridiscens*). During prophase I, gene regions of this duplicated chromosome decondensed into thousands of loops. A *red* fluorescent dye labeled one of the components of ribonucleoproteins. The presence of ribonucleoproteins here is evidence of gene activity. (A *white* fluorescent dye labeled the proteins making up the axis of the duplicated chromosome's structural framework.)

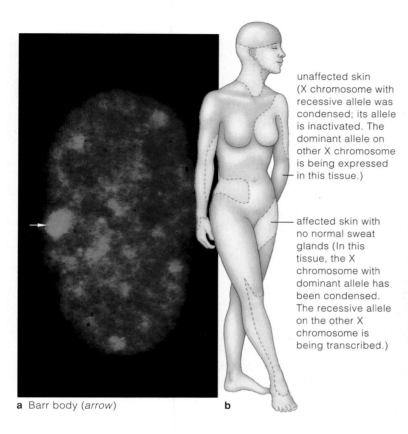

unaffected skin (X chromosome with recessive allele was condensed; its allele is inactivated. The dominant allele on other X chromosome is being expressed in this tissue.)

affected skin with no normal sweat glands (In this tissue, the X chromosome with dominant allele has been condensed. The recessive allele on the other X chromosome is being transcribed.)

a Barr body (*arrow*)

b

Figure 15.7 (**a**) Micrograph of an inactivated X chromosome, called a Barr body, as it appears in a human female's somatic cell during interphase. The X chromosome is not condensed this way in a human male's cells. (**b**) Anhidrotic ectodermal dysplasia, a mosaic pattern of gene expression. The condition arises as a result of random X chromosome inactivation.

Figure 15.8 Why is this female calico cat "calico"? In her cells, one X chromosome carries a dominant allele for the brownish-black pigment melanin. The allele on her other X chromosome specifies yellow fur. At an early stage of the cat's embryonic development, one of the two X chromosomes was inactivated at random in each cell that had formed by then. In all descendants of those cells, the same chromosome also became inactivated, leaving only one functional allele for the coat-color trait. We see patches of different colors, depending on which allele was inactivated in cells that formed a given tissue region. (The white patches result from a gene interaction involving the "spotting gene," which blocks melanin synthesis entirely.)

X Chromosome Inactivation

A mammalian zygote destined to become a female has two X chromosomes (one from the mother, one from the father). As it develops, *one* of the two condenses in each cell. The condensation is a programmed event, but the outcome is random. *One or the other chromosome may become inactivated.* You may observe such a condensed X chromosome in the interphase nucleus; it shows up as a dark spot (Figure 15.7*a*). It is called a **Barr body** after its discoverer, Murray Barr.

When the maternal (or paternal) X chromosome is inactivated in a cell, it also becomes inactivated in all of the cell's descendants. By adulthood, each adult female is "mosaic" for the X chromosomes. *She has patches of tissues where maternal genes of the X chromosome are being expressed—and patches of tissue in which paternal genes of the X chromosome are being expressed.* Because any pair of alleles on her two X chromosomes may or may not be identical, the tissue patches may or may not have the same characteristics.

Mary Lyon discovered the mosaic tissue effect that arises from random X chromosome inactivation. We see this effect in human females who are heterozygous for a recessive allele on the X chromosome that prevents sweat glands from forming. The lack of sweat glands is one symptom of *anhidrotic ectodermal dysplasia*. In the affected females, the X chromosome with the dominant

allele has condensed, and genes on the one bearing the mutant allele are being transcribed in the patches of skin with no sweat glands. Figure 15.7*b* is a diagram of the mosaic tissue effect. The same effect is apparent in female calico cats. Such cats are heterozygous for black and yellow coat-color alleles on their X chromosomes. The coat color in a given body region depends on which X chromosome's genes are transcribed (Figure 15.8).

In a typical cell of multicelled organisms, controls repress all but about 5 to 10 percent of the genes at a given time.

The remaining fraction of genes is selectively expressed at different stages of growth and development. Lampbrush chromosome formation, X chromosome inactivation, and other events provide evidence of this.

In the chapter introduction, we introduced the idea that a great variety of signals influence gene activity. The following examples from animals and plants will give you an idea of their effects at the molecular level.

Hormonal Signals

Hormones, a major category of signaling molecules, can stimulate or inhibit gene activity in target cells. Any cell with receptors for a given hormone is a target. Animal cells secrete hormones into tissue fluid. Most hormone molecules are picked up by the bloodstream, which distributes them to cells some distance away. In plants, hormones do not travel far from cells that secrete them.

Certain hormones bind to membrane receptors at the surface of target cells. Others enter target cells, bind to regulatory proteins, and so help initiate transcription. In many cases, a hormone molecule must first touch bases, so to speak, with an enhancer. **Enhancers** are base sequences that serve as binding sites for suitable activator proteins. They may or may not be adjacent to a promoter in the same DNA molecule. When some distance separates the two, a loop forms in the DNA and brings them

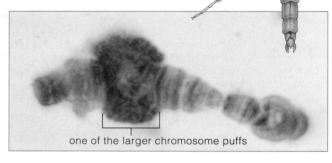

one of the larger chromosome puffs

Figure 15.9   Visible evidence of transcription in a polytene chromosome from a larva of a midge (*Chironomus*). Midges are flies, one to ten millimeters long. The short-lived, winged adults often congregate in great swarms, which help them find mates in a hurry. They might seem to be pests merely because of their large numbers. They also look somewhat like mosquitoes, but they don't bite.

Most midge larvae develop in aquatic habitats. To sustain their rapid growth, they feed continuously, as on decaying organic material. They require a lot of saliva, and they must continuously transcribe genes for saliva's protein components. Those genes have undergone amplification. Ecdysone, a hormone, serves as a regulatory protein that helps promote their transcription. Midge chromosomes loosen and puff out in regions where the genes are being transcribed in response to the hormonal signal. The puffs become large and appear quite diffuse when transcription is most intense. Staining techniques reveal banding patterns in the chromosomes, as evident in the micrograph.

together. RNA polymerase binds avidly to the bound complex at the base of the loop.

Consider the effect of **ecdysone**, a hormone with key roles in the life cycle of many insects. Immature stages of the insects, called larvae, grow rapidly. They feed continually on organic matter, such as decaying leaves, and require copious amounts of saliva to prepare food for digestion. DNA in their salivary gland cells has been replicated repeatedly. The copies of DNA molecules have remained together in parallel array, forming what is known as a *polytene* chromosome. When ecdysone binds to a receptor on the gland cells, its signal triggers very rapid transcription of multiple copies of genes in the DNA. The gene regions affected by this hormonal signal puff out during transcription, as in Figure 15.9. Afterward, translation of the mRNA transcripts from the genes produces the protein components of saliva.

In vertebrates, certain hormones have widespread effects on gene expression because many types of cells have receptors for them. As one example, the pituitary gland secretes somatotropin, or growth hormone. This hormonal signal stimulates synthesis of all the proteins required for cell division and, ultimately, the body's growth. Most cells have receptors for somatotropin.

Other vertebrate hormones signal only certain cells at certain times. Prolactin, a secretion from the pituitary gland, is like this. Beginning a few days after a female mammal gives birth, researchers can detect prolactin in the blood. This hormone activates genes in mammary gland cells that have receptors for it. Those genes have responsibility for milk production. Liver cells and heart cells have the same genes but do not have the receptors necessary to respond to signals from prolactin.

Explaining hormonal control of gene activity is like explaining a full symphony orchestra to someone who has never seen one or heard it perform. Many separate parts must be defined before their interactions can be understood! We will return to this topic, starting with Chapter 37 on the endocrine system. As you will see in Chapters 44 and 45, some of the most elegant examples of hormonal controls are drawn from studies of animal reproduction and development.

Sunlight as a Signal

Plant a few seeds from a corn or bean plant in a pot that contains moist, nutrient-rich soil. Next, let the seeds germinate, but keep them in total darkness. After eight days have passed, they will develop into seedlings that are spindly and notably pale, owing to the absence of chlorophyll (Figure 15.10). Now expose the seedlings to a single burst of dim light from a flashlight. Within ten minutes, they will start converting stockpiles of certain molecules to the activated form of chlorophylls, the

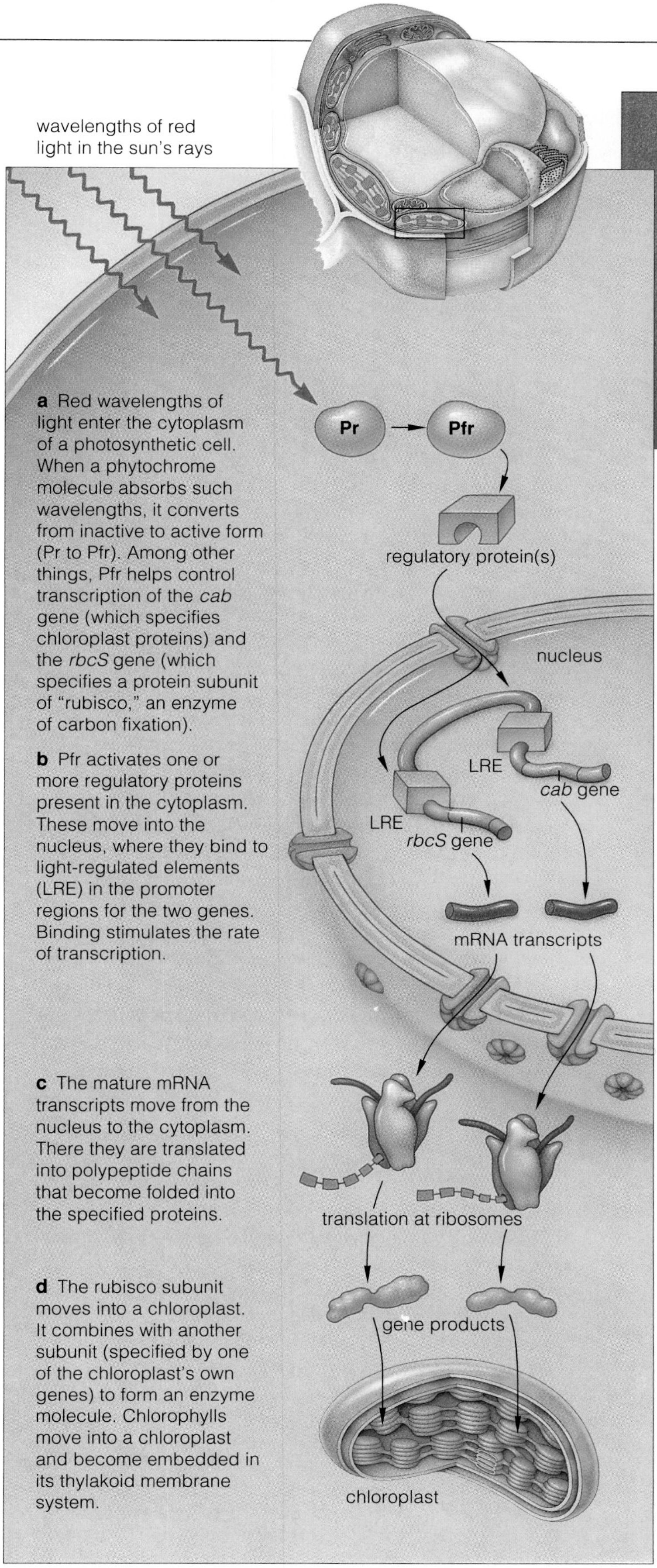

wavelengths of red
light in the sun's rays

a Red wavelengths of light enter the cytoplasm of a photosynthetic cell. When a phytochrome molecule absorbs such wavelengths, it converts from inactive to active form (Pr to Pfr). Among other things, Pfr helps control transcription of the *cab* gene (which specifies chloroplast proteins) and the *rbcS* gene (which specifies a protein subunit of "rubisco," an enzyme of carbon fixation).

b Pfr activates one or more regulatory proteins present in the cytoplasm. These move into the nucleus, where they bind to light-regulated elements (LRE) in the promoter regions for the two genes. Binding stimulates the rate of transcription.

c The mature mRNA transcripts move from the nucleus to the cytoplasm. There they are translated into polypeptide chains that become folded into the specified proteins.

d The rubisco subunit moves into a chloroplast. It combines with another subunit (specified by one of the chloroplast's own genes) to form an enzyme molecule. Chlorophylls move into a chloroplast and become embedded in its thylakoid membrane system.

Pr → Pfr

regulatory protein(s)

nucleus

LRE

cab gene

LRE

rbcS gene

mRNA transcripts

translation at ribosomes

gene products

chloroplast

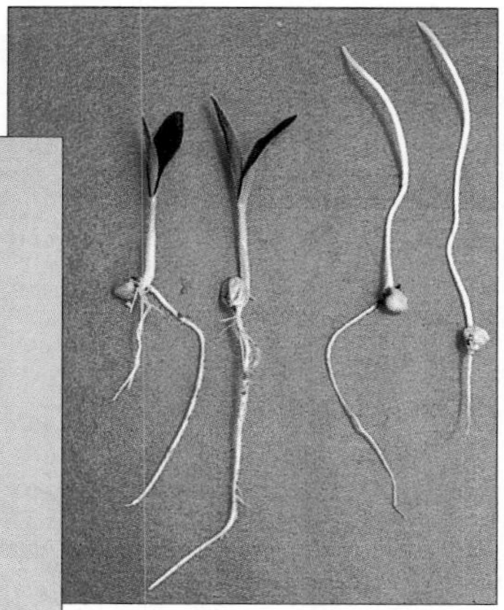

Figure 15.10 Sunlight as a signal for gene activity. The photograph shows the effect of the absence of light on corn seedlings. The two seedlings to the left were the control group; they were grown in sunlight, in a greenhouse. The two seedlings positioned next to them were grown in total darkness for eight days. The dark-grown plants were not able to convert their stockpiled precursors of chlorophyll molecules to active form, and they never did green up. The diagram illustrates one model for the mechanism by which phytochrome helps control gene transcription in plants. Red wavelengths can convert the phytochrome molecule from inactive form (here designated Pr) to active form (Pfr). In this form, the phytochrome can serve as a regulator of transcription.

light-trapping pigment molecules that are absolutely central to photosynthesis.

Phytochrome, a blue-green pigment, is a signaling molecule that helps plants adapt over the short term to changes in light conditions. Section 32.4 takes a close look at this molecule. For now, simply be aware that it alternates between active and inactive forms. At sunset, during the night, or in the shade, far-red wavelengths predominate, and the phytochrome in cells is inactive. It becomes activated at sunrise, when red wavelengths dominate the sky. In addition, the quantity of incoming red or far-red wavelengths varies as days alternate with nights, and as days grow shorter and then longer with the changing seasons. Such variations serve as signals that help regulate phytochrome activity, which in turn influences transcription of certain genes at certain times of day and year. The genes specify enzymes and other proteins with key roles in germination, stem elongation, branching, leaf expansion, and the formation of flowers, fruits, and seeds.

Figure 15.10 shows a model of phytochrome control of transcription. Experiments by Elaine Tobin and her coworkers at the University of California, Los Angeles, provide evidence in favor of such a model. Working with dark-grown seedlings of duckweed (*Lemna*), they discovered a marked increase in the number of certain mRNA transcripts after a one-minute exposure to red light. Exposure enhanced transcription of the genes for proteins that bind chlorophylls and for an enzyme that mediates carbon fixation. Both proteins are required for the development and greening of chloroplasts.

Hormones and other signaling molecules, including diverse kinds that respond to environmental stimuli, have profound influence on gene expression.

15.6 CANCER AND THE CELLS THAT FORGET TO DIE

THE CELL CYCLE REVISITED   Every second, millions of cells in your skin, gut lining, liver, and other body parts divide and replace their worn-out, dead, and dying predecessors. They do not divide willy-nilly. Controls govern the expression of genes that specify enzymes and other proteins required for cell growth, DNA replication, spindle formation, chromosome movements, and the division of the cytoplasm. Regulatory proteins control the synthesis and use of these gene products, and they control when the division machinery is put to rest.

Controls over the cell cycle are extensive. At various interrelated points, gene activity can be stepped up or slowed down, and gene products can advance, delay, or block the cycle. Some genes directly regulate passage through the cycle and thereby are its primary controllers. Among the most crucial are **protein kinases**, a class of enzymes that attach phosphate groups to proteins. You will find molecules of them operating at the boundary between G1 and S of the cycle, and at the transition from G2 to mitosis. (Here you might refer to Section 9.2.)

Other genes encode proteins that modify the activity of the primary control genes or their products. Among them are genes that specify **growth factors**, which are transcriptional signals sent by one cell to trigger growth in other cells. The products include CSF (short for *Colony Stimulating Factor*), which normally stimulates growth of certain white blood cells. EFG (for *Epidermal Growth Factor*) stimulates skin cells to grow and divide; NFG stimulates growth of neurons. Still other genes specify cell receptors for particular growth factors.

Also, certain genes specify enzymes and other factors required to maintain DNA replication, stockpile the ICE-like proteases, and perform other vital activities. They do not act directly as controls. But if they become altered, as by mutations, a cell may be deprived of important enzymes or proteins, and some controls over the cycle may be lost.

CHARACTERISTICS OF CANCER   When controls are lost, cell division will not stop as long as conditions for growth stay favorable. When cells are not responding to normal controls over growth and division, they form a tissue mass called a **tumor** (Figure 15.11). The cells of common skin warts and other *benign* tumors grow slowly, in an unprogrammed way. They still have surface recognition proteins that can hold them together in their home tissue. Surgically remove a benign tumor, and you remove its potential threat to the surrounding tissues.

In a *malignant* tumor, abnormal cells grow and divide more rapidly, with destructive physical and metabolic effects on surrounding tissues. These cells are disfigured, grossly so. They cannot construct a normal cytoskeleton or plasma membrane, and they cannot synthesize normal versions of recognition proteins. Such cells can break loose from their home tissue, enter lymph or blood vessels, travel through the body, and become lodged where they do not belong. This process of abnormal cell migration and invasion is called **metastasis** (Figure 15.12).

The photographs in Section 13.4 show merely 3 of the more than 200 types of malignant tumors that have been identified so far. **Cancer** is the general category in which all malignant tumors are grouped. Each year in developed countries alone, 15 to 20 percent of all deaths result from cancer. And it is not just a human affliction. Cancer has also been observed in most of the animals that have been studied. Similar abnormalities have even been observed in many kinds of plants.

At the minimum, all cancer cells show the following characteristics:

1. *Profound changes in the plasma membrane and cytoplasm.* Membrane permeability increases. Membrane proteins are lost or altered, and different ones form. The cytoskeleton becomes disorganized, shrinks, or both. Enzyme activity shifts, as in amplified reliance on glycolysis.

2. *Abnormal growth and division.* Controls that prevent overcrowding in tissues are lost. Cell populations reach high densities. New proteins trigger abnormal increases in small blood vessels that service the growing cell mass.

3. *Weakened capacity for adhesion.* Recognition proteins are altered or lost; cells can't stay anchored in proper tissues.

4. *Lethality.* Unless cancer cells are eradicated, they will kill the individual.

Any gene having the potential to induce a cancerous transformation is called an **oncogene**. Oncogenes were first identified in retroviruses, which are a class of RNA viruses. They are altered forms of normal genes, now called **proto-oncogenes**, that specify certain proteins required for normal cell functioning. In other words, normal expression of proto-oncogenes is vital, which helps explain why their *abnormal* expression is lethal.

Transformation may start with mutations in control elements that govern proto-oncogenes or with the

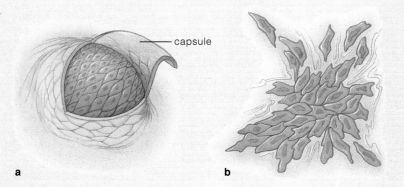

— capsule

a   b

Figure 15.11   Evidence of loss of gene control: (**a**) A benign tumor, a mass of cells with an outwardly normal appearance that are enclosed in a capsule of connective tissue. (**b**) A malignant tumor, a disorganized collection of cancer cells.

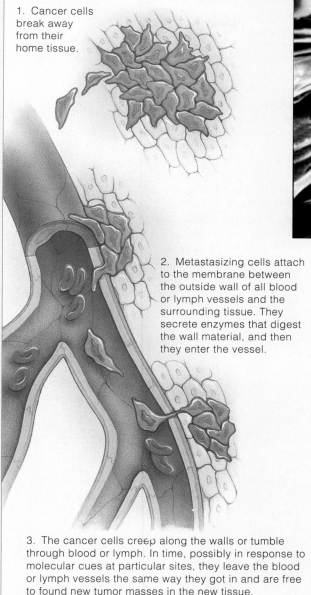

1. Cancer cells break away from their home tissue.

2. Metastasizing cells attach to the membrane between the outside wall of all blood or lymph vessels and the surrounding tissue. They secrete enzymes that digest the wall material, and then they enter the vessel.

3. The cancer cells creep along the walls or tumble through blood or lymph. In time, possibly in response to molecular cues at particular sites, they leave the blood or lymph vessels the same way they got in and are free to found new tumor masses in the new tissue.

a

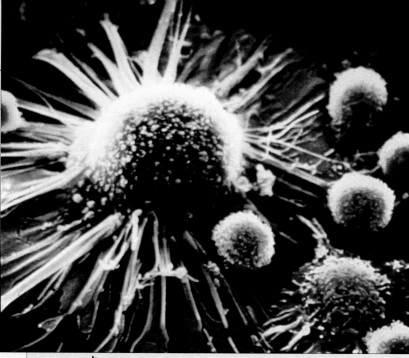

b

Figure 15.12 (**a**) Metastasis, a migration of cancer cells from the tissue in which they originated and their invasion of other tissues in the body. (**b**) The scanning electron micrograph shows a few of the body's defenders—white blood cells—flanking a cancer cell. The defenders may or may not be able to destroy it. Cancer arises through loss of controls that govern specific genes, many of which influence cell growth and division.

Consider the *myc* gene, which is located in human chromosome 8. It specifies a regulatory protein that helps control the cell cycle. Sometimes a break occurs in this gene region and in a region of chromosome 14. The two chromosomes swap segments. The *myc* gene ends up in a new location, where it may escape normal controls over its transcription and come under the influence of new ones. Whatever the case, a transformation begins that converts *myc* from a proto-oncogene to an oncogene— and Burkitt's lymphoma may be the outcome.

Burkitt's lymphoma is a malignant tumor of the white blood cells called B lymphocytes. This is the only type of cell that produces antibody molecules, which are protein weapons against specific agents of disease (Section 40.7). In activated B cells, a gene region that specifies antibody production is transcribed at an extremely high rate. The translocation between chromosomes 8 and 14 puts the *myc* gene next to this region and makes it abnormally active.

Yet cancer is a multistep process, involving more than one oncogene. Researchers have already identified three such genes with roles in nearly all colon cancers. In 1995, they identified a key gene in ovarian and breast cancers. Through such discoveries, it may be possible to diagnose carriers early and frequently. A cancer detected early enough may still be curable by surgery.

Perhaps more than any other example, cancerous transformations bring home the extent to which you and all other organisms depend on controls over transcription, translation, and when and where gene products will be used. This example alone reinforces the vital nature of the topics introduced in this chapter.

genes themselves. For instance, this happens with certain insertions of viral DNA into cellular DNA. It can happen when carcinogens (cancer-inducing agents) cause changes in the DNA. Ultraviolet radiation and ionizing radiation (x-rays and gamma rays) are common carcinogens. So are many natural and synthetic compounds, including asbestos and certain components of tobacco smoke.

Some cancers arise if base substitutions or deletions alter a proto-oncogene or one of the controls over its transcription. Others arise if a gene becomes abnormally amplified. And still others arise if a whole gene or part of it moves to a new location in the same chromosome or to an entirely different chromosome.

SUMMARY

1. Cells control gene expression; that is, which gene products appear, when, and in what amounts. When the control mechanisms come into play depends on the type of cell, prevailing chemical conditions, and outside signals that can change the cell's activities.

2. Regulatory proteins, enzymes, hormones, and other control elements interact with one another, with DNA and RNA, and with the gene products. They operate before, during, and after transcription and translation.

3. In all cells, two of the most common types of control systems operate to block or enhance transcription. Their effects are reversible.
   a. In negative control systems, a regulatory protein binds at a specific DNA sequence and prevents one or more genes from being transcribed.
   b. In positive control systems, a regulatory protein binds to DNA and promotes transcription.

4. Most prokaryotic cells (bacteria) do not require many genes to grow and reproduce. Most of their control systems affect transcription rates. Control of operons (groupings of related genes and control elements) are examples.

5. Promoters are DNA binding sites that signal the start of a gene. In bacteria alone, operators are binding sites for regulatory proteins that can inhibit transcription. Enhancers are binding sites for activator proteins that promote transcription.

6. Eukaryotic cells, particularly those of multicelled organisms, require more complex gene controls. Gene activity must change rapidly in response to short-term shifts in the surroundings, as in prokaryotic cells. But it also must be adjusted in intricate ways during long-term growth and development, when great numbers of cells multiply, make physical contact with one another in tissues, and interact chemically.

7. Being descended from the same cell, all of the cells in a multicelled organism inherit the same assortment of genes. However, different types of cells activate and suppress some fraction of the genes in different ways. This behavior is called selective gene expression.

8. One outcome of selective gene expression is called cell differentiation. By this process, different lineages of cells in the developing multicelled organism become specialized in appearance, composition, and function.

9. Cancer results when controls over the cell cycle and the cell's death machinery are lost. Some genes encode proteins that are the primary regulators of the cycle. Others encode proteins that modify the activity of the primary controllers or their products. Others encode enzymes and other proteins that can, when altered, indirectly affect the cycle.

Review Questions

1. In what fundamental way do negative and positive control of transcription differ? Is the effect of one or the other form of control (or both) reversible? *15.1*

2. Distinguish between: *15.2*
   a. promoter and operator
   b. repressor protein and activator protein

3. Describe one type of control over transcription of the lactose operon in *E. coli*, a prokaryotic cell. *15.2*

4. A plant, fungus, or animal is composed of diverse cell types. How might this diversity arise, given that the body cells in each organism inherit the same set of genetic instructions? As part of your answer, define cell differentiation and explain how selective gene expression brings it about. *15.3*

5. Review Figure 15.4. Using the diagram below, define five types of gene controls in eukaryotic cells and indicate where they take effect. *15.3*

6. If a polytene chromosome and a lampbrush chromosome are both evidence of transcriptional activity, in what respect do they differ? *15.4, 15.5*

7. What is a Barr body? Does it appear in the cells of males, females, or both? Explain your answer. *15.4*

8. Define hormones. Why do hungry midge larvae depend on the hormone ecdysone? *15.5*

9. What are the characteristics of cancer cells? Explain the difference between a benign tumor and one that is malignant. *15.6*

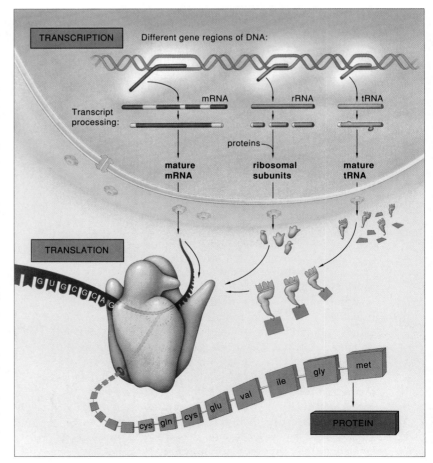

Self-Quiz (Answers in Appendix IV)

1. Apoptosis is _____ .
   a. cell death by severe tissue damage
   b. cell death by suicide
   c. a popping sound in mutated toes

2. ICE-like proteases are _____ .
   a. regulatory proteins   c. environmental signals
   b. lethal weapons   d. low-temperature enzymes

3. The expression of a given gene depends on _____ .
   a. cell type and functions   c. environmental signals
   b. chemical conditions   d. all of the above

4. Regulatory proteins interact with _____ .
   a. DNA   c. gene products
   b. RNA   d. all of the above

5. In prokaryotic but not eukaryotic cells, a(n) _____ precedes the genes of an operon.
   a. lactose molecule   c. operator
   b. promoter   d. both b and c

6. A base sequence signaling the start of a gene is a(n) _____ .
   a. promoter   c. enhancer
   b. operator   d. activator protein

7. An operon most typically governs _____ .
   a. bacterial genes   c. genes of all types
   b. a eukaryotic gene   d. DNA replication

8. Prokaryotic cells rely most heavily on _____ controls over transcription.
   a. slow, continuous   c. slow, on-off
   b. rapid, continuous   d. rapid, on-off

9. Cell differentiation _____ .
   a. occurs in all multicelled organisms
   b. requires different genes in different cells
   c. involves selective gene expression
   d. both a and c
   e. all of the above

10. Some gene controls in eukaryotic cells operate by _____ .
    a. amplifying genes   d. rearranging DNA
    b. processing RNA transcripts   e. all of the above
    c. chemically modifying DNA

11. X chromosome inactivation in mammalian females may result in a _____ for some traits.
    a. male phenotype   c. rise in transcription rates
    b. mosaic tissue effect   d. rise in translation rates

12. Hormones interact with _____ .
    a. membrane receptors   c. enhancers
    b. regulatory proteins   d. all of the above

13. Match the terms with their most suitable descriptions.
    ____ phytochrome   a. inhibits gene transcription
    ____ Barr body   b. gene with the potential to induce
    ____ oncogene   cancerous transformation
    ____ repressor   c. normally required, lethal when
    ____ proto-oncogene   mutated
                         d. helps plants adapt to changes
                            in light
                         e. inactivated X chromosome

Critical Thinking

1. Kristen and Dan have isolated a strain of *E. coli* in which a mutation has affected the capacity of CAP to bind to a region of the lactose operon, as it would do normally. State how the mutation affects transcription of the lactose operon when cells of this bacterial strain are subjected to the following conditions:
   a. Lactose and glucose are available.
   b. Lactose is available but glucose is not.
   c. Both lactose and glucose are absent.

2. *Duchenne muscular dystrophy*, a genetic disorder, affects boys almost exclusively. Early in childhood, muscles begin to atrophy (waste away) in affected individuals, who typically die in their teens or early twenties as a result of respiratory failure. Muscle biopsies of women who carry a gene associated with the disorder reveal some regions of atrophied muscle tissue. Yet muscle tissue adjacent to these regions was normal or even larger and more chemically active, as if to compensate for the weakness of the adjoining region. How can you explain these observations?

3. Unlike most rodents, guinea pigs are already well developed at the time of birth. Within a few days, they are able to eat grass, vegetables, and other plant material. Suppose a breeder decides to separate the baby guinea pigs from their mothers after three weeks. He wants to keep the males and females in different cages, but it is quite difficult to determine the gender of guinea pigs when they are so young. Suggest a simple test that the breeder can perform to determine gender.

4. Individuals affected by *pituitary dwarfism* cannot synthesize somatotropin (growth hormone). Children with this genetic abnormality will not grow to normal height unless they receive injections of somatotropin. Develop a hypothesis to explain why this hormone therapy is effective.

5. In this chapter, you have read that transcription is generally controlled by the binding of a protein to a DNA sequence that is "upstream" from a gene, rather than by modification of RNA polymerase. Develop a hypothesis to explain why this is so.

Selected Key Terms

activator protein 15.2	oncogene 15.6
apoptosis CI	operator 15.2
Barr body 15.4	operon 15.2
cancer 15.6	phytochrome 15.5
cell differentiation 15.3	positive control system 15.1
ecdysone 15.5	promoter 15.2
enhancer 15.5	protein kinases 15.6
growth factor 15.6	proto-oncogene 15.6
hormone 15.5	regulatory protein 15.1
ICE-like proteases CI	repressor 15.2
metastasis 15.6	tumor 15.6
negative control system 15.1	

Readings

Duke, R. D. Ojcius, and J. Ding-E Young. December 1996. "Cell Suicide in Health and Disease." *Scientific American* 80–87.

Feldman, M., and L. Eisenbach. November 1988. "What Makes a Tumor Cell Metastatic?" *Scientific American* 259(5): 60–85.

Murray, A., and M. Kirschner. March 1991. "What Controls the Cell Cycle?" *Scientific American* 264(3): 56–63.

Tijan, R. February 1995. "Molecular Machines That Control Genes." *Scientific American* 54–61.

Web Site See *http://www.wadsworth.com/biology* for practice quiz questions, hypercontents, BioUpdates, and critical thinking. The Wadsworth Biology Resource Center provides a wealth of information fully organized and integrated by chapter.

16

RECOMBINANT DNA AND GENETIC ENGINEERING

Mom, Dad, and Clogged Arteries

Butter! Bacon! Eggs! Ice cream! Cheesecake! Possibly you think of such foods as enticing, off-limits, or both. After all, who among us doesn't know about animal fats and the dreaded cholesterol?

Soon after you feast on these fatty foods, cholesterol enters the bloodstream. Cholesterol is important. It is a structural component of animal cell membranes, and without membranes, there would be no cells. Cells also remodel cholesterol into various molecules, including the vitamin D that is necessary for the development of good bones and teeth. Normally, however, your liver synthesizes enough cholesterol for your cells.

Some proteins circulating in the blood combine with cholesterol and other substances to form lipoprotein particles. The *HDLs* (high-density lipoproteins) collect cholesterol and transport it to the liver, where it can be metabolized. *LDLs* (low-density lipoproteins) normally end up in cells that store or use cholesterol.

Sometimes too many LDLs form, and the excess infiltrates the elastic walls of arteries. There they promote formation of abnormal masses called atherosclerotic plaques (Figure 16.1). These interfere with blood flow and narrow the arterial diameter. If the plaques clog one of the tiny coronary arteries that deliver blood to the heart, the resulting symptoms can range from mild chest pains to a heart attack.

How your body handles dietary cholesterol depends on what you inherited from your parents. Consider the gene for a protein that serves as the cell's receptor for LDLs. Inherit two "good" alleles of that gene, and your blood level of cholesterol will tend to remain so low that your arteries will never get clogged, even with a high-fat diet. Inherit two copies of a certain mutated allele, however, and you are destined to develop a rare genetic disorder called *familial cholesterolemia*. With this disorder, cholesterol builds up to abnormally high levels. Many affected individuals die of heart attacks during childhood or their teens.

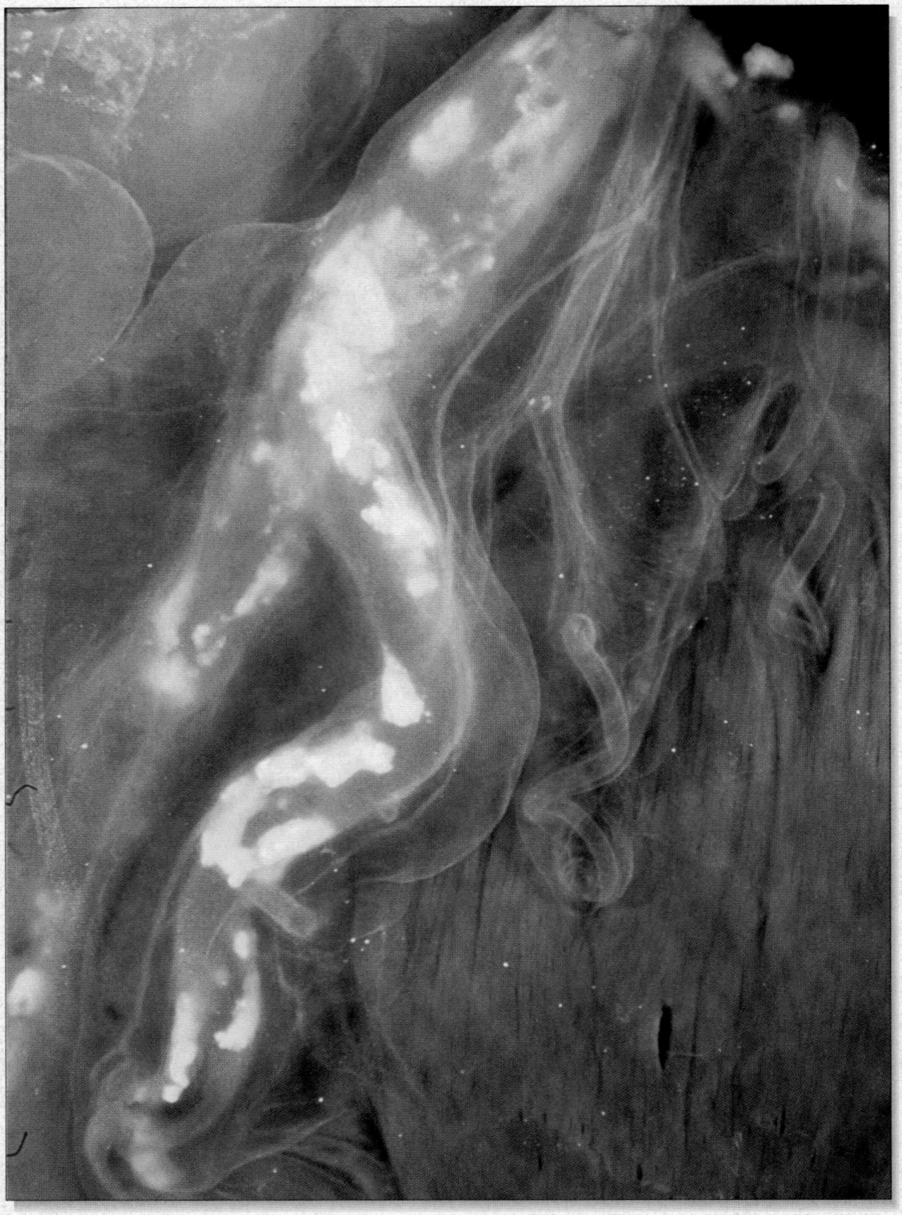

Figure 16.1 Potentially life-threatening plaques (*bright yellow*) inside one of the coronary arteries, which are small vessels that deliver blood to the heart. The plaques are the legacy of abnormally high levels of cholesterol.

In 1992 a woman from Quebec, Canada, became a milestone in the history of genetics. She was thirty years old. Like two of her younger brothers who had died from heart attacks in their early twenties, she inherited the defective gene for the LDL receptor. She herself survived a heart attack when she was sixteen. At twenty-six, she had coronary bypass surgery.

At the time, people were hotly debating the risks and promise of **gene therapy**—the transfer of one or more normal or modified genes into an individual's body cells to correct a genetic defect or boost resistance to disease. Even so, the woman consented to undergo an untried, physically wrenching procedure designed to give her body working copies of the good gene.

Medical researchers removed about 15 percent of the woman's liver. They placed liver cells in a nutrient-rich medium that promoted growth and division. *And they spliced the good gene into the genetic material of a harmless virus.* That modified virus served roughly the same function as a hypodermic needle. The researchers allowed it to infect the cultured liver cells and thereby insert copies of the good gene into them.

Later, the researchers infused about a billion of the modified cells into the woman's portal vein, a major blood vessel that leads directly to the liver. There, at least some cells took up residence, and they started to produce the missing cholesterol receptor. Two years after this, between 3 and 5 percent of the woman's liver cells were behaving normally and sponging up cholesterol from the blood. Her blood levels of LDLs had declined nearly 20 percent. Scans of her arteries showed no evidence at all of the progressive clogging that had nearly killed her. At a recent press conference, the woman announced she is active and doing well.

Her cholesterol levels do remain more than twice as high as normal, and it is too soon to know whether the gene therapy will prolong her life. Yet the intervention provides solid proof that the concept of gene therapy is sound, and hopes are high.

As you might gather from this pioneering clinical application, recombinant DNA technology has truly staggering potential for medicine. It also has great potential for agriculture and industry. The technology does not come without risks. With this chapter, we consider some basic aspects of the new technology. At the chapter's end, we also address some ecological, social, and ethical questions related to its application.

KEY CONCEPTS

1. Genetic experiments have been proceeding in nature for billions of years, through gene mutations, crossing over and recombination, and other natural events.

2. Humans are now purposefully bringing about genetic changes by way of recombinant DNA technology. Such enterprises are called genetic engineering.

3. With this technology, researchers isolate, cut, and splice together gene regions from different species, then greatly amplify the number of copies of the genes that interest them. The genes, and in some cases the proteins they specify are produced in quantities that are large enough to use for research and for practical applications.

4. Three activities are at the heart of recombinant DNA technology. First, procedures based on specific types of enzymes are used to cut DNA molecules into fragments. Second, the fragments are inserted into cloning tools, such as plasmids. Third, the fragments containing the genes of interest are identified, then copied rapidly and repeatedly.

5. Genetic engineering involves isolating, modifying, and inserting genes back into the same organism or into a different one. The goal is to beneficially modify traits that the genes influence. Human gene therapy, which focuses on controlling or curing genetic disorders, is an example.

6. The new technology raises social, legal, ecological, and ethical questions regarding its benefits and risks.

RECOMBINATION IN NATURE—AND IN THE LABORATORY

For at least 3 billion years, nature has been conducting uncountable numbers of genetic experiments, through mutation, crossing over, and other events that introduce changes in genetic messages. This is the source of life's diversity.

For many thousands of years, we humans have been changing numerous genetically based traits of species. By artificial selection practices, we produced new crop plants and breeds of cattle, birds, dogs, and cats from wild ancestral stocks. We developed meatier turkeys and sweeter oranges, larger corn, seedless watermelons, flamboyant ornamental roses, and other useful plants. We produced splendid hybrids, including the tangelo (tangerine × grapefruit) and mule (horse × donkey).

Researchers now use **recombinant DNA technology** to analyze genetic changes. With this technology, they cut and splice DNA from different species, then insert the modified molecules into bacteria or other types of cells that engage in rapid replication and cell division. The cells copy the foreign DNA right along with their own. In short order, huge populations produce useful quantities of recombinant DNA molecules. The new technology also is the basis of **genetic engineering**, by which genes are isolated, modified, and inserted back into the same organism or into a different one.

Plasmids, Restriction Enzymes, and the New Technology

Believe it or not, this astonishing technology originated with the innards of bacteria. Bacterial cells have a single chromosome, a circular DNA molecule that has all the genes they require to grow and reproduce. But many species also have **plasmids**, or small, circular molecules of "extra" DNA that contain a few genes (Figure 16.2).

Usually, plasmids are not essential for survival, but some of the genes they carry may benefit the bacterium. For instance, some plasmid genes confer resistance to antibiotics. (*Antibiotics*, remember, are toxic metabolic products of microorganisms that can kill or inhibit the growth of competing microorganisms.) The bacterium's replication enzymes copy and reproduce plasmid DNA, just as they copy chromosomal DNA.

In nature, many bacteria are able to transfer plasmid genes to a bacterial neighbor of the same species or a different one (Section 22.2). Replication enzymes may even integrate a transferred plasmid into the bacterial chromosome of a recipient cell. A recombinant DNA molecule is the result.

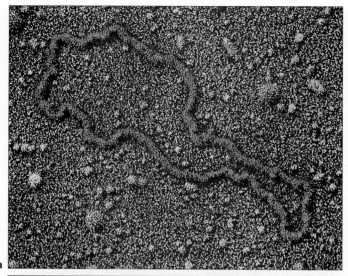

a

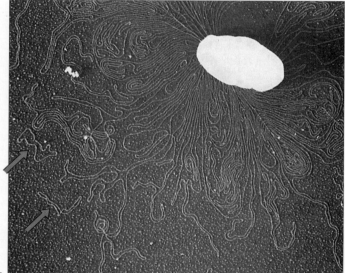

b

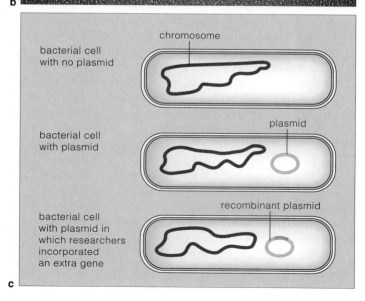

c

Figure 16.2 (**a**) A plasmid at high magnification. (**b**) Plasmids (*blue* arrows) released from a ruptured *Escherichia coli* cell. (**c**) Naturally occurring and modified plasmids.

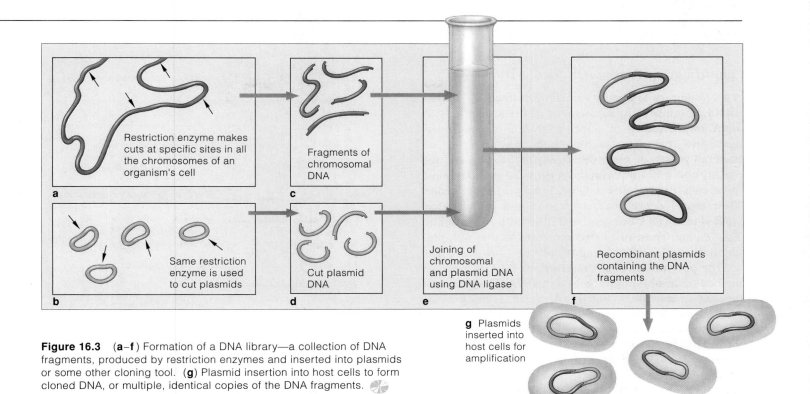

Figure 16.3 (**a–f**) Formation of a DNA library—a collection of DNA fragments, produced by restriction enzymes and inserted into plasmids or some other cloning tool. (**g**) Plasmid insertion into host cells to form cloned DNA, or multiple, identical copies of the DNA fragments.

Infectious particles called viruses as well as bacteria dabble in gene transfers and recombinations. And so do most eukaryotic species. As you might imagine, viral infection does a bacterium no good. Over evolutionary time, bacteria developed an arsenal against invasion by harmful genes. They became equipped with many types of **restriction enzymes**, which are able to recognize and cut apart foreign DNA that may enter a cell. Eventually, researchers learned how to use plasmids *and* restriction enzymes for genetic recombination in the laboratory.

Producing Restriction Fragments

Each type of restriction enzyme makes a cut wherever it recognizes a specific, very short nucleotide sequence in the DNA. Cuts at two identical sequences in the same DNA molecule produce a fragment. Because some types of enzymes make *staggered* cuts, some fragments have single-stranded portions at both ends. Sometimes these are referred to as sticky ends:

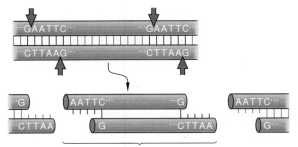

liberated DNA fragment with sticky ends

By "sticky," we mean the short, single-stranded ends of a DNA fragment will have the chemical capacity to base-pair with any other DNA molecule that also has been cut by the same restriction enzyme.

For example, suppose you use the same restriction enzyme to cut plasmids *and* DNA molecules that you have isolated from a human cell. When you mix the cut molecules together, they base-pair at the cut sites. After this, you add **DNA ligase** to the mixture. DNA ligase is an enzyme that seals DNA's sugar-phosphate backbone at the cut sites, just as it does during DNA replication. In this way, you create "recombinant plasmids," which have pieces of DNA from another organism inserted into them (Figure 16.2*c*).

You now have a **DNA library**. It is a collection of DNA fragments, produced by restriction enzymes, that have been incorporated into plasmids, as illustrated in Figure 16.3.

With recombinant DNA technology, DNA from different species is cut and spliced together; then the recombinant molecules are amplified, by way of copying mechanisms, to produce useful quantities of the genes of interest.

Recombination is made possible by the use of restriction enzymes that make specific cuts in DNA molecules and of DNA ligases that seal the cut ends.

Recombinant DNA technology has uses in basic research.

The new technology also has uses in genetic engineering—the deliberate modification of genes, followed by their insertion into the same individual or a different one.

WORKING WITH DNA FRAGMENTS

Amplification Procedures

A DNA library is almost vanishingly small. To obtain useful amounts of it, biochemists resort to methods of **DNA amplification**, by which a DNA library is copied again and again. One such method uses "factories" of bacteria, yeasts, or some other cells that can reproduce rapidly and take up plasmids. A growing population of these cells can amplify a DNA library in short order. Their repeated cycles of replication and cell division yield cloned DNA that has been inserted into plasmids. The "cloned" part of this name refers to the multiple, identical copies of DNA fragments.

The **polymerase chain reaction**, or **PCR**, is a newer method of amplifying fragments of DNA. The reactions proceed in test tubes, not in microbial factories. First, researchers identify short nucleotide sequences located just before and just after a region of DNA from a cell of the organism that interests them. Then they synthesize **primers**. These short nucleotide sequences, recall, will base-pair with any complementary sequences in DNA. And the replication enzymes called **DNA polymerases** recognize them as START tags (Section 13.3).

For PCR, a DNA polymerase from a bacterium that lives in hot springs, even water heaters, is the enzyme of choice, because it remains functional at the elevated temperatures necessary to unwind DNA and also at the lower temperatures necessary for base pairing. Researchers mix together the primers, the polymerases, all the DNA from one of the organism's cells, and free nucleotides. Next, they expose the mixture to precise temperature cycles. During each cycle, the two strands of all the DNA molecules unwind from each other. And primers become positioned on exposed nucleotides at the targeted sites according to base-pairing rules (Figure 16.4). With each round of reactions, the number of DNA molecules doubles. For example, if there are 10 such molecules in the test tube, there soon will be 20, then 40, 80, 160, 320, 640, 1,280, and so on. Very quickly, a target region from a single DNA molecule can be amplified to *billions* of molecules.

In short, *PCR amplifies samples that contain even tiny amounts of DNA.* As you will see in the next section, such samples can be obtained from fossils—even from a single hair or drop of blood left at the scene of a crime.

Sorting Out Fragments of DNA

When restriction enzymes cut DNA, the fragments they produce are not all the same length. Researchers can use **gel electrophoresis** to separate fragments from one another according to length. This laboratory procedure employs an electric field to force molecules through a viscous gel and to separate them according to physical

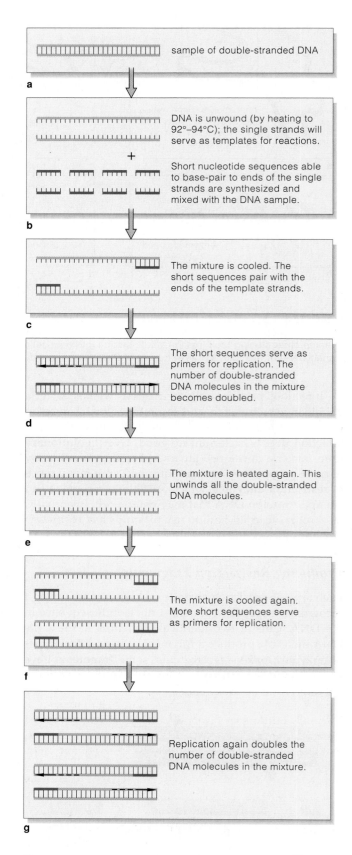

a sample of double-stranded DNA

b DNA is unwound (by heating to 92°–94°C); the single strands will serve as templates for reactions.

Short nucleotide sequences able to base-pair to ends of the single strands are synthesized and mixed with the DNA sample.

c The mixture is cooled. The short sequences pair with the ends of the template strands.

d The short sequences serve as primers for replication. The number of double-stranded DNA molecules in the mixture becomes doubled.

e The mixture is heated again. This unwinds all the double-stranded DNA molecules.

f The mixture is cooled again. More short sequences serve as primers for replication.

g Replication again doubles the number of double-stranded DNA molecules in the mixture.

Figure 16.4 The polymerase chain reaction (PCR).

Figure 16.5 One of the methods used for sequencing DNA, as first developed by Frederick Sanger. The method is employed to determine the nucleotide sequence of specific DNA fragments, as this example illustrates.

a Single-stranded DNA fragments are added to a solution in four different test tubes:

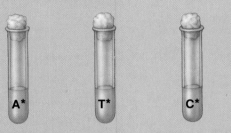

All four tubes contain DNA polymerases, short nucleotide sequences that can serve as primers for replication, and the nucleotide subunits of DNA (A, T, C, and G). Each tube contains a modified, labeled version of only one of the four kinds of nucleotides. We can show these as A^*, T^*, C^*, and G^*. The labeled form is present in low concentration, along with a generous supply of the unmodified form of the same nucleotide. Let's follow what happens in the tube with the A^* subunits.

b As expected, the DNA polymerase recognizes a primer that has become attached to a fragment, which it uses as a template strand. The enzyme assembles a complementary strand according to base-pairing rules (A only to T, and C only to G). Sooner or later, the enzyme picks up an A^* subunit for pairing with a T on the template strand. The modified nucleotide is a chemical roadblock—it prevents the enzyme from adding more nucleotides to the growing complementary strand. In time, the tube contains labeled strands of different lengths, as dictated by the location of each A^* in the sequence:

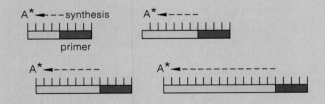

The same thing happens in the other three tubes, with strand lengths dictated by the location of T^*, C^*, and G^*.

c DNA from each of the four tubes is placed in four parallel lanes in the same gel. Then the DNA can be subjected to electrophoresis. The resulting nucleotide sequence can be read off the resulting bands in the gel. Look at the numbers running down the side of this diagram:

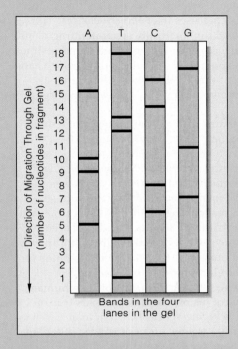

Bands in the four
lanes in the gel

Start with "1" and read across the four lanes (A, T, C, G). As you can see, T is the closest to the start of the nucleotide sequence; it has migrated farthest through the gel. At "2," the next nucleotide is C, and so on. The entire sequence, read from the first nucleotide to the last, is

T C G T A C G C A A G T T C A C G T

And now, by applying the rules of base pairing, you can deduce the sequence of the DNA fragment that served as the template.

---

and chemical properties. For DNA, size alone affects how far the molecules move. (For proteins, remember, a molecule's size, shape, and net surface charge are the determining factors.)

A gel that contains DNA fragments is immersed in a buffered solution, and electrodes connect the solution to a power source. Apply voltage, and negatively charged phosphate groups of the fragments respond to it. The fragments move toward the positively charged pole at different rates and thereby become separated into bands according to their lengths. Again, how far they migrate depends only on their size; the larger fragments cannot move as fast through it. After a predetermined period of electrophoresis, fragments of different length can be identified by staining the gel.

DNA Sequencing

Once DNA fragments from a sample have been sorted out according to length, researchers can work out the nucleotide sequence of each type. The Sanger method of DNA sequencing, as detailed in Figure 16.5, will give you a sense of one of the ways this can be done.

Restriction fragments can be rapidly amplified by PCR and by large populations of bacterial cells or yeast cells.

Restriction fragments can be sorted out according to length, as by gel electrophoresis. Then sequencing methods can be employed to determine the nucleotide sequences of the DNA fragments.

16.3

RIFF-LIPS AND DNA FINGERPRINTS

Suppose a researcher has isolated a good-size sample of your DNA molecules. She cuts them up with restriction enzymes, then subjects the restriction fragments to gel electrophoresis. Large ones cannot move as fast as small ones, so the fragments separate from one another in the gel. Now the researcher uses the *Southern blot method.* She transfers the electrophoretically separated fragments from the gel to a membrane filter and then immerses the filter in a solution containing a labeled probe, which can bind to the fragment that interests her. Afterward, she applies the same method to DNA samples from several other people.

When she compares the banding patterns, she finds small variations among them. Why? Although the DNA molecules of any two people are alike in some regions, they differ significantly in others. *The slight molecular differences shift the number and location of sites where restriction enzymes can make their cuts.* As a result, some fragments from corresponding regions of DNA from two or more individuals differ in length. The differences in DNA electrophoresis patterns among individuals are called **RFLPs** (pronounced RIFF-lips). The term stands for **restriction fragment length polymorphisms**.

As you know, the set of fingerprints of each human is unique, a marker of his or her identity. As is true of other sexually reproducing species, each human also has a **DNA fingerprint**—a unique array of RFLPs, inherited from each parent in a Mendelian pattern.

RFLP analysis is a wonderful new procedure for basic research. It also has startling applications in society at large. Consider the human genome project. As described in Section 16.6, this is an ambitious effort to decipher the nucleotide sequence of human DNA, with its 3.2 billion base pairs. Or consider how evolutionary biologists are analyzing RFLPs to decipher DNA from mummies, from fossilized insects and plants, and from mammoths and humans that were preserved many thousands of years ago in glacial ice. Investigators even used RFLP analysis to confirm that bones exhumed from a shallow pit in Siberia belonged to five members of the Russian imperial family, all shot to death in secrecy in 1918.

RFLP analysis has spectacular potential for medicine. Researchers have already pinpointed unique restriction sites in several mutated genes, including the ones that cause sickle-cell anemia, cystic fibrosis, and other genetic disorders. They can use the sites to find out whether an individual carries a mutated gene, as clinicians are now doing in prenatal diagnosis (Section 12.11).

What about social issues? Think of all the paternity and maternity cases against celebrities and ordinary folks alike. They now can be resolved by comparing the child's DNA fingerprints with those of the disputed parent. Or think about forensic medicine. A few drops of blood, semen, or even cells from a hair follicle at a crime scene or on a suspect's clothing often yield enough DNA to identify the perpetrator (Figure 16.6).

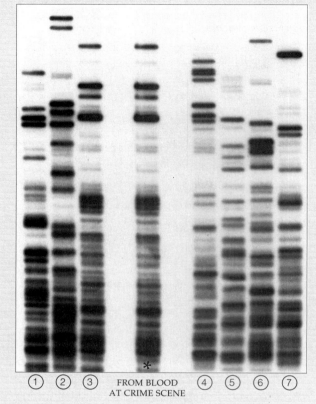

Figure 16.6 DNA fingerprint from a bloodstain discovered at a crime scene, with DNA fingerprints from blood samples of seven suspects (*circled numbers*). Which one of the seven was a match?

Britain, which started the first DNA database in the world, is at the forefront of solving crimes with DNA fingerprinting. Also, police in Britain can lawfully collect samples of hair or saliva from crime suspects, even if by force, for RFLP analysis. At this writing, detectives in Cardiff, Wales, are engaged in a massive "blood drive" to locate the man who raped and killed a local teenager. They expect to contact as many as 5,000 "donors." The detectives have let it be known that anyone who declines to donate blood voluntarily risks calling attention to himself—and inviting exceedingly fine scrutiny. Mass DNA screening has already helped solve other murders in Britain and Germany.

Recently in the United States, defense lawyers in a high-profile and seemingly interminable criminal case hammered away at DNA evidence. Someone nearly cut off the head of a celebrity's wife, slashed her friend to death, and apparently left a trail of blood leading to the celebrity's house. RFLP analysis seems to implicate him. His lawyers pounded on the competence and character of just about everybody who found blood at the crime scene and the house, collected samples, ran the PCRs, and did the RFLP analyses. And one can only wonder if sensationalized efforts of defense attorneys alone will send DNA research back to the twelfth century.

Use of DNA Probes

Recombinant plasmids are not much use in themselves. They must be mixed with living cells that can take them up. So how do you find out which cells do this? You can utilize **DNA probes**: short DNA sequences synthesized from radioactively labeled nucleotides. Part of a probe must be designed to base-pair with some portion of the DNA of interest. Any base pairing between nucleotide sequences (that is, RNA as well as DNA) from different sources is called **nucleic acid hybridization**.

The challenge is to "select" the cells that have taken up recombinant plasmids and the gene of interest. One way to do this is to use a plasmid containing a gene that confers resistance to a particular antibiotic. You put prospective host cells on a culture medium that has the antibiotic added to it. The antibiotic prevents growth of all cells *except* the ones that house plasmids with the antibiotic-resistance gene.

As the cells divide, they form colonies. Assume the colonies are growing on agar, a gel-like substance, in a petri dish. You blot the agar against a nylon filter. Some cells stick to the filter at sites that mirror the locations of the original colonies (Figure 16.7). You use solutions to rupture the cells, fix the released DNA onto the filter, and make the double-stranded DNA unwind. Then you add DNA probes, which hybridize only with the gene region having a complementary base sequence. Probe-hybridized DNA emits radioactivity and allows you to tag the colonies that harbor the gene of interest.

Use of cDNA

Researchers study genes to find out about the protein products and how they are put to use. However, even if a host cell takes up a gene, the cell may not be able to synthesize the protein. For example, recall that new mRNA transcripts of human genes contain noncoding sequences (introns). They cannot be translated until the introns are snipped out and coding regions (exons) are spliced together into mature form. Bacterial enzymes do not recognize the splice signals, so bacterial host cells cannot always properly translate human DNA.

Sometimes researchers get around the problem by using **cDNA**, which is a DNA strand "copied" from a *mature* mRNA transcript for the gene. By a backwards process called **reverse transcription**, a complementary DNA strand is assembled on mRNA. The outcome is a hybrid mRNA-cDNA molecule (Figure 16.8). Enzymes remove the RNA and then synthesize a complementary DNA strand, thus making double-stranded cDNA.

Double-stranded cDNA can be further modified, as by attaching signals for transcription and translation. The modified cDNA can be inserted into a plasmid for

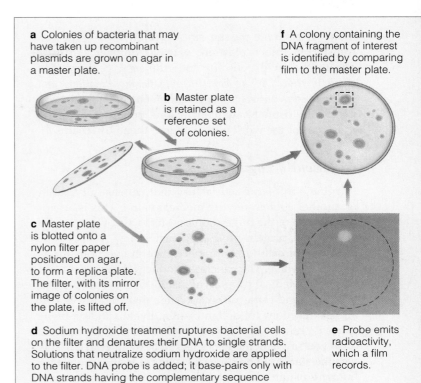

a Colonies of bacteria that may have taken up recombinant plasmids are grown on agar in a master plate.

b Master plate is retained as a reference set of colonies.

c Master plate is blotted onto a nylon filter paper positioned on agar, to form a replica plate. The filter, with its mirror image of colonies on the plate, is lifted off.

d Sodium hydroxide treatment ruptures bacterial cells on the filter and denatures their DNA to single strands. Solutions that neutralize sodium hydroxide are applied to the filter. DNA probe is added; it base-pairs only with DNA strands having the complementary sequence

e Probe emits radioactivity, which a film records.

f A colony containing the DNA fragment of interest is identified by comparing film to the master plate.

Figure 16.7 Use of a DNA probe to identify a colony of bacterial cells that have taken up plasmids with a specific DNA fragment.

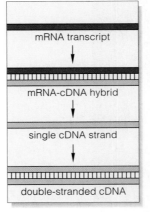

a A mature mRNA transcript of a gene of interest is used as a template to assemble a single strand of DNA. The enzyme reverse transcriptase catalyzes the assembly. An nRNA-cDNA hybrid molecule is the result.

b By enzyme action, the mRNA is removed and a second strand of DNA is assembled on the DNA strand remaining. The outcome is a double-stranded cDNA, as "copied" from an mRNA template.

mRNA transcript

mRNA-cDNA hybrid

single cDNA strand

double-stranded cDNA

Figure 16.8 Formation of cDNA from an mRNA transcript.

amplification. Recombinant plasmids can be inserted into bacterial cells, which may use the cDNA directions for synthesizing the protein of interest.

Host cells that take up modified genes may be identified by procedures involving DNA probes.

Modified eukaryotic genes may not be expressed in host bacterial cells unless first introduced into the cells as cDNA.

Many years have passed since foreign DNA was first transferred into a plasmid, yet that transfer started a debate that will continue into the next century. The issue is this: *Do the benefits of gene modifications and gene transfers outweigh the potential dangers?* Before you personally come to any conclusions, reflect upon a few examples of the work with bacteria, then with plants.

Genetically Engineered Bacteria

Imagine a miniaturized factory that churns out insulin or another protein having medical value. This is an apt description of the huge stainless steel vats of genetically engineered, protein-producing bacteria.

Think of *diabetics*, who need insulin injections for as long as they live. The pancreas produces insulin, a protein hormone. Medical supplies of insulin once were obtained from pigs and cattle, although some diabetics developed allergic reactions to foreign insulin. Eventually, synthetic genes for human insulin were transferred into *E. coli* cells. (Can you say why the genes had to be synthesized?) This was the start of huge bacterial factories that manufacture human insulin, hemoglobin, somatotropin, interferon, albumin, and other valuable proteins.

In addition, in research laboratories around the world, certain strains of bacteria are now being engineered to degrade oil

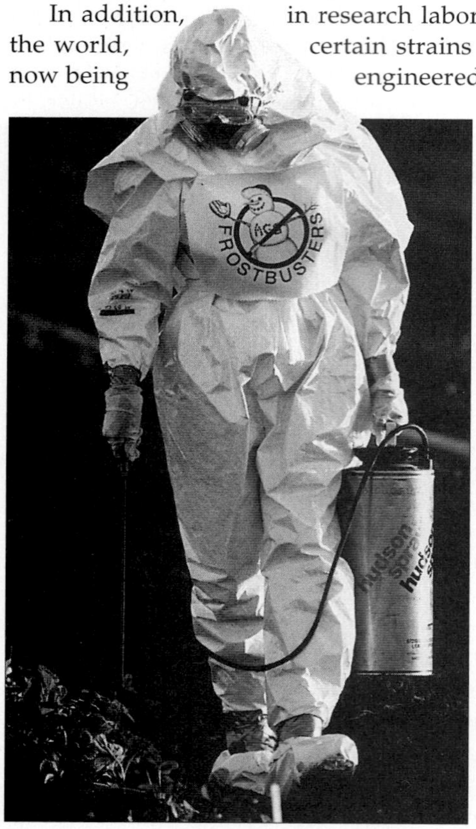

Figure 16.9 Spraying an experimental strawberry patch in California with "ice-minus" bacteria. The sprayer used elaborate protective gear to meet government regulations in effect then.

spills from tankers, to manufacture alcohol and other chemicals, to process minerals, or to leave crop plants alone. The strains are harmless to begin with. The ones meant to be confined to the laboratory are "designed" to prevent escape. As added precautions, the foreign DNA usually includes "fail-safe" genes. Such genes are silent *unless* the engineered bacteria somehow are exposed to conditions that occur in the environment. Exposure will activate the genes, with lethal results for the bacteria.

For example, foreign DNA may include a *hok* gene next to a promoter of the lactose operon (Section 15.2). If an engineered bacterium does manage to escape to the environment, where sugars are common, the gene will trip into action. The protein the gene specifies will destroy membrane function and so destroy the cell.

Even so, some people worry about the possible risks of introducing genetically engineered bacteria into the environment. Consider how Steven Lindow engineered a bacterium that can make many crop plants resist frost. A protein at the bacterial surface promotes formation of ice crystals. Lindow excised the ice-forming gene from some cells. As he hypothesized, spraying his "ice-minus bacteria" on strawberry plants in an isolated field prior to a frost would make plants less vulnerable to freezing. Even though Lindow deleted a *harmful* gene from an organism, it triggered a bitter legal battle over releasing engineered bacteria into the environment. In time, the courts did rule in favor of allowing such a release, and researchers sprayed a small strawberry patch (Figure 16.9). Nothing bad ever happened. Since then, the rules governing the release of genetically engineered species have become less restrictive.

Genetically Engineered Plants

A HUNT FOR BENEFICIAL GENES Currently, botanists are combing the world for seeds and other living tissues from wild ancestors of potatoes, corn, and other plants. They send their prizes—genes from a plant's lineage—to a safe storage laboratory in Colorado. Why? *Farmers now rely on only a few strains of high-yield crop plants that feed most of the human population.* The near-absence of genetic diversity means our food base is dangerously vulnerable to many disease-causing viruses, bacteria, and fungi. It is a race against time. Our population has reached an astounding size. People, bulldozers, and power saws are encroaching on previously uninhabited areas; hundreds of wild plant species disappear weekly.

The danger is real. Recently a new fungal strain that rots potatoes entered the United States from Mexico. In 1994, crop losses from the resulting *potato blight* topped $100 million. Pennsylvania farmers had little to harvest. Possibly laboratory-stored, fungus-resistant strains of potato plants will save crops in the future.

PLANT REGENERATION Botanists also search for good genes in the laboratory. Years ago, Frederick Steward and his coworkers cultured cells of carrot plants and induced them to grow into small embryos, some of which grew into whole plants (Section 31.7). Other species, including many crop plants, are regenerated today from cultured cells. The methods raise mutation rates, so cultures are a source of genetic modifications.

Researchers can pinpoint a useful mutation among millions of cells. Suppose a culture medium contains a toxic product of a disease agent. If a few cells have a mutated gene that confers resistance to the toxin, only they will live in the culture. If cells regenerate whole plants that can be hybridized with other varieties, the hybrids may get the good gene.

METHODS OF GENE TRANSFER Now genetic engineers insert genes into cultured plant cells. For example, they insert DNA fragments into the "Ti" plasmid from the bacterium *Agrobacterium tumefaciens*, which can infect many species of flowering plants. Some plasmid genes invade a plant's DNA and induce formation of abnormal tissue masses called crown gall tumors (Figure 16.10*a*). Before introducing the Ti plasmid into plants, researchers first remove the tumor-inducing genes and insert desired genes into it. They grow the modified bacterial cells with cultured plant cells, then regenerate plants from the cells that take up the genes. In some instances, the foreign genes have been expressed in plant tissues, with observable effects (Figure 16.10*b*).

A. tumefaciens can only infect beans, peas, potatoes, and other dicots. However, wheat, corn, rice, and many major food crops are monocots, which the bacterium cannot infect. In some cases, genetic engineers can use electric shocks or chemicals to deliver modified DNA into protoplasts. (Protoplasts are plant cells stripped of their walls.) At this writing, regenerating some species of plants from protoplast cultures is not yet possible. But researchers have had some success in delivering genes into cultured plant cells by actually shooting them with pistols. Instead of bullets, they use blanks to drive DNA-coated, microscopic particles into the cells.

ON THE HORIZON Despite many obstacles, improved varieties of crop plants have been developed or are in the works. For example, genetically engineered cotton

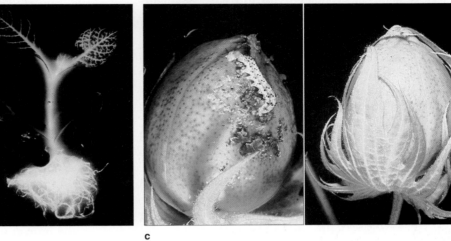

Figure 16.10 Gene transfers in plants. (**a**) Genes in a plasmid from a common bacterium (*Agrobacterium tumefaciens*) cause crown gall tumors on woody plants. Scientists use this plasmid to move desirable genes into cultured plant cells. (**b**) Evidence of a successful gene transfer: a modified tobacco plant that glows in the dark. A firefly gene became incorporated into the plant's DNA and is being translated. The protein product is luciferase, an enzyme required for bioluminescence (Section 6.7). (**c**) Example of a valuable plant that benefits from genetic engineering. *Left:* Buds of cotton plants are vulnerable to worm attack. *Right:* Buds of a genetically modified cotton plant resist attack.

plants now resist worm attacks without any help from pesticides, which have the disadvantage of also killing off beneficial pest-eating insects (Figure 16.10*c*).

Also on the horizon are engineered plants that can serve as factories for pharmaceuticals. A few years ago, genetically engineered tobacco plants that make human hemoglobin, melanin, and other proteins were planted in a field in North Carolina. Ecologists later found no trace of foreign genes or proteins in the soil or in other plants or animals in the vicinity. Recently, mustard plant cells made plastic beads in a Stanford University laboratory. (Plastics are long-chain polymers of identical subunits.) The plant DNA incorporates three bacterial enzymes that are able to convert carbon dioxide, water, and nutrients into inexpensive, biodegradable plastic.

Fail-safe measures must be built into genetically engineered organisms that may end up where they shouldn't be.

Genetically engineered plants are one of the few options available for protecting our vulnerable food supply.

GENETIC ENGINEERING OF ANIMALS

Supermice and Biotech Barnyards

The first mammals enlisted for experiments in genetic engineering were laboratory mice. Consider an example of this work. R. Hammer, R. Palmiter, and R. Brinster corrected a hormone deficiency that leads to dwarfism in mice. Insufficient levels of somatotropin give rise to the abnormality. The researchers used a microneedle to inject the gene for rat somatotropin into fertilized mouse eggs. After the eggs were implanted in an adult female, the gene was successfully integrated into the mouse DNA. Later, the baby mice in which the foreign gene was expressed were 1-1/2 times larger than their dwarf littermates. In other experiments, researchers transferred the gene for human somatotropin into a mouse embryo, where it became integrated into the DNA. The modified embryo grew up to be "supermouse" (Figure 16.11).

Today, as part of research into the molecular basis of Alzheimer's disease and other genetic disorders, a few human genes are being inserted into mouse embryos. Besides microneedles, microscopic laser beams are used to open up temporary holes in the plasma membrane of cultured cells, although such methods have varying degrees of success. Retroviruses also are used to insert genes into cultured cells, which may incorporate the foreign genes into their DNA. But the genetic material of retroviruses can undergo rearrangements, deletions, and other alterations that shut down the introduced genes. And if the virus particles escape from laboratory isolation, they may infect organisms.

Animals of "biotech barnyards" are competing with bacterial factories as genetically engineered sources of proteins. For example, goats produce CFTR protein (for treating cystic fibrosis) and TPA (which diminishes the severity of heart attacks). Cattle may soon produce human collagen for repairing cartilage, bone, and skin. Also, in 1996 researchers made a genetic duplicate of an adult ewe. They reprogrammed one of her mammary gland cells so all of its genes could be expressed. They fused that cell with an egg (from another ewe) from which the nucleus had been removed. Signals from the egg cytoplasm triggered the development of a cluster of embryonic cells, which was implanted into a surrogate mother. The result was a clone, a lamb named *Dolly*. At this writing, the clone is thriving. The experiment opens up the possibility of producing genetically engineered clones of sheep, cattle, and other farm animals that can supply consistently uniform quantities of proteins, even tissues and organs, for medical uses and for research.

Applying the New Technology to Humans

Researchers around the world are working their way through the 3.2 billion base pairs of the twenty-three

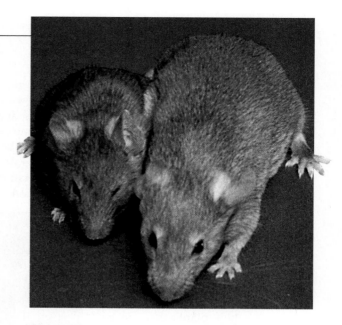

Figure 16.11 Ten-week-old mouse littermates. The mouse to the left weighs 29 grams and the one to the right, 44 grams. The larger mouse grew from a fertilized egg into which the gene for human somatotropin (growth hormone) had been inserted.

pairs of human chromosomes. This ambitious effort is the **human genome project**. Some researchers focus on specific chromosomes. Others are deciphering certain gene regions only. For example, rather than studying the noncoding sequences (introns), J. Venter and Sidney Brenner isolate mRNAs from brain cells and use them to make cDNA. By sequencing only cDNAs, they have identified hundreds of previously unknown genes.

About 99.9 percent of the nucleotide sequence is the same in all humans on Earth. The remaining 0.1 percent —about 3,200,000 base pairs—are mutations and other sequence variations sprinkled throughout the genome. They account for *all* genetic differences in the human population. Therefore, completion of the project should give us the ultimate reference book of human biology and genetic disorders. What will we do with all of that information? Certainly we will use it in the search for treatments and cures for genetic disorders. Of the 2,000 or so genes studied so far, 400 have already been linked to genetic disorders. And the knowledge opens doors to gene therapy, the transfer of one or more normal or modified genes into the body cells of an individual to correct a genetic defect or boost resistance to disease.

But what about forms of human gene expression that are neither disabling nor life-threatening? Will we tinker with them, also? We now leave the chapter with a *Focus* essay that looks at these questions.

As with any new technology, the potential benefits of recombinant DNA technology and genetic engineering should be weighed carefully against the potential risks.

REGARDING GENE THERAPY

This chapter opened with a historic, inspiring case, the first proof that gene therapy can help save human lives. It closes with questions that invite you to consider some social and ethical issues related to the application of recombinant DNA technology to our rapidly advancing knowledge of the human genome.

To most of us, human gene therapy to correct genetic abnormalities seems like a socially acceptable goal. Let's take this idea one step further. Is it also socially desirable or acceptable to change certain genes of a normal human individual (or sperm or egg) to alter or enhance traits?

The idea of selecting desirable human traits is called *eugenic engineering*. Yet who decides which forms of a trait are most "desirable"? For example, what happens if prospective parents start picking the sex of a child by way of genetic engineering? In one survey group, three-fourths of the people who were asked this question said they would choose a boy. What would be the long-term social implications of a drastic shortage of girls?

As other examples, would it be okay to engineer taller or blue-eyed or fair-skinned boys and girls? Would it be okay to engineer "superhumans" with amazing strength or intelligence? Suppose a person of average intelligence moved into a town of 800 Einsteins. Would the response go beyond mutterings of "There goes the neighborhood"? Are there people narcissistic enough to commission a clone of themselves, one that has just a few genetically engineered "improvements"? Researchers can alter genes now, and they have already cloned a sheep from a fully differentiated cell.

There are those who say that the DNA of any organism must never be altered. Put aside the fact that nature itself alters DNA much of the time, and has done so for nearly all of life's history. The concern is that *we* do not have the wisdom to bring about beneficial changes without causing great harm to ourselves or to the environment.

When it comes to manipulating human genes, one is reminded of our human tendency to leap before we look. When it comes to restricting genetic modifications of any sort, one also is reminded of the old saying, "If God had wanted us to fly, he would have given us wings." Yet something about the human experience has given us a capacity to imagine wings of our own making—and that capacity has carried us to the frontiers of space.

Where are we going from here with recombinant DNA technology, this new product of our imagination? To gain perspective on the question, spend some time reading the history of the human species. It is a history of survival in the face of all manner of new challenges, threats, bumblings, and sometimes disasters on a grand scale. It is also a story of our connectedness with the environment and with one another.

The basic questions confronting you today are these: Should we be more cautious, believing that one day the risk takers may go too far? And what do we as a species stand to lose if the risks are not taken?

SUMMARY

1. Uncountable gene mutations and other forms of genetic "experiments" have been proceeding in nature for at least 3 billion years.

2. Through artificial selection practices, humans have been manipulating the genetic character of species for many thousands of years. Today, recombinant DNA technology has enormously expanded our capacity to genetically modify organisms.

3. With recombinant DNA technology, researchers can isolate, cut, and then splice together gene regions from different species. They can greatly amplify the number of copies of those regions into usefully large quantities. Certain enzymes make the cuts and do the splicing.

4. Researchers employ a variety of restriction enzymes and DNA ligase to cut and insert DNA into plasmids from bacterial cells.

   a. Plasmids are small, circular DNA molecules that contain extra genes—that is, genes in addition to those of the circular bacterial chromosome.

   b. A DNA library is a collection of DNA fragments produced by restriction enzymes and later incorporated into plasmids.

5. It is possible to amplify the quantity of restriction fragments by enlisting a population of rapidly dividing cells, such as bacteria or yeasts. Short DNA fragments can be amplified hugely and far more rapidly in a test tube by the polymerase chain reaction, or PCR.

6. Following amplification, the DNA fragments can be sorted out, according to length, and their nucleotide sequences can be determined.

7. Each individual of a species has a DNA fingerprint, which is a unique array of restriction fragment length polymorphisms (RFLPs). Small molecular variations in the base sequence of the DNA of different individuals lead to small, identifiable differences in the restriction fragments cut from the DNA.

8. Cells that take up foreign genes can be identified by DNA probes, which base-pair with the genes of interest. The base pairing of nucleotide sequences from different sources is called nucleic acid hybridization.

9. In genetic engineering, genes are isolated, modified, and inserted into the same organism or a different one. In gene therapy, copies of normal or modified genes are inserted into an individual in order to correct a genetic defect or boost resistance to disease.

10. Both recombinant DNA technology and genetic engineering have enormous potential for research and applications in medicine, agriculture, the home, and industry. As with any new technology, the potential benefits must be weighed against the potential risks, including ecological and social repercussions.

Review Questions

1. Distinguish these terms from one another:
   a. recombinant DNA technology and genetic engineering *16.1*
   b. restriction enzyme and DNA ligase *16.1*
   c. cloned DNA and cDNA *16.2, 16.4*
   d. PCR and DNA sequencing *16.2*

2. Define RFLPs. Briefly explain one method by which they are produced, then name some applications for RFLP analysis. *16.3*

3. Explain how DNA probes and nucleic acid hybridization can be used to identify genetically modified host cells. *16.4*

Self-Quiz *(Answers in Appendix IV)*

1. _____ are small circles of bacterial DNA that are separate from the circular bacterial chromosome.

2. DNA fragments result when _____ cut DNA molecules at specific sites.
   a. DNA polymerases   c. restriction enzymes
   b. DNA probes   d. RFLPs

3. PCR stands for _____ .
   a. polymerase chain reaction
   b. polyploid chromosome restrictions
   c. polygraphed criminal rating
   d. politically correct research

4. A _____ is a collection of DNA fragments, produced by restriction enzymes and incorporated into plasmids.
   a. DNA clone   c. DNA probe
   b. DNA library   d. gene map

5. A _____ is multiple, identical copies of a collection of DNA fragments inserted into plasmids.
   a. DNA clone   c. DNA probe
   b. DNA library   d. gene map

6. Gel electrophoresis, a standard laboratory procedure, can be used to separate DNA restriction fragments according to _____ .
   a. shape   c. net surface charge
   b. size   d. all of the above

7. In reverse transcription, _____ is assembled on _____ .
   a. mRNA; DNA   c. DNA; enzymes
   b. cDNA; mRNA   d. DNA; agar

8. _____ is the transfer of normal genes into body cells to correct a genetic defect.
   a. Reverse transcription   c. Gene mutation
   b. Nucleic acid hybridization d. Gene therapy

9. Match the terms with the most suitable description.
   ____ DNA fingerprint   a. selecting "desirable" traits
   ____ Ti plasmid   b. deciphering 3.2 billion base
   ____ nature's genetic   pairs of 23 human chromosomes
        experiments   c. used in some gene transfers
   ____ nucleic acid   d. unique array of RFLPs
        hybridization   e. base pairing of nucleotide
   ____ human genome   sequences from different
        project   DNA or RNA sources
   ____ eugenic engineering   f. mutations, crossovers

Critical Thinking

1. Besides this chapter's examples, list what you believe might be some potential benefits and risks of genetic engineering.

2. Ryan, a forensic scientist, obtained a very small DNA sample from a crime scene. In order to examine the sample by DNA

fingerprinting, he must first amplify the sample by the polymerase chain reaction. He estimates there are 50,000 copies of the DNA in his sample. Derive a simple formula and calculate the number of copies he will have after fifteen cycles of PCR.

3. A game warden in Africa confiscated eight ivory tusks from elephants. Some tissue is still attached to the tusks. Now he must determine whether the tusks were taken illegally from northern populations of endangered elephants, or from other elephants from populations to the south that can be hunted legally. How can he use DNA fingerprinting to find the answer?

4. A restriction enzyme dubbed EcoRI is specified by a gene on an R plasmid in *E. coli*. In nature, it cleaves the circular DNA of a small virus that infects mammals. Lisa, a biotechnologist, has cloned an EcoRI fragment into a plasmid. The cloned EcoRI fragment is 850 base pairs (bp) long. And 300 bp from one end of the fragment is a site where a different restriction enzyme (BamHI) can cleave it. The plasmid is 1,900 bp long and has no BamHI sites.
   a. When EcoRI and the plasmid are mixed together in solution, what length does Lisa expect the resulting fragments to be? What can she expect when she uses BamHI alone to cut it? When she uses both EcoRI and BamHI?
   b. Draw a circular map of the recombination plasmid, showing the relative distance between the two restriction enzyme sites.

5. Last winter, Lunardi's Market put out a bin of tomatoes having splendid vine-ripened redness, flavor, and texture. The sign posted above the bin identified them as genetically engineered produce. Most shoppers selected unmodified tomatoes in the adjacent bin, even though those tomatoes were pale pink, mealy-textured, and tasteless. Which ones would you pick? Why?

Selected Key Terms

cDNA *16.4*
DNA amplification *16.2*
DNA fingerprint *16.3*
DNA library *16.1*
DNA ligase *16.1*
DNA polymerase *16.2*
DNA probe *16.4*
gel electrophoresis *16.2*
gene therapy *CI*
genetic engineering *16.1*

human genome project *16.6*
nucleic acid hybridization *16.4*
plasmid *16.1*
polymerase chain reaction (PCR) *16.2*
primer *16.2*
recombinant DNA technology *16.1*
restriction enzyme *16.1*
restriction fragment length
   polymorphism (RFLP) *16.3*
reverse transcription *16.4*

Readings

Anderson, W. F. 1992. "Human Gene Therapy." *Science* 256: 808–813.

Gasser, C. S., and R. T. Fraley. 1989. "Genetically Engineering Plants for Crop Improvement." *Science* 244: 1293–1299.

Joyce, G. December 1992. "Directed Molecular Evolution." *Scientific American* 267(6): 90–97.

10 March 1997. "Send in the Clones." *Newsweek.*

Watson, J. D. 1990. "The Human Genome Project: Past, Present, and Future." *Science* 248: 44–49.

Web Site See *http://www.wadsworth.com/biology* for practice quiz questions, hypercontents, BioUpdates, and critical thinking. The Wadsworth Biology Resource Center provides a wealth of information fully organized and integrated by chapter.

APPENDIX I. BRIEF CLASSIFICATION SCHEME

The classification scheme that follows is a composite of several that microbiologists, botanists, and zoologists use. The major groupings are agreed upon, more or less. There is not always agreement, however, on what to call a given grouping or where it might fit within the overall hierarchy. There are several reasons for the lack of total consensus.

First, the fossil record varies in its quality and in its completeness. Therefore, the phylogenetic relationship of one group to others is sometimes open to interpretation. Comparative studies at the molecular level are firming up the picture, but this work is still under way.

Second, ever since the time of Linnaeus, classification schemes have been based on the perceived morphological similarities and differences among organisms. Although some original interpretations are now open to question, we are so used to thinking about organisms in certain ways that reclassification often proceeds slowly. Traditionally, for example, birds and reptiles have been considered to be separate classes (Reptilia and Aves). And yet there are now compelling arguments for grouping the lizards and snakes together in one class, and the crocodilians, dinosaurs, and birds in a different class.

Third, researchers in microbiology, mycology, botany, zoology, and the other fields of biological inquiry have inherited a wealth of literature, based on classification schemes that were developed over time in each of those fields. Many see no good reason to give up the established terminology and thereby disrupt access to the past. Until recently, for example, microbiologists and botanists have been using *division*, and zoologists *phylum*, for taxa that are equivalent in the hierarchy of classification. Many still do. Opinions are still polarized with respect to the kingdom Protista, certain members of which could just as easily be grouped in the kingdoms of plants, fungi, or animals. Indeed, the term protozoan is a holdover from an earlier scheme in which amoebas and certain other single-celled organisms were ranked as simple animals.

Given the problems, why do we even bother imposing artificial frameworks on the history of life? We do this for the same reason that a writer might decide to break up the history of civilization into several volumes, a number of chapters, and many paragraphs. Both efforts are attempts to impart obvious structure to what might otherwise be an overwhelming body of knowledge and to enhance the retrieval of information from it.

Finally, bear in mind that we include this classification scheme primarily for your reference purposes. Besides being open to revision, it also is by no means complete. Numerous existing and extinct organisms of the so-called lesser phyla are not represented here. Our strategy is to focus mainly on organisms mentioned in the text. A few examples of organisms also are listed under the entries.

SUPERKINGDOM PROKARYOTA. Prokaryotes. Almost all microscopic species with DNA concentrated in a region of cytoplasm, not inside a membrane-bounded nucleus. All are bacteria, either single cells or simple associations of cells. Autotrophs and heterotrophs (Table 22.2). Reproduce by prokaryotic fission, sometimes by budding and by bacterial conjugation. *Bergey's Manual of Systematic Bacteriology*, the authoritative reference in the field, calls this "a time of taxonomic transition." It groups bacteria mostly by numerical taxonomy (Section 22.6), not on phylogeny. The scheme presented here reflects strong evidence of evolutionary relationships for at least some bacterial groupings.

KINGDOM EUBACTERIA. Gram-negative and gram-positive forms. Peptidoglycan in cell wall. Photosynthetic autotrophs, chemosynthetic autotrophs, and heterotrophs.

PHYLUM GRACILICUTES.   Typical Gram-negative, thin wall. Autotrophs (photosynthetic and chemosynthetic) and heterotrophs. *Anabaena* and other cyanobacteria. *Escherichia, Pseudomonas, Neisseria, Myxococcus.*
PHYLUM FIRMICUTES.   Typical Gram-positive, thick wall. Heterotrophs. *Bacillus, Staphylococcus, Streptococcus, Clostridium, Actinomycetes.*
PHYLUM TENERICUTES.   Gram-negative, wall absent. Heterotrophs (saprobes, pathogens). *Mycoplasma.*

KINGDOM ARCHAEBACTERIA. Methanogens, halophiles, thermophiles. Strict anaerobes, distinct from other bacteria in cell wall, membrane lipids, ribosomes, and RNA sequences. *Methanobacterium, Halobacterium, Sulfolobus.*

SUPERKINGDOM EUKARYOTA. Eukaryotes. Single-celled and multicelled species. Cells start out life with a nucleus (encloses the DNA) and usually other membrane-bound organelles. Chromosomes with numerous proteins attached.

KINGDOM PROTISTA. Diverse single-celled, colonial, and multicelled eukaryotic species, currently most easily classified by what they are *not* (not bacteria, fungi, plants, or animals). Autotrophs, heterotrophs, or both (Table 23.3). Reproduce sexually and asexually (by meiosis, mitosis, or both). Many related evolutionarily to plants, fungi, and possibly animals.

PHYLUM CHYTRIDIOMYCOTA.   Chytrids. Heterotrophs; saprobic decomposers or parasites. *Chytridium.*
PHYLUM OOMYCOTA.   Water molds. Heterotrophs. Decomposers, some parasites. *Saprolegnia, Phytophthora, Plasmopara.*
PHYLUM ACRASIOMYCOTA.   Cellular slime molds. Heterotrophs with free-living, phagocytic amoeboid cells and spore-bearing stages. *Dictyostelium.*
PHYLUM MYXOMYCOTA.   Plasmodial slime molds. Heterotrophs with free-living, phagocytic amoeboid cells and spore-bearing stages. Aggregate into streaming mass of cells that discard plasma membranes. *Physarum.*
PHYLUM SARCODINA.   Amoeboid protozoans. Heterotrophs, free-living or endosymbiotic, some pathogens. Soft-or shelled bodies, locomotion by pseudopods. The rhizopods (naked amoebas, foraminiferans), *Amoeba proteus, Entomoeba.* Also the actinopods (radiolarians, heliozoans).
PHYLUM MASTIGOPHORA.   Animal-like flagellated protozoans. Heterotrophs, free-living, many internal parasites. All with one to several flagella. *Trypanosoma, Trichomonas, Giardia.*
APICOMPLEXA. Heterotrophs, many parasitic. Complex of rings, tubules, other structures at head end. Most familiar members called sporozoans. *Plasmodium, Toxoplasma.*

PHYLUM CILIOPHORA. Ciliated protozoans. Heterotrophs, predators or symbionts, some parasitic. All have cilia. Free-living, sessile, or motile. *Paramecium*, hypotrichs.

PHYLUM EUGLENOPHYTA. Euglenoids. Mostly heterotrophs, some autotrophs (photosynthetic). Flagellated. *Euglena*.

PHYLUM PYRRHOPHYTA. Dinoflagellates. Photosynthetic, mostly, but some heterotrophs. *Gymnodinium breve*.

PHYLUM CHRYSOPHYTA. Golden algae, yellow-green algae, diatoms. Photosynthetic. Some flagellated, others not. *Mischococcus, Synura, Vaucheria*.

PHYLUM RHODOPHYTA. Red algae. Photosynthetic, some parasitic. Nearly all marine, some freshwater. *Porphyra. Bonnemaisonia, Euchema*.

PHYLUM PHAEOPHYTA. Brown algae. Photosynthetic, nearly all in temperate or marine waters. *Macrocystis, Fucus, Sargassum, Ectocarpus, Postelsia*.

PHYLUM CHLOROPHYTA. Green algae. Mostly photosynthetic, some parasitic. Most freshwater, some marine or terrestrial. *Chlamydomonas, Spirogyra, Ulva, Volvox, Codium, Halimeda*.

KINGDOM FUNGI. Nearly all multicelled eukaryotic species. Heterotrophs; mostly saprobic decomposers, some parasites. Nutrition based on extracellular digestion of organic matter and absorption of nutrients by individual cells. Multicelled species form absorptive mycelium within substrates and structures that produce asexual spores (and sometimes sexual spores).

PHYLUM ZYGOMYCOTA. Zygomycetes. Zygosporangia (zygote inside thick wall) formed by sexual reproduction. Bread molds, related forms. *Rhizopus, Philobolus*.

PHYLUM ASCOMYCOTA. Ascomycetes. Sac fungi. Sac-shaped cells form sexual spores (ascospores). Most yeasts and molds, morels, truffles. *Saccharomycetes, Morchella, Neurospora, Sarcoscypha. Claviceps, Ophiostoma*.

PHYLUM BASIDIOMYCOTA. Basidiomycetes. Club fungi. Most diverse group. Produce basidiospores inside club-shaped structures. Mushrooms, shelf fungi, stinkhorns. *Agaricus, Amanita, Puccinia, Ustilago*.

IMPERFECT FUNGI. Sexual spores absent or undetected. The group has no formal taxonomic status. If better understood, a given species might be grouped with sac fungi or club fungi. *Verticillium, Candida, Microsporum, Histoplasma*.

LICHENS. Mutualistic interactions between fungal species and a cyanobacterium, green alga, or both. *Usnea, Cladonia*.

KINGDOM PLANTAE. Multicelled eukaryotes. Nearly all photosynthetic autotrophs with chlorophylls *a* and *b*. Some parasitic. Nonvascular and vascular species, generally with well-developed root and shoot systems. Nearly all adapted in form and function to survive dry conditions in land habitats; a few in aquatic habitats. Sexual reproduction predominant; also asexual reproduction by vegetative propagation.

PHYLUM RHYNIOPHYTA. Earliest known vascular plants; muddy habitats. Extinct. *Cooksonia, Rhynia*.

PHYLUM PROGYMNOSPERMOPHYTA. Progymnosperms. Ancestral to early seed-bearing plants; extinct. *Archaeopteris*.

PHYLUM PTERIDOSPERMOPHYTA. Seed ferns. Fernlike gymnosperms; extinct. *Medullosa*

PHYLUM CHAROPHYTA. Stoneworts.

PHYLUM BRYOPHYTA. Bryophytes: mosses, liverworts, hornworts. Seedless, nonvascular, haploid dominance. *Marchantia, Polytrichum, Sphagnum*.

PHYLUM PSILOPHYTA. Whisk ferns. Seedless, vascular. No obvious roots, leaves on sporophyte. *Psilotum*.

PHYLUM LYCOPHYTA. Lycophytes, club mosses. Seedless, vascular. Leaves, branching rhizomes, vascularized roots and stems. *Lycopodium, Selaginella*.

PHYLUM SPHENOPHYTA. Horsetails. Seedless, vascular. Some sporophyte stems photosynthetic, others nonphotosynthetic, spore-producing. *Equisetum*.

PHYLUM PTEROPHYTA. Ferns. Largest group of seedless vascular plants (12,000 species), mainly tropical, temperate habitats.

PHYLUM CYCADOPHYTA. Cycads. Type of gymnosperm (vascular, bears "naked" seeds). Tropical, subtropical. Palm-shaped leaves, simple cones on male and female plants. *Zamia*.

PHYLUM GINKGOPHYTA. Ginkgo (maidenhair tree). Type of gymnosperm. Seeds with fleshy outer layer. *Ginkgo*.

PHYLUM GNETOPHYTA. Gnetophytes. Only gymnosperms with vessels in xylem and double fertilization (but endosperm does not form). *Ephedra, Welwitschia*.

PHYLUM CONIFEROPHYTA. Conifers. Most common and familiar gymnosperms. Generally cone-bearing with needle-like or scale-like leaves.
  Family Pinaceae. Pines, firs, spruces, hemlock, larches, Douglas firs, true cedars. *Pinus*.
  Family Cupressaceae. Junipers, cypresses. *Juniperus*.
  Family Taxodiaceae. Bald cypress, redwoods, Sierra bigtree, dawn redwood. *Sequoia*.
  Family Taxaceae. Yews.

PHYLUM ANTHOPHYTA. Angiosperms (flowering plants). Largest group of vascular seed-bearing plants. Only organisms that produce flowers, fruits.
  Class Dicotyledonae. Dicotyledons (dicots). Some families of several different orders are listed:
    Family Nymphaeaceae. Water lilies.
    Family Papaveraceae. Poppies.
    Family Brassicaceae. Mustards, cabbages, radishes.
    Family Malvaceae. Mallows, cotton, okra, hibiscus.
    Family Solanaceae. Potatoes, eggplant, petunias.
    Family Salicaceae. Willows, poplars.
    Family Rosaceae. Roses, apples, almonds, strawberries.
    Family Fabaceae. Peas, beans, lupines, mesquite.
    Family Cactaceae. Cacti.
    Family Euphorbiaceae. Spurges, poinsettia.
    Family Cucurbitaceae. Gourds, melons, cucumbers, squashes.
    Family Apiaceae. Parsleys, carrots, poison hemlock.
    Family Aceraceae. Maples.
    Family Asteraceae. Composites. Chrysanthemums, sunflowers, lettuces, dandelions.
  Class Monocotyledonae. Monocotyledons (monocots). Some families of several different orders are listed:
    Family Liliaceae. Lilies, hyacinths, tulips, onions, garlic.
    Family Iridaceae. Irises, gladioli, crocuses.
    Family Orchidaceae. Orchids.
    Family Arecaceae. Date palms, coconut palms.
    Family Cyperaceae. Sedges.
    Family Poaceae. Grasses, bamboos, corn, wheat, sugarcane.
    Family Bromeliaceae. Bromeliads, pineapples, Spanish moss.

KINGDOM ANIMALIA. Multicelled eukaryotes, nearly all with tissues, organs, and organ systems and with motility during at least part of the life cycle. Heterotrophs; predators (herbivores, carnivores, omnivores), parasites, detritivores. Reproduce sexually (and asexually in many species). Continuous stages of embryonic development.

PHYLUM PLACOZOA. Marine. Simplest known animal. Two cell layers, no mouth, no organs. *Trichoplax*.

PHYLUM MESOZOA. Ciliated, wormlike parasites, about the same level of complexity as *Trichoplax*.

PHYLUM PORIFERA. Sponges. No symmetry, tissues, or organs.

PHYLUM CNIDARIA. Radial symmetry, tissues, nematocysts.
  Class Hydrozoa. Hydrozoans. *Hydra, Obelia, Physalia*.
  Class Scyphozoa. Jellyfishes. *Aurelia*.
  Class Anthozoa. Sea anemones, corals. *Telesto*.

PHYLUM CTENOPHORA. Comb jellies. Modified radial symmetry.

PHYLUM PLATYHELMINTHES. Flatworms. Bilateral, cephalized; simplest animals with organ systems. Saclike gut.
  Class Turbellaria. Triclads (planarians), polyclads. *Dugesia*.
  Class Trematoda. Flukes. *Schistosoma*.
  Class Cestoda. Tapeworms. *Taenia*.

PHYLUM NEMERTEA. Ribbon worms.

PHYLUM NEMATODA. Roundworms. *Ascaris, Trichinella.*

PHYLUM ROTIFERA. Rotifers.

PHYLUM MOLLUSCA. Mollusks.
  Class Polyplacophora. Chitons.
  Class Gastropoda. Snails (periwinkles, whelks, limpets, abalones, cowries, conches, nudibranchs, tree snails, garden snails), sea slugs, land slugs.
  Class Bivalvia. Clams, mussels, scallops, cockles, oysters, shipworms.
  Class Cephalopoda. Squids, octopuses, cuttlefish, nautiluses. *Loligo.*

PHYLUM BRYOZOA. Bryozoans (moss animals).

PHYLUM BRACHIOPODA. Lampshells.

PHYLUM ANNELIDA. Segmented worms.
  Class Polychaeta. Mostly marine worms.
  Class Oligochaeta. Mostly freshwater and terrestrial worms, but many marine. *Lumbricus* (earthworms).
  Class Hirudinea. Leeches.

PHYLUM TARDIGRADA. Water bears.

PHYLUM ONYCHOPHORA. Onychophorans. *Peripatus.*

PHYLUM ARTHROPODA.
  Subphylum Trilobita. Trilobites; extinct.
  Subphylum Chelicerata. Chelicerates. Horseshoe crabs, spiders, scorpions, ticks, mites.
  Subphylum Crustacea. Shrimps, crayfishes, lobsters, crabs, barnacles, copepods, isopods (sowbugs).
  Subphylum Uniramia.
    Superclass Myriapoda. Centipedes, millipedes.
    Superclass Insecta.
      Order Ephemeroptera. Mayflies.
      Order Odonata. Dragonflies, damselflies.
      Order Orthoptera. Grasshoppers, crickets, katydids.
      Order Dermaptera. Earwigs.
      Order Blattodea. Cockroaches.
      Order Mantodea. Mantids.
      Order Isoptera. Termites.
      Order Mallophaga. Biting lice.
      Order Anoplura. Sucking lice.
      Order Homoptera. Cicadas, aphids, leafhoppers, spittlebugs.
      Order Hemiptera. Bugs.
      Order Coleoptera. Beetles.
      Order Diptera. Flies.
      Order Mecoptera. Scorpion flies. *Harpobittacus.*
      Order Siphonaptera. Fleas.
      Order Lepidoptera. Butterflies, moths.
      Order Hymenoptera. Wasps, bees, ants.

PHYLUM ECHINODERMATA. Echinoderms.
  Class Asteroidea. Sea stars.
  Class Ophiuroidea. Brittle stars.
  Class Echinoidea. Sea urchins, heart urchins, sand dollars.
  Class Holothuroidea. Sea cucumbers.
  Class Crinoidea. Feather stars, sea lilies.
  Class Concentricycloidea. Sea daisies.

PHYLUM HEMICHORDATA. Acorn worms.

PHYLUM CHORDATA. Chordates.
  Subphylum Urochordata. Tunicates, related forms.
  Subphylum Cephalochordata. Lancelets.
  Subphylum Vertebrata. Vertebrates.
    Class Agnatha. Jawless vertebrates (lampreys, hagfishes).
    Class Placodermi. Jawed, heavily armored fishes; extinct.
    Class Chondrichthyes. Cartilaginous fishes (sharks, rays, skates, chimaeras).
    Class Osteichthyes. Bony fishes.
      Subclass Dipnoi. Lungfishes.
      Subclass Crossopterygii. Coelacanths, related forms.
      Subclass Actinopterygii. Ray-finned fishes.
        Order Acipenseriformes. Sturgeons, paddlefishes.
        Order Salmoniformes. Salmon, trout.
        Order Atheriniformes. Killifishes, guppies.

Order Gasterosteiformes. Seahorses.
Order Perciformes. Perches, wrasses, barracudas, tunas, freshwater bass, mackerels.
Order Lophiiformes. Angler fishes.
Class Amphibia. Mostly tetrapods; embryo enclosed in amnion.
  Order Caudata. Salamanders.
  Order Anura. Frogs, toads.
  Order Apoda. Apodans (caecilians).
Class Reptilia. Skin with scales, embryo enclosed in amnion.
  Subclass Anapsida. Turtles, tortoises.
  Subclass Lepidosaura. *Sphenodon*, lizards, snakes.
  Subclass Archosaura. Dinosaurs (extinct), crocodiles, alligators.
Class Aves. Birds. (In more recent schemes, dinosaurs, crocodilians, and birds are often grouped in the same category.)
  Order Struthioniformes. Ostriches.
  Order Sphenisciformes. Penguins.
  Order Procellariiformes. Albatrosses, petrels.
  Order Ciconiiformes. Herons, bitterns, storks, flamingoes.
  Order Anseriformes. Swans, geese, ducks.
  Order Falconiformes. Eagles, hawks, vultures, falcons.
  Order Galliformes. Ptarmigan, turkeys, domestic fowl.
  Order Columbiformes. Pigeons, doves.
  Order Strigiformes. Owls.
  Order Apodiformes. Swifts, hummingbirds.
  Order Passeriformes. Sparrows, jays, finches, crows, robins, starlings, wrens.
Class Mammalia. Skin with hair; young nourished by milk-secreting glands of adult.
  Subclass Prototheria. Egg-laying mammals (duckbilled platypus, spiny anteaters).
  Subclass Metatheria. Pouched mammals or marsupials (opossums, kangaroos, wombats).
  Subclass Eutheria. Placental mammals.
    Order Insectivora. Tree shrews, moles, hedgehogs.
    Order Scandentia. Insectivorous tree shrews.
    Order Chiroptera. Bats.
    Order Primates.
      Suborder Strepsirhini (prosimians). Lemurs, lorises.
      Suborder Haplorhini (tarsioids and anthropoids).
        Infraorder Tarsiiformes. Tarsiers.
        Infraorder Platyrrhini (New World monkeys).
          Family Cebidae. Spider monkeys, howler monkeys, capuchin.
        Infraorder Catarrhini (Old World monkeys and hominoids).
          Superfamily Cercopithecoidea. Baboons, macaques, langurs.
          Superfamily Hominoidea. Apes and humans.
            Family Hylobatidae. Gibbon.
            Family Pongidae. Chimpanzees, gorillas, orangutans.
            Family Hominidae. Existing and extinct human species (*Homo*) and australopiths.
    Order Carnivora. Carnivores.
      Suborder Feloidea. Cats, civets, mongooses, hyenas.
      Suborder Canoidea. Dogs, weasels, skunks, otters, raccoons, pandas, bears.
    Order Proboscidea. Elephants; mammoths (extinct).
    Order Sirenia. Sea cows (manatees, dugongs).
    Order Perissodactyla. Odd-toed ungulates (horses, tapirs, rhinos).
    Order Artiodactyla. Even-toed ungulates (camels, deer, bison, sheep, goats, antelopes, giraffes).
    Order Edentata. Anteaters, tree sloths, armadillos.
    Order Tubulidentata. African aardvarks.
    Order Cetacea. Whales, porpoises.
    Order Rodentia. Most gnawing animals (squirrels, rats, mice, guinea pigs, porcupines).

Metric-English Conversions

Length

English		Metric
inch	=	2.54 centimeters
foot	=	0.30 meter
yard	=	0.91 meter
mile (5,280 feet)	=	1.61 kilometer

To convert	multiply by	to obtain
inches	2.54	centimeters
feet	30.00	centimeters
centimeters	0.39	inches
millimeters	0.039	inches

Weight

English		Metric
grain	=	64.80 milligrams
ounce	=	28.35 grams
pound	=	453.60 grams
ton (short) (2,000 pounds)	=	0.91 metric ton

To convert	multiply by	to obtain
ounces	28.3	grams
pounds	453.6	grams
pounds	0.45	kilograms
grams	0.035	ounces
kilograms	2.2	pounds

Volume

English		Metric
cubic inch	=	16.39 cubic centimeters
cubic foot	=	0.03 cubic meter
cubic yard	=	0.765 cubic meters
ounce	=	0.03 liter
pint	=	0.47 liter
quart	=	0.95 liter
gallon	=	3.79 liters

To convert	multiply by	to obtain
fluid ounces	30.00	milliliters
quart	0.95	liters
milliliters	0.03	fluid ounces
liters	1.06	quarts

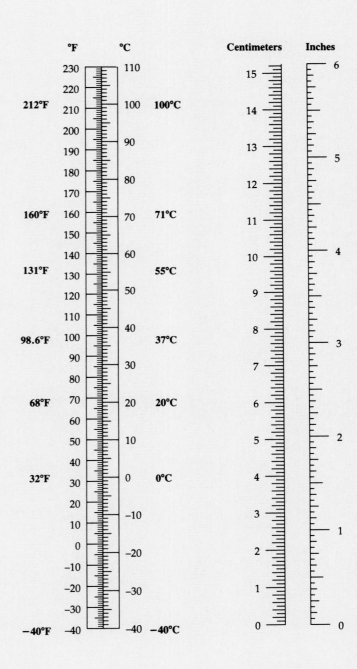

APPENDIX III. ANSWERS TO GENETIC PROBLEMS

CHAPTER 11

1. a. *AB*
   b. *AB* and *aB*
   c. *Ab* and *ab*
   d. *AB*, *aB*, *Ab*, and *ab*

2. a. All offspring will be *AaBB*.
   b. 25% *AABB*; 25% *AaBB*; 25% *AABb*; 25% *AaBb*
   c. 25% *AaBb*; 25% *Aabb*; 25% *aaBb*; 25% *aabb*

   d.
1/16	*AABB*	(6.25%)
1/8	*AaBB*	(12.5%)
1/16	*aaBB*	(6.25%)
1/8	*AABb*	(12.5%)
1/4	*AaBb*	(25%)
1/8	*aaBb*	(12.5%)
1/16	*AAbb*	(6.25%)
1/8	*Aabb*	(12.5%)
1/16	*aabb*	(6.25%)

3. Yellow is recessive. Because the first-generation plants have a green phenotype and must be heterozygous, green must be dominant over the recessive yellow.

4. a. *ABC*
   b. *ABc* and *aBc*
   c. *ABC*, *aBC*, *ABc*, and *aBc*
   d. *ABC*, *aBC*, *AbC*, *abC*, *ABc*, *aBc*, *Abc*, and *abc*

5. The F_1 plants must all be heterozygous for both genes. When these plants self-pollinate, 1/4 (or 25%) of the F_2 plants will be heterozygous for both genes.

6. a. The mother must be heterozygous for both genes; the father is homozygous recessive for both of the genes. The first child is homozygous recessive for both genes.

   b. The probability that a second child will not be a tongue roller and will have detached earlobes is 1/4 (25%).

7. a. Bill, *Aa Ee Ss* ; and Marie, *AA EE SS* .

   b. The probability that each child will have Bill's phenotype is 100 percent. Because Marie's phenotype is the same as Bill's, the probability that each child will have Marie's phenotype is also 100 percent.

   c. Zero.

8. a. The mother must be heterozygous ($I^A i$). The male with type B blood could have fathered the child if he, too, were heterozygous ($I^B i$).

   b. If the male were heterozygous, then he *could* be the father. However, *any* male who carried the *i* allele could be the father, so one can't say this particular individual absolutely must be. This would include the males with type O blood (*ii*), the heterozygotes with type A blood ($I^A i$), and the heterozygotes with type B blood ($I^B i$).

9. The most direct way to do this would be to allow mice from a true-breeding white-furred strain to mate with mice from a true-breeding brown-furred strain. (True-breeding strains typically are homozygous for alleles influencing the trait being studied.) All F_1 offspring from this cross will be heterozygous for the alleles that influence brown or white fur color. The phenotype of the F_1 offspring should be recorded, then the offspring should be allowed to mate with one another to produce an F_2 population. Assuming that only a single gene locus is involved, several outcomes are possible, and they include the following:

   All F_1 offspring are brown. The F_2 offspring segregate 3 brown: 1 white. *Conclusion*: Segregation is occurring at a single locus where the allele specifying brown is dominant to the allele specifying white.

   All F_1 offspring are white. The F_2 offspring segregate 3 white: 1 brown. *Conclusion*: Segregation is occurring at a single locus where the allele specifying white is dominant to the allele specifying brown.

   All F_2 offspring are tan. The F_2 offspring segregate 1 brown: 2 tan: 1 white. *Conclusion*: Segregation is occurring at a single gene locus where the alleles show incomplete dominance.

10. If the allele that causes the hip disorder is recessive, you *cannot* guarantee that Dandelion's puppies will not carry the allele, inasmuch as Dandelion may be a heterozygous carrier of that allele.

11. Fred should allow his black guinea pig to mate with a white one. This would be a testcross.

    The white one has the genotype *ww*, and the black one might be homozygous dominant (*WW)* or heterozygous (*Ww*). If any offspring from the cross are white, then his black guinea pig is *Ww*. If all offspring are black after Fred allows enough matings to produce a significant number of offspring, then there is a good probability that his guinea pig is *WW*.

    (If ten of the offspring are all black and none is white, then there is about a 99.9 percent chance that his guinea pig is *WW*. The greater the number of offspring, the greater the degree of confidence in the conclusion.)

12. a. $R^1R^1 \times R^1R^2$ yields 1/2 red flowers, and 1/2 pink.
    b. $R^1R^1 \times R^2R^2$ yields all pink flowers.
    c. $R^1R^2 \times R^1R^2$ yields 1/4 red, 1/2 pink, and 1/4 white.
    d. $R^1R^2 \times R^2R^2$ yields 1/2 pink and 1/2 white.

13. 9/16 walnut (pea-rose); 3/16 rose; 3/16 pea; 1/16 single.

14. Both parents are heterozygotes ($Hb^A Hb^S$).
    a. 1/4
    b. 1/4
    c. 1/2

15. The mating is M^LM. Genotypes of the offspring will be 1/4 MM, 1/2 M^LM, and 1/4 M^LM^L. Phenotypes will be:

    1/4 homozygous dominant (MM) survivors

    1/2 homozygous (M^LM) survivors

    1/4 lethal (M^LM^L) nonsurvivors

The probability that any one kitten among the surviving progeny will be heterozygous is 2/3.

16. a. Both parents must be heterozygous (Aa). They may produce albino (aa) and unaffected (AA or Aa) children.

    b. Both parents and all their children must be aa.

    c. The albino male must be aa. Because there is an albino child, the female must be heterozygous Aa. If she were AA, they would have no albino children. The one albino child must be aa; the three unaffected children must be Aa. There is a 50 percent chance that any child will be albino. The 3 : 1 ratio is not surprising, given the small number of progeny.

17. a. All offspring of the mating will have an intermediate amount of pigmentation corresponding to genotype $A^1A^2B^1B^2$.

    b. All possible genotypes could occur among offspring of this mating, in these expected proportions:

1/16	$A^1A^1B^1B^1$	(dark red)
1/8	$A^1A^1B^1B^2$	(medium-dark red)
1/16	$A^1A^1B^2B^2$	(medium red)
1/8	$A^1A^2B^1B^1$	(medium-dark red)
1/4	$A^1A^2B^1B^2$	(medium red)
1/8	$A^1A^2B^2B^2$	(light red)
1/16	$A^2A^2B^1B^1$	(medium red)
1/8	$A^2A^2B^1B^2$	(light red)
1/16	$A^2A^2B^2B^2$	(white)

CHAPTER 12

1. a. Human males inherit their X chromosome from their mother.

    b. With respect to an X-linked gene, a human male can produce two kinds of gametes. One kind will have a Y chromosome without the gene. The other kind of gamete will contain an X chromosome—hence it will contain the X-linked gene.

    c. All gametes of a human female who is homozygous for an X-linked gene will have an X chromosome, hence it will contain that gene.

    d. A female who is heterozygous for an X-linked gene will produce two kinds of gametes. One kind will have an X chromosome with one type of allele. The other kind will have an X chromosome with the other allele.

2. a. This gene is only carried on Y chromosomes. So we would not expect females to have hairy pinnae, because females normally do not have Y chromosomes.

    b. Because sons always inherit a Y chromosome from their father and daughters never do, a male having hairy pinnae will always transmit the gene for the trait to his sons and never to his daughters.

3. A 0 percent crossover frequency means that 50 percent of the gametes will be AB and 50 percent will be ab.

4. X rays might have induced a deletion of the dominant allele on one of the fruit fly chromosomes and led to the rare vestigial-wing condition. An alternative explanation is that the irradiation might have induced a mutation in the dominant allele.

5. The likelihood that crossing over will occur between the genes is low if they are very close together, and it is higher if the distance between them is greater.

6. a. Either anaphase I or anaphase II of meiosis.

    b. If a translocation resulted in the attachment of most of human chromosome 21 to the end of a normal chromosome 14, then the new individual would develop Down syndrome. The the cells of that individual would contain the attached chromosome 21 in addition to two normal chromosomes 21. The chromosome number would still be 46.

7. By using c as the symbol for color blindness and C for normal color vision, the cross between these animals can be diagrammed as follows:

$$C(Y) \text{ female} \times Cc \text{ male}$$

In mugwumps, a son receives one sex-linked allele from each of his parents, but a daughter inherits her unpaired sex-linked allele from her father. In this cross, half of the sons will be CC and half Cc, but none will be color blind. Of the daughters, half will be $C(Y)$ and half $c(Y)$. There is a 50 percent chance that a daughter will be color blind. Note: this sex chromosome designation is not the same as it is in humans. But this is the way it is not only for mugwumps, but also for all birds, moths, butterflies, and some other organisms; the Y chromosome is associated with females, not males.

8. To have a daughter who will be affected by childhood muscular dystrophy, the prospective father and the mother must carry the allele for the disorder. Males with the allele have muscular dystrophy and rarely father children.

APPENDIX IV. ANSWERS TO SELF-QUIZZES

CHAPTER 1
1. cell
2. Metabolism
3. Homeostatsis
4. adaptive
5. mutation
6. d
7. d
8. b
9. c, e, d, b, a

CHAPTER 2
1. b
2. c
3. f
4. e
5. f
6. c, a, b, d

CHAPTER 3
1. d
2. e
3. f
4. b
5. d
6. d
7. d
8. c, e, b, d, a

CHAPTER 4
1. c
2. d
3. d
4. False; many cell types have walls that surround their plasma membrane, which is the outermost *membrane.*
5. c
6. e, d, a, b, c

CHAPTER 5
1. c
2. c
3. a
4. e
5. a
6. centrifuge
7. d
8. b

CHAPTER 6
1. c
2. a
3. a
4. d
5. b
6. e
7. a
8. a
9. d
10. c, g, a, d, e, b, f

CHAPTER 7
1. carbon dioxide, water
2. d
3. c
4. b
5. d
6. c
7. c
8. c, d, e, b, f, a

CHAPTER 8
1. d
2. c
3. b
4. c
5. d
6. c
7. b
8. d
9. b, c, a, d

CHAPTER 9
1. a
2. b
3. b
4. c
5. a
6. d
7. b
8. d, b, c, a

CHAPTER 10
1. d
2. d
3. a
4. b
5. d
6. c
7. c
8. d, a, c, b,

CHAPTER 11
1. a
2. b
3. a
4. c
5. b
6. a
7. d
7. b, d, a, c

CHAPTER 12
1. c
2. c
3. e
4. c
5. d
6. d
7. d
8. c, e, d, b, a

CHAPTER 13
1. c
2. d
3. d
4. c
5. a
6. d
7. d, b, c, a

CHAPTER 14
1. d
2. d
3. b
4. c
5. c
6. a
7. a
8. d
9. c
10. e, c, a, f, d, g , b

CHAPTER 15
1. b
2. b
3. d
4. d
5. d
6. a
7. a
8. d
9. e
10. e
11. b
12. d
13. d, e, b, a, c

CHAPTER 16
1. Plasmids
2. c
3. a
4. b
5. a
6. b
7. b
8. d
9. d, c, f, e, b, a

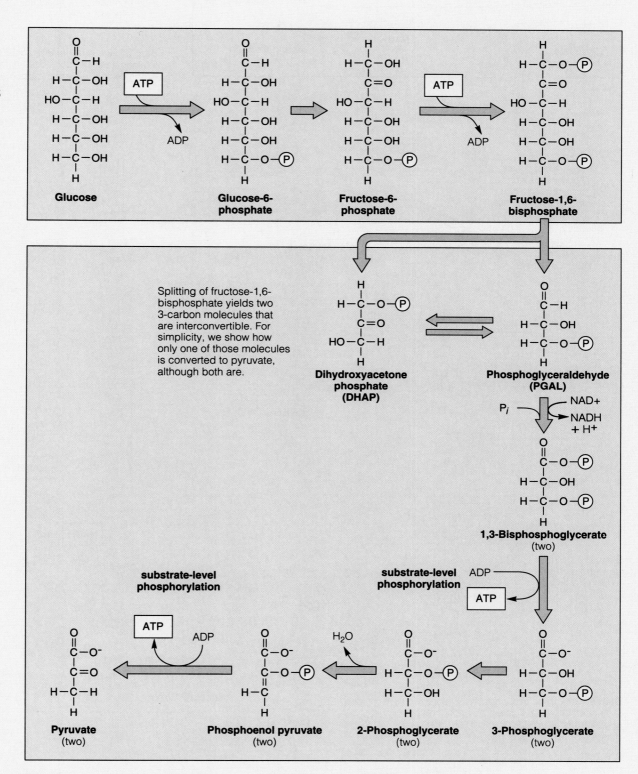

ENERGY-REQUIRING STEPS OF GLYCOLYSIS

(two ATP invested)

ENERGY-RELEASING STEPS OF GLYCOLYSIS

(four ATP produced)

Splitting of fructose-1,6-bisphosphate yields two 3-carbon molecules that are interconvertible. For simplicity, we show how only one of those molecules is converted to pyruvate, although both are.

Figure a Glycolysis, ending with two 3-carbon pyruvate molecules for each 6-carbon glucose entering the reactions. The *net* energy yield is two ATP molecules (two invested, four produced).

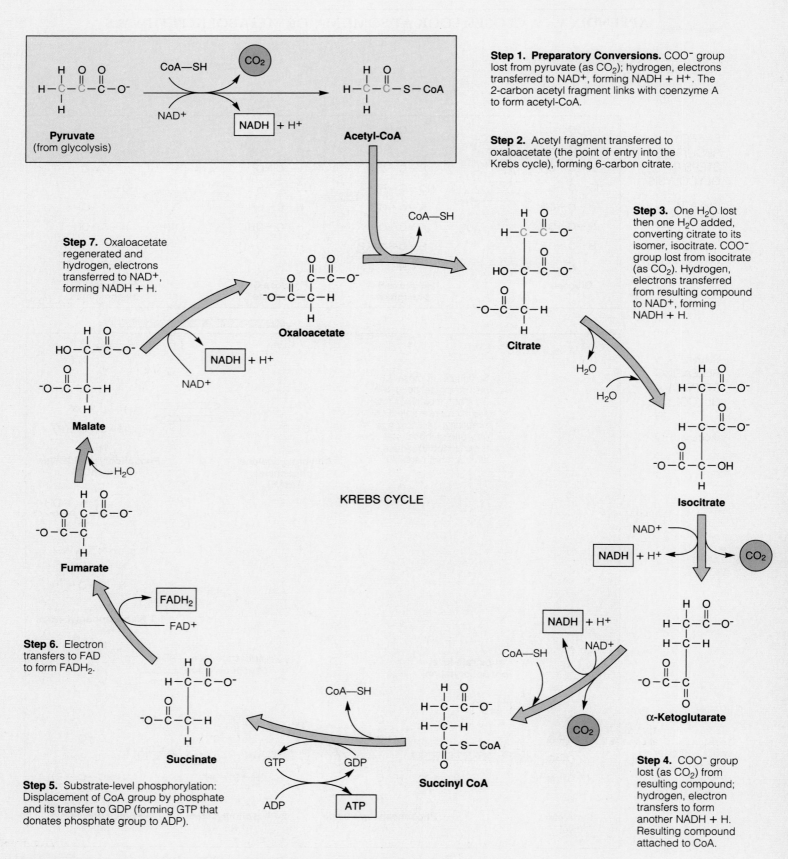

Step 1. Preparatory Conversions. COO^- group lost from pyruvate (as CO_2); hydrogen, electrons transferred to NAD^+, forming $NADH + H^+$. The 2-carbon acetyl fragment links with coenzyme A to form acetyl-CoA.

Step 2. Acetyl fragment transferred to oxaloacetate (the point of entry into the Krebs cycle), forming 6-carbon citrate.

Step 3. One H_2O lost then one H_2O added, converting citrate to its isomer, isocitrate. COO^- group lost from isocitrate (as CO_2). Hydrogen, electrons transferred from resulting compound to NAD^+, forming $NADH + H$.

Step 7. Oxaloacetate regenerated and hydrogen, electrons transferred to NAD^+, forming $NADH + H$.

Step 6. Electron transfers to FAD to form $FADH_2$.

Step 5. Substrate-level phosphorylation: Displacement of CoA group by phosphate and its transfer to GDP (forming GTP that donates phosphate group to ADP).

Step 4. COO^- group lost (as CO_2) from resulting compound; hydrogen, electron transfers to form another $NADH + H$. Resulting compound attached to CoA.

Pyruvate (from glycolysis)

Acetyl-CoA

Oxaloacetate

Citrate

Isocitrate

α-Ketoglutarate

Succinyl CoA

Succinate

Fumarate

Malate

KREBS CYCLE

Figure B Krebs cycle (citric acid cycle). *Red* identifies the carbon atoms entering the cyclic pathway by way of acetyl-CoA.

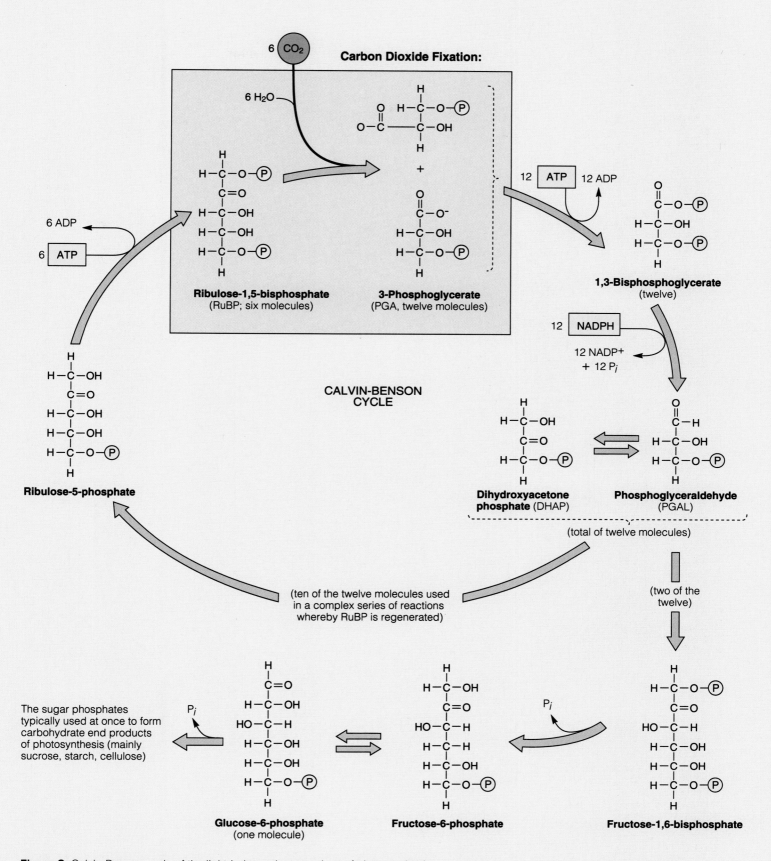

Figure C Calvin-Benson cycle of the light-independent reactions of photosynthesis.

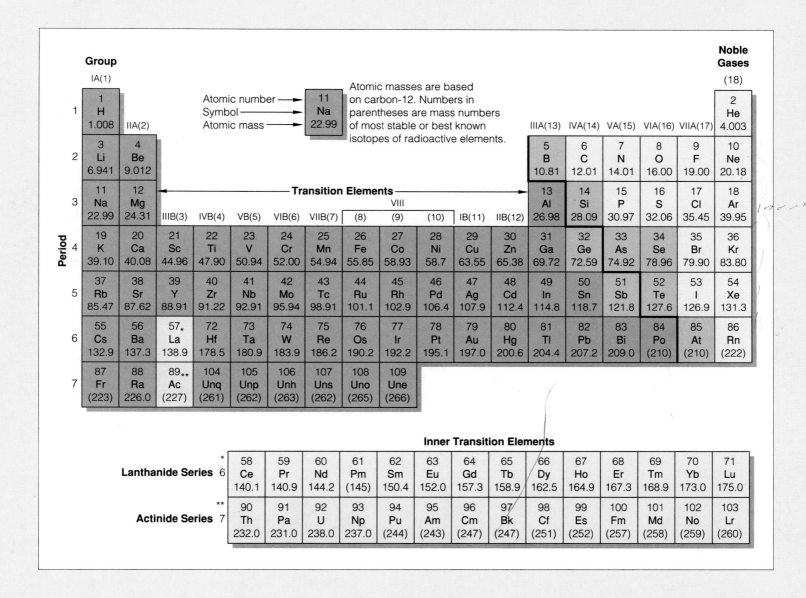

A GLOSSARY OF BIOLOGICAL TERMS

ABO blood typing Method of using two surface proteins (A, B, or both) of red blood cells to characterize an individual's blood. O signifies the absence of both proteins.

abortion The spontaneous expulsion or the dislodging of an embryo or a fetus from the uterus.

abscisic acid (ab-SISS-ik) Plant hormone that promotes stomatal closure, bud dormancy, and seed dormancy.

abscission (ab-SIH-zhun) [L. *abscindere*, to cut off] The dropping of leaves, flowers, fruits, or other plant parts due to hormonal action.

absorption Of most animals, movement of nutrients, fluid, and ions across the gut lining and into the internal environment.

accessory pigment Light-trapping pigment molecule; it contributes to photosynthesis by extending the range of usable wavelengths beyond those absorbed by the chlorophylls.

acid [L. *acidus*, sour] A substance that releases hydrogen ions when dissolved in water.

acid rain The falling to Earth of rain (or snow) that contains sulfur and nitrogen oxides. Also called wet acid deposition (as opposed to dry acid deposition of airborne particles of sulfur and nitrogen oxides).

acoelomate (ay-SEE-luh-mate) Of some of the invertebrates, having no fluid-filled cavity between the gut and body wall.

acoustical signal Sounds that are normally used as a form of communication between animals of the same species.

actin (AK-tin) One of the motor proteins with roles in contraction. Interacts with myosin in muscle cells.

action potential Abrupt, brief reversal in the steady voltage difference across the plasma membrane (the resting membrane potential) of a neuron and other excitable cells.

activation energy The minimum amount of collision energy necessary to drive reactant molecules to an activated state (the transition state) at which a given chemical reaction will proceed spontaneously.

activator A regulatory protein having a role in a positive control system that promotes gene transcription.

active site A crevice in the surface of an enzyme molecule where a specific reaction is catalyzed, or made to proceed far faster than it would spontaneously.

active transport A pumping of one or more specific solutes through the interior of a transport protein that spans the lipid bilayer of a cell membrane. The solute is transported against its concentration gradient. An energy boost, as from ATP, activates the protein.

adaptation [L. *adaptare*, to fit] Of evolution, being adapted (or becoming more adapted) to a given set of environmental conditions. Of a sensory neuron, decreasing frequency of action potentials (or their cessation) even if a stimulus is maintained at constant strength.

adaptive behavior A behavior that promotes propagation of an individual's genes and that tends to increase in frequency in the population over time.

adaptive radiation A burst of divergences from a single lineage that gives rise to many new species, each adapted to an unoccupied or new habitat or to using a novel resource.

adaptive trait Any aspect of form, function, or behavior that helps an individual survive and reproduce under prevailing conditions.

adaptive zone A way of life available for organisms that are physically, ecologically, and evolutionarily equipped to live it, such as "catching insects in the air at night."

adenine (AH-de-neen) A purine; a nitrogen-containing base in certain nucleotides.

adenosine diphosphate (ah-DEN-uh-seen die-FOSS-fate) ADP, an organic compound that can make energy transfers in cells; typically formed by hydrolysis of ATP.

adenosine phosphate Any of a number of relatively small organic compounds, some of which function as chemical messengers within and between cells, and others (such as ATP) that function as energy carriers.

ADH Antidiuretic hormone. Hypothalamic hormone that induces water conservation as required during control of extracellular fluid volume and solute concentrations.

adhesion protein A protein that helps cells of the same type locate one another during development, adhere, and remain in the proper position in the proper tissue.

adipose tissue A type of connective tissue having an abundance of fat-storing cells and blood vessels for transporting fats.

ADP Adenosine diphosphate. A nucleotide coenzyme that accepts unbound phosphate or a phosphate group to become ATP.

ADP/ATP cycle In cells, a mechanism of ATP renewal. During a phosphate-group transfer, ATP reverts to ADP, then it forms again by phosphorylation of ADP.

aerobic respiration (air-OH-bik) [Gk. *aer*, air, + *bios*, life] The main pathway of ATP formation, for which oxygen is the final acceptor of electrons stripped from glucose or another organic compound. It proceeds from glycolysis through the Krebs cycle and electron transport phosphorylation. For each glucose molecule, a typical net yield is 36 ATP.

age structure The number of individuals in each of several or many age categories for a population.

agglutination (ah-glue-tin-AY-shun) In this defensive response, antibodies circulating in blood act against foreign cells (such as transfused red blood cells of the wrong type) and cause them to clump together.

aging A range of processes, including the breakdown of cell structure and function, by which the body gradually deteriorates. All multicelled species that show extensive cell differentiation undergo aging.

AIDS Short for acquired immunodeficiency syndrome. A set of chronic disorders that arises following infection by the human immunodeficiency virus (HIV), which destroys key cells of the immune system.

alcohol An organic compound that has one or more hydroxyl groups (—OH) and readily dissolves in water. Sugars are examples.

alcoholic fermentation Anaerobic pathway of ATP formation. Pyruvate from glycolysis is degraded to acetaldehyde, which accepts electrons from NADH to form ethanol with a net yield of two ATP. NAD$^+$ is regenerated.

aldosterone (al-DOSS-tuh-rohn) Hormone secreted by the adrenal cortex that helps regulate sodium reabsorption.

allantois (ah-LAN-twahz) [Gk. *allas*, sausage] One of four extraembryonic membranes that functions in respiration and in storing the metabolic wastes of embryos of reptiles, birds, and some mammals. In humans, it gives rise to blood vessels for the placenta and to the urinary bladder.

allele (uh-LEEL) At a given gene locus on a chromosome, one of two or more slightly different molecular forms of a gene that arise through mutation and that code for different versions of the same trait.

allele frequency The abundance of each kind of allele in the population as a whole.

allergen Normally harmless substance that provokes inflammation, excessive mucus secretion, and other defense responses.

allergy A response to an allergen.

allopatric speciation [Gk. *allos*, different, L *patria*, native land] Speciation that follows the end of gene flow between populations or subpopulations of a species as a result of their geographic isolation from each other.

allosteric control (AL-oh-STARE-ik) Form of control over a metabolic reaction or pathway that operates by binding a specific substance at a control site on a specific enzyme.

altruism (al-true-ISS-tik) Self-sacrificing behavior; an individual behaves in a way that helps others but decreases its own chances of reproductive success.

alveolus (ahl-VEE-uh-lus), plural **alveoli** [L. *alveus*, small cavity] One of the cupped, thin-walled outpouchings of respiratory bronchioles. A site where oxygen diffuses from air in the lungs to blood, and carbon dioxide diffuses from blood to the lungs.

amino acid (uh-MEE-no) A small organic molecule with a hydrogen atom, an amino group, an acid group, and an R group bonded covalently to a central carbon atom; the subunit of polypeptide chains.

ammonification (uh-moan-ih-fih-KAY-shun) A process by which certain soil bacteria and fungi break down nitrogenous wastes and remains of organisms; part of the nitrogen cycle.

amnion (AM-nee-on) Of land vertebrates, one of four extraembryonic membranes; the boundary layer of a fluid-filled sac that allows the embryo to grow in size, move freely, and be protected from sudden impacts and temperature shifts.

amniote egg An egg, often with a leathery or calcified shell, that has extraembryonic membranes, including the amnion.

amphibian A type of vertebrate somewhere between fishes and reptiles in body plan and reproductive mode; salamanders, frogs and toads, and caecilians are existing groups.

anaerobic pathway (an-uh-ROW-bik) [Gk. *an*, without, + *aer*, air] Metabolic pathway in which a substance other than oxygen serves as the final acceptor of electrons that have been stripped from substrates.

analogous structures (ann-AL-uh-gus) [Gk. *analogos*, similar to one another] Body parts that once differed in evolutionarily distant lineages, then converged in structure and function as the lineages responded to similar environmental pressures.

anaphase (AN-uh-faze) Of anaphase I of meiosis, the stage when each homologous chromosome separates from its partner and both move to opposite spindle poles. Of mitosis and of anaphase II of meiosis, sister chromatids of each chromosome separate and move to opposite poles.

aneuploidy (AN-yoo-ploy-dee) Having one more chromosome or one less relative to the parental chromosome number.

angiosperm (AN-gee-oh-spurm) [Gk. *angeion*, vessel, and *spermia*, seed] A flowering plant.

animal A multicelled heterotroph that feeds on other organisms, that is motile for at least part of the life cycle, that develops by a series of embryonic stages, and that usually has tissues, organs, and organ systems.

annelid A type of invertebrate classified as a segmented worm; an oligochaete (such as an earthworm), leech, or polychaete.

annual A flowering plant that completes its life cycle in one growing season.

anther [Gk. *anthos*, flower] A pollen-bearing part of a stamen.

antibiotic One of many metabolic products of certain microorganisms that can kill their bacterial competitors for nutrients in soil.

antibody [Gk. *anti*, against] One of a diverse array of antigen-binding receptors. Only B cells make antibody molecules and position them at their surface or secrete them.

anticodon A sequence of three nucleotide bases in a tRNA molecule that can base-pair with a codon in an mRNA molecule.

antigen (AN-tih-jen) [Gk. *anti*, against, + *genos*, race, kind] A molecular configuration that white blood cells recognize as foreign and that triggers an immune response. Most antigens are proteins at the surface of pathogens or tumor cells.

antigen-MHC complex Processed antigen fragments bound with a suitable MHC molecule and displayed at the cell surface; basis of antigen recognition that promotes lymphocyte cell divisions.

antigen-presenting cell Any cell displaying antigen-MHC complexes at its surface.

aorta (ay-OR-tah) [Gk. *airein*, to lift, heave] Main artery of systemic circulation; carries oxygenated blood away from the heart to all body regions except the lungs.

apical dominance Inhibitory influence of a terminal bud on growth of lateral buds.

apical meristem (AY-pih-kul MARE-ih-stem) [L. *apex*, top, + Gk. *meristos*, divisible] A mass of self-perpetuating cells responsible for primary growth at root and shoot tips.

apoptosis (APP-oh-TOE-sis) Of multicelled organisms, a form of cell death; molecular signals activate weapons of self-destruction already stockpiled in target cells. It occurs when a cell has completed its prescribed function or becomes altered, as by infection or cancerous transformation.

appendicular skeleton (ap-en-DIK-yoo-lahr) Bones of the limbs, hips, and shoulders.

Archaebacteria A kingdom of prokaryotes; encompasses methanogens, halophiles, and thermophiles, all of which differ from the eubacteria in chemical composition and in cell wall and membrane characteristics.

archipelago An island chain some distance away from a continent.

area effect Larger islands tend to support more species than smaller ones at equivalent distances from sources of colonizing species.

arteriole (ar-TEER-ee-ole) A blood vessel between an artery and capillary; a control point where the blood volume delivered to a given body region can be adjusted.

artery A large-diameter, rapid-transport blood vessel with a thick, muscular wall that smooths out the pulsations in blood pressure caused by heart contractions.

arthropod An invertebrate with a hardened exoskeleton, specialized body segments, and jointed appendages. Spiders, crabs, and insects are examples.

artificial selection Selection of traits among individuals of a population that occurs in an artificial environment, under contrived, manipulated conditions.

asexual reproduction Any of a number of modes of reproduction by which offspring arise from a single parent and inherit the genes of that parent only.

atmosphere A volume of gases, airborne particles, and water vapor that envelops the Earth; 80 percent of its mass is distributed within seventeen miles of the Earth's surface.

atmospheric cycle A biogeochemical cycle in which the atmosphere is the largest reservoir of an element. The carbon cycle and nitrogen cycle are examples.

atom The smallest particle unique to a given element; it has one or more positively charged protons, electrons, and (except for hydrogen), neutrons.

atomic number The number of protons in the nucleus of each atom of an element; the number differs for each element.

ATP Adenosine triphosphate (ah-DEN-uh-seen try-FOSS-fate). A nucleotide of adenine, ribose, and three phosphate groups that acts as an energy carrier. Its phosphate-group transfers drive nearly all energy-requiring metabolic reactions.

australopith (OHSS-trah-low-pith) [L. *australis*, southern, + Gk. *pithekos*, ape] Any of the earliest known hominids; a primate species on or near the evolutionary road that led to modern humans.

autoimmune response Misdirected immune response in which lymphocytes mount an attack against normal body cells.

autonomic nervous system (auto-NOM-ik) All nerves from the central nervous system to the smooth muscle, cardiac muscle, and glands of the viscera (internal organs and structures) of the vertebrate body.

autosome Any of the pairs of chromosomes that are the same in both males and females of the species.

autotroph (AH-toe-trofe) [Gk. *autos*, self, + *trophos*, feeder] Organism that synthesizes its own organic compounds using carbon dioxide (as the carbon source) and energy from the physical environment (such as sunlight energy). *Compare* heterotroph.

auxin (AWK-sin) A plant hormone that influences growth, such as stem elongation.

axial skeleton (AX-ee-uhl) Of a vertebrate skeleton, the skull, backbone, ribs, and breastbone (sternum).

axon A cylindrical extension from the cell body of a neuron, often with finely branched endings, that is specialized for the rapid propagation of action potentials.

B lymphocyte (B cell) The only white blood cell that produces antibodies, then positions them at the cell surface or secretes them as weapons in immune responses.

bacterial conjugation A transfer of plasmid DNA from one bacterial cell to another.

bacterial flagellum Of many bacterial cells, a whiplike motile structure that does not contain a core of microtubules.

bacteriophage (bak-TEER-ee-oh-fahj) [Gk. *baktērion*, small staff, rod, + *phagein*, to eat] Category of viruses that infect bacterial cells.

balancing selection All forms of selection that are maintaining two or more alleles for a trait in a population. When the resulting genetic variation persists, this is a case of balanced polymorphism.

Barr body In cells of female mammals, one of two X chromosomes that was randomly condensed so that its genes are inactivated.

basal body A centriole which, after giving rise to microtubules of a flagellum or cilium, remains attached to its base in the cytoplasm.

base Any substance that accepts hydrogen ions when dissolved in water.

base pair Two nucleotide bases that are located in two adjacent strands of DNA or RNA and that are hydrogen-bonded to each other.

base sequence The particular order in which one nucleotide base follows the next in a strand of DNA or RNA. The order is unique in at least some regions for each species.

basophil Fast-acting white blood cell that secretes histamine and other substances to maintain an inflammatory response.

behavior, animal A response to external and internal stimuli that requires sensory, neural, endocrine, and effector components. Behavior has a genetic basis and can evolve; it also can be modified through learning.

benthic province All sediments and rocky formations of the ocean bottom.

biennial (bi-EN-yul) A flowering plant that lives through two growing seasons.

bilateral symmetry Body plan in which the left and right halves of an animal are mirror-images of each other.

binary fission Of flatworms and some other animals, a mode of asexual reproduction; growth by way of mitotic cell divisions is followed by division of the whole body into two parts of the same or different sizes. Not the same as prokaryotic fission, by which bacteria reproduce.

biogeochemical cycle The movement of an element from the environment to organisms, then back to the environment.

biogeographic realm [Gk. *bios*, life, + *geographein*, to describe the Earth's surface] One of six major land divisions, each with distinguishing plants and animals; it retains its general identity because of climate and geographic barriers to gene flow.

biological clock Internal time-measuring mechanism that helps adjust an organism's daily activities, seasonal activities, or both in response to environmental cues.

biological magnification The increasing concentration of a nondegradable or slowly degradable substance in body tissues as it is passed along food chains.

biological species concept A species is one or more populations of individuals that are interbreeding under natural conditions and producing fertile offspring, and that are reproductively isolated from other such populations. The concept applies only to sexually reproducing species.

bioluminescence A flashing of light that emanates from an organism when excited electrons of luciferins (highly fluorescent substances) return to a lower energy level.

biomass Combined weight of all organisms at a given trophic level in an ecosystem.

biome A broad, vegetational subdivision of a biogeographic realm; shaped by climate, topography, and composition of regional soils.

biosphere [Gk. *bios*, life, + *sphaira*, globe] All regions of the Earth's waters, crust, and atmosphere in which organisms live.

biosynthetic pathway A metabolic pathway by which organic compounds necessary for life are synthesized.

biotic potential Of population growth for a given species, the maximum rate of increase per individual under ideal conditions.

bipedalism Habitually walking on two feet, as by ostriches and humans.

bird The only vertebrate that produces feathers and that has strong resemblances and evolutionary connections to reptiles.

blastocyst (BLASS-tuh-sist) [Gk. *blastos*, sprout, + *kystis*, pouch] Outcome of cleavage of a fertilized mammalian egg; a surface layer of blastomeres around a cavity filled with their secretions (the blastocoel), and an inner cell mass (several blastomeres huddled together against the cavity's inner surface).

blastomere One of the small, nucleated cells produced during the cleavage stage of animal development.

blastula (BLASS-chew-lah) Outcome of one type of cleavage pattern; blastomeres formed by the successive cuts surround a fluid-filled cavity (blastocoel).

blood A fluid connective tissue composed of water, solutes, and formed elements (blood cells and platelets); it carries substances to and from cells and helps maintain an internal environment favorable for cell activities.

blood pressure Fluid pressure, generated by heart contractions, that circulates blood.

blood-brain barrier Mechanism that exerts some control over which solutes enter the cerebrospinal fluid and thus helps protect the brain and spinal cord.

bone Mineral-hardened connective tissue of bones; one of the organs of the vertebrate skeleton that help move the body and its parts, protect other organs, store minerals, and (in some) produce blood cells.

bottleneck A severe reduction in population size, as brought about by intense selection pressure or a natural calamity.

Bowman's capsule The cup-shaped portion of a nephron that receives water and solutes being filtered from the blood in the kidneys.

brain The most complex integrating center of most nervous systems; it receives, processes, and integrates sensory input and issues coordinated commands for response by muscles and glands.

brain stem The vertebrate nervous tissue that evolved first and that still persists in the hindbrain, midbrain and forebrain.

bronchiole A component of the finely branched bronchial tree inside each lung.

bronchus, plural **bronchi** (BRONG-CUSS, BRONG-kee) [Gk. *bronchos*, windpipe] Tube-like branchings of the trachea that lead into the lungs of most vertebrates.

brown alga A photoautotrophic protistan with a notable abundance of xanthophyll pigments; such algae occupy marine habitats.

bryophyte A nonvascular land plant, such as a moss.

bud An undeveloped shoot of meristematic tissue, primarily; often covered and protected by scales (modified leaves).

buffer system A partnership between a weak acid and the base that forms when it dissolves in water. The two work as a pair to counter slight shifts in pH.

bulk A volume of fiber and other undigested material that absorption processes in the small intestine cannot decrease.

bulk flow In response to a pressure gradient, the movement of more than one kind of molecule in the same direction in the same medium (as in blood, sap, or air).

C4 pathway A pathway of photosynthesis in which carbon dioxide is fixed twice, in two different cell types. Carbon dioxide accumulates in the leaf and helps counter photorespiration. The first compound formed is the 4-carbon oxaloacetate.

Calvin-Benson cycle Cyclic reactions that are the *synthesis* part of the light-independent reactions of photosynthesis. In plants, RuBP or some other compound to which carbon has been affixed undergoes rearrangements; a sugar phosphate forms and the RuBP is regenerated. The cycle runs on ATP and NADPH from the light-dependent reactions.

CAM plant A plant that conserves water by opening stomata only at night, when it fixes carbon dioxide by way of a C4 pathway.

cambium (KAM-bee-um), plural **cambia** One of two types of meristems responsible for secondary growth (increases in stem and root diameter). Vascular cambium gives rise to secondary xylem and secondary phloem; cork cambium gives rise to periderm.

camouflage Adaptations in body color, form, or patterning, and in behavior that function in predator avoidance; they help an organism hide in the open (blend with its surroundings) and escape detection.

cancer A malignant tumor; its cells show gross abnormalities in the plasma membrane and cytoplasm, skewed growth and division, and weakened capacity for adhesion within the parent tissue (leading to metastasis). Unless eradicated, cancer is lethal.

capillary, blood [L. *capillus*, hair] A thin-walled vessel that functions in the exchange of carbon dioxide, oxygen, and some other substances between blood and interstitial fluid, which bathes living cells.

capillary bed A diffusion zone, consisting of a great number of capillaries, where substances are exchanged between blood and interstitial fluid.

carbohydrate [L. *carbo*, charcoal, + *hydro*, water] A molecule that consists of carbon, hydrogen, and oxygen in a 1:2:1 ratio (there are exceptions). All cells use carbohydrates as structural materials, energy reservoirs, and transportable forms of energy. The monosaccharides, oligosaccharides, and polysaccharides are three classes.

carbon cycle A biogeochemical cycle in which carbon moves from the atmosphere (its largest reservoir), through the ocean and organisms, then to the atmosphere.

carbon dioxide fixation Of photosynthesis, the first enzyme-mediated step of the light-independent reactions. Carbon (from CO_2) is affixed to RuBP or another compound for entry into the Calvin-Benson cycle.

carcinogen (kar-SIN-uh-jen) A substance or an agent, such as ultraviolet radiation, that can trigger cancer.

cardiac cycle (KAR-dee-ak) [Gk. *kardia*, heart, + *kyklos*, circle] The sequence of muscle contraction and relaxation for one heartbeat.

cardiac pacemaker Sinoatrial (SA) node; the basis of the normal rate of heartbeat. The self-excitatory cardiac muscle cells that spontaneously generate rhythmic waves of excitation over the heart chambers.

cardiovascular system Of most animals, an organ system of blood, one or more hearts, and blood vessels that functions in the rapid transport of substances to and from cells.

carnivore [L. *caro, carnis*, flesh, + *vovare*, to devour] An animal that eats other animals; a type of heterotroph.

carotenoid (kare-OTT-en-oyds) A light-sensitive, accessory pigment that transfers absorbed energy to chlorophylls. Different types absorb violet and blue wavelengths and transmit red, orange, and yellow.

carpel (KAR-pul) The female reproductive part of a flower; sometimes called a pistil. The lower portion of a single carpel (or of a structure composed of two or more) is an ovary. The upper portion has a stigma (a pollen-capturing surface tissue) and often a style (slender extension of the ovary wall).

carrying capacity The maximum number of individuals in a population (or species) that can be sustained indefinitely by a given environment.

cartilage A type of connective tissue with solid yet pliable intercellular material that resists compression.

Casparian strip A waxy band that is an impermeable barrier between the walls of abutting cells making up the endodermis (and exodermis, if present) inside roots.

cDNA Any DNA molecule copied from a mature mRNA transcript by way of reverse transcription.

cell [L. *cella*, small room] The smallest living unit; an organized unit that can survive and reproduce on its own, given suitable DNA instructions and environmental resources— notably energy and raw materials.

cell count The number of cells of a given type in a microliter of blood.

cell cycle Events by which a cell increases in mass, roughly doubles its number of cytoplasmic components, duplicates its DNA, then undergoes nuclear and cytoplasmic division. It extends from the time a new cell is produced until it completes division.

cell differentiation Developmental process in which different cell populations activate and suppress a fraction of their genes in different ways and so become specialized in composition, structure, and function.

cell junction Of multicelled organisms, a point of contact that links two adjoining cells physically, functionally, or both.

cell plate A disklike structure that forms from remnants of a microtubular spindle when a plant cell divides; it develops into a crosswall that partitions the cytoplasm.

cell theory A theory in biology stating that (1) all organisms are composed of one or more cells, (2) the cell is the smallest unit that retains a capacity for independent life, and (3) all cells arise from preexisting cells.

cell wall A semirigid, permeable structure that helps a cell hold its shape and resist rupturing if internal fluid pressure rises.

central nervous system The brain and spinal cord of vertebrates.

central vacuole A fluid-filled organelle in mature, living plant cells that stores amino acids, sugars, ions, and toxic wastes. As it enlarges, it forces increases in cell surface area that improve nutrient uptake.

centriole (SEN-tree-ohl) A cylinder of triplet microtubules that gives rise to microtubules of cilia and flagella.

centromere (SEN-troh-meer) [Gk. *kentron*, center, + *meros*, a part] A small, constricted region of a chromosome having attachment sites for the microtubules that move the chromosome during nuclear division.

cephalization (sef-ah-lah-ZAY-shun) [Gk. *kephalikos*, head] During the evolution of bilateral animals, the concentration of sensory structures and nerve cells in a head.

cerebellum (ser-ah-BELL-um) [L. diminutive of *cerebrum*, brain] Hindbrain region with reflex centers for maintaining posture and smoothing out limb movements.

cerebral cortex Thin surface layer of the cerebral hemispheres. Some parts receive sensory input, others integrate information and coordinate suitable responses.

cerebrospinal fluid Clear extracellular fluid surrounding and cushioning the brain and spinal cord.

cerebrum (suh-REE-bruhm) Forebrain region that first evolved to integrate olfactory input and select motor responses to it. In mammals, it evolved into the most complex integrating center.

channel protein A transport protein that acts as a channel through which specific ions and other water-soluble substances cross the plasma membrane. Some channels remain open; others are gated, and these open and close in controlled ways.

chemical bond A union between the electron structures of two or more atoms or ions.

chemical synapse (SIN-aps) [Gk. *synapsis*, union] A small cleft between a presynaptic neuron and a postsynaptic cell (another neuron, a muscle cell, or a gland cell) that is bridged by neurotransmitter molecules released from the presynaptic neuron.

chemiosmotic theory (kim-ee-OZ-MOT-ik) An electrochemical gradient across a cell membrane drives ATP formation. Hydrogen ions accumulate in a compartment formed by the membrane. The combined force of the H+ concentration and electric gradients propels ions through transport proteins (ATP synthases) spanning the membrane. By enzyme action at these proteins, ADP and inorganic phosphate combine to form ATP.

chemoreceptor (KEE-moe-ree-sep-tur) A sensory receptor that detects chemical energy (ions or molecules) dissolved in the fluid that bathes it.

chemosynthetic autotroph (KEE-moe-sin-THET-ik) One of a few kinds of bacteria able to synthesize its own organic compounds by using carbon dioxide as the carbon source and certain inorganic substances (such as sulfur) as the energy source.

chlorofluorocarbon (KLORE-oh-FLOOR-oh-car-bun), or **CFC** One of the odorless, invisible compounds of chlorine, fluorine, and carbon, widely used in commercial products, that are contributing to the thinning of the ozone layer above the Earth's surface.

chlorophyll (KLOR-uh-fills) [Gk. *chloros*, green, + *phyllon*, leaf] A light-sensitive pigment that absorbs violet-to-blue and red wavelengths but that transmits green. The main pigments in all but one small group of photoautotrophs. Certain chlorophylls donate electrons to the light-dependent reactions of photosynthesis.

chloroplast (KLOR-uh-plast) An organelle that specializes in photosynthesis in plants and photosynthetic protistans.

chordate An animal having a notochord, a dorsal hollow nerve cord, a pharynx, and gill slits in the pharynx wall for at least part of the life cycle.

chorion (CORE-ee-on) Of placental mammals, one of four extraembryonic membranes; it becomes a key component of the placenta. Absorptive structures (villi) develop at its surface and enhance the rapid exchange of substances between the embryo and mother.

chromatid (CROW-mah-tid) Of a duplicated eukaryotic chromosome, one of two DNA molecules (and associated proteins) that remain attached at their centromere region until separated by mitosis or meiosis; after this, each is a separate chromosome.

chromosome (CROW-moe-some) [Gk. *chroma*, color, + *soma*, body] Of eukaryotes, a DNA molecule with many associated proteins. Of prokaryotes, a DNA molecule without a comparable profusion of proteins.

chromosome number The sum total of chromosomes in cells of a given type. *See* haploidy; diploidy.

cilium (SILL-ee-um), plural **cilia** [L. *cilium*, eyelid] Of eukaryotic cells, a short, hairlike projection with an internal, regular array of microtubules. Cilia can serve as motile or sensory structures or help create currents of fluids. Typically more profuse than flagella.

circadian rhythm (ser-KAYD-ee-un) [L. *circa*, about, + *dies*, day] A cycle of physiological events that is completed every twenty-four hours or so independently of environmental change.

circulatory system An organ system having a muscular pump (heart, most often), blood vessels, and blood; the system transports materials to and from cells and often helps stabilize body temperature and pH.

cladogram [Gk. *clad-*, branch] Evolutionary tree diagram that arranges groups by branch points to show their relative relationships. Groups closer together share a more recent common ancestor than those farther apart.

classification system A way of organizing and retrieving information about species.

cleavage Third stage of animal embryonic development. Mitotic cell divisions divide the volume of egg cytoplasm into a number of smaller, nucleated cells (blastomeres). The number of cells increases, but the original volume of egg cytoplasm does not.

cleavage furrow A ringlike depression that forms during cytoplasmic division of an animal cell and that defines the cleavage plane for the cell. Microfilaments attached to the plasma membrane contract and draw it inward to cut the cell in two.

cleavage reaction A molecule splits into two smaller ones. Hydrolysis is an example.

climate Prevailing weather conditions for an ecosystem, such as temperature, humidity, wind speed, cloud cover, and rainfall.

climax community A self-perpetuating, stable array of species in equilibrium with one another and with their habitat.

climax pattern model Idea that one climax community may extend into another along gradients of environmental conditions, such as variations in climate, topography, and species interactions.

cloaca Of some vertebrates, the last part of a gut that receives feces, urine, and sperm or eggs; of some invertebrates, an excretory, respiratory, or reproductive duct.

cloned DNA Multiple, identical copies of restriction fragments that have been inserted into plasmids or some other cloning vector.

club fungus A fungus with reproductive structures having microscopic, club-shaped cells that produce and bear spores.

cnidarian A radial invertebrate at the tissue level of organization and the only organism to produce nematocysts. Two body forms (medusae and polyps) are common.

coal A nonrenewable source of energy that formed more than 280 million years ago from submerged, undecayed plant remains.

codominance A pair of nonidentical alleles that specify two phenotypes are expressed at the same time in heterozygotes.

codon One of the base triplets in an mRNA molecule, the linear sequence of which corresponds to a linear sequence of amino acids in a polypeptide chain. Of 64 codons, 61 specify different amino acids, and 3 of these also are start signals for translation; 1 serves as a stop signal for translation.

coelom (SEE-lum) [Gk. *koilos*, hollow] A cavity, lined with peritoneum, between the gut and body wall of most animals.

coenzyme A nucleotide; an enzyme helper that accepts electrons and hydrogen atoms stripped from substrates at a reaction site and transfers them elsewhere.

coevolution The joint evolution of two or more closely interacting species; when one species evolves, the change affects selection pressures operating between the two, so the other also evolves.

cofactor A metal ion or coenzyme; it helps an enzyme catalyze a reaction or transfers electrons, atoms, or functional groups from one substrate to another.

cohesion Capacity to resist rupturing when placed under tension (stretched).

cohesion theory of water transport Theory that water moves up through plants due to hydrogen bonding among water molecules confined as narrow columns in xylem. The collective cohesive strength of the bonds allows water to be pulled up in response to transpiration (evaporation from leaves).

collenchyma (coll-ENG-kih-mah) A simple plant tissue that offers flexible support for primary growth, as in lengthening stems.

colon (CO-lun) The large intestine.

commensalism [L. *com*, together, + *mensa*, table] An ecological interaction between species that directly benefits one but does not affect other much, if at all.

communication display A pattern of behavior, often ritualized with intended changes in the function of common behavior patterns, that serves as a social signal.

communication signal A social cue encoded in stimuli that holds unambiguous meaning for individuals of the same species. Specific odors, sounds, coloration and patterning, postures, and movements are examples.

community All populations living in the same habitat. Also, a group of organisms with similar life-styles in a habitat, such as a community of birds.

companion cell A specialized parenchyma cell that helps load organic compounds into conducting cells of phloem.

comparative morphology [Gk. *morph*, form] Study of comparable body parts of adults or embryonic stages of major lineages.

competitive exclusion Theory that species that require identical resources cannot coexist indefinitely.

complement system A set of about twenty proteins circulating in inactive form within vertebrate blood; different kinds induce lysis of pathogens, promote inflammation, and stimulate phagocytes to act during both nonspecific defenses and immune responses.

compound A substance consisting of two or more elements in unvarying proportions.

concentration gradient A difference in the number of molecules or ions of a substance between adjoining regions. Energy inherent in their constant molecular motion makes them collide and careen outward from the region of higher to lower concentration. Barring other forces, all substances tend to diffuse down their concentration gradient.

condensation reaction Through covalent bonding, two molecules combine to form a larger molecule, often with the formation of water as a by-product.

cone cell In a vertebrate eye, a photoreceptor that responds to intense light and contributes to sharp daytime vision and color perception.

conifer A pollen- and seed-bearing plant of the dominant group of gymnosperms; mostly evergreen, woody trees and shrubs with needle-like or scale-like leaves.

conjugation, bacterial Of bacteria only, a mechanism by which a donor cell transfers plasmid DNA to a recipient cell.

connective tissue proper A category of animal tissues, all having mostly the same components but in different proportions. They incorporate fibroblasts and other cells, the secretions of which form fibers (mostly of collagen and elastin) and a ground substance of modified polysaccharides.

consumer [L. *consumere*, to take completely] A heterotroph that obtains energy and carbon by feeding on the tissues of other organisms. Herbivores, carnivores, and parasites are examples.

continuous variation Of a population, a more or less continuous range of small differences in a given trait among all of its individuals.

contractile vacuole (kun-TRAK-till VAK-you-ohl) [L. *contractus*, to draw together] Of some protistans, such as a paramecium, an organelle that takes up excess water in the cell body, then contracts; the contractile force is enough to expel the water outside the cell through a pore to its surface.

control group Of an experimental test, a group used to evaluate possible side effects of a test involving an experimental group. Ideally, the control group is identical to the experimental group in all respects except for the variable being studied.

cork cambium A lateral meristem that gives rise to a corky replacement for the epidermis of woody plant parts.

corpus callosum (CORE-pus ka-LOW-sum) A band of axons (200 million in humans) that functionally link two cerebral hemispheres.

corpus luteum (CORE-pus LOO-tee-um) A glandular structure that develops from cells of a ruptured ovarian follicle and secretes progesterone and estrogen.

cortex [L. *cortex*, bark] In general, a rindlike layer such as the kidney or adrenal cortex. In vascular plants, the ground tissue that makes up most of the primary plant body, supports plant parts, and stores food.

cotyledon A seed leaf, which develops as part of the embryo of monocots and dicots; cotyledons provide nourishment for the seedling at the time of germination and initial growth.

courtship display A pattern of ritualized social behavior between potential mates. It may include frozen postures as well as movements that are exaggerated and yet simplified. It may include visual signals, such as body parts that are conspicuously enlarged, distinctively colored or patterned, or some combination of these.

covalent bond (koe-VAY-lunt) [L. *con*, together, + *valere*, to be strong] A sharing of one or more electrons between atoms or groups of atoms. If electrons are shared equally, the bond is nonpolar. If shared unequally, it is polar (slightly positive at one end, slightly negative at the other).

continuous variation Of the individuals of a population, a range of small differences in one or more traits.

cross-bridge formation Of a muscle cell, a reversible interaction between its many actin and myosin filaments that is the basis of contraction.

crossing over During prophase I of meiosis, the breakage and exchange of corresponding segments between nonsister chromatids of a pair of homologous chromosomes; a form of genetic recombination that breaks up old combinations of alleles and puts new ones together in chromosomes.

culture The sum of behavior patterns of a social group, passed between generations by learning and by symbolic behavior, especially language.

cuticle (KEW-tih-kull) A body covering. Of land plants, a transparent cover of waxes and lipid-rich cutin deposited on the outer surface of epidermal cell walls. Of annelids, a thin, flexible surface coat. Of arthropods, a hardened, lightweight cover with protein and chitin components that functions as an exoskeleton.

cyclic AMP (SIK-lik) A nucleotide; cyclic adenosine monophosphate. It functions in intercellular communication, as when it is a second messenger (a cytoplasmic mediator of a cell's response to signaling molecules).

cyclic pathway of ATP formation Ancient photosynthetic pathway occurring at the plasma membrane of some bacteria and at the thylakoid membrane of chloroplasts. A photosystem embedded in the membrane gives up electrons to a transport system, which gives them back to the photosystem. The electron flow sets up concentration and electric gradients across the membrane that drive ATP formation at nearby membrane sites.

cyst Of many microorganisms, a resistant resting stage with thick, tough outer layers that typically forms in response to adverse conditions; of skin, any abnormal, fluid-filled sac without an external opening.

cytochrome (SIGH-toe-krome) [Gk. *kytos*, hollow vessel, + *chrōma*, color] Iron-containing protein molecule; a component of the electron transport systems used in photosynthesis and aerobic respiration.

cytokinesis (SIGH-toe-kih-NEE-sis) [Gk. *kinesis*, motion] Cytoplasmic division; the splitting of a parent cell into daughter cells.

cytokinin (SIGH-tow-KY-nin) Any of the class of plant hormones that stimulate cell division, promote leaf expansion, and retard leaf aging.

cytological marker One or more observable, unusual differences between chromosomes of the same type.

cytomembrane system [Gk. *kytos*, hollow vessel] Organelles functioning as a system to modify, package, and distribute newly formed proteins and lipids. Endoplasmic reticulum, Golgi bodies, lysosomes, and a variety of vesicles are its components.

cytoplasm (SIGH-toe-plaz-um) [Gk. *plassein*, to mold] All cellular parts, particles, and semifluid substances enclosed within the plasma membrane except for the nucleus (or nucleoid, in bacterial cells).

cytoplasmic localization When cleavage divides an animal zygote, each resulting blastomere receives a localized portion of maternal messages in the egg cytoplasm.

cytosine (SIGH-toe-seen) A pyrimidine; one of the nitrogen-containing bases in nucleotides.

cytoskeleton The internal "skeleton" of eukaryotic cells. Its microtubules and other components structurally support the cell and organize and move its internal components. The cytoskeleton also helps free-living cells move through their environment.

cytotoxic T cell A T lymphocyte that uses touch-killing to eliminate infected body cells or tumor cells. When it contacts targets, it delivers cell-killing chemicals into them.

decomposer [partly fr. L. *dis-*, to pieces, + *companere*, arrange] Of ecosystems, a heterotroph that gets energy and carbon by chemically breaking down the remains, products, or wastes of other organisms and helps cycle nutrients back to producers. Certain fungi and bacteria are examples.

deforestation The removal of all trees from a large tract of land, such as the Amazon Basin and the Pacific Northwest.

degradative pathway A metabolic pathway by which organic compounds are broken down in stepwise reactions that lead to products of lower energy.

deletion Loss of a chromosome segment.

denaturation (deh-NAY-chur-AY-shun) Of any molecule, the loss of three-dimensional shape following disruption of hydrogen bonds and other weak bonds.

dendrite (DEN-drite) [Gk. *dendron*, tree] A short, slender extension from the cell body of a neuron; commonly an input zone.

denitrification (DEE-nite-rih-fih-KAY-shun) Conversion of nitrate or nitrite by certain bacteria to gaseous nitrogen (N_2) and a small amount of nitrous oxide (N_2O).

density-dependent control A factor that limits population growth by reducing the birth rate, increasing the rates of death and dispersal, or all of these. Predation, parasitism, disease, and competition for resources are examples.

density-independent factor A factor that tends to cause a population's death rate to increase independently of its density. Storms and floods are examples.

dentition (den-TIH-shun) The type, size, and number of an animal's teeth.

derived trait A novel feature that evolved only once and is shared only by descendants of the ancestral species in which it evolved.

dermal tissue system All the tissues that cover and protect the surfaces of a plant.

dermis The layer of skin underlying the epidermis; consists primarily of dense connective tissue.

desert A biome that typically forms where the potential for evaporation greatly exceeds rainfall and vegetation cover is limited.

desertification (dez-urt-ih-fih-KAY-shun) Conversion of a grassland or an irrigated or rain-fed cropland to a desertlike condition, with a drop in agricultural productivity of 10 percent or more.

detrital food web A network of food chains in which energy flows mainly from plants through arrays of detritivores and decomposers.

detritivore (dih-TRY-tih-vorez) [L. *detritus*; after *deterere*, to wear down] A heterotroph that consumes decomposing particles of organic matter. Earthworms, crabs, and nematodes are examples.

deuterostome (DUE-ter-oh-stome) [Gk. *deuteros*, second, + *stoma*, mouth] A bilateral animal for which the first indentation that forms in the early embryo develops into an anus. An echinoderm or a chordate.

development Of multicelled organisms, the programmed emergence of specialized, morphologically different body parts.

diaphragm (DIE-uh-fram) [Gk. *diaphragma*, to partition] Muscular partition between the thoracic and abdominal cavities; its contraction and relaxation contribute to breathing. Also, a contraceptive device used temporarily to prevent sperm from entering the uterus during sexual intercourse.

dicot (DIE-kot) [Gk. *di*, two, + *kotylēdōn*, cup-shaped vessel] A dicotyledon. In general, a flowering plant characterized by seeds having embryos with two cotyledons (seed leaves); net-veined leaves; and floral parts arranged in fours, fives, or multiples of these.

diffusion Net movement of like molecules (or ions) down their concentration gradient. In the absence of other forces, the energy inherent in molecules makes them move constantly and collide at random. Their collisions are most frequent where they are most crowded together; thus they show a net outward movement from regions of higher to lower concentration.

digestive system An internal sac or tube from which ingested food is absorbed into the internal environment.

dihybrid cross An experimental cross in which true-breeding F_1 offspring inherit two gene pairs, each consisting of two nonidentical alleles.

diploidy (DIP-loyd-ee) The presence of two of each type of chromosome (that is, pairs of homologous chromosomes) in the interphase nucleus of somatic cells and germ cells. *Compare* haploidy.

directional selection A mode of natural selection by which the range of variation for some trait shifts in a consistent direction in response to directional change in the environment or to new environmental conditions.

disaccharide (die-SAK-uh-ride) [Gk. *di*, two, + *sakcharon*, sugar] A simple carbohydrate; one of the oligosaccharides consisting of two covalently bonded sugar monomers.

disease Outcome of an infection when the body's defenses cannot be mobilized fast enough; the pathogen's activities interfere with normal body functions.

disruptive selection A mode of natural selection by which forms of a trait at both ends of a range of variation are favored and intermediate forms are selected against.

distal tubule The tubular portion of a nephron farthest from the glomerulus; a region of water and sodium reabsorption.

distance effect Only species adapted for long-distance dispersal are potential colonists of islands far from their home range.

diversity, organismic Sum total of all the variations in form, function, and behavior that have accumulated in different lineages. Variations in traits generally are adaptive to prevailing conditions or were adaptive to conditions that existed in the past.

DNA Deoxyribonucleic acid (dee-OX-ee-RYE-bow-new-CLAY-ik). For all cells and many viruses, a nucleic acid that is the molecule of inheritance. It consists of two nucleotide strands twisted together helically and held together by numerous hydrogen bonds. The nucleotide sequence encodes instructions for synthesizing proteins and, ultimately, new individuals of a particular species.

DNA amplification Any of several methods by which a DNA library is copied again and again to yield multiple, identical copies of DNA fragments (cloned DNA).

DNA-DNA hybridization *See* nucleic acid hybridization.

DNA fingerprint A unique array of RFLPs, inherited in a Mendelian pattern from each parent, that gives each individual a unique identity.

DNA library A collection of DNA fragments produced by restriction enzymes and later incorporated into plasmids.

DNA ligase (LYE-gaze) An enzyme that seals together the new base-pairings during DNA replication; also used by technologists to seal base-pairings between DNA fragments and cut plasmid DNA.

DNA polymerase (poe-LIM-uh-raze) An enzyme that assembles a new strand on a parent DNA strand during replication; also takes part in DNA repair.

DNA probe A short DNA sequence that is synthesized from radioactively labeled nucleotides. Part of the probe is designed to base-pair with part of a gene under study.

DNA repair Following an alteration in the base sequence of a DNA strand, a process that may restore the original sequence, as carried out by DNA polymerases, DNA ligases, and other enzymes.

DNA replication Of cells, the process by which hereditary material is duplicated for distribution to daughter nuclei. Occurs prior to mitosis and meiosis in eukaryotic cells and during prokaryotic fission in bacterial cells.

dominance hierarchy A social organization in which some members of the group have adopted a subordinate status to others.

dominant allele In a diploid cell, an allele that masks the expression of its partner on the homologous chromosome.

dormancy [L. *dormire*, to sleep] A hormone-mediated time of inactivity during which metabolic activities idle. Perennials, seeds, many spores, cysts, and some animals go through dormancy.

double fertilization Of flowering plants only, the fusion of one sperm nucleus with the egg nucleus (to produce a zygote), *and* the fusion of a second sperm nucleus with nuclei of the endosperm mother cell, which gives rise to a nutritive tissue (endosperm).

doubling time The length of time it takes for a population to double in size.

drug addiction Chemical dependence on a drug following habituation and tolerance of it; in time the drug assumes an "essential" biochemical role in the body.

dry shrubland A biome that typically forms where annual rainfall is less than 25 to 60 centimeters; short, multibranched woody shrubs (e.g., chaparral) predominate.

dry woodland A biome that typically forms where annual rainfall is about 40 to 100 centimeters; there may be tall trees, but these do not form a dense canopy.

duplication A repeat of the same linear stretch of an individual's DNA in the same chromosome or in a different one.

ecdysone A hormone with major influence over the development of many insects.

echinoderm A type of invertebrate that has calcified spines, needles, or plates on the body wall. Although radially symmetrical, it has some bilateral features. Sea stars and sea urchins are examples.

ecology [Gk. *oikos*, home, + *logos*, reason] Study of the interactions of organisms with one another and with their physical and chemical environment.

ecosystem [Gk. *oikos*, home] An array of organisms and their physical environment, all of which are interacting through a flow of energy and a cycling of materials.

ecosystem modeling An analytical method, based on computer programs and models, of predicting unforeseen effects of specific disturbances to an ecosystem.

ectoderm [Gk. *ecto*, outside, + *derma*, skin] The first-formed, outermost primary tissue layer of animal embryos; forerunner of cell lineages that give rise to nervous system tissues and the integument's outer layer.

effector Of homeostatic systems, a muscle (or gland) that responds to signals from an integrator, such as the brain, by producing movement (or chemical change) that helps adjust the body to changing conditions.

effector cell A differentiated cell of one of the subpopulations of lymphocytes that form during an immune response; it acts at once to engage and destroy the antigen-bearing agent that triggered the response.

egg A type of mature female gamete; also called an ovum.

El Niño A recurring, massive eastward displacement of warm surface waters of the western equatorial Pacific, which displaces cooler waters off the South American coast. Causes global disruptions in climate.

electromagnetic spectrum The entire range of wavelengths, from the forms of radiant energy less than 10^{-5} nanometer long to radio waves more than 10 kilometers long.

electron A negatively charged unit of matter, with both particulate and wavelike properties, that occupies one of the orbitals around the atomic nucleus. Atoms can gain, lose, or share electrons with other atoms.

electron transfer The donation of one or more electrons stripped from one molecule to another molecule.

electron transport phosphorylation (FOSS-for-ih-LAY-shun) Final stage of aerobic respiration, when electrons from reaction intermediates flow through a membrane transport system that gives them up to oxygen. The flow sets up electrochemical gradients that drive ATP formation at other sites in the membrane.

electron transport system Organized array of enzymes and cofactors, bound in a cell membrane, that accept and donate electrons in series. When it operates, hydrogen ions flow across the membrane, and the flow drives ATP formation and other reactions.

element A substance that cannot be broken down to substances with different properties.

embryo (EM-bree-oh) [Gk. *en*, in, + probably *bryein*, to swell] Of animals generally, a stage formed by cleavage, gastrulation, and other early developmental events. Of seed plants, the young sporophyte, from the first cell divisions after fertilization until germination.

embryonic induction A change in the developmental fate of an embryonic tissue, as brought about by exposure to a gene product released from an adjacent tissue.

embryo sac Common name of the female gametophyte of flowering plants.

emerging pathogen A deadly pathogen, either a newly mutated strain of an existing species or one that evolved long ago and is only now taking great advantage of the increased presence of human hosts.

emulsification Of the chyme in the small intestine, a suspension of droplets of fat coated with bile salts.

end product A substance present at the end of a metabolic pathway.

endangered species A species at the brink of extinction owing to the extremely small size and severely limited genetic diversity of its remaining populations.

endergonic reaction (en-dur-GONE-ik) A chemical reaction having a net gain in energy.

endocrine gland A ductless gland that secretes hormones, which usually enter interstitial fluid and then the bloodstream.

endocrine system System of cells, tissues, and organs, functionally linked to the nervous system, that exerts control by way of its hormones and other chemical secretions.

endocytosis (EN-doe-sigh-TOE-sis) Transport of a substance into a cell by a vesicle, the membrane of which is a patch of plasma membrane that forms around the substance and sinks into the cytoplasm. Phagocytes also engulf prey or pathogens this way.

endoderm [Gk. *endon*, within, + *derma*, skin] The innermost primary tissue layer of animal embryos; gives rise to the inner lining of the gut and organs derived from it.

endodermis A sheetlike wrapping of single cells around the vascular cylinder of a root that functions in controlling the uptake of water and dissolved nutrients.

endometrium (EN-doh-MEET-ree-um) [Gk. *metrios*, of the womb] Innermost lining of the uterus, consisting of connective tissues, glands, and blood vessels.

endoplasmic reticulum or **ER** (EN-doe-PLAZ-mik reh-TIK-yoo-lum) An organelle that begins at the nucleus and curves through the cytoplasm. In rough ER (with many ribosomes on its cytoplasmic side), many new polypeptide chains acquire specialized side chains. Smooth ER (with no attached ribosomes) is a site of lipid synthesis.

endoskeleton [Gk. *endon*, within, + *skleros*, hard, stiff] An internal framework of bone, cartilage, or both in chordates. Together with skeletal muscle, supports and protects other body parts, helps maintain posture, and moves the body.

endosperm (EN-doe-sperm) Nutritive tissue that surrounds a flowering plant embryo and becomes food for the young seedling.

endospore A resting structure that forms around a copy of the chromosome and part of the cytoplasm of certain bacteria.

endosymbiosis In general, a mutually beneficial interdependence between two species, one of which resides permanently inside the other's body.

energy A capacity to do work.

energy carrier A molecule that delivers energy from one metabolic reaction site to another. ATP is the most common energy carrier in all cells.

energy flow pyramid A pyramid-shaped representation of an ecosystem's trophic structure, illustrating the energy losses at each transfer to a different trophic level.

enhancer A base sequence in DNA that is a binding site for an activator protein.

entropy (EN-trow-pee) A measure of the degree of disorder in a system (how much energy has become so disorganized and dispersed, usually as heat, that it is no longer readily available to do work). Any organized system tends toward entropy without energy inputs to make up for the flow of energy out of it.

enzyme (EN-zime) One of a class of proteins that enormously speed (catalyze) reactions between specific substances, usually at their functional groups.

eosinophil Fast-acting white blood cell; its enzyme secretions digest holes in parasitic worms during an inflammatory response.

epidermis The outermost tissue layer of a multicelled plant and of nearly all animals.

epinephrine (ep-ih-NEF-rin) Hormone of the adrenal medulla; raises blood levels of sugar and fatty acids; increases heart rate and the force of contraction.

epiglottis A flaplike structure at the start of the larynx, the position of which directs the movement of air into the trachea or of food into the esophagus.

epistasis (eh-PISS-tah-sis) An interaction between gene pairs. Two alleles of one gene mask expression of another gene's alleles, so expected phenotypes may not appear.

epithelium (EP-ih-THEE-lee-um) An animal tissue of one or more layers of adhering cells that covers the body's external surfaces and lines its internal cavities and tubes. It has one free surface; the opposite surface rests on a basement membrane between it and an underlying connective tissue. Epidermis is an example.

equilibrium, dynamic [Gk. *aequus*, equal, + *libra*, balance] The point at which a chemical reaction runs forward as fast as in reverse; the concentrations of reactant molecules and product molecules show no net change.

erosion The movement of land under the force of wind, running water, and ice.

erythrocyte (eh-RITH-row-site) [Gk. *erythros*, red, + *kytos*, vessel] Red blood cell.

esophagus (ee-SOF-uh-gus) Tubular portion of a digestive system that receives ingested food and leads to the stomach.

essential amino acid An amino acid that an organism cannot synthesize for itself and must obtain from a food source.

essential fatty acid A fatty acid that an organism cannot synthesize for itself and must obtain from food source.

estrogen (ESS-trow-jen) A sex hormone that helps oocytes mature, induces changes in the uterine lining during the menstrual cycle and pregnancy, and helps maintain secondary sexual traits; also influences bodily growth and development.

estrus (ESS-truss) [Gk. *oistrus*, frenzy] For mammals generally, the cyclic period of a female's sexual receptivity to the male.

estuary (EST-you-ehr-ee) A partly enclosed coastal region where seawater mixes with freshwater and runoff from the surrounding land, as by streams and rivers.

ethylene (ETH-il-een) Plant hormone that stimulates fruit ripening and abscission.

Eubacteria Kingdom of the most common species of bacterial cells.

eukaryotic cell (yoo-CARRY-oh-tic) [Gk. *eu*, good, + *karyon*, kernel] A cell having a "true nucleus" and other distinguishing membrane-bound organelles. *Compare* prokaryotic cell.

eutrophication Nutrient enrichment of a body of water, such as a lake, that typically results in reduced transparency and a phytoplankton-dominated community.

evaporation [L. *e-*, out, + *vapor*, steam] Heat energy converts a substance from the liquid to the gaseous state.

evolution, biological [L. *evolutio*, unrolling] Genetic change in a line of descent over time; brought about by microevolutionary processes (gene mutation, natural selection, genetic drift, and gene flow).

evolutionary systematics The branch of biology that applies evolutionary theory to the task of identifying patterns of diversity over time and in the environment.

evolutionary tree A treelike diagram in which the branches represent separate lines of descent from a common ancestor and branch points represent divergences.

excitatory postsynaptic potential (or EPSP) One of two competing signals at an input zone of a neuron; a graded potential that brings the neuron's plasma membrane closer to threshold.

excretion Any of several processes by which excess water, excess or harmful solutes, or waste materials leave the body by way of a urinary system or certain glands.

exergonic reaction (EX-ur-GONE-ik) A chemical reaction that shows a net loss in energy.

exocrine gland (EK-suh-krin) [Gk. *es*, out of, + *krinein*, to separate] Glandular structure that secretes products, usually through ducts or tubes, to a free epithelial surface.

exocytosis (EK-so-sigh-TOE-sis) Transport of a substance out of a cell by means of a vesicle, the membrane of which fuses with the plasma membrane, so that the vesicle's contents are released outside.

exodermis Layer of cells just inside the root epidermis of most flowering plants; helps control the uptake of water and solutes.

exon Any of the nucleotide sequences of a pre-mRNA molecule that become spliced together to form a mature mRNA transcript and ultimately get translated into protein.

exoskeleton [Gk. *exo*, out, + *sklēros*, hard, stiff] An external skeleton, as in arthropods.

experiment A test of potentially falsifiable hypotheses about some aspect of nature. Its premise is that any aspect of the natural world has one or more underlying causes.

exponential growth (EX-po-NEN-shul) A pattern of population growth in which the population size expands by ever increasing increments during successive time intervals because the reproductive base becomes ever larger. The plot of population size against time has a characteristic J-shaped curve.

extinction, background A steady rate of species turnover that characterizes lineages through most of their histories.

extinction, mass An abrupt increase in the rate at which major taxa disappear, with several taxa being affected simultaneously.

extracellular fluid In animals generally, all the fluid not inside cells; includes plasma (the liquid portion of blood) and interstitial fluid (occupying the spaces between cells and tissues).

extracellular matrix A matrix that helps impart shape to many animal tissues; its ground substance contains fibrous proteins and other materials (mostly cell secretions).

FAD Flavin adenine dinucleotide, one of the nucleotide coenzymes that transfers electrons and unbound protons (H^+) from one reaction site to another. At such times it is abbreviated $FADH_2$.

fall overturn The vertical mixing of a body of water in autumn. Its upper layer cools, increases in density, and sinks; dissolved oxygen moves down and nutrients from bottom sediments move up.

family pedigree A chart of the genetic relationship of the individuals in a family through successive generations.

fat A lipid with a glycerol head and one, two, or three fatty acid tails. Tryglycerides (neutral fats) have three. Unsaturated tails have single covalent bonds in their carbon backbone; saturated tails also have one or more double bonds.

fate map A map of the surface of an animal embryo that shows the origin of each kind of differentiated cell in the adult.

fatty acid A molecule with a backbone of up to thirty-six carbon atoms, a carboxyl group (—COOH) at one end, and hydrogen atoms at most or all of the remaining bonding sites.

feedback inhibition Of cells or multicelled organisms, a control mechanism by which the output of a substance changes a specific condition or activity, which then triggers a decrease in further output of the substance or further activity.

fermentation [L. *fermentum*, yeast] A type of anaerobic pathway of ATP formation. It starts with glycolysis, ends with a transfer of electrons back to one of the breakdown products or intermediates, and regenerates NAD^+ required for the reaction. Its has a net yield of two ATP per glucose molecule.

fertilization [L. *fertilis*, to carry, to bear] The fusion of a sperm nucleus with the nucleus of an egg, which thus becomes a zygote.

fever A body temperature higher than a set point in the hypothalamic region that acts as the body's thermostat.

fibrous root system Of most monocots, all the lateral branchings of adventitious roots, which arose earlier from the young stem.

filter feeder An animal that filters food from a current of water that is directed through a body part, such as a sea squirt's pharynx.

filtration Of vertebrates, a process by which blood pressure forces water and solutes from capillaries into interstitial fluid. Filtration occurs in Bowman's capsule of a nephron.

fin Of fishes generally, an appendage that helps propel, stabilize, and guide the body through water.

first law of thermodynamics [Gk. *therme*, heat, + *dynamikos*, powerful] A law of nature stating the total amount of energy in the universe remains constant. Energy cannot be created from nothing and existing energy cannot be destroyed.

fish An aquatic animal of the most ancient, diverse vertebrate lineage, which includes jawless, cartilaginous, and bony fishes.

fitness An increase in adaptation to the environment, as brought about by genetic change.

fixation Of the individuals of a population, only one kind of allele remains at a specified locus; all individuals are homozygous for it.

fixed action pattern Of animals, a program of coordinated, stereotyped muscle activity that runs to completion independently of feedback from the environment.

flagellum (fluh-JELL-um), plural **flagella** [L. whip] Tail-like motile structure of many free-living eukaryotic cells; its core has a 9 + 2 array of microtubules.

flower A reproductive structure that distinguishes angiosperms from other seed plants and often attracts pollinators.

fluid mosaic model All cell membranes consist of a lipid bilayer and proteins. The lipids (phospholipids, mainly) impart basic structure, impermeability to water-soluble molecules, and (through packing variations and movements) fluidity. Diverse proteins spanning the bilayer or attached to one of its surfaces perform most of the membrane functions, including transport, enzyme activity, and reception of molecular signals or substances.

follicle (FOLL-ih-kul) A small sac, pit, or cavity, as around a hair; also a mammalian oocyte with its surrounding layer of cells.

food chain A straight-line sequence of who eats whom in an ecosystem.

food web A network of cross-connecting, interlinked food chains with some number of producers and consumers, as well as decomposers, detritivores, or both.

forebrain Most complex part of a vertebrate brain; it includes the cerebrum (and cerebral cortex), olfactory lobes, and hypothalamus.

forest A biome where tall trees grow together closely enough to form a fairly continuous canopy over a broad region.

fossil Recognizable, physical evidence of an organism that lived in the distant past.

fossil fuel Coal, petroleum, or natural gas; a nonrenewable source of energy formed in sediments by the compression of plant remains over hundreds of millions of years.

fossilization How fossils form. First an organism or traces of it becomes buried in sediments or volcanic ash. Water infiltrates the remains, which become infused with dissolved inorganic compounds. Sediments accumulate and exert pressure above the burial site. Over great spans of time, the pressure and chemical changes transform the remains to stony hardness.

founder effect A form of bottlenecking. By chance, allele frequencies of founders of a new population may not be the same as those of the original population. If there is no gene flow between the two, then natural selection will influence gene frequencies in drastically different ways through its interaction with genetic drift.

free radical A highly reactive, unbound molecular fragment that has the wrong number of electrons.

fruit [L. after *frui*, to enjoy] Of flowering plants, the expanded and ripened ovary of one or more carpels, some with accessory floral structures incorporated.

FSH Follicle-stimulating hormone; one of the hormones produced and secreted by the anterior lobe of the pituitary gland; serves reproductive roles in both sexes.

functional group An atom or group of atoms that is covalently bonded to the carbon backbone of an organic compound and that influences its chemical behavior.

functional-group transfer Donation of a functional group by one molecule to another.

Fungi The kingdom of fungi which, as a group, are major decomposers.

fungus A eukaryotic heterotroph that uses extracellular digestion and absorption; it secretes enzymes that break down an external food source into molecules small enough to be absorbed by its cells. Saprobes feed on nonliving organic matter, parasites feed on living organisms.

gall bladder Organ that stores bile secreted from the liver and that is connected by way of a duct to the small intestine.

gamete (GAM-eet) [Gk. *gametēs*, husband, and *gametē*, wife] A haploid cell, formed by meiosis and cytoplasmic division of a germ cell; required for sexual reproduction. Eggs and sperm are examples.

gamete formation Of animals, the first stage of development, in which sperm or eggs form and mature within reproductive tissues or organs of parents.

gametophyte (gam-EET-oh-fite) [Gk. *phyton*, plant] A haploid, multicelled, gamete-producing body that forms during the life cycle of most plants.

ganglion (GANG-lee-un), plural **ganglia** [Gk. *ganglion*, a swelling] A distinct clustering of cell bodies of neurons in regions other than the brain or spinal cord.

gastrulation (gas-tru-LAY-shun) The fourth stage of animal development; a time of major cellular reorganization when newly formed cells become arranged into two or three primary tissues, or germ layers.

gene [short for German *pangan*, after Gk. *pan*, all + *genes*, to be born] A unit of information about a heritable trait that is passed on from parents to offspring. Each gene has a specific location (locus) on a chromosome.

gene flow A microevolutionary process; the movement of alleles into and out of populations as a result of immigration and emigration.

gene frequency More precisely, allele frequency; the relative abundances of all the different alleles at a given gene locus that are carried by all individuals of a population.

gene locus The particular location of a gene along the length of a chromosome.

gene mutation [L. *mutatus*, a change] A heritable change in a DNA molecule by the deletion, addition, or substitution of one to several bases in its nucleotide sequence.

gene pair Of diploid cells, the two alleles at a particular gene locus (that is, on a pair of homologous chromosomes).

gene pool Sum total of all genotypes in a population.

gene therapy Generally, the transfer of one or more normal genes into an organism to correct or lessen the adverse effects of a genetic disorder.

genetic code [After L. *genesis*, to be born] The correspondence between the nucleotide triplets in DNA (then in RNA) and the specific sequences of amino acids in the resulting polypeptide chains; the basic language of protein synthesis in cells.

genetic disease An illness in which the expression of one or more genes increased the susceptibility of an individual to an infection or weakened its immune response.

genetic disorder An inherited condition that results in mild to severe medical problems.

genetic divergence Build-up of differences in gene pools of two or more populations of a species after a geographic barrier arises and separates them, because gene mutation, natural selection, and genetic drift are free to operate independently in each one.

genetic drift A random change in allele frequencies over the generations brought about by chance alone. The magnitude of its effect on genetic diversity and on the range of phenotypes relates to population size.

genetic engineering Altering the information content of DNA through use of recombinant DNA technology.

genetic equilibrium A state in which a population is not evolving. It occurs only if there is no mutation, if the population is large and isolated from other populations of the species, and if there is no natural selection (all members survive and reproduce equally by random mating).

genetic recombination A nonparental combination of some number of alleles in offspring; an outcome of gene mutation, crossing over, changes in chromosome structure or number, or recombinant DNA technology.

genome All the DNA in a haploid number of chromosomes of a given species.

genotype (JEEN-oh-type) Genetic constitution of an individual; a single gene pair or the sum total of an individual's genes. *Compare* phenotype.

genus, plural **genera** (JEEN-US, JEN-er-ah) [L. *genus*, race, origin] A taxon into which all species with phenotypic similarities and evolutionary relationship are grouped.

geographic dispersal Directional movement in which some residents of an established community leave their home range and take up residence elsewhere; they are considered to be exotic species in the new location.

geologic time scale A time scale for Earth history, the subdivisions of which are based on boundaries marked by episodes of mass extinction and which have been refined by radiometric dating.

germ cell Of animals, a cell of a lineage set aside for sexual reproduction; germ cells give rise to gametes. *Compare* somatic cell.

germ layer Of animal embryos, one of two or three primary tissue layers that form at gastrulation as forerunners of adult tissues. *Compare* ectoderm; endoderm; mesoderm.

germination (jur-min-AY-shun) Of resting spores and seeds, a resumption of activity following a period of arrested development.

gibberellin (JIB-er-ELL-un) A type of plant hormone that promotes stem elongation.

gill A respiratory organ, typically with a moist, thin vascularized layer of epidermis that functions in gas exchange.

gland A secretory cell or structure derived from epithelium and often connected to it.

glomerular capillaries Set of blood capillaries inside Bowman's capsule of the nephron.

glomerulus (glow-MARE-you-luss) [L. *glomus*, ball] First portion of the nephron, where water and solutes are filtered from blood.

glucagon (GLUE-kuh-gone) A hormone that stimulates cells to convert glycogen and amino acids to glucose; secreted by alpha cells of the pancreas when the blood level of glucose decreases.

glyceride (GLISS-er-eyed) One of the fats or oils; a molecule with one, two, or three fatty acid tails attached to a glycerol backbone.

glycerol (GLISS-er-oh) [Gk. *glykys*, sweet, + L. *oleum*, oil] A three-carbon molecule with three hydroxyl groups attached; one of the components of fats and oils.

glycogen (GLY-kuh-jen) A highly branched polysaccharide that is the main storage carbohydrate of animals; cleavage reactions at its many branchings yield an abundance of glucose monomers when required.

glycocalyx A sticky mesh of polysaccharides, polypeptides, or both around the cell wall of many bacteria.

glycolysis (gly-CALL-ih-sis) [Gk. *glykys*, sweet, + *lysis*, loosening or breaking apart] Initial energy-releasing reactions of aerobic and anaerobic pathways by which enzymes break down glucose (or some other organic compound) to pyruvate. It proceeds in the cytoplasm of all cells, it has a net yield of two ATP, and oxygen has no role in it.

glycoprotein A protein having linear or branched oligosaccharides covalently bonded to it. Nearly all surface proteins of animal cells and many proteins circulating in blood are glycoproteins.

gnetophyte Only gymnosperm known to have vessels in its xylem.

Golgi body (GOHL-gee) Organelle of lipid assembly, polypeptide chain modification, and packaging of both in vesicles for export or for transport to locations in the cytoplasm.

gonad (GO-nad) Primary reproductive organ in which animal gametes are produced.

graded potential Of neurons, a local signal that slightly alters the voltage difference across a patch of plasma membrane and that varies in magnitude according to the stimulus. With prolonged or intense stimulation, such signals may spread to a trigger zone of the membrane and initiate an action potential.

granum, plural **grana** In many chloroplasts, any of the stacks of flattened, membranous compartments with chlorophyll and other light-trapping pigments and reaction sites for ATP formation.

grassland A biome with flat or rolling land, 25-100 centimeters of annual rainfall, warm summers, grazing animals, and periodic fires that regenerate dominant species.

gravitropism (GRAV-ih-TROPE-izm) [L. *gravis*, heavy, + Gk. *trepein*, to turn] Tendency of a plant to grow directionally in response to the Earth's gravitational force.

gray matter Inside the brain and spinal cord, the unmyelinated axons, dendrites, and nerve cell bodies and neuroglial cells.

grazing food web A network of food chains in which energy flows from plants to an array of herbivores, then carnivores.

green alga An aquatic protistan with chlorophylls *a* and *b*; early members of its lineage may have given rise to plants.

green revolution In developing countries, the use of improved crop varieties, modern agricultural practices (including massive inputs of fertilizers and pesticides), and equipment to increase crop yields.

greenhouse effect Warming of the lower atmosphere as a result of the presence of increasing levels of greenhouse gases (such as carbon dioxide and methane).

ground meristem (MARE-ih-stem) [Gk. *meristos*, divisible] A primary meristem that produces the ground tissue system, hence the bulk of the plant body.

ground substance Of certain animal tissues, intercellular material made of cell secretions and other noncellular components.

ground tissue system Tissues making up the bulk of a plant body, the most common being parenchyma.

growth Of multicelled organisms, increases in the number, size, and volume of cells.

guanine A nitrogen-containing base; present in one of the four nucleotide building blocks of DNA and RNA.

guard cell Either of two adjacent cells with roles in the movement of carbon dioxide, oxygen, and water vapor across leaf or stem epidermis. An opening (stoma) forms when both swell with water and move apart; it closes when they lose water and collapse against one another.

gut A body region where food is digested and absorbed; of complete digestive systems, the gastrointestinal tract (the portions from the stomach onward).

gymnosperm (JIM-noe-sperm) [Gk. *gymnos*, naked, + *sperma*, seed] A vascular plant that bears seeds at exposed surfaces of reproductive structures, such as cone scales.

habitat [L. *habitare*, to live in] The type of place where an organism normally lives, as characterized by physical and chemical features and by its array of species.

hair cell A mechanoreceptor that may give rise to action potentials when bent or tilted.

half-life The time it takes for half of a given quantity of any radioisotope to decay into a different, less unstable daughter isotope.

halophile A type of archaebacterium that lives in extremely saline habitats.

haploidy (HAP-loyd) The presence of half the parental number of chromosomes in a gamete, as brought about by meiosis; the gamete has one of each pair of homologous chromosomes. *Compare* diploidy.

Hardy-Weinberg rule Allele frequencies will stay the same through the generations if there is no mutation, if the population is infinitely large and is isolated from other populations of the same species, if mating is random, and if all individuals survive and reproduce equally.

HCG Human chorionic gonadotropin. A hormone that helps maintain the lining of the uterus during the menstrual cycle and during the first trimester of pregnancy.

heart Muscular pump that keeps blood circulating through the animal body.

helper T cell A T lymphocyte which, when activated, produces and secretes chemicals that induce responsive T or B lymphocytes to divide and give rise to large populations of effector cells and memory cells.

hemoglobin (HEEM-oh-glow-bin) [Gk. *haima*, blood, + L. *globus*, ball] Iron-containing, oxygen-transporting protein that gives red blood cells their color.

hemostasis (HEE-mow-STAY-sis) [Gk. *haima*, blood, + *stasis*, standing] Stopping of blood loss from a damaged blood vessel through coagulation, blood vessel spasm, platelet plug formation, and other mechanisms.

herbivore [L. *herba*, grass, + *vovare*, to devour] Plant-eating animal.

hermaphrodite An individual with both male and female gonads; two individuals sexually reproduce by the mutual transfer of sperm.

heterocyst (HET-er-oh-sist) A self-modified cyanobacterial cell that makes a nitrogen-fixing enzyme when nitrogen is scarce.

heterotroph (HET-er-oh-trofe) [Gk. *heteros*, other, + *trophos*, feeder] Organism unable to synthesize its own organic compounds; it feeds on autotrophs, other heterotrophs, or organic wastes. *Compare* autotroph.

heterozygous condition (HET-er-oh-ZYE-guss) [Gk. *zygoun*, join together] Of a specified trait, having a pair of nonidentical alleles at a gene locus (on a pair of homologous chromosomes).

higher taxon (plural, **taxa**) One of the ever more inclusive groupings meant to reflect relationships among species. Family, order, class, phylum, and kingdom are examples.

hindbrain One of three divisions of the vertebrate brain; the medulla oblongata, cerebellum, and pons. Has reflex centers for respiration, blood circulation, and other basic functions; also coordinates motor responses and many complex reflexes.

histone Any of a class of proteins that are intimately associated with eukaryotic DNA and largely responsible for the organization of eukaryotic chromosomes.

homeostasis (HOE-me-oh-STAY-sis) [Gk. *homo*, same, + *stasis*, standing] Of animals, a physiological state in which physical and chemical aspects of the internal environment (blood and interstitial fluid) are maintained within ranges suitable for cell activities.

homeotic gene One of a class of master genes that specify the development of specific body part in animals.

hominid [L. *homo*, man] All species on or near the evolutionary road leading to modern humans.

hominoid Apes, humans, and their recent ancestors.

homologous chromosome (huh-MOLL-uh-gus) [Gk. *homologia*, correspondence] Of cells having a diploid chromosome number, one of a pair of chromosomes that are identical in size, shape, and gene sequence, and that interact during meiosis. Nonidentical sex chromosomes in a cell also interact during meiosis and are considered homologues also.

homology Similarity in one or more body parts in different species that is attributable to their descent from a common ancestor.

homologous structures The same body parts, modified in different ways, in different lines of descent from a common ancestor.

homozygous condition (HOE-moe-ZYE-guss) For a specified trait, having a pair of identical alleles at a gene locus (on a pair of homologous chromosomes).

homozygous dominant condition Having a pair of dominant alleles at a gene locus (on a pair of homologous chromosomes).

homozygous recessive condition Having a pair of recessive alleles at a gene locus (on a pair of homologous chromosomes).

hormone [Gk. *hormon*, to stir up, set in motion] Signaling molecule that stimulates or inhibits gene transcription in nonadjacent target cells (any cell having receptors for it).

horsetail A seedless vascular plant with photosynthetic stems that look like horsetails.

human genome project Worldwide basic research project to sequence the estimated 3 billion nucleotides present in the DNA of human chromosomes.

humus Decomposing organic matter in soil.

hydrogen bond A weak interaction between a small, highly electronegative atom of a molecule and a neighboring hydrogen atom that is taking part in a polar covalent bond.

hydrogen ion A free (unbound) proton; a hydrogen atom that has lost its electron and so bears a positive charge (H^+).

hydrologic cycle A biogeochemical cycle, driven by solar energy, in which water moves slowly through the atmosphere, on or through surface layers of land masses, to the ocean, and back again.

hydrolysis (high-DRAWL-ih-sis) [L. *hydro*, water, + Gk. *lysis*, loosening or breaking apart] Cleavage reaction in which covalent bonds break, splitting a molecule into two or more parts. Often H^+ and OH^- (derived from a water molecule) become attached to the exposed bonding sites.

hydrophilic substance [Gk. *philos*, loving] A polar substance that is attracted to the polar water molecule and dissolves easily in water. Sugars are examples.

hydrophobic substance [Gk. *phobos*, dreading] A nonpolar substance that is repelled by the polar water molecule and thus resists being dissolved in water. Oil is an example.

hydrosphere All liquid or frozen water on or near the Earth's surface.

hydrostatic pressure Any volume of fluid that exerts a force directed against a wall, membrane, or another structure enclosing that fluid. The greater its concentration of solutes, the greater will be the pressure that the fluid exerts.

hydrothermal vent ecosystem Ecosystem, near a fissure in the ocean floor, based on chemosynthetic bacteria that use dissolved minerals as their energy source.

hypha (HIGH-fuh), plural **hyphae** [Gk. *hyphe*, web] Of fungi, a filament with chitin-reinforced walls and, often, reinforcing cross-walls; component of the mycelium.

hypodermis A subcutaneous layer having stored fat that helps insulate the body; although not part of skin, it anchors skin and allows it some freedom of movement.

hypothalamus [Gk. *hypo*, under, + *thalamos*, inner chamber or possibly *tholos*, rotunda] Of the forebrain, a major center for homeostatic control of visceral activities (such as salt-water balance, temperature control, and reproduction), related forms of behavior (as in hunger, thirst, and sex), and emotional expression, such as sweating with fear.

hypothesis In science, a possible explanation of a specific phenomenon in nature, one that has the potential to be proved false by experimental tests.

immune response Events by which B and T lymphocytes recognize antigen, undergo cell divisions that form huge populations of lymphocytes, which differentiate into subpopulations of effector and memory cells. The effector cells destroy cells bearing antigen-MHC complexes. Memory cells are not activated until subsequent encounters with the same antigen.

immunization Various processes, including vaccination, that promote increased immunity against specific diseases.

immunoglobulin (Ig) One of five classes of antibodies, each with antigen-binding sites and other sites with specialized functions.

implantation Process by which a blastocyst adheres to the endometrium and establishes connections by which the mother and embryo will exchange substances during pregnancy.

imprinting A time-dependent form of learning that is triggered by exposure to sign stimuli and that usually occurs during a sensitive period when the animal is young.

inbreeding Nonrandom mating among close relatives, which have many identical alleles in common. Inbreeding is a form of genetic drift in a small group of relatives that are preferentially interbreeding.

incomplete dominance One allele of a pair is not fully dominant over its partner, so a heterozygous phenotype in between the two homozygous phenotypes emerges.

independent assortment, theory of By the end of meiosis in a germ cell, each pair of homologous chromosomes—hence the genes that they carry—have been sorted for shipment into gametes independently of how the other pairs were sorted out.

indirect selection A theory in evolutionary biology that self-sacrificing individuals can pass on their genes indirectly by helping relatives survive and reproduce.

induced-fit model A substrate induces change in the shape of an enzyme's active site when bound to it, the result being a more precise molecular fit between the two that promotes reactivity.

infection Invasion and multiplication of a pathogen in host cells or tissues. Disease follows if defenses cannot be mobilized fast enough to prevent the pathogen's activities from interfering with normal functions.

inflammation, acute Important aspect of nonspecific defenses and immune responses when cells of a local tissue are damaged or killed, as by infection; requires action of fast-acting phagocytes and plasma proteins, including complement proteins.

inheritance The transmission, from parents to offspring, of structural and functional patterns that have a genetic basis and are characteristic of their species.

inhibiting hormone A signaling molecule produced and secreted by the hypothalamus that suppresses a particular secretion by the anterior lobe of the pituitary gland.

inhibitor A substance that can bind with an enzyme and interfere with its functioning.

inhibitory postsynaptic potential (IPSP) Of neurons, one of two competing types of graded potentials at an input zone; it tends to drive the resting membrane potential away from threshold.

instinctive behavior A behavior performed without having been learned by experience in the environment. The nervous system of a newly born or hatched animal is prewired to recognize one or two sign stimuli (simple, well-defined cues in the environment) that can trigger a suitable response.

insulin Hormone secreted by beta cells of the pancreas that lowers the glucose level in blood by stimulating cells to take up glucose; also promotes protein and fat synthesis and inhibits protein conversion to glucose.

integration, neural [L. *integrare*, coordinate] Moment-by-moment summation of all the excitatory and inhibitory synapses acting on the neuron; it takes place at each level of synapsing in a nervous system.

integrator Of homeostatic systems, a control point such as a brain where information is pulled together in the selection of responses to stimuli.

integument Of animals, a protective body cover such as skin. Of seed-bearing plants, one or more layers around an ovule that harden, thicken, and form a seed coat.

integumentary exchange (in-teg-you-MEN-tuh-ree) A mode of respiration in which oxygen and carbon dioxide diffuse across a thin, moist, vascularized layer at the body surface of certain animals.

interleukin One of several chemical mediator molecules secreted by helper T cells that fan mitotic cell divisions and differentiation of responsive T and B cells.

intermediate A compound formed between the start and end of a metabolic pathway.

intermediate filament In different types of animal cells, a cytoskeletal element made of particular proteins.

interneuron Any neuron of the vertebrate brain and spinal cord.

internode In vascular plants, the stem region between two successive nodes.

interphase Of the cell cycle, an interval in between nuclear divisions when a cell increases in mass, roughly doubles the number of its cytoplasmic components, and then duplicates its chromosomes (replicates its DNA). The interval differs among species.

interspecific competition Individuals of different species compete with one another for a share of resources in their habitat.

interstitial fluid (IN-ter-STISH-ul) [L. *interstitus*, to stand in the middle of something] That portion of extracellular fluid occupying the spaces between cells and tissues of complex animals.

intertidal zone Generally, part of a rocky or sandy shoreline above the low water mark and below the high water mark; organisms that inhabit it are alternately submerged, then exposed, by tides.

intervertebral disk One of a number of disk-shaped structures containing cartilage that act as shock absorbers and flex points between bony segments of the vertebral column.

intraspecific competition All individuals of a population compete with one another for a share of resources in their habitat.

intron A noncoding portion of a newly formed mRNA molecule.

inversion A linear stretch of DNA within a chromosome that has become oriented in the reverse direction, with no molecular loss.

invertebrate Animal without a backbone.

in vitro fertilization Conception outside the body (literally, "in glass" petri dishes or test tubes).

ion, negatively charged (EYE-on) An atom or a compound that acquired an overall negative charge by gaining one or more electrons.

ion, positively charged An atom or a compound that acquired an overall positive charge by losing one or more electrons.

ionic bond Ions of opposite charge have attracted each other and are staying together.

isotonic condition Equality in the relative concentrations of solutes in two fluids; for two fluids separated by a cell membrane, there is no net osmotic (water) movement across the membrane.

isotope (EYE-so-tope) Of an element, an atom with more or fewer neutrons than the atoms having the most common number.

J-shaped curve The type of curve that emerges when population size is plotted against time; it represents unrestricted, exponential growth.

joint An area of contact or near-contact between bones.

juvenile Of many animals, a miniaturized form between the embryo and adult; it simply changes in size and proportion until it reaches sexual maturity.

karyotype (CARRY-oh-type) For an individual (or a species), a preparation of metaphase chromosomes sorted by length, centromere location, and other defining features.

keratin A tough, water-insoluble protein made by most epidermal cells that becomes concentrated in skin's outermost layers.

keratinization (care-AT-in-iz-AY-shun) Process by which keratin-producing epidermal cells die and accumulate as keratinized bags at the skin surface to form a barrier against dehydration, bacteria, many toxins, and ultraviolet radiation.

keystone species A species that dominates a community and dictates its structure.

key innovation A modified structure or function that allows a lineage to exploit the environment in more efficient or novel ways.

kidney One of a pair of vertebrate organs that filter mineral ions, organic wastes, and other substances from the blood; it controls the amounts returned to blood and thereby helps maintain the volume and solute levels of extracellular fluid.

kilocalorie 1,000 calories of heat energy; the amount of energy required to raise the temperature of 1 kilogram of water by 1°C; unit of measure for the caloric value of foods.

kinase One of a class of enzymes that catalyze phosphate-group transfers.

kinetochore Group of proteins and DNA at a chromosome's centromere where spindle microtubules become attached at mitosis or meiosis. Each chromatid of a duplicated chromosome has its own kinetochore.

Krebs cycle A cyclic pathway that occurs in mitochondria; together with a few preparatory steps, the stage of aerobic respiration in which pyruvate is completely broken down to carbon dioxide and water. Coenzymes accept unbound protons (H^+) and electrons stripped from intermediates and deliver them to the next stage.

lactate fermentation Anaerobic pathway of ATP formation; pyruvate from glycolysis is converted to the three-carbon compound lactate, and NAD^+ is regenerated. The net energy yield is two ATP.

lactation Milk production by hormone-primed mammary glands.

lake A body of standing freshwater with littoral, limnetic, and profundal zones.

large intestine Colon; a gut region that receives unabsorbed food residues from the small intestine and concentrates and stores feces until they are expelled from the body.

larva, plural **larvae** Of many animals, an immature developmental stage between the embryo and adult.

larynx (LARE-inks) A tubular airway to and from the lungs. In humans, it contains vocal cords.

lateral meristem A type of meristem in plants that show secondary growth; either vascular cambium or cork cambium.

lateral root Of taproot systems, a lateral branching from the first, primary root.

leaching The removal of some nutrients in soil as water percolates through it.

leaf For most vascular plants, a structure having chlorophyll-containing tissue that is the major region of photosynthesis.

learned behavior Variation or change in responses to stimuli after an animal has processed and integrated information gained from specific experiences.

lek Of some birds and other animals, a type of communal display ground occupied during times of courtship.

lethal mutation A mutation with drastic effects on phenotype that usually cause the individual's death.

LH Luteinizing hormone. Hormone secreted by the anterior lobe of the pituitary gland, with roles in male and female reproduction.

lichen (LY-kun) A symbiotic interaction between a fungus and a photoautotroph, such as a green alga.

life cycle A recurring pattern of genetically programmed events by which individuals grow, develop, maintain themselves, and reproduce.

life table Tabulation of age-specific patterns of birth and death for a population.

ligament A strap of dense connective tissue that bridges a joint.

light-dependent reactions The first stage of photosynthesis, in which sunlight energy is trapped and converted to the chemical energy of ATP alone (by a cyclic pathway) or ATP and NADPH (by a noncyclic pathway).

light-independent reactions The second stage of photosynthesis, in which ATP makes phosphate-group transfers required to build sugar phosphates. Often NADPH delivers electrons and hydrogen atoms for the synthesis reactions, which also require carbon from carbon dioxide. The sugar phosphates enter other reactions by which starch, cellulose, and other end products of photosynthesis are assembled.

lignification Of land plants, a process by which lignin is deposited in secondary cell walls. Lignin imparts strength and rigidity by anchoring cellulose strands in the walls, stabilizes and protects other components of the wall, and forms a waterproof barrier around the cellulose. A key factor in the evolution of vascular plants.

lignin A complex organic compound that strengthens and waterproofs cell walls in certain tissues of vascular plants.

limbic system Brain centers, located in the middle of the cerebral hemispheres, that interact to govern emotions and influence memory. Distantly related to the olfactory lobes; it still deals with the sense of smell.

limiting factor Any essential resource that, in short supply, limits population growth.

lineage (LIN-ee-age) A line of descent.

linkage group A quantitative measure of the relative positions of the genes along the length of a chromosome.

linkage mapping An investigative approach by which the relative linear positions of a chromosome's genes are deduced by tracking the percentage of recombinants in gametes.

lipid Mainly a hydrocarbon, greasy or oily, that strongly resists dissolving in water but quickly dissolves in nonpolar substances. In nearly all cells, certain lipids are the main reservoirs of stored energy. Other lipids are structural materials (as in membranes) and cell products (such as surface coatings).

lipid bilayer Structural basis of all cell membranes, with two layers of mostly phospholipid molecules. The hydrophobic tails of the molecules are sandwiched in between the hydrophilic heads, which are dissolved in the fluid that bathes them.

lipoprotein A molecule that forms when proteins in blood combine with cholesterol, triglycerides, and phospholipids that were absorbed from the gut, after a meal.

liver A large gland in vertebrates and many invertebrates that stores and interconverts organic compounds and helps maintain their blood concentrations. It additionally inactivates most hormone molecules that have served their functions as well as ammonia and other compounds that can be toxic at high concentrations.

local signaling molecules Secretions from cells of many local tissues that quickly alter chemical conditions in the vicinity, then are degraded before they can travel elsewhere.

logistic population growth (low-JISS-tik) A pattern of population growth in which a low-density population slowly increases in size, then enters a rapid growth phase that levels off once the carrying capacity has been reached.

loop of Henle The hairpin-shaped, tubular region of a nephron that functions in the reabsorption of water and solutes.

lung An internal respiratory surface in the shape of a cavity or sac that can be filled with air; a pair occur in a few fishes and in amphibians, birds, reptiles, and mammals.

lycophyte A seedless vascular plant of mostly wet or shaded habitats; requires free water to complete its life cycle.

lymph (LIMF) [L. *lympha*, water] Tissue fluid that has drained into the vessels of the lymphatic system.

lymph capillary A small-diameter vessel of the lymph vascular system that has no pronounced entrance; tissue fluid moves inward by passing between overlapping endothelial cells at the vessel's tip.

lymph node A lymphoid organ that is a battleground for immune responses; packed, organized arrays of lymphocytes inside cleanse lymph before it reaches the blood.

lymph vascular system [L. *lympha*, water, + *vasculum*, a small vessel] Of the lymphatic system, vessels that take up and transport excess tissue fluid and reclaimable solutes as well as fats absorbed from the digestive tract.

lymphatic system An organ system that supplements the circulatory system. Its vessels deliver fluid and solutes from interstitial fluid to the bloodstream; its lymphoid organs have roles in immunity.

lymphocyte Any of various T and B cells with roles in immune responses that follow recognition of antigen.

lysis [Gk. *lysis*, a loosening] Gross damage to a plasma membrane, cell wall, or both that allows the cytoplasm to leak out and thereby leads to cell death.

lysogenic pathway A common pathway in which a viral replication cycle enters a latent period and viral genes are integrated with a host chromosome. The recombinant molecule is a miniature time bomb that is passed on to all of the host cell's progeny.

lysosome (LYE-so-sohm) The main organelle of intracellular digestion, with enzymes that can break down nearly all polysaccharides, proteins, and nucleic acids, and some lipids.

lysozyme An infection-fighting enzyme in mucous membranes.

lytic pathway A common pathway in which a viral replication cycle proceeds rapidly and ends with lysis of the host cell.

macroevolution The large-scale patterns, trends, and rates of change among families and other more inclusive groups of species.

macrophage A leukocyte (white blood cell) that phagocytizes worn-out cells and tissue debris and anything bearing antigen. It may become an antigen-presenting cell and thus trigger immune responses.

mammal The only vertebrate whose females nourish offspring by milk, produced by and secreted from mammary glands.

mass extinction A sudden rise in rates of extinction above the background level; a catastrophic, global event during which a number of major taxa are wiped out simultaneously.

mass number The total number of protons and neutrons in an atom's nucleus.

mast cell In connective tissues, a basophil-like cell that releases histamine during an inflammatory response.

mechanoreceptor Sensory cell or nearby cell that detects mechanical energy (changes in pressure, position, or acceleration).

medulla oblongata Part of the hindbrain; it has reflex centers for vital tasks, such as respiration and circulation. It coordinates motor responses with complex reflexes, such as coughing. It influences brain regions concerned with sleep and arousal.

medusa (meh-DOO-sah) [Gk. *Medousa*, one of three sisters in Greek mythology having snake-entwined hair] Free-swimming, bell-shaped stage in cnidarian life cycles; oral arms and tentacles extend from the bell.

megaspore A haploid spore that forms in the ovary of seed-bearing plants; one of its cellular descendants develops into an egg.

meiosis (my-OH-sis) [Gk. *meioun*, to diminish] Two-stage nuclear division process that reduces the chromosome number of a germ cell by half, to the haploid number; each daughter nucleus receives one of each type of chromosome. It is the basis of gamete formation and (in plants) spore formation. *Compare* mitosis.

membrane excitability A special membrane property of any cell that can generate action potentials in response to stimulation.

memory The capacity of the brain to store and retrieve information about past sensory experience.

memory cell One of the subpopulations of B or T cells that form during an immune response to antigen but that do not act at once; it enters a resting phase, from which it is released during a secondary immune response.

menopause (MEN-uh-pozz) [L. *mensis*, month, + *pausa*, stop] End of the period of a human female's reproductive potential.

menstrual cycle Cyclic release of a secondary oocyte and priming of the endometrium to receive it, should it become fertilized; the complete cycle averages about twenty-eight days in female humans.

menstruation Cyclic sloughing of the blood-enriched endometrium (lining of the uterus) when pregnancy does not occur.

mesoderm (MEH-zoe-derm) [Gk. *mesos*, middle, + *derm*, skin] Of most animals, intermediate primary tissue layer that forms in the embryo and gives rise to muscle, most of the skeleton; circulatory, reproductive, and excretory organs; and connective tissue layers of the gut and integument.

mesophyll (MEH-zoe-fill) Of vascular plants, a tissue of photosynthetic parenchyma cells with an abundance of air spaces.

messenger RNA (mRNA) A single strand of ribonucleotides transcribed from DNA and translated into a polypeptide chain; the only RNA with protein-building instructions.

metabolic pathway (MEH-tuh-BALL-ik) One of many orderly sequences of enzyme-mediated reactions by which cells maintain, increase, or decrease the concentrations of substances. Different pathways are linear or circular, and one typically interconnects with others.

metabolism (meh-TAB-oh-lizm) [Gk. *meta*, change] All of the controlled, enzyme-mediated chemical reactions by which cells acquire and use energy to synthesize, store, degrade, and eliminate substances in ways that contribute to growth, survival, and reproduction.

metamorphosis (me-tuh-MOR-foe-sis) [Gk. *meta*, change, + *morphe*, form] Major change in body form during the transition from an embryo to the adult owing to hormonally controlled increases in size, reorganization of tissues, and remodeling of body parts.

metaphase Of meiosis I, the stage when all pairs of homologous chromosomes have become positioned midway between the spindle poles. Of mitosis or meiosis II, a stage when each duplicated chromosome has become positioned midway between the spindle poles.

metazoan A multicelled animal.

methanogen A type of archaebacterium that lives in oxygen-free habitats and produces methane gas as a metabolic by-product.

MHC marker Any of a variety of proteins that are self-markers. Some occur on all body cells of an individual; others are unique to the macrophages and lymphocytes.

micelle (my-SELL) Of fat digestion, a small droplet that is assembled from bile salts, fatty acids, and monoglycerides and that assists in fat absorption from the small intestine.

microevolution Change in allele frequencies resulting from mutation, genetic drift, gene flow, and natural selection.

microfilament [Gk. *mikros*, small, + L. *filum*, thread] Thin cytoskeletal element of two twisted polypeptide chains; it has roles in cell movement, especially at the cell surface, and in producing and maintaining cell shapes.

microorganism An organism, usually single-celled, that is far too small to be observed without the aid of a microscope.

microspore Of gymnosperms and flowering plants, a walled haploid spore that develops into a pollen grain.

microtubular spindle Of eukaryotic cells, a temporary bipolar structure composed of organized arrays of microtubules that forms during nuclear division and that moves the chromosomes to prescribed destinations.

microtubule (my-crow-TUBE-yool) A hollow, cylindrical cytoskeletal element, consisting mainly of tubulin, with roles in cell shape, motion, and growth and in the structure of cilia and flagella.

microtubule organizing center (MTOC) In the cytoplasm of eukaryotic cells, a small mass of proteins and other substances. The number, type, and location of MTOCs dictate the particular organization and orientation of microtubules inside the cell.

microvillus (MY-crow-VILL-us) [L. *villus*, shaggy hair] Of many cells, one of a number of very slender cylindrical extensions of the plasma membrane that serve in absorption or secretion.

midbrain Part of the vertebrate brain with coordination centers for reflex responses to visual and auditory input; also swiftly relays sensory signals to forebrain.

migration A recurring pattern of movement between two or more locations in response to environmental rhythms, such as circadian rhythms and seasonal changes in daylength. It requires activation or suppression of internal timing mechanisms that govern physiological and behavioral functions.

mimicry (MIM-ik-ree) In form, behavior, or both, a prey organism (the mimic) closely resembles a dangerous, unpalatable, or hard-to-catch species (its model).

mineral An element or inorganic compound formed by natural geologic processes and required for normal cell functioning.

mitochondrion (MY-toe-KON-dree-on), plural **mitochondria** Of eukaryotic cells, the organelle that specializes in formation of ATP; the second and third stages of aerobic respiration, an oxygen-requiring pathway, occur only in mitochondria.

mitosis (my-TOE-sis) [Gk. *mitos*, thread] Type of nuclear division that maintains the parental chromosome number for daughter cells; for eukaryotes, the basis of growth in size (and often of asexual reproduction).

mixture Two or more elements intermingled in proportions that can and usually do vary.

molar One of the cheek teeth; a tooth with a platform having cusps (surface bumps) that help crush, grind, and shear food.

molecular clock An accumulation of neutral mutations in a lineage that can be measured as a series of predictable "ticks" back through time; a way of calculating the time of origin of one lineage or species relative to others.

molecule A unit of matter in which chemical bonding holds together two or more atoms of the same or different elements.

mollusk An invertebrate with a tissue fold (mantle) draped around a soft, fleshy body; snails, clams, and squids are examples.

molting Shedding of hair, feathers, horns, epidermis, a shell, or cuticle in a process of increases in size or periodic renewal.

Monera In some traditional classification schemes, a kingdom that encompasses both archaebacteria and eubacteria.

monocot (MON-oh-kot) A monocotyledon; a flowering plant with seeds bearing one cotyledon, floral parts generally in threes (or multiples of three), and leaves typically parallel-veined. *Compare* dicot.

monohybrid cross [Gk. *monos*, alone] An experimental cross in which offspring inherit a pair of nonidentical alleles for a single trait being studied, so that they are heterozygous.

monomer Small molecule that is commonly a subunit of polymers, such as the sugar monomers of starch.

monosaccharide (MON-oh-SAK-ah-ride) [Gk. *monos*, alone, single, + *sakharon*, sugar] A simple carbohydrate, with only one sugar monomer. Glucose is an example.

monosomy A chromosome abnormality; the presence of a chromosome that has no homologue in a diploid cell.

morphogenesis (MORE-foe-JEN-ih-sis) [Gk. *morphe*, form, + *genesis*, origin] A program of orderly changes in an animal embryo's size, shape, and proportions, the outcome being specialized tissues and early organs. As part of the program, cells divide, grow, migrate, and change in size. Tissues expand and fold, and cells in some die in controlled ways at prescribed locations.

morphological convergence Lineages only remotely related evolved in response to similar environmental pressures, and they became similar in appearance, functions, or both. Analogous structures are evidence of this macroevolutionary pattern.

morphological divergence Of two or more lineages, change in appearance, functions, or both as they evolve from a shared ancestor. Homologous structures are evidence of this macroevolutionary pattern.

motor neuron A type of neuron that swiftly delivers signals from the brain or spinal cord to muscle cells, gland cells, or both.

multicelled organism An organism with differentiated cells arranged into tissues, organs, and often organ systems.

multiple allele system Three or more different molecular forms of the same gene (alleles) among individuals of a population.

muscle fatigue A decline in tension of a muscle kept in a state of tetanic contraction as a result of continuous, high-frequency stimulation.

muscle tension A mechanical force exerted by a contracting muscle; it resists opposing forces such as gravity and the weight of an object being lifted.

muscle tissue A tissue having cells able to contract under stimulation, then passively lengthen and return to the resting position.

mutagen (MEW-tuh-jen) Any environmental agent that can alter the molecular structure of DNA. Ultraviolet radiation and certain viruses are examples.

mutation [L. *mutatus*, a change, + *-ion*, result or a process or an act] A heritable change in the molecular structure of DNA. Mutations are the source of all alleles and, ultimately, of life's diversity. *See also* lethal mutation; neutral mutation.

mutation frequency The number of times a gene mutation has occurred in a population, as in 1 million gametes from which 500,000 individuals were produced.

mutation rate Of a gene, the probability of it undergoing spontaneous mutation during some specified interval, such as each DNA replication cycle.

mutualism [L. *mutuus*, reciprocal] A two-species, symbiotic ecological interaction that directly benefits both participants.

mycelium (my-SEE-lee-um), plural **mycelia** [Gk. *mykes*, fungus, mushroom, + *helos*, callus] A mesh of tiny, branching filaments (hyphae) that is the food-absorbing part of most fungi.

mycorrhiza (MY-coe-RIZE-uh) "Fungus-root"; a form of mutualism between the hyphae of a fungus and young roots of many plants. The fungus obtains carbohydrates from the plant and in turn releases dissolved mineral ions that plant roots can take up.

myelin sheath Of many sensory and motor neurons, an axonal sheath that enhances the propagation of action potentials; the plasma membranes of Schwann cells are wrapped repeatedly around the axon, and only a small node separates one from the other.

myofibril (MY-oh-FY-brill) One of the many threadlike structures inside a muscle cell; it is divided into many sarcomeres, the basic units of contraction.

myosin (MY-uh-sin) A motor protein with roles in contraction. In muscles, it interacts with actin to bring about contraction.

NAD+ Nicotinamide adenine dinucleotide; a nucleotide coenzyme. When carrying electrons and unbound protons (H+) between reaction sites, it is abbreviated NADH.

NADP+ Nicotinamide adenine dinucleotide phosphate; a phosphorylated nucleotide coenzyme. When carrying electrons and unbound protons (H+) between reaction sites, it is abbreviated NADPH$_2$.

natural killer cell A cytotoxic lymphocyte that reconnoiters for tumor cells and virus-infected cells, then touch-kills them.

natural selection A microevolutionary process; outcome of differences in survival and reproduction among individuals that show variation in heritable traits. Over the generations, it can lead to increased fitness (increased adaptation to the environment).

necrosis (neh-CROW-sis) Of multicelled organisms, the passive death of many cells that results from severe tissue damage.

negative feedback mechanism Homeostatic feedback mechanism in which an activity changes some condition in the internal environment; the altered condition triggers a response that reverses the change.

nematocyst (NEM-ad-uh-sist) [Gk. *nema*, thread, + *kystis*, pouch] A capsule formed only by cnidarians; it houses dischargeable, tubular threads that have prey-piercing barbs and that dispense toxins or sticky substances.

nephridium (neh-FRID-ee-um), plural **nephridia** Of earthworms and some other invertebrates, a system of regulating water and solute levels.

nephron (NEFF-ron) [Gk. *nephros*, kidney] Of the vertebrate kidney, a slender tubule into which water and solutes are filtered from blood, then selectively reabsorbed, and in which urine forms.

nerve Cordlike communication line of the peripheral nervous system, composed of axons of sensory neurons, motor neurons, or both bundled in connective tissue. In the brain and spinal cord, similar cord-like bundles are called nerve tracts.

nerve cord Of many animals, a cordlike communication line of axons of neurons.

nerve impulse *See* action potential.

nerve net A simple nervous system of a diffuse mesh of nerve cells that interact with contractile and often sensory cells in an epithelium.

nervous system An organ system in which nerve cells, such as neurons, are oriented relative to one another in signal-conducting and information-processing pathways. It detects and processes information about changes outside and inside the body and elicits responses from muscle and gland cells.

nervous tissue A type of connective tissue composed of neurons.

net energy Of energy resources available to the human population, the amount of energy that is left over after subtracting the energy used to locate, extract, transport, store, and deliver energy to consumers.

net population growth rate per individual (*r*) Of population growth equations, a single variable combining birth and death rates, which are assumed to remain constant.

neural tube Embryonic forerunner of the brain and spinal cord.

neuroglial cell (NUR-oh-GLEE-uhl) One of the cells that structurally and metabolically support neurons. Neuroglia represent about half the volume of the vertebrate nervous system.

neuromodulator A signaling molecule that can magnify or reduce the effects of a given neurotransmitter on neighboring or distant neurons.

neuromuscular junction A site of chemical synapsing between the axonal endings of a motor neuron and a muscle cell.

neuron (NUR-on) A nerve cell; the basic unit of communication in nervous systems. *See* motor neuron; interneuron; sensory neuron.

neurotransmitter Any of a class of signaling molecules that are secreted from neurons and act on immediately adjacent cells, then are rapidly degraded or recycled.

neutral mutation A mutation with little or no effect on phenotype; natural selection cannot increase or decrease its frequency in a population, for it cannot influence an individual's survival and reproduction. *Compare* molecular clock.

neutron A unit of matter, one or more of which occupies the atomic nucleus and has mass but no electric charge.

neutrophil The most abundant, fast-acting white blood cell; it phagocytizes bacteria during an inflammatory response.

niche (NITCH) [L. *nidas*, nest] The sum total of all activities and relationships in which individuals of a species engage as they secure and use the resources necessary to survive and reproduce.

nitrification (nye-trih-fih-KAY-shun) Of the nitrogen cycle, a process by which certain bacteria strip electrons from ammonia or ammonium in soil. The end product, nitrite (NO_2^-), is broken down to nitrate (NO_3^-) by other bacteria.

nitrogen cycle An atmospheric cycle; a biogeochemical cycle in which the largest nitrogen reservoir is the atmosphere.

nitrogen fixation Process by which a few kinds of bacteria convert gaseous nitrogen (N_2) to ammonia. This dissolves rapidly in their cytoplasm to form ammonium, which can be used in biosynthetic pathways.

nociceptor (NO-SEE-sep-tur) A pain receptor, such as a free nerve ending that detects tissue damage as by burns or distortions.

node A site where one or more leaves are attached to a plant stem.

noncyclic pathway of ATP formation (non-SIK-lik) [L. *non*, not, + Gk. *kylos*, circle] Photosynthetic pathway of ATP formation in which electrons from water molecules flow through two photosystems and two electron transport chains, and end up in NADPH.

nondisjunction The failure of two sister chromatids or of a pair of homologous chromosomes to separate at meiosis or mitosis, so that daughter cells end up with too many or too few chromosomes.

notochord (KNOW-toe-kord) Of chordates, a rod of stiffened tissue (not cartilage or bone) that serves as a supporting structure for the body.

nuclear envelope A double membrane (two lipid bilayers and associated proteins) that is the outermost portion of a cell nucleus.

nucleic acid (new-CLAY-ik) A single-stranded or double-stranded chain of four different kinds of nucleotides joined one after the other at their phosphate groups. They differ in which nucleotide base follows the next in sequence. DNA and RNA are examples.

nucleic acid hybridization Single strands of DNA or RNA from different species are recombined into double-stranded, hybrid molecules, which are then heated to break hydrogen bonds between them. The amount of heat required to separate the strands is a measure of biochemical similarity, which is greatest among closely related species and weakest among distantly related ones.

nucleoid (NEW-KLEE-oid) Of bacteria, the region of the cell in which the bacterial chromosome is physically organized (but not bounded by membrane, as with the nuclear envelope of eukaryotic cells).

nucleolus (new-KLEE-oh-lus) [L. *nucleolus*, a little kernel] In the nucleus of a nondividing cell, a site where the protein and RNA subunits of ribosomes are assembled.

nucleosome (NEW-KLEE-oh-sohm) One of the organizational units of the eukaryotic chromosome; a stretch of DNA looped twice around a "spool" of histone molecules.

nucleotide (NEW-KLEE-oh-tide) A small organic compound with a five-carbon sugar (deoxyribose), nitrogen-containing base, and phosphate group. Nucleotides are the structural units of adenosine phosphates, nucleotide coenzymes, and nucleic acids.

nucleotide coenzyme A protein that assists an enzyme by transporting electrons and hydrogen atoms released at one reaction site to another site.

nucleus (NEW-klee-us) [L. *nucleus*, a kernel] Of atoms, the central core of one or more positively charged protons and (in all but hydrogen) electrically neutral neutrons. In eukaryotic cells, a membranous organelle that physically isolates and organizes DNA out of the way of cytoplasmic machinery.

numerical taxonomy Practice by which traits of an unidentified organism are compared with those of a known group. The greater the total number of traits that it has in common with the known group, the closer is their inferred relatedness.

nutrient An element essential for growth and survival of a given organism; directly or indirectly, it has one or more roles in metabolism that no other element fulfills.

nutrition All of those processes by which food is selectively taken in, digested, absorbed, and later converted to the body's own organic compounds.

obesity An excess of fat in the body's adipose tissues, caused by imbalances between caloric intake and energy output.

oligosaccharide A carbohydrate consisting of a short chain of two or more covalently bonded sugar monomers. The type called a disaccharide has two sugar monomers. *Compare* monosaccharide; polysaccharide.

omnivore [L. *omnis*, all, + *vovare*, to devour] An animal able to obtain energy and carbon from more than one food source rather than being limited to one trophic level.

oncogene (ON-koe-jeen) Any gene having the potential to induce cancerous transformation.

oocyte An immature egg.

oogenesis (oo-oh-JEN-uh-sis) Formation of a female gamete, from a germ cell to a haploid ovum (mature egg).

operator A short base sequence intervening between a promoter and bacterial genes.

operon A promoter-operator sequence that services more than one bacterial gene.

orbital One of a number of volumes of space around the atomic nucleus in which electrons are likely to be at any instant.

organ A structure having definite form and function that consists of more than one tissue.

organ formation A stage of embryonic development in animals in which primary tissue layers split into subpopulations of cells, and different lines of cells become unique in structure and function; basis of growth and tissue specialization.

organ system Two or more organs that interact chemically, physically, or both in performing a common task.

organelle Of cells, an internal, membrane-bounded sac or compartment having one or more specific, specialized metabolic functions.

organic compound A molecule consisting of one or more elements covalently bonded to some number of carbon atoms.

osmoreceptor A sensory receptor that detects changes in water volume (hence solute concentration) in fluid that bathes it.

osmosis (oss-MOE-sis) [Gk. *osmos*, act of pushing] The diffusion of water in response to a water concentration gradient between two regions that are separated from each other by a selectively permeable membrane. The greater the number of molecules and ions dissolved in a solution, the lower its water concentration will be.

osmotic pressure At some point following the development of hydrostatic pressure in a cell, an amount of force that counters the inward diffusion of water and prevents any further increase in fluid volume.

ovary (oh-vuh-ree) A type of primary reproductive organ in which eggs form. In seed-bearing plants, the part of the carpel where eggs develop, fertilization takes place, and seeds mature; at maturity, it and sometimes other plant parts is a fruit.

oviduct (OH-vih-dukt) One of a pair of ducts through which eggs travel from an ovary to the uterus. Formerly called Fallopian tube.

ovulation (AHV-you-LAY-shun) The release of a secondary oocyte from an ovary during a menstrual cycle.

ovule (OHV-youl) [L. *ovum*, egg] Of seed-bearing plants, a female gametophyte with egg cell, surrounding tissue, and one or two protective layers (integuments). After fertilization, it matures into a seed.

ovum (OH-vum) Of vertebrates, the mature female gamete (egg).

oxidation-reduction reaction An electron transfer from one atom or molecule to another. Often hydrogen is also transferred along with an electron or electrons.

ozone thinning A pronounced seasonal thinning of the ozone layer in the lower stratosphere, especially above polar regions.

pancreas (PAN-cree-us) Gland that secretes enzymes and bicarbonate into the small intestine during digestion, and that secretes the hormones insulin and glucagon.

pancreatic islet Any of the 2 million clusters of endocrine cells in the pancreas, including alpha cells, beta cells, and delta cells.

parapatric speciation Neighbor populations become distinct species while maintaining contact along their common border.

parasite [Gk. *para*, alongside, + *sitos*, food] An organism that lives in or on other living organisms (hosts), feeds on specific tissues during part of its life cycle, and usually does not kill its host outright.

parasitism A two-species interaction in which one species lives in or on a host species and uses its tissues for nutrients. The parasite benefits; the host does not.

parasitoid An insect larva that grows and develops inside a host organism (usually another insect), eventually consuming the soft tissues and killing it.

parasympathetic nerve Of the autonomic nervous system, any of the nerves carrying signals that tend to slow down the body's activities and divert energy to basic tasks; such nerves work continually in opposition with sympathetic nerves to make small adjustments in internal organ activity.

parenchyma (par-ENG-kih-mah) A simple tissue that makes up the bulk of the plant body. Its cells take part in photosynthesis, storage, secretion, and other tasks.

parthenogenesis Development of an embryo from an unfertilized egg.

passive transport A transport protein that spans the lipid bilayer of a cell membrane passively allows a solute to diffuse through its interior, down its concentration gradient.

pathogen (PATH-oh-jen) [Gk. *pathos*, suffering, + *-genēs*, origin] A virus, bacterium, fungus, protozoan, parasitic worm, or some other infectious agent that can invade organisms, multiply in them, and cause disease.

pattern formation Of animals, a sculpting of nondescript clumps of embryonic cells into specialized tissues and organs according to an ordered, spatial pattern.

PCR Polymerase chain reaction; a method of enormously amplifying the quantity of DNA fragments cut by restriction enzymes.

peat bog An accumulation of the remains of peat mosses in a compressed, excessively moist, and highly acidic mat.

pedigree A chart of genetic connections among related individuals, constructed according to standardized methods.

pelagic province The full volume of ocean water; it is subdivided into a neritic zone (relatively shallow water over continental shelves) and an oceanic zone (water over ocean basins).

penis A male copulatory organ by which sperm can be deposited inside a female reproductive tract.

peptide hormone A protein hormone or some other water-soluble hormone unable to cross the lipid bilayer of a target cell. It enters by receptor-mediated endocytosis or activates cell surface receptors.

perennial [L. *per-*, throughout, + *annus*, year] A flowering plant that lives for more than two growing seasons.

pericycle (PARE-ih-sigh-kul) [Gk. *peri-*, around, + *kyklos*, circle] Of a root vascular cylinder, one or more layers just inside the endodermis that gives rise to lateral roots and contributes to secondary growth.

periderm Of plants that show extensive secondary growth, a protective covering that replaces epidermis.

peripheral nervous system (per-IF-ur-uhl) [Gk. *peripherein*, to carry around] The nerves leading into and out from a spinal cord and brain and the ganglia along those communication lines.

peristalsis (pare-ih-STAL-sis) Waves of contraction and relaxation of muscles in the wall of a tubular or saclike organ.

peritoneum (pare-ih-tah-NEE-um) A lining of the coelom that also covers and helps maintain the position of internal organs.

peritubular capillary One of the set of blood capillaries around the tubular parts of a nephron; it functions in reabsorption of water and solutes back into the body and in secretion of hydrogen ions and some other substances in the forming urine.

permafrost A permanently frozen, water-impenetrable layer beneath the soil surface in arctic tundra.

peroxisome Enzyme-filled vesicle in which fatty acids and amino acids are digested into hydrogen peroxide and converted to harmless products.

pest resurgence Insecticide resistance arises among a population of pests; the insecticide acts as an agent of directional selection.

PGA Phosphoglycerate (FOSS-foe-GLISS-er-ate) A key intermediate of glycolysis and of the Calvin-Benson cycle.

PGAL Phosphoglyceraldehyde. A key intermediate of glycolysis and of the Calvin-Benson cycle.

pH scale A measure of the concentration of free hydrogen ions in blood, water, and other solutions; pH 0 is the most acidic, 14 the most basic, and 7, neutral.

phagocyte (FAG-uh-sight) [Gk. *phagein*, to eat, + *kytos*, hollow vessel] A cell that obtains nutrients or destroys foreign cells and cellular debris by phagocytosis. The macrophages and amoeboid protozoans are examples.

phagocytosis (FAG-uh-sigh-TOE-sis) [Gk. *phagein*, to eat, + *kytos*, hollow vessel] The engulfment of foreign cells or substances by means of pseudopod formation and then endocytosis.

pharynx (FARE-inks) A muscularized tube by which food enters the gut. Also an organ of gas exchange in some chordates; in land vertebrates, a dual entrance for the tubular part of the digestive tract and windpipe.

phenotype (FEE-no-type) [Gk. *phainein*, to show, + *typos*, image] Observable trait or traits of an individual that arise from the interactions between genes, and between genes and the environment.

pheromone (FARE-oh-moan) [Gk. *phero*, to carry, + *-mone*, as in hormone] A nearly odorless, hormone-like, exocrine gland secretion that is a communication signal between individuals of the same species and that helps integrate social behavior.

phloem (FLOW-um) Of vascular plants, a tissue with living cells that interconnect and form the tubes through which sugars and other dissolved organic compounds are conducted.

phospholipid An organic compound with a glycerol backbone, two fatty acid tails, and a hydrophilic head (a phosphate group and another polar group). Phospholipids are the main structural materials of cell membrane.

phosphorus cycle Movement of phosphorus from rock or soil through organisms, then back to soil.

phosphorylation (FOSS-for-ih-LAY-shun) The attachment of unbound (inorganic) phosphate to a molecule; also the transfer of a phosphate group from one molecule to another, as when ATP phosphorylates glucose.

photoautotroph An organism able to synthesize all organic molecules it requires using carbon dioxide as the carbon source and sunlight as the energy source. All plants, some protistans, and a few bacteria are photoautotrophs.

photolysis (foe-TALL-ih-sis) [Gk. *photos*, light, + *-lysis*, breaking apart] A reaction sequence of the noncyclic pathway of photosynthesis, triggered by photon energy, in which water is split into oxygen, hydrogen, and electrons.

photoperiodism A biological response to a change in the relative length of daylight and darkness.

photoreceptor Light-sensitive sensory cell.

photosynthesis The trapping of energy from the sun and its conversion to chemical energy (ATP, NADPH, or both), followed by synthesis of sugar phosphates that become converted to sucrose, cellulose, starch, and other products. The main pathway by which energy and carbon enter the web of life.

photosystem One of the clusters of light-trapping pigments that are embedded in photosynthetic membranes and that donate electrons to transport systems that are involved in the formation of ATP, NADPH, or both. Examples are photosystems II and I of the thylakoid membrane of chloroplasts.

phototropism [Gk. *photos*, light, + *trope*, turning, direction] An adjustment in the direction and rate of plant growth in response to a source of light.

photovoltaic cell A device that can convert sunlight energy into electricity.

phycobilin (FIE-koe-BY-lins) One of the light-sensitive, accessory pigments that transfer energy they absorb to chlorophylls. The phycobilins are especially abundant in red algae and cyanobacteria.

phylogeny Evolutionary relationships among species, starting with most ancestral forms and including the branches leading to their descendants.

phytochrome Light-sensitive pigment, the activation and inactivation of which triggers plant hormone activities governing leaf expansion, stem branching, stem length and often seed germination and flowering.

phytoplankton (FIE-toe-PLANK-tun) [Gk. *phyton*, plant, + *planktos*, wandering] A community of floating or weakly swimming photoautotrophs in saltwater and freshwater habitats.

pigment A light-absorbing molecule.

pioneer species An opportunistic colonizer of barren or disturbed habitats, notable for its high dispersal rate and rapid growth.

pituitary gland An endocrine gland that interacts with the hypothalamus to control diverse physiological functions, including activity of many other endocrine glands. The posterior lobe stores and secretes hypothalamic hormones; the anterior lobe produces and secretes its own hormones.

placenta (play-SEN-tuh) A blood-engorged organ, consisting of endometrial tissue and extraembryonic membranes, that develops during pregnancy in the female placental mammal. It permits exchanges between the mother and her fetus without intermingling their bloodstreams, thus sustaining the new individual and allowing its blood vessels to develop apart from the mother's.

plankton [Gk. *planktos*, wandering] Any community of floating or weakly swimming organisms, mostly microscopic, living in freshwater and saltwater habitats. *See also* phytoplankton; zooplankton.

plant One of the eukaryotic organisms, nearly all of which are photoautrophic, that have chlorophylls *a* and *b* and that generally have well-developed root and shoot systems.

Plantae The kingdom of plants.

plasma (PLAZ-muh) Liquid component of blood; water and dissolved ions, diverse proteins, sugars, gases, and other substances.

plasma membrane Of cells, the outermost membranous boundary between cytoplasm and external fluid bathing the cell. Its lipid bilayer is the basic structural part of the membrane; diverse proteins embedded in the bilayer or attached to its surfaces carry out most functions, such as transport and reception of extracellular signals.

plasmid Of many bacteria, a small, circular molecule of extra DNA that carries only a few genes and that replicates independently of the bacterial chromosome.

plasmodesma, plural **plasmodesmata** (PLAZ-moe-DEZ-muh) A plant cell junction; a membrane-lined channel across walls of adjacent cells that connects their cytoplasm.

plasmolysis An osmotically induced shrinkage of a cell's cytoplasm.

plate tectonics Theory that slabs, or plates, of the Earth's outer layer (lithosphere) float on a hot, plastic layer of the underlying mantle. All plates are in motion and they raft the continents to new positions over great spans of geologic time.

platelet (PLAYT-let) Any of the fragments of megakaryocytes that release substances necessary for clot formation.

pleiotropy (PLEE-oh-troe-pee) [Gk. *pleon*, more, + *trope*, direction] As a result of expression of alleles at a single gene locus, positive or negative effects on two or more traits. The effects may or may not be manifest at the same time in the individual.

polar body One of four cells that form during the meiotic cell division of an oocyte but that does not become the ovum.

pollen grain [L. *pollen*, fine dust] Depending on the species, the immature or mature, sperm-bearing male gametophyte of gymnosperms and angiosperms.

pollen sac In anthers of flowers, any of the chambers in which pollen grains develop.

pollen tube A tube formed after a pollen grain germinates; grows through carpel tissues, and carries sperm to the ovule.

pollination Of flowering plants, the arrival of a pollen grain on the landing platform (stigma) of a carpel.

pollutant Any substance with which an ecosystem has had no prior evolutionary experience, in terms of kinds or amounts, and that can accumulate to disruptive or harmful levels. Can be naturally occurring or synthetic.

polymer (POH-lih-mur) [Gk. *polus*, many, + *meris*, part] A large molecule with three to millions of subunits.

polymerase (puh-LIM-ur-aze) An enzyme that catalyzes a polymerization reaction. Examples are DNA polymerase (of DNA replication and repair), RNA polymerase (of gene transcription), and the enzyme that catalyzes cellulose synthesis.

polypeptide chain Organic compound with nitrogen atoms arrayed in a regular pattern in its backbone (-N-C-C-N-C-C-). A protein consists of one or more of these chains.

polymorphism (poly-MORE-fizz-um) [Gk. *polus*, many, + *morphe*, form] Of populations, the persistence of two or more distinctive forms of a trait (morphs).

polyp (POH-lip) Vase-shaped, sedentary stage of cnidarian life cycles.

polypeptide chain Three or more amino acids joined by peptide bonds.

polyploidy (POL-ee-PLOYD-ee) Having three or more of each type of chromosome in the interphase nucleus, compared to a diploid chromosome number.

polysaccharide [Gk. *polus*, many, + *sakcharon*, sugar] A straight or branched chain of many covalently linked sugar units, of the same or different kinds. The most common are cellulose, starch, and glycogen.

polysome A number of ribosomes translating the same mRNA molecule one after the other, at the same time, during protein synthesis.

pons Part of the hindbrain; a traffic center for signals passing between the cerebellum and integrating centers of the forebrain.

population Individuals of the same species occupying the same area.

population density A count of individuals of a population occupying a specified area or volume of a habitat.

population distribution The general pattern of dispersion of individuals of a population throughout their habitat.

population size The number of individuals that make up the gene pool of a population.

positive feedback mechanism Homeostatic mechanism by which a chain of events is set in motion that intensifies a change from an original condition; after a limited time, the intensification reverses the change.

predation An ecological interaction in which a predatory species eats (and so directly harms) a prey species.

predator [L. *prehendere*, to grasp, seize] An organism that feeds on other living organisms (its prey), that does not live in or on them, and that may or may not kill them.

prediction A statement about what you can expect to observe in nature if a theory or hypothesis is not false.

pressure flow theory Of vascular plants, a theory that organic compounds move through phloem because of gradients in solute concentrations and pressure between sources (such as photosynthetically active leaves) and sinks (such as growing parts).

primary growth Plant growth originating at root tips and shoot tips and resulting in increases in length.

primary immune response Defensive acts by white blood cells and their secretions, as elicited by a first-time encounter with an antigen; includes both antibody-mediated and cell-mediated activities.

primary productivity, gross Of ecosystems, the rate at which producers capture and store a given amount of energy in their cells and tissues during a specified interval.

primary productivity, net Of ecosystems, the rate of energy storage in the cells and tissues of producers in excess of their rate of aerobic respiration.

primate A mammalian lineage that includes the prosimians, tarsioids, and anthropoids (monkeys, apes, and humans).

primer A short nucleotide sequence that will base-pair with any complementary DNA sequence; later, DNA polymerases recognize them as start tags for replication.

prion A small protein linked with eight rare, fatal degenerative diseases of the nervous system. Prions are altered products of a gene that is present in normal as well as infected individuals.

probability The chance that each outcome of a given event will occur is proportional to the number of ways it can be reached.

procambium (pro-KAM-bee-um) Of vascular plants, a primary meristem that gives rise to primary vascular tissues.

producer Of ecosystems, an autotrophic (self-feeding) organism; it nourishes itself using sources of energy and carbon from its physical environment. Photoautorophs and chemoautotrophs are examples.

progesterone (pro-JESS-tuh-rown) A type of sex hormone secreted by the pair of ovaries in female mammals.

prokaryotic cell (pro-CARRY-oh-tic) [L. *pro*, before, + Gk. *karyon*, kernel] A bacterium; a single-celled organism, usually walled, that does not have the profusion of membrane-bound organelles seen in eukaryotic cells.

prokaryotic fission Division mechanism by which a bacterial cell reproduces.

promoter A short base sequence in DNA that signals the start of a gene; the site where RNA polymerase binds to start transcription.

prophase Of mitosis, the stage when each duplicated chromosome starts to condense, microtubules form a spindle apparatus, and the nuclear envelope starts to break up.

prophase I Of meiosis, the stage when the microtubular spindle starts to form, the nuclear envelope starts to break up, and each duplicated chromosome condenses and pairs with its homologous partner. Nonsister chromatids typically undergo crossing over and genetic recombination.

prophase II Brief first stage of meiosis II when chromosomes start moving toward spindle equator; each has already been separated from its homologue but still consists of two chromatids.

protein Organic compound of one or more chains of amino acids, which peptide bonds join in a linear sequence that is the basis of the protein's three-dimensional structure and chemical behavior.

Protista The kingdom of protistans.

protistan (pro-TISS-tun) [Gk. *prōtistos*, primal, very first] One of the staggeringly diverse eukaryotes, single-celled species of which may resemble the first eukaryotic cells. At present categorized in part by what they are *not* (not bacteria, fungi, plants, or animals).

proto-oncogene A gene sequence similar to an oncogene but that codes for a protein required for normal cell function. When a mutation alters its structure, it may trigger cancerous transformation.

proton Positively charged particle, one or more of which resides in the nucleus of each type of atom; when unbound, a proton is the same as a hydrogen ion (H^+).

protostome (PRO-toe-stome) [Gk. *proto*, first, + *stoma*, mouth] One of the lineage of coelomate, bilateral animals in which the first indentation to form in the embryo develops into a mouth. Includes mollusks, annelids, and anthropods.

protozoan A group of protistans, some predatory, others parasitic, that in certain respects resemble single-celled heterotrophs that presumably gave rise to animals. The amoeboid and animal-like protozoans, as well as sporozoans, are examples.

proximal tubule Of a nephron, the tubular portion into which water and solutes enter after being filtered from the blood at the preceding portion (Bowman's capsule).

pulmonary circuit Blood circulation route leading to and from the lungs.

Punnett-square method A way to predict the probable outcome of a genetic cross in simple diagrammatic form.

purine A type of nucleotide base with a double ring structure. Examples are adenine and guanine.

pyrimidine (phi-RIM-ih-deen) A type of nucleotide bases with a single ring structure. Examples are cytosine and thymine.

pyruvate (PIE-roo-vate) A small organic compound with a backbone of three carbon atoms. Glycolysis produces two molecules of pyruvate as end products.

r A variable used in population growth equations to signify net population growth rate; the assumption is that birth and death rates remain constant and therefore may be combined into this one variable.

radial symmetry Of animals, a body plan having four or more roughly equivalent parts arranged around a central axis.

radiometric dating A method of measuring the proportions of (1) a radioisotope in a mineral trapped long ago in a newly formed rock and (2) a daughter isotope that formed from it by radioactive decay in the same rock sample. Used to assign absolute dates to fossil-containing rocks and to a geologic time scale.

radioisotope An unstable atom, with an uneven number of protons and neutrons. It spontaneously emits particles and energy, and over a predictable time span becomes transformed (decays) into a different atom.

rain shadow A reduction in rainfall on the leeward side of high mountains, resulting in arid or semiarid conditions.

reabsorption In kidneys, diffusion or active transport of water and reclaimable solutes from a nephron into peritubular capillaries under the control of ADH and aldosterone. At a capillary bed, osmotic movement of some interstitial fluid into a capillary when there is a difference in the concentration of water between plasma and interstitial fluid.

rearrangement, molecular A conversion of one type of organic compound to another through a juggling of its internal bonds.

receptor, molecular Of cells, a membrane protein that binds a hormone or some other extracellular substance that triggers change in cell activities.

receptor, sensory Of nervous systems, a sensory cell or specialized cell adjacent to it that can detect a particular stimulus.

recessive allele [L. *recedere*, to recede] In heterozygotes, an allele whose expression is fully or partially masked by expression of its partner; only in the homozygous recessive condition is it fully expressed.

recognition protein Of mammalian cell membranes, a glycolipid or glycoprotein that functions as a molecular fingerprint; it identifies a cell as being of a specific type. Self-proteins of the body's own cells are an example.

recombinant technology Procedures by which DNA (genes) from different species may be isolated, cut, spliced together, and the new recombinant molecules multiplied in quantity, as by PCR or by a population of rapidly dividing bacterial cells.

red alga A protistan, most of which are multicelled, aquatic, and photosynthetic, with an abundance of phycobilins.

blood cell An erythrocyte; an oxygen-transporting cell in blood.

red marrow A site of blood cell formation in the spongy tissue of many bones.

reflex [L. *reflectere*, to bend back] A simple, stereotyped movement elicited directly by sensory stimulation.

reflex pathway [L. *reflectere*, to bend back] The simplest route by which stimulation of nerve cells leads to contractile or glandular action. In complex animals, sensory neurons directly stimulate or inhibit motor neurons, without intervention by interneurons.

refractory period Of neurons, the period following an action potential at a given patch of membrane when sodium gates are shut and potassium gates are open, so that the patch is insensitive to stimulation.

regulatory protein A protein involved in a positive or negative control system that operates during transcription or translation or after translation. The protein interacts with DNA, RNA, or a gene product.

releasing hormone A signaling molecule produced by the hypothalamus that enhances or slows down secretions from target cells in the anterior lobe of the pituitary gland.

repressor Regulatory protein that can block gene transcription.

reproduction Of nature, any of a number of processes by which a new generation of cells or multicelled individuals is produced. For eukaryotes, it may be sexual or it may be asexual (as by binary fission, budding, or vegetative propagation). For bacteria only, it occurs by prokaryotic fission.

reproductive isolating mechanism Any heritable feature of body form, functioning, or behavior that prevents interbreeding between one or more genetically divergent populations.

reproductive success The survival and production of offspring by an individual.

reptile A type of carnivorous vertebrate; its ancestors were the first vertebrates to escape dependency on standing water, largely by means of internal fertilization and amniote eggs. Examples are dinosaurs (extinct), crocodiles, lizards, and snakes.

resource partitioning Coexistence among competing species; they share the same resource differently or at different times.

respiration [L. *respirare*, to breathe] Of animals, the exchange of oxygen from the environment for carbon dioxide wastes from cells by way of integumentary exchange, a respiratory system, or both. *Compare* aerobic respiration.

respiratory surface An epithelium or some other surface tissue that functions in gas exchange in animals; it is thin enough to allow oxygen to diffuse into the body and carbon dioxide to diffuse out of it.

resting membrane potential Of a neurons and other excitable cells, a steady voltage difference across the plasma membrane in the absence of outside stimulation.

restoration ecology Attempts to reestablish biodiversity in abandoned farmlands and other large areas altered by agriculture, mining, and other severe disturbances.

restriction enzyme Of bacteria, one of a class of enzymes that can cut apart foreign DNA injected into the cell body, as by viruses; also used in recombinant DNA technology.

reticular formation A mesh of interneurons extending from the upper spinal cord, through the brain stem, and into the cerebral cortex; a low-level pathway through the vertebrate nervous system.

reverse transcriptase A viral enzyme that uses viral RNA as a template to synthesize either DNA or mRNA in a host cell.

reverse transcription A process by which reverse transcriptase assembles DNA on an RNA template. Used by RNA viruses as well as by technologists who use mature mRNA transcripts as the templates for synthesizing cDNA.

RFLP Short for restriction fragment length polymorphism. Such fragments are formed by using restriction enzymes to cut DNA; for each individual, they show slight but unique differences in their banding pattern.

Rh blood typing A laboratory method of characterizing red blood cells on the basis of a protein that serves as a self-marker at their surface; Rh^+ signifies its presence, and Rh^- its absence.

rhizoid A rootlike absorptive structure of some fungi and nonvascular plants.

ribosomal RNA (rRNA) A type of RNA molecule that combines with proteins to form ribosomes, on which the polypeptide chains of proteins are assembled.

ribosome In all cells, the structure at which amino acids are strung together in specified sequence to form the polypeptide chains of proteins. An intact ribosome consists of two subunits, each composed of ribosomal RNA and protein molecules.

RNA Ribonucleic acid. A general category of single-stranded nucleic acids that function in processes by which genetic instructions encoded in DNA are used to build proteins.

rod cell A vertebrate photoreceptor that is sensitive to very dim light and contributes to coarse perception of movement.

root hair Of vascular plants, an extension of a specialized root epidermal cell; root hairs collectively enhance the surface area available for absorbing water and solutes.

root nodule A localized swelling on roots of certain legumes and other plants that contain symbiotic, nitrogen-fixing bacteria.

root A part of a plant that typically grows belowground, absorbs water and dissolved nutrients, helps anchor aboveground parts, and often stores food.

RuBP Ribulose bisphosphate. A compound with a backbone of five carbon atoms; used for carbon fixation by C3 plants and also regenerated in the Calvin-Benson cycle of photosynthesis.

S-shaped curve A curve, obtained when population size is plotted against time, that is characteristic of logistic growth.

salination A salt buildup in soil as a result of evaporation, poor drainage, and heavy irrigation.

salivary gland A gland that secretes saliva, a fluid that initially mixes with food in the mouth and starts the breakdown of starch.

salt A compound that releases ions other than H^+ and OH^- in solution.

saltatory conduction Of myelinated neurons, a rapid form of action potential propagation by a hopping of excitation from one node to another between the jellyrolled membranes of cells making up the myelin sheath.

sampling error A rule of probability: The fewer times that a chance event occurs, the greater will be the variance from the expected outcome of that occurrence. Of experimental groups, large sampling tends to lessen the chance that differences among individuals will distort the results.

saprobe A heterotroph that obtains energy and carbon from nonliving organic matter and so causes its decay. Saprobic fungi are examples.

sarcomere (SAR-koe-meer) Of muscles, the basic unit of contraction; a region of myosin and actin filaments organized in parallel arrays between two Z lines of one of the many myofibrils inside a muscle cell.

sarcoplasmic reticulum (sar-koe-PLAZ-mik reh-TIK-you-lum) A membrane system around a muscle cell's myofibrils that takes up, stores, and releases the calcium ions required for cross-bridge formation in sarcomeres, hence for contraction.

Schwann cell One of a series of neuroglial cells wrapped like jellyrolls around an axon in a nerve and forming a myelin sheath.

sclerenchyma (skler-ENG-kih-mah) A simple plant tissue in which most cells have thick, lignin-impregnated walls. Sclerenchyma supports mature plant parts and commonly protects seeds.

sea-floor spreading Molten rock erupts from great, continuous ridges on the ocean floor, flows laterally in both directions, and hardens to form new crust. As the seafloor spreads out, it forces older crust down into great trenches elsewhere in the seafloor.

second law of thermodynamics A law of nature stating that the spontaneous direction of energy flow is from organized to less organized forms. Overall, the total amount of energy in the universe is spontaneously flowing from forms of higher to lower quality; with each conversion, some energy gets randomly dispersed in a form (usually heat) not as readily available to do work.

second messenger A molecule within a cell that mediates a hormonal signal by triggering a response to it.

secondary immune response An immune response to previously encountered antigen that is more rapid and prolonged than the first owing to the swift participation of memory cells.

secondary sexual trait A trait associated with maleness or femaleness but one with no direct role in reproduction.

secretion A product released across the plasma membrane of a cell that may act singly or as part of glandular tissue.

sedimentary cycle A biogeochemical cycle with no gaseous phase; the element moves from land to the seafloor, then returns only through long-term geological uplifting.

seed Of gymnosperms and angiosperms, a fully mature ovule (contains the plant embryo) with integuments that form the seed coat.

segmentation Of animals, a body plan with a repeating series of units that may or may not be similar to one another in appearance. Also, an oscillating movement created by rings of circular muscle in a tubular wall. In a small intestine, such movement constantly mixes and forces the contents of the lumen against the absorptive wall surface.

segregation, theory of [L. se-, apart, + grex, herd] A Mendelian theory that diploid organisms inherit pairs of genes for traits (on pairs of homologous chromosomes) and that the two genes segregate during meiosis and end up in separate gametes.

selective gene expression Of a multicelled organism, the activation and suppression of some fraction of the same inherited genes in different populations of cells; it leads to cell differentiation.

selective permeability Of a cell membrane, a capacity to allow some substances but not others to cross it at certain sites, at certain times, owing to its molecular structure.

selfish behavior A behavior by which an individual protects or increases its own chance of producing offspring, regardless of the consequences to the group to which it belongs.

selfish herd A society held together simply by reproductive self-interest.

semen (SEE-mun) [L. serere, to sow] Sperm-bearing fluid expelled from a penis during male orgasm.

semiconservative replication [Gk. hēmi, half, + L. conservare, to keep] How a DNA molecule duplicates itself. A DNA double helix unzips and a complementary strand is assembled on the exposed bases of each strand. Each conserved strand and its new partner are wound together into a double helix. Thus the outcome is two "half-old, half-new" molecules.

senescence (sen-ESS-cents) [L. senescere, to grow old] Sum of processes leading to the natural death of an organism or some of its parts.

sensation A conscious awareness of a stimulus. Not the same thing as perception, which is an understanding of the meaning of a sensation.

sensory neuron Any of the nerve cells or cells adjacent to them that detect specific stimuli (such as light energy) and relay signals to the brain and spinal cord.

sensory system Of animals, the "front door" of the nervous system; the components that detect external and internal stimuli and relay information about them into the central nervous system.

sessile animal (SESS-ihl) An animal that remains attached to a substrate during some stage (often the adult) of its life cycle.

sex chromosome Of animals and some plants, one of two types of chromosomes, the combinations of which govern gender. *Compare* autosome.

sexual dimorphism Of a species, having individuals with distinctively male or female phenotypes.

sexual reproduction Production of offspring by way of meiosis, gamete formation, and fertilization.

sexual selection A microevolutionary process; natural selection favors a trait that gives the individual a competitive edge in attracting and holding onto mates, hence in reproductive success.

shifting cultivation The cutting and burning of trees, followed by tilling of ashes into the soil; once called slash-and-burn agriculture.

shell model Model of electron distribution in atoms, in which all orbitals available to electrons occupy a nested series of shells.

shoot system The aboveground parts of a plant, such as stems, leaves, and flowers.

sieve-tube member Of flowering plants, a cellular component of the interconnecting sugar-conducting tubes in phloem.

sign stimulus A simple but important cue in the environment that triggers a suitable response to a stimulus that the nervous system is prewired to recognize.

sink region Of plants, any region using or stockpiling organic compounds for growth and development.

sister chromatid One of two DNA molecules (and associated proteins) of a duplicated chromosome that remain attached at their centromere region until separated at mitosis or meiosis; each ends up as a chromosome in a different daughter nucleus.

skeletal muscle An organ with hundreds to many thousands of muscle cells arrayed in bundles. Connective tissue surrounds the bundles and extends beyond the muscle, in the form of tendons that attach it to bone.

sliding filament model Model of muscle contraction in which each myosin filament in a sarcomere physically slides along and pulls actin filaments toward the center of the sarcomere, which shortens. The sliding requires ATP energy and formation of cross-bridges between the actin and myosin.

small intestine Of vertebrates, the portion of the digestive system where digestion is completed and most nutrients absorbed.

smog, industrial Polluted air, gray colored, that predominates in industrialized cities with cold, wet winters.

smog, photochemical Polluted air, brown and smelly, that occurs in large cities with many gas-burning vehicles and a warm climate.

social behavior Intraspecific interactions that require mixes of the instinctive and learned behaviors by which individuals send and respond to communication signals.

social parasite An animal that exploits the social behavior of another species to gain food, care for young, or some other factor necessary to survive, reproduce, or both.

sodium-potassium pump A membrane transport protein; when activated by ATP, it selectively transports potassium ions across the membrane, against its concentration gradient, and allows the crossing of sodium ions in the opposite direction.

soil A variable mixture of mineral particles and decomposing organic material; air and water occupy spaces between the particles.

solute (SOL-yoot) [L. *solvere*, to loosen] Any substance dissolved in a solution. In water, this means that hydration spheres surround the charged parts of individual ions or molecules and keep them dispersed.

solvent A fluid, such as water, in which one or more substances is dissolved.

somatic cell (so-MAT-ik) [Gk. *somā*, body] Of animals, any body cell that is not a germ cell (which gives rise to gametes).

somatic nervous system Nerves leading from the central nervous system to skeletal muscles.

somite One of many paired segments in a vertebrate embryo that will give rise to most bones, skeletal muscles of the head and trunk, and the overlying dermis.

source region Of plants, a site where organic compounds form by photosynthesis.

speciation (spee-cee-AY-shun) Evolutionary process by which a daughter species forms from a population or subpopulation of the parent species. The process can vary in its details and in the length of time it takes before the required reproductive isolation from each other is complete.

species (SPEE-sheez) [L. *species*, a kind] In general, one kind of organism. Of sexually reproducing organisms only, one or more populations of individuals that are interbreeding under natural conditions and producing fertile offspring, and that are reproductively isolated from other such groups.

sperm [Gk. *sperma*, seed] A type of mature male gamete.

spermatogenesis (sperm-AT-oh-JEN-ih-sis) Process by which mature sperm form from a germ cell in males.

sphere of hydration Through positive or negative interactions, a clustering of water molecules around individual molecules of a substance placed in water. *Compare* solute.

spinal cord The portion of the central nervous system that threads through a canal inside the vertebral column and that affords direct reflex connections between sensory and motor neurons, and that has tracts to and from the brain.

spleen One of the lymphoid organs; it is a filtering station for blood, a reservoir of red blood cells, and a reservoir of macrophages.

spore A single-celled reproductive structure or resistant, resting body; often walled, and often adapted to survive adverse conditions. An asexual spore may directly give rise to a new stage of the life cycle; a sexual spore must unite with another sexual spore to produce a new stage.

sporophyte [Gk. *phyton*, plant] Of plant life cycles, a vegetative body that grows by way of mitosis from a zygote and that produces spore-bearing structures.

spring overturn Of some bodies of water, movement during spring of dissolved oxygen from the surface layer to the depths and movement of nutrients from bottom sediments to the surface.

stabilizing selection A mode of natural selection by which intermediate phenotypes are favored, and extremes at both ends of the range of variation are eliminated.

stamen (STAY-mun) A male reproductive structure in a flower, commonly consisting of a pollen-bearing structure (anther) on a single stalk (filament).

start codon A base triplet in a strand of mRNA that serves as the start signal for the translation stage of protein synthesis.

stem cell A type of self-perpetuating cell that remains unspecialized. Some of its daughter cells also are self-perpetuating; others differentiate into cells of specialized structure and function. Stem cells in bone marrow that give rise to daughter stem cells and to blood cells are an example.

sterol (STAIR-all) A type of lipid with a rigid backbone of four fused carbon rings. Sterols differ in the number, position, and type of functional groups. They occur in eukaryotic cell membranes; cholesterol is the main type in animal tissues.

steroid hormone A lipid-soluble hormone, synthesized from cholesterol, that diffuses directly across the lipid bilayer of a target cell's plasma membrane and binds with a receptor inside the cell.

stigma Of many flowers, a sticky or hairy surface tissue on upper portion of ovary; it captures pollen grains and favor their germination.

stimulus [L. *stimulus*, goad] A specific form of energy, such as mechanical energy, light, and heat, that activates a sensory receptor having the capacity to detect it.

stoma (STOW-muh), plural **stomata** [Gk. *stoma*, mouth] A controllable gap between two guard cells in stems and leaves; any of the tiny passageways across the epidermis by which carbon dioxide moves into a plant and water vapor and oxygen move out.

stomach A muscular, stretchable sac that receives ingested food; of vertebrates, an organ between the esophagus and intestine in which much protein digestion occurs.

stop codon A base triplet in a strand of mRNA that serves as a stop signal during the translation stage of protein synthesis; it blocks further additions of amino acids to a new polypeptide chain.

strain One of two or more organisms with differences that are too minor to classify it as a separate species. An example is a strain of *Escherichia coli* or some other bacterium.

stratification Ancient layers of sedimentary rock resulting from the slow deposition of volanic ash, silt, and other materials, one above the other.

stream A flowing-water ecosystem that starts out as a freshwater spring or seep.

stroma [Gk. *strōma*, bed] The semifluid interior between a thylakoid membrane system and the two outer membranes of a chloroplast; the chloroplast zone where sucrose, starch, cellulose, and other end products of photosynthesis are assembled.

stromatolite A formation of sediments and matted remains of photosynthetic species that lived shallow seas and gradually accumulated in thin, cakelike layers.

substrate A reactant or precursor for a metabolic reaction; a specific molecule or molecules that an enzyme can chemically recognize, bind reversibly to itself, and modify rapidly in a predictable way.

substrate-level phosphorylation A direct, enzyme-mediated transfer of a phosphate group from a substrate of a reaction to a different molecule, as when an intermediate of glycolysis gives up a phosphate group to ADP, thus forming ATP.

succession, primary (suk-SESH-un) [L. *succedere*, to follow after] An ecological pattern by which a community develops in sequence, from the time pioneer species colonize a new, barren habitat to the climax community (an end array of species that remain in equlibrium over the region).

succession, secondary A pattern by which some disturbed area within a community recovers and moves back toward the climax state; typical of abandoned fields, burned forests, and storm-battered intertidal zones.

surface-to-volume ratio A mathematical relationship in which volume increases with the cube of the diameter, but surface area increases only with the square. Of growing cells, the volume of cytoplasm increases faster than the surface area of the plasma membrane that must service the cytoplasm. This constraint generally keeps cells small, elongated, or with membrane foldings.

survivorship curve A plot of age-specific survival of a group of individuals in a given environment, from the time of their birth until the last one dies.

swim bladder An adjustable flotation device that changes in volume as it exchanges gases with blood and that allows many fishes to maintain neutral buoyancy in water.

symbiosis (sim-by-OH-sis) [Gk. *sym*, together, + *bios*, life, mode of life] Literally, living together. For at least part of the life cycle, individuals of one species live near, in, or on individuals of another species. Commensalism, mutualism, and parasitism are examples of symbiotic interactions.

sympathetic nerve Of autonomic nervous systems, one of the nerves dealing mainly with increasing overall body activities at times of heightened awareness, excitement, or danger; also work continually in opposition with parasympathetic nerves to make minor adjustments in internal organ activity.

sympatric speciation [Gk. *sym*, together, + *patria*, native land] Species form within the home range of an existing species, in the absence of a physical barrier. It may happen instantaneously, as by polyploidy.

synaptic integration (sin-AP-tik) Of an individual neuron, the moment-by-moment combining of all excitatory and inhibitory signals arriving at its trigger zone.

syndrome A set of symptoms that may not individually be a telling clue but that collectively characterize a particular genetic disorder or disease.

systemic circuit (sis-TEM-ik) Of a closed circulatory system, a route by which oxygen-enriched blood flows from the lungs to the left half of the heart, through the rest of the body (where it gives up oxygen and takes up carbon dioxide), then back to the right half of the heart.

T lymphocyte One of a class of white blood cells that carry out immune responses. The helper T and cytotoxic T cells are examples.

taproot system A primary root together with its lateral branchings.

target cell Of a signaling molecule such as a hormone, any cell having the type of receptor that can bind with that molecule.

tectum The midbrain's roof. In fishes and amphibians, a center for coordinating most sensory input and initiating motor responses. In most vertebrates (not mammals), a reflex center that swiftly relays sensory input to higher integrative centers in the forebrain.

telophase (TEE-low-faze) Of meiosis I, the stage when one of each pair of homologous chromosomes has arrived at a spindle pole. Of mitosis and of meiosis II, the final stage when chromosomes decondense into threadlike structures and two daughter nuclei form.

temperature A measure of how rapidly the molecules or ions of a substance are moving.

tendon A cord or strap of dense connective tissue that attaches muscle to bones.

territory An area that one or more animals defend against competitors for food, water, mates, or suitable living space.

test A way to determine the accuracy of a prediction, as by conducting experimental or observational tests and by developing models. In science, such predictions must be based on potentially falsifiable hypotheses; that is, they must be tested in the natural world in ways that might disprove them.

testcross For an individual of unknown genotype that shows dominance for a trait, a type of experimental cross that may reveal whether the individual is homozygous dominant or heterozygous.

testis, plural **testes** Male gonad; a primary reproductive organ in which male gametes and sex hormones are produced.

testosterone (tess-TOSS-tuh-rown) Of male vertebrates, a sex hormone with key roles in the development and functioning of the male reproductive system.

tetanus Of muscles, a large contraction in which repeated stimulation of a motor unit causes muscle twitches to mechanically run together. In a disease by the same name, toxins block muscles from relaxing.

thalamus Of the forebrain, a coordinating center for sensory input and a relay station for signals to the cerebrum.

theory, scientific A testable explanation of a broad range of related phenomena, one that has been extensively tested and can be used with a high degree of confidence. A scientific theory remains open to tests, revision, and tentative acceptance or rejection.

thermal inversion The trapping of a layer of dense, cool air beneath a layer of warm air; can cause air pollutants to accumulate to high levels close to the ground.

thermophile A type of archaebacterium of unusually hot aquatic habitats, as in hot springs or near hydrothermal vents.

thermoreceptor A sensory cell or specialized cell by it that detects radiant energy (heat).

thigmotropism (thig-MOE-truh-pizm) [Gk. *thigm*, touch] An orientation of the direction of growth in response to physical contact with a solid object, as when a vine curls around a fencepost.

threshold Of a neuron and other excitable cells, the minimum amount by which the resting membrane potential must change to trigger an action potential.

thylakoid membrane system Of chloroplasts, an internal membrane system commonly folded into flattened channels and disks (grana) that have light-absorbing pigments and enzymes used in the formation of ATP, NADPH, or both during photosynthesis.

thymine A nitrogen-containing base of one of the nucleotides in DNA.

thymus gland A lymphoid organ that has endocrine functions; lymphocytes of the immune system multiply, differentiate, and mature in its tissues, and its hormone secretions affect their functioning.

thyroid gland An endocrine gland, the hormones of which affect overall metabolic rates, growth, and development.

tissue Of multicelled organisms, a group of cells and intercellular substances that function together in the performance of one or more specialized tasks.

tonicity The relative concentrations of solutes in two fluids, such as inside and outside a cell. When solute concentrations are isotonic (equal in both fluids), water shows no net osmotic movement in either direction. When one fluid is hypotonic (has less solutes), the other is hypertonic (has more solutes) and water tends to move into it.

toxin A normal metabolic product of one species with chemical effects that can harm or kill individuals of a different species that encounter it.

trace element Any element that represents less than 0.01 percent of body weight.

tracer A substance that has a radioisotope attached to it so that its pathway or destination in a cell, organism, ecosystem, or some other system can be tracked, as by scintillation counters that detect its emissions.

trachea (TRAY-kee-uh), plural **tracheae** An air-conducting tube of respiratory systems; of land vertebrates, the windpipe, which carries air between the larynx and bronchi.

tracheal respiration Of insects, spiders, and some other animals, a respiratory system consisting of finely branching tracheae that extend from openings in the integument and that dead-end in body tissues.

tracheid (TRAY-kid) Of flowering plants, one of two types of cells in xylem that conduct water and dissolved minerals.

tract A communication line inside the brain and spinal cord; comparable to a nerve of the peripheral nervous system.

transcription [L. *trans*, across, + *scribere*, to write] The first stage of protein synthesis, when an RNA strand is assembled on one of the two strands of a DNA double helix; the base sequence of the resulting transcript is complementary to the DNA template.

transfer RNA (tRNA) An RNA molecule that binds and delivers an amino acid to a ribosome and that pairs with an mRNA codon during the translation stage of protein synthesis.

translation The stage of protein synthesis when the encoded sequence of information in mRNA becomes converted to a sequence of particular amino acids, the outcome being a polypeptide chain; rRNA, tRNA, and mRNA interact to bring this about.

translocation Of cells, a stretch of DNA that physically moved to a different location in the same chromosome or in a different one, with no molecular loss. Of vascular plants, the process by which organic compounds are distributed by way of phloem.

transpiration Evaporative water loss from aboveground plant parts, leaves especially.

transport protein A membrane protein that allows water-soluble substances to move through their interior, which spans the lipid bilayer of a cell membrane.

transposable element A DNA region that moves spontaneously from one location to another in the same DNA molecule or a different one. Often such regions inactivate genes into which they become inserted and cause changes in phenotype.

triglyceride (neutral fat) A lipid having three fatty acid tails attached to a glycerol backbone. Triglycerides are the body's most abundant lipids and richest energy source.

trisomy (TRY-so-mee) Of cells, the presence of three chromosomes of a given type rather than the two characteristic of the parental diploid chromosome number.

trophic level (TROE-fik) [Gk. *trophos*, feeder] All of the organisms in an ecosystem that are the same number of transfer steps away from the energy input into the system.

tropical rain forest A biome where rainfall is regular and heavy, the annual mean temperature is 25°C, humidity is 80 percent or more, and biodiversity is spectacular.

tropism (TROE-pizm) Of plants, a directional growth response to an environmental factor, such as growth toward light.

true breeding Of a sexually reproducing species, a lineage in which the offspring of successive generations are just like the parents in one or more traits being studied.

tumor A tissue mass composed of cells that are dividing at an abnormally high rate.

turgor pressure (TUR-gore) [L. *turgere*, to swell] Internal fluid pressure applied to a cell wall when water moves into the cell by way of osmosis.

ultrafiltration Bulk flow of a small amount of protein-free plasma from a blood capillary when the outward-directed force of blood pressure is greater than the inward-directed osmotic force of interstitial fluid.

uniformity Various theories that mountain building, erosion, and other forces of nature have worked over the Earth's surface in the same repetitive ways through time.

upwelling An upward movement of deep, nutrient-rich water along coasts; it replaces surface waters that move away from shore when the direction of prevailing wind shifts.

uracil (YUR-uh-sill) Nitrogen-containing base of a nucleotide found in RNA molecules; like thymine, it can base-pair with adenine.

ureter A tubular channel for urine flow between the kidney and urinary bladder.

urethra A tubular channel for urine flow between the urinary bladder and an opening at the body surface.

urinary bladder A distensible sac in which urine is stored before being excreted.

urinary excretion A mechanism by which excess water and solutes are removed by way of a urinary system.

urinary system An organ system that adjusts the volume and composition of blood, and so helps maintain extracellular fluid.

urine A fluid formed in kidneys by filtration, reabsorption, and secretion; urine consists of wastes, excess water, and solutes.

uterus (YOU-tur-us) [L. *uterus*, womb] A chamber in which the developing embryo is contained and nurtured during pregnancy.

vaccination An immunization procedure against a specific pathogen.

vaccine An antigen-containing preparation, swallowed or injected, that increases immunity to certain diseases by inducing formation of armies of effector and memory B and T cells.

vagina Part of a reproductive system of mammalian females that receives sperm, forms part of the birth canal, and channels menstrual flow to the exterior.

variable Of an experimental test, a specific aspects of an object or event that may differ over time and among individuals.

vascular bundle Of vascular plants, the arrangement of primary xylem and phloem into multistranded, sheathed cords that extend lengthwise through the ground tissue system.

vascular cambium A lateral meristem that increases stem or root diameter (girth).

vascular cylinder Arrangement of vascular tissues as a central cylinder in roots.

vascular plant A plant with xylem and phloem, and well-developed roots, stems, and leaves.

vascular tissue system Xylem and phloem; conducting tissues that distribute water and solutes through the body of vascular plants.

vein Of the circulatory system, any of the large-diameter vessels that lead back to the heart; of leaves, one of the vascular bundles that thread through photosynthetic tissues.

ventricle (VEN-tri-kuhl) Of the vertebrate heart, one of two chambers from which blood is pumped out. Blood circulation is driven by ventricular contraction.

venule A small blood vessel that accepts blood from capillaries and delivers it to a vein.

vernalization Of flowering plants, stimulation of flowering by exposure to low temperatures.

vertebra, plural **vertebrae** One of a series of hard bones organized, with intervertebral disks, into a backbone.

vertebrate Animal with a backbone.

vesicle (VESS-ih-kul) [L. *vesicula*, little bladder] In the cytoplasm of cells, one of a variety of small membrane-bound sacs that function in the transport, storage, or digestion of substances or in some other activity.

vessel member A cell in xylem; although dead at maturity, its wall becomes part of a water-conducting pipeline.

villus (VIL-us), plural **villi** Any of several fingerlike absorptive structures projecting from the free surface of an epithelium.

viroid An infectious particle consisting only of very short, tightly folded strands or circles of RNA. Viroids may have evolved from introns, which they resemble.

virus A noncellular infectious agent that consists of DNA or RNA and a protein coat; it can replicate only after its genetic material enters a host cell and subverts the host's metabolic machinery.

vision Perception of visual stimuli; requires focusing of light precisely onto a layer of photoreceptive cells that is dense enough to sample details of a light stimulus, followed by image formation in a brain.

visual signal An observable action or cue that functions as a communication signal.

vitamin Any of more than a dozen organic substances that animals require in small amounts for metabolism but that they generally cannot synthesize for themselves.

vocal cord One of the thickened, muscular folds of the larynx that help produce sound waves for speech.

water potential Sum of two opposing forces (osmosis and turgor pressure) that can cause a directional movement of water into or out of a walled cell.

water table The upper limit at which the ground in a specified area has become fully saturated with water.

watershed Any specified region in which all precipitation drains into a single stream or river.

wavelength A wavelike form of energy in motion. The horizontal distance between two crests of every two successive waves.

wax A molecule with long-chain fatty acids packed together and linked to long-chain alcohols or to carbon rings. Waxes have a firm consistency and repel water.

white blood cell One of the eosinophils, neutrophils, macrophages, T and B cells, and other leukocytes which, together with their chemical weapons and mediators, defend the vertebrate body against attacks by pathogens and against tissue damage.

white matter Inside the brain and spinal cord, axons that have glistening white sheaths and that specialize in rapid signal transmission.

wild-type allele For a given gene locus, the allele that occurs normally or with the greatest frequency among individuals of a population.

wing A body part that functions in flight, as among birds, bats, and many insects. A bird wing is a forelimb with feathers, strong muscles, and extremely lightweight bones. An insect wing develops as a lateral fold of the exoskeleton.

X chromosome A sex chromosome with genes that, in humans, causes an embryo to develop into a female, provided that it inherits a pair of these.

X-linked gene Any gene located on an X chromosome.

X-linked recessive inheritance Recessive condition in which the responsible, mutated gene occurs on the X chromosome.

xylem (ZYE-lum) [Gk. *xylon*, wood] Of vascular plants, a tissue that transports water and solutes through the plant body.

Y chromosome A distinctive chromosome present in males or females of many species, but not both. In human males, one is paired with an X chromosome (XY); and in human females, the pairing is XX.

Y-linked gene One of the genes located on a Y chromosome.

yellow marrow A fatty tissue in the cavities of most mature bones that produces red blood cells when blood loss from the body is severe.

yolk sac Of land vertebrates, one of four extraembryonic membranes. In most shelled eggs, it holds nutritive yolk; in humans, part becomes a site of blood cell formation and some of its cells give rise to the forerunners of gametes.

zero population growth A population for which the number of births is balanced by the number of deaths over a specified period, assuming that immigration and emigration also are balanced.

zooplankton A freshwater or marine community of floating or weakly swimming heterotrophs, mostly microscopic, such as rotifers and copepods.

zygote (ZYE-goat) The first cell of a new individual, formed by the fusion of a sperm nucleus with the nucleus of an egg at fertilization; also called a fertilized egg.

CREDITS AND ACKNOWLEDGMENTS

FRONT MATTER **Pages iv-v**, David Macdonald. **vi-vii**, James M. Bell/Photo Researchers. **viii-ix** S. Stammers/SPL/Photo Researchers. **x-xi** © 1990 Arthur M. Greene. **xii-xiii** Gary Head. **xiv-xv** Lennart Nilsson from *Behold Man*, © 1974 Albert Bonniers Forlag and Little, Brown and Company, Boston. **xvi-xvii** John Alcock. **xviii-xix**, Lennart Nilsson from *A Child Is Born*, © 1966, 1977 Dell Publishing Company, Inc. **xx-xxi**, Jim Doran. **xxxii** and **page 1** © 1990 Tom Van Sant/The GeoSphere Project, Santa Monica, California.

CHAPTER 1 **1.1** Frank Kaczmarek. **1.2** Art, American Composition & Graphics (ACG). **1.3, 1.4** Jack deConingh. **1.5** (a) Walt Anderson/ Visuals Unlimited; (b) Gregory Dimijian/ Photo Researchers; (c) Alan Weaving/Ardea London. **1.6** (a) R. Robinson, Visuals Unlimited; (b) Tony Brain/SPL/Photo Researchers; (b) M. Abbey, Visuals Unlimited; (d, f) Edward S. Ross; (e) Dennis Brokaw; (g,h) Pat & Tom Leeson/Photo Researchers. **1.7** J. A. Bishop, L. M. Cook. **1.8** Levi Publishing Company. **1.9** Jack deConingh; art, Raychel Ciemma. **1.11** Gary Head. **1.12** James Carmichael, Jr./NHPA. **Page 19** James M. Bell/Photo Researchers

CHAPTER 2 **2.1** Gary Head for Norman Terry, University of California, Berkeley. **2.2** (a) Lee Newman, University of Washington; (b) Dr. Slavik Dushenkov PhD., Phytotech, Inc. **2.3** Jack Carey. **2.6** (a) Kingsley R. Stern; (b) Chip Clark; (c) Gary Byerly, LSU. **2.9** (a) Hank Morgan/Rainbow; (b) Art, R. Ciemma; (c) Harry T. Chugani, M.D., UCLA School of Medicine. **2.12** Maris and Cramer. **2.14** Photo Researchers. **2.16** (a) Richard Riley/FPG; (b) art, R. Ciemma; (c) copyright © Kennan Ward/The Stock Market. **2.17** H. Eisenbeiss/ Frank Lane Picture Agency. **2.18** Art, R. Ciemma. **2.20** Michael Grecco/Picture Group.

CHAPTER 3 **3.1** Dave Schiefelbein, NASA. **3.4** Tim Davis/Photo Researchers. **3.6** *Above,* Gary Head; *below,* Martin Rogers/FPG. **3.10** David Scharf/Peter Arnold, Inc. **3.12** Clem Haagner/Ardea, London. **3.13** (a) Art, Precision Graphics; (c) Micrograph Lewis L. Lainey; (d) Larry Lefever/Grant Heilman; (e) Kenneth Lorenzen. **3.19** Art, Palay/Beaubois. **3.20** CNRI/ SPL/Photo Researchers; art, Robert Demarest. **3.23** Art, Precision Graphics; A. Lesk/SPL/ Photo Researchers.

CHAPTER 4 **4.1** (a) Corbis-Bettmann. (b) Armed Forces Institute of Pathology; (c) National Library of Medicine; (e) The Francis A. Countway Library of Medicine. **4.2** Art, R. Ciemma. **4.3** George S. Ellmore; art, R. Demarest. **4.5** Driscoll, Youngquist, and Baldeschwieler/Cal Tech/ Science Source/Photo Researchers; art, R. Ciemma. **4.6** Jeremy Pickett-Heaps, School of Botany, University of Melbourne. **4.7** Art, R. Ciemma and ACG. **4.8** M. C. Ledbetter, Brookhaven National Laboratory; art, R. Ciemma and ACG. **4.9** G. L. Decker; art, R. Ciemma and ACG. **4.10** Stephen L. Wolfe. **4.11** (a) *Left,* Don W. Fawcett/Visuals Unlimited; *right,* A. C. Faberge, *Cell and Tissue Research,* 151:403-415, 1974. **4.12** Art, R. Ciemma and ACG. **4.13** (a,b) Micrographs, Don W.Fawcett/ Visuals Unlimited; *below,* art, R. Ciemma. **4.14** Gary W. Grimes; art R. Demarest after a model by J. Kephart. **4.15** Keith R. Porter. **4.16** L. K.

Shumway; *below right* art, Palay/Beaubois. **4.17** J. Victor Small, Gottfried Rinnerthaler. **4.18** Art, Precision Graphics / **4.19** (a) C. J. Brokaw; (b) Sidney L. Tamm. **4.20** Art, Precision Graphics after Stephen L. Wolfe, *Molecular and Cellular Biology,* Wadsworth. 1993. **4.21** Art, Precision Graphics after Stephen L. Wolfe, *Molecular and Cellular Biology,* Wadsworth. **4.22** Ron Hoham, Dept. of Biology, Colgate University. **4.23** Art, R. Ciemma; (b) Photo Researchers. **4.24** (a) George S. Ellmore and art, R. Ciemma; (b) Ed Reschke and art, Joel Ito; . **4.25** Art, R. Ciemma. **4.26** (a) Art, R. Ciemma; (b) G. Cohen-Bazire; (c) K. G. Murti/Visuals Unlimited; (d) R. Calentine/Visuals Unlimited; (e) Gary Gaard and Arthur Kelman.

CHAPTER 5 **5.1** (a) Runk/Schoenberger/Grant Heilman; (b) Inigo Everson/Bruce Coleman Ltd. **5.2** (a,b) Art, Precision Graphics; (c) Art, R. Ciemma. **5.3** Art, Raychel Ciemma. **5.4** Charles D. Winters/Photo Researchers; art, Precision Graphics after Stephen L. Wolfe, *Molecular and Cellular Biology,* Wadsworth; **5.5** P. Pinto da Silva, D. Branton, *Journal of Cell Biology,* 45:598, by copyright permission of The Rockefeller University Press; art, Palay/ Beaubois. **5.6** Art, Palay/Beaubois. **5.7** Art, Precision Graphics. **5.9** M. Sheetz, R. Painter, and S. Singer, *Journal of Cell Biology,* 70:193, by copyright permission of The Rockefeller University Press. **5.12** Art, R. Ciemma. **5.15** Art, Leonard Morgan. **5.16** M. M. Perry and A. B. Gilbert. **5.17** M. Abbey/Visuals Unlimited. **5.19** Frieder Sauer/Bruce Coleman Ltd.

CHAPTER 6 **6.1** Gary Head; molecular models, Protein Data Bank. **6.2** Evan Cerasoli. **6.3** *Above* NASA; *below* Manfred Kage/Peter Arnold, Inc. **6.8** Art, Nadine Sokol. **6.10** Art, R. Ciemma; (b,c) art, Precision Graphics. **6.11** (a,b) Thomas A. Steitz; art, Palay/Beaubois. **6.12** Douglas Faulkner/Sally Faulkner Collection. **6.16** Art, R. Ciemma, ACG after B. Alberts et al., *Molecular Biology of the Cell,* Garland Publishing, 1983. **6.17** Art, R. Ciemma and ACG. **6.18** (a) © Oxford Scientific Films/ Animals Animals; (b) © Raymond Mendez/Animals Animals; (c,d) Keith V. Wood. **6.19** (a,b) Photographs, Christopher C. Contag, *Molecular Microbiology,* November 1995, Vol. 18, No. 4, pp. 593-603. "Photonic Detection of Bacterial Pathogens in Living Hosts." Reprinted by permission of Blackwell Science.

CHAPTER 7 **7.1** David R. Frazier/Photo Researchers. **7.2** (a) Hans Reinhard/Bruce Coleman Ltd.; (e,f) David Fisher; art, R. Ciemma and ACG. **7.3** (a) Barker-Blakenship/FPG; (b) Art, Precision Graphics after Govindjee. **7.4** Carolina Biological Supply Company; art, R. Ciemma. **7.5** Art, Precision Graphics after Stephen L. Wolfe, *Molecular and Cellular Biology,* Wadsworth. **7.6** Art, Precision Graphics. **7.7** Larry West/FPG. **7.8** (a) Barker-Blakenship/ FPG; (b) Art, Precision Graphics after Govindjee. **7.9** Art, R. Ciemma. **7.10, 7.11** Art, Precision Graphics. **7.12** E. R. Degginger. **7.13** Art, R. Ciemma and Precision Graphics. **7.14** Art, Precision Graphics. **7.15** (a) Art, R. Ciemma. **7.16** (a) Grant Heilman Inc.; (b) Dick Davis/Photo Researchers; (c) Martin Grosnick/Ardea London. **7.17** NASA. **7.18** Art, R. Ciemma and Precision Graphics. **Page 129** art, R. Ciemma.

CHAPTER 8 **8.1** Stephen Dalton/Photo Researchers. **8.3** (a,b) Gary Head; (c) Janeart/ Image Bank. **8.4** *Right,* art; Palay/Beaubois. **8.5** (a) Keith R. Porter; (b) Art, L. Calver; (c) Art, R. Ciemma. **8.7, 8.8** Art, R. Ciemma. **8.11** (b) Adrian Warren/Ardea London; (c) David M. Phillips/Visuals Unlimited. **8.12** Gary Head. **Page 144** R. Llewellyn/Superstock, Inc. **Page 147** © Lennart Nilsson

CHAPTER 9 **9.1** (*left and right, above*) Chris Huss; (*left inset and right, below*) Tony Dawson. **9.2** C. J. Harrison et al., *Cytogenetics and Cell Genetics* 35:21-27, © 1983 S. Karger A.G., Basel. **9.4** A. S. Bajer, University of Oregon. **9.5** Ed Reschke; art, R. Ciemma. **9.6** (a) *Left,* C. J. Harrison et al., *Cytogenetics and Cell Genetics,* 35:21-27, © 1983 S. Karger A. G., Basel; (b) B. Hamkalo; (c) O. L. Miller, Jr., Steve L. McKnight; (b-d) art, Nadine Sokol. **9.8** B.A. Palevitz and E.H. Newcomb, University of Wisconsin/BPS/Tom Stack & Associates. **9.9** H. Beams, R. G. Kessel, *American Scientist,* 64:279-290, 1976. **9.10** (a-c, e) Lennart Nilsson from *A Child Is Born* © 1966, 1967 Dell Publishing Company, Inc.; (d) L. Nilsson from *Behold Man,* © 1974 by Albert Bonniers Forlag and Little, Brown and Company, Boston.

CHAPTER 10 **10.1** (a) Jane Burton/Bruce Coleman Ltd.; (b) Dan Kline/Visuals Unlimited. **10.2** Art, R. Ciemma. **10.3** CNRI/SPL/Photo Researchers. **10.4** Art, R. Ciemma. **10.5** Art, R. Ciemma and ACG. **10.6** Art, R. Ciemma. **10.9** David M. Phillips/Visuals Unlimited. **10.10** Art, R. Ciemma. **10.12** © Richard Corman/ Outline

CHAPTER 13 **11.1** *Left* Frank Trapper/Sygma; *center* Focus on Sports; *right, above* Fabian/ Sygma; *right, below* Moravian Museum, Brno. **11.2** Jean M. Labat/Ardea London; art, Jennifer Wardrip. **11.5** Art, Hans & Cassady, Inc. **11.7** Art, Hans & Cassady, Inc. **11.8** Art, R. Ciemma. **11.10** William E. Ferguson. **11.12** Micrographs Stanley Flegler/Visuals Unlimited. **11.13** (a,b) Michael Stuckey/Comstock Inc.; (c) Russ Kinne/ Comstock Inc. **11.14** David Hosking. **11.15** Tedd Somes. **11.16** *Top to bottom* Frank Cezus; Frank Cezus; Michael Keller; Ted Beaudin; Stan Sholik/all FPG. **11.17** (a) Dan Fairbanks, Brigham Young University. **11.18** Jane Burton/Bruce Coleman Ltd.; art, D. & V. Hennings. **11.19** (a) William E. Ferguson; (b) Eric Crichton/Bruce Coleman Ltd. **11.20** Evan Cerasoli. **11.21** Leslie Faltheisek/Clacritter Manx. **11.22** Joe McDonald/Visuals Unlimited.

CHAPTER 12 **12.1** Eddie Adams/AP Photo. **12.2** (b) Photograph Omikron/Photo Researchers. **12.4** (a) From Lennart Nilsson, *A Child Is Born,* © 1966, 1977 Dell Publishing Company, Inc.; (b) Redrawn by R. Demarest by permission from page 126 of Michael Cummings, *Human Heredity: Principles and Issues,* Third Edition. © 1994 by West Publishing. All rights reserved; (c) Art, R. Demarest after Patten, Carlson, and others. **12.5** Photograph Carolina Biological Supply Company / **12.8** Art, R. Ciemma. **12.10** (b) Photo Dr. Victor A. McKusick; (c) Steve Uzzell. **12.14** Giraudon/Art Resource, NY. **12.16** After Victor A. McKusick, *Human Genetics,* Second edition, © 1969. Reprinted by permission of Prentice-Hall, Inc., Engelwood Cliffs, NJ; photo Corbis-Bettmann. **12.17** (a,b) W. M. Carpenter. **12.18** Art, R. Ciemma.

12.19 (a) Cytogenetics Laboratory, University of California, San Francisco; (b) After Collman, Stoller, *American Journal of Public Health*, 52, 1962; photographs*above* courtesy of Peninsula Association for Retarded Children and Adults, San Mateo Special Olympics, Burlingame, CA; *below* used by permission of Carole Lafrate. 12.21 (a,b) Courtesy G. H. Valentine. 12.23 C. J. Harrison. 12.26 Art, R. Ciemma. 12.27 (a) Fran Heyl Associates © Jacques Cohen, computer enhanced by © Pix Elation; (b) Art, R. Ciemma. 12.29 Bonnie Kamin/Stuart Kenter Associates. 12.31 Carolina Biological Supply Company.

CHAPTER 13   13.1, 13.2 *Above,* A. Lesk/SPL/ Photo Researchers; *below,* A. C. Barrington Brown © 1968 J. D. Watson. 13.3 Art, R. Ciemma. 13.4 (b) Lee D. Simon/Science Source/Photo Researchers. 13.6 Biophoto Associates/SPL/ Photo Researchers. 13.7 Art, Precision Graphics. 13.10 (a) Ken Greer/Visuals Unlimited; (b) Biophoto Associates/Science Source/Photo

Researchers; (c) James Stevenson/SPL/Photo Researchers; (d) Gary Head.

CHAPTER 14   14.1 *Above,* Kevin Magee/Tom Stack & Associates; *below,* Dennis Hallinan/FPG. 14.3 Gary Head. 14.4 From Stephen L. Wolfe, *Molecular and Cellular Biology,* Wadsworth. 14.5 Art, Palay/Beaubois and Precision Graphics. 14.8 Art, Hans & Cassady, Inc. 14.12 (a) Courtesy of Thomas A. Steitz from *Science,* 246:1135-1142, December 1, 1989. 14.14 Art, R. Ciemma and ACG. 14.15 Dr. John E. Heuser, Washington University School of Medicine, St. Louis, MO; art, Palay/Beaubois. 14.18 *Right* Nik Kleinberg; *left,* Peter Starlinger. 14.19 Art, R. Ciemma.

CHAPTER 15   15.1 (a) Slim Films; (b) Dr. Brian V. Harmon, Queensland University of Technology. From *Methods in Cell Biology,* Volume. 46, 1995, "Anatomical Methods in Cell Death," Academic Press. Reprinted by Permission. 15.2 Brian Matthews, University of Oregon. 15.3 Art, Palay/Beaubois and Hans

& Cassady. 15.4 Art, R. Ciemma. 15.5 (a) M. Roth and J. Gall. 15.6 (a) C. J. Harrison et al., *Cytogenetics and Cell Genetics,* 35:21-27, © 1983 S. Karger A. G., Basel; (b) art, Palay/Beaubois. 15.7 (a) Dr. Karen Dyer Montgomery. 15.8 Jack Carey. 15.9 W. Beerman; art, R. Ciemma. 15.10 Art, Precision Graphics; Frank B. Salisbury. 15.11 Art, Betsy Palay/Artemis. 15.12 (a) Art, Betsy Palay/Artemis; (b) Lennart Nilsson © Boehringer Ingelheim International GmbH.

CHAPTER 16   16.1 Lewis L. Lainey. 16.2 (a) Stanley N. Cohen/Science Source/Photo Researchers; (b) Dr. Huntington Potter and Dr. David Dressler. 16.3 Art, R Ciemma. 16.6 Cellmark Diagnostics, Abingdon, U.K. 16.9 Michael Maloney, *San Francisco Chronicle.* 16.10 (a) Stephen L. Wolfe, *Molecular and Cellular Biology;* (b) Keith V. Wood; (c,d) Monsanto Company. 16.11 R. Brinster, R. E. Hammer, School of Veterinary Medicine, University of Pennsylvania. **Page 269** S. Stammers/SPL/ Photo Researchers.

INDEX

Numbers followed by i refer to illustrations; those followed by t refer to tables.